From Spinors to Quantum Mechanics

From Spinors to Quantum Mechanics

Gerrit Coddens

Commissariat à l'énergie atomique et aux énergies alternatives,
France

Imperial College Press

Published by

Imperial College Press
57 Shelton Street
Covent Garden
London WC2H 9HE

Distributed by

World Scientific Publishing Co. Pte. Ltd.
5 Toh Tuck Link, Singapore 596224
USA office: 27 Warren Street, Suite 401-402, Hackensack, NJ 07601
UK office: 57 Shelton Street, Covent Garden, London WC2H 9HE

Library of Congress Cataloging-in-Publication Data
Coddens, Gerrit, author.
From spinors to quantum mechanics / Gerrit Coddens, Commissariat à l'énergie atomique et aux énergies alternatives, France
pages cm
Includes bibliographical references and index.
ISBN 978-1-78326-636-4 (hardcover : alk. paper)
1. Group theory. 2. Spinor analysis. 3. Quantum theory. I. Title.
QC174.17.G7C63 2015
530.1201'5122--dc23
2015000300

British Library Cataloguing-in-Publication Data
A catalogue record for this book is available from the British Library.

Typeset by Stallion Press
Email: enquiries@stallionpress.com

"I came to offer thee a flower,
But thou must have all my garden,-
It is thine."
(*Rabindranath Tagore*)

To the memory of my parents
To Claude, Isabelle, Felice and Chantal
To José, Poesje, Guy and Kenzie
To Théo and Sarah

Preface

This book has a two-fold objective: giving the reader a non-conventional introduction to the representation theory for the rotation and the homogeneous Lorentz groups, that will allow them to understand these topics better, and to show how the insights gained this way can lead to a better understanding of quantum mechanics. Contrary to what a knowledgeable reader may expect on the basis of this statement, it will not be something he has seen elsewhere, neither for group theory nor for quantum mechanics.

This alternative approach to the representation theory is geometrical. Rotations are just Euclidean geometry; it should be a piece of cake. But textbooks keep it abstract, only covering the algebraic aspects of the representation theory while not explaining what the geometric counterpart of that algebra is. One can readily check that the SU(2) matrices behave like they should do. But somebody must have discovered this, based on insight, and that geometrical insight is *not* explained in textbooks. When one tries to figure it out oneself, one will have to pay an unreasonable price in amount of time spent and frustration endured. The reason for this is that a radical change in approach is required. The underlying idea is not difficult but it can take the unwary completely off guard: one has to modify the definition of a rotation as a function, by changing both its domain and range.

We are used to seeing the rotations as functions g from $\mathbb{R}^3$ to $\mathbb{R}^3$, that rotate vectors $\mathbf{r} \in \mathbb{R}^3$ to other vectors $g(\mathbf{r}) = \mathbf{r}' \in \mathbb{R}^3$. But one can see a rotation g also as a function on the group of rotations G, that transforms group elements $g_j \in G$ into other group elements $g^\circ g_j = g'_j \in G$. Instead of operating with rotations on vectors coded in the form of 3×1 column matrices (rotating vectors), one operates then with rotations on (other) rotations coded in the form of 2×1 column matrices (rotating rotations). The explicit mention of this prerequisite change in imagery is the only link that is missing; the rest is straightforward.

One can derive the representation theory for the Lorentz group following the same principles. This is more difficult because there a few more quirky mathematical twists to it. But once these have been ironed out, group theory will no longer appear as a concatenation of tedious and mysterious algebraic calculations. The algebraic quantities will have acquired a recognizable geometrical meaning, just as one recognizes a circle in the equation of a circle.

In learning quantum mechanics, one goes through the same feelings of alienation and dismay as with learning group theory. Here is this intricate set of rules with this very disorienting explanation for it. It just comes out of the blue, and it looks ever so hard to make sense of it. Much as with group theory, the reader is invited to just stick to the algebra without asking any further questions as to what it means. Certain claims, for instance that a particle cannot have a well-defined position and a well-defined momentum at the same time, render the subject even more impervious. The Hungarian philosopher Imre Lakatos summarized it wittily as follows: *"When a particle is accelerated in Brookhaven, it is not in Brookhaven"*. Even more puzzling is that this is being derived from a mathematical formalism wherein the momentum and position vectors $\mathbf{p}$ and $\mathbf{r}$ appear as very well-defined quantities in the equations.

The narrative appears thus to run a bit as follows: these quantities do not exist simultaneously, but by starting from an incorrect theory based on the assumption that they *do* exist simultaneously, one can derive mathematically another, correct theory wherein they *do not* exist simultaneously. It just happens that there exists some magic that can be used to find the right starting from the wrong. It is hard to understand how this could be. Quantum mechanics is full of such mysteries.

What is proposed in this book is that the geometrical insights from group representation theory can be very helpful in making sense of quantum mechanics. It will take even some more surprising mathematical leaps, but ultimately many mysterious aspects of quantum mechanics become clear when one bases the reasoning on the true geometrical meaning of a spinor. From the results I have obtained up to now, I am convinced that this method is the only one that might permit us to eventually understand the whole of quantum mechanics. I think that this book could function as a welcome complement to anyone who wishes to obtain a better understanding of the group representation theory and of quantum mechanics.

The contents of this book have all been rethought from scratch; it has been a long and winding road. None of the results derived are novel; they have all been well known within the traditional approach for a long time. But the work cannot be assessed using a criterion of novelty of results. What counts in this book is the additional insight that can be gained from an alternative approach.

I would like to thank my employer, the Commissariat à l'Energie Atomique et aux Energies Alternatives, and my directors Guillaume Petite, Martine Soyer and Kees van der Beek for having offered me the opportunity to carry out this work, and my colleagues for their moral support. Finally I would like to thank Sébastien Ceste for his help with the figures.

Palaiseau,
August 2011

Gerrit Coddens

List of Symbols

$\mathbb{1}$	unit matrix
$\otimes$	tensor product
$\boldsymbol{\nabla}$	gradient
$\Box$	d'Alembertian operator $\frac{1}{c^2}\frac{\partial^2}{\partial t^2} - \frac{\partial^2}{\partial x^2} - \frac{\partial^2}{\partial y^2} - \frac{\partial^2}{\partial z^2}$
$\forall$	for all
$\exists$	there exists
$\exists!$	there exists a unique
$\parallel$	parallel
$\perp$	perpendicular
$*$	scalar product in space-time: $(a_{ct}, \mathbf{a}) * (b_{ct}, \mathbf{b}) = a_{ct}b_{ct} - \mathbf{a} \cdot \mathbf{b}$
$\subset$	subset
$\Rightarrow$	logical implication
$\Leftrightarrow$	logical equivalence
&	logical "and" operator
$\vee$	logical "or" operator
$\neg$	logical negation
/	not
$\mathbf{A}^\dagger$	hermitian conjugated matrix of matrix $\mathbf{A}$
$\mathbf{A}^\top$	transposed of matrix $\mathbf{A}$
A_n	alternating group of n elements (even permutations)
$a\|b$	a is replaced by b
$\{a, b, \cdots\}$	set containing $a, b, \cdots$
$\mathbf{a} \wedge \mathbf{b}$	vector product of vectors $\mathbf{a} \in \mathbb{R}^3$ and $\mathbf{b} \in \mathbb{R}^3$
$< \mathbf{a}, \mathbf{b} >$	Hermitian in-product $\sum_{j=1}^{n} a_j^* b_j$ of $\mathbf{a} \in \mathbb{C}^n$ and $\mathbf{b} \in \mathbb{C}^n$
$A \cap B$	intersection of sets A and B
$A \cup B$	union of sets A and B
$A \backslash B$	difference of sets A and B

$\mathbf{a}\cdot\hat{\mathscr{L}}$	angular-momentum operator $a_x\hat{\mathrm{L}}_x\sigma_x + a_y\hat{\mathrm{L}}_y\sigma_y + a_z\hat{\mathrm{L}}_z\sigma_z$
$\mathbf{a}\cdot\boldsymbol{\sigma}$	notation for the 2×2 matrix $a_x\sigma_x + a_y\sigma_y + a_z\sigma_z$
α	fine-structure constant $\alpha = \frac{q^2}{c\hbar}$
c	speed of light in vacuum
$\mathscr{C}$	light cone in Minkowski space-time $\mathbb{R}^4$
$\mathbb{C}$	set of complex numbers
$\mathbb{C}^n$	set of n-dimensional complex vectors
$\hat{\mathrm{D}}$	$\mathbf{r}\cdot\boldsymbol{\nabla} = x\frac{\partial}{\partial x} + y\frac{\partial}{\partial y} + z\frac{\partial}{\partial z}$
$\mathbf{D}(g)$	representation matrix of group element g
$\det(\mathbf{A})$	determinant of the matrix $\mathbf{A}$
Δ	Laplacian $\frac{\partial^2}{\partial x^2} + \frac{\partial^2}{\partial y^2} + \frac{\partial^2}{\partial z^2}$
δ_{jk}	Kronecker symbol
$\hat{\mathrm{E}}$	energy operator $-\frac{\hbar}{\imath}\frac{\partial}{\partial t}$
$(\mathbf{E}, \mathbf{B})$	electromagnetic field
$\mathbf{e}_j$	unit vector
ε	eccentricity of an orbit
$F(A, B)$	set of functions with domain A and values in B
φ	rotation angle
$g \in G$	group element g belongs to the group G
$g_2 \circ g_1$	abstract group operation in group $(G, \circ)$, g_2 "after" g_1
$g_{\mu\nu}$	metric tensor
γ, β	Lorentz factor and velocity parameter
γ_μ	the four Dirac matrices $\gamma_{ct}, \gamma_x, \gamma_y, \gamma_z$
$\boldsymbol{\gamma}$	the four Dirac matrices $(\gamma_{ct}, \gamma_x, \gamma_y, \gamma_z)$ in "vector" notation
h	Planck's constant
$\hbar$	$h/2\pi$
$\mathscr{I}$	isotropic cone in $\mathbb{C}^3$
J	total angular momentum, from coupling ℓ and S, $J = \ell + S$
J_z	spin quantum number for z-component of $\hat{\mathbf{J}}$
$\mathbf{L}$	Lorentz transformation matrix
ℓ	quantum number of $\hat{\mathrm{L}}$
$\hat{\mathscr{L}}$	the three operators $(\hat{\mathrm{L}}_x\sigma_x,\ \hat{\mathrm{L}}_y\sigma_y,\ \hat{\mathrm{L}}_z\sigma_z)$ in "vector notation"
L(n,$\mathbb{C}$)	linear group of $n\times n$ complex matrices
L(n,$\mathbb{R}$)	linear group of $n\times n$ real matrices
$\hat{\mathrm{L}}_z$	$-\imath\hbar(x\frac{\partial}{\partial y} - y\frac{\partial}{\partial x})$, sometimes just $x\frac{\partial}{\partial y} - y\frac{\partial}{\partial x}$
$\hat{\mathrm{L}}_z$	$-\imath\frac{\partial}{\partial\phi}$ in spherical coordinates
λ	eigenvalue or wavelength (depending on the context)

m	quantum number of $\hat{\mathrm{L}}_z$
m	relativistic mass of a moving electron
m_0	electron rest mas
m_*	modified electron rest mass (within a potential)
$\mathbb{M}_n$	Riemann manifold constructed from n copies of $\mathbb{R}^3$
μ	gyromagnetic ratio $\frac{q}{2m_0c}$
$\boldsymbol{\mu}$	magnetic dipole
$\mu_B = \hbar\mu$	Bohr magneton
$\mathbf{n}$	mathematical rotation axis specified by a unit vector $\mathbf{n}$
$\mathbb{N}$	set of positive integer numbers
ω, Ω	angular frequencies
ω_0	angular frequency in rest frame
$\hat{\mathrm{p}}_j$	momentum operator $\frac{\hbar}{\imath}\frac{\partial}{\partial x_j}$
$P_{\ell,m}$	spherical harmonic, belonging to $F(\mathbb{C}^3, \mathbb{C})$ or $F(\mathbb{R}^3, \mathbb{R})$
ψ	wave function (Schrödinger equation), spinor in SU(2) or SL(2,$\mathbb{C}$)
Ψ	2×2 (SL(2,$\mathbb{C}$)) or 4×1 (Dirac representation) spinor
ψ^c	conjugated spinor of spinor ψ in SU(2)
q	electron charge
$\mathbb{Q}$	set of rational numbers
r	radius, length of position vector $\mathbf{r}$
$\mathbf{R}$	rotation matrix
$\mathbb{R}$	set of real numbers
$\mathbb{R}^n$	set of n-dimensional real vectors
$\mathbb{R}P^2$	projective space, set of directions of $\mathbb{R}^3$
$\mathbf{s}$	physical rotatation axis as specified by a unit vector $\mathbf{s} = \mathbf{e}'_z$
$\mathbf{S}$	spin vector $\frac{\hbar}{2}\mathbf{s}$
S	spin quantum number $\frac{1}{2}$, used in coupling schemes $J = \ell + S$
$S_1 \rightsquigarrow S_2$	the calculations on set S_1 lead to the same result as on set S_2
SL(2,$\mathbb{C}$)	two-dimensional complex special linear group
S_n	symmetric group of n elements (permutations)
SO(3)	three-dimensional special orthogonal group (rotations in $\mathbb{R}^3$)
SU(2)	two-dimensional special unitary group (rotations in $\mathbb{R}^3$)
S_z	spin quantum number for z-component $\pm\frac{1}{2}$ of $\hat{\mathbf{S}}$
$\boldsymbol{\sigma}$	the three Pauli matrices $(\sigma_x, \sigma_y, \sigma_z)$ in "vector" notation
σ_j	the three Pauli matrices $\sigma_x, \sigma_y, \sigma_z$
ς	curvilinear path length

(θ, ϕ)	spherical coordinates
τ	proper time
$\mathbf{V}^\star$	$ct\mathbb{1} - \mathbf{r}\cdot\boldsymbol{\sigma}$ for $\mathbf{V} = ct\mathbb{1} + \mathbf{r}\cdot\boldsymbol{\sigma}$
$(V, \mathbf{A})$	electromagnetic four-potential
(ξ_0, ξ_1)	components of a spinor of SU(2)
$(\dot{\xi}_0, \dot{\xi}_1)$	dotted spinor
Y	icosahedral group
$Y_{\ell,m}$	complex spherical harmonic as a function belonging to $F(\mathbb{R}^2, \mathbb{C})$
$\mathbb{Z}$	set of integer numbers
ζ	$\zeta = \xi_1/\xi_0$ (only used in the stereographic projection)

Contents

Chapter 1

Introduction

1.1 Motivation

Cartan's work is by all reckoning fundamental for modern physics, both within the realms of general relativity and quantum mechanics. In his book about spinors [Cartan (1981)], Cartan writes that physicists "use spinors like vectors". One has to understand what a spinor is to understand that this is a terrible blame. Unfortunately, Cartan's book is not very helpful for a physicist who would want to understand what a spinor is.

As described by the prominent mathematician Jean Dieudonné in his book *Pour l'honneur de l'esprit humain* [Dieudonné (1987)], even a top mathematician can have a very hard time in trying to decipher what the underlying ideas of a mathematical work are. He may be able to verify the proofs mechanically but still feel very perplexed and mystified. Mathematical presentations are written that way for the sake of rigour. As reported by Blumenthal [Blumenthal (1935)], Hilbert explained that in mathematics we could rephrase the axiom "two points define a single straight line"as a statement "two chairs define a single table", and that it would not matter, as the intuitive meaning of the mathematical objects that occur within a theorem should not interfere with its proof. Knowing the intuitive meaning of a mathematical object could have the consequence that one smuggles something taken as granted into a proof while it is not, and this must be avoided. But all this is hardly of any help for a physicist who wants to use these mathematics just as a tool. A physicist cannot afford to attempt to reinvent the mathematics; it may take months of running around in circles before one sees the light. He will therefore find himself forced to skip the true learning process described by Dieudonné.

What Dieudonné describes is exactly what happens when one opens a textbook about group theory. It is ever so easy to verify the mathematical proofs, but what it all means remains a mystery. This has led to the nickname *Gruppenpest.* Abdus Salam's comment [Salam (1963)], about the dismay he experienced after attending a set of lectures by Racah, is also worth citing: "After attending these lectures I thought this is really too hard. I can never learn this. One is hardly ever likely to need all this complicated matter." To which he added: "I was wrong."

I have therefore tried to make a book that should allow the reader to gain a good understanding of group theory without going through the fight and loss of time involved in the true learning process. By presenting this as research, I would expose myself to polemics. Do we really need another text about a subject that is well known? To avoid this I will simply answer by presenting a paradox in Section 1.2.

The results of Chapters 3 and 4 of this book were obtained in a personal attempt [Coddens (2002, 2008)] to gain an understanding of what spinors are. The original motivation of this work was not to derive new results about the mathematics of spinors. In this respect, the existing literature [Cartan (1981); Chaichian and Hagedorn (1998); Cornwell (1984); Hladik (1996); Inui *et al.* (1990); Jones (1990); Misner *et al.* (1970); Penrose and Rindler (1984); Smirnov (1972); Sternberg (1994)] contains very complete accounts that we cannot aim to surpass. The goal was rather to obtain a better understanding of the underlying mathematical ideas.

The motivation for this was a profound conviction that a thorough understanding of the ideas behind the mathematics is absolutely necessary if one wants to overcome the conceptual difficulties inherent to quantum mechanics. As both the mathematics and the physics involved are difficult, one might at a certain stage not even be able to figure out which difficulties are purely mathematical (and therefore not mysterious) and which ones belong to physics. *It would be futile to try to search for a physical explanation of some difficult point if the difficulty in question is in reality purely mathematical.* If one wants to avoid such situations, one has to know exactly where the demarcation line lies between the physics and the mathematics, i.e. one must master all the underlying ideas of the mathematics. As described by Dieudonné [Dieudonné (1987)], in an axiomatic presentation, some of these ideas can remain totally hidden, and it can be quite an effort to figure them out. It is this effort the author of this book wanted to save the reader.

Insofar as the initial motivation for this book is concerned, it is thus entirely personal, and as the reader will notice, the presentation is also entirely personal and non-standard, because I was not concerned about an axiomatic presentation or about deriving theorems.

At a certain stage one cannot remain blind to the fact that all the results derived resemble a lot to standard quantum mechanics. It is in fact quite surprising to see that a lot of "physics" turns out to be just pure mathematics. For example, we are able to derive the Dirac equation with very few additional assumptions. What comes out is a clear geometrical description, of which it is perfectly possible to make simple mental pictures. In a more conventional algebraic approach these geometrical ideas remain hidden. The visual pictures offer interesting new viewpoints and a good insight into the mathematical language that is used in the formalism of quantum mechanics. *What I have found is that in many instances our difficulties to understand quantum mechanics reside within the mathematics rather than in the physics.*

What is described above is a kind of communication problem. Due to the austerity of certain presentations it becomes difficult to decode the full message. The problem does not only reside in the lack of clarity that might result from the austerity of mathematical presentations. An additional problem resides in the lack of rigour in the way physicists use the mathematics. This can be illustrated by the way Dirac introduced his famous "delta function" δ_a for $a \in \mathbb{R}$. He defined it[1] as a function $\delta_a \in F(\mathbb{R}, \mathbb{R})$, with the properties $\forall x \in \mathbb{R} : (x \neq a \Rightarrow \delta(x-a) = 0)$, $(x = a \Rightarrow \delta(x-a) = \infty)$, and:

$$\forall f \in F(\mathbb{R}, \mathbb{R}) : \int_{-\infty}^{+\infty} f(x)\delta(x-a)\, dx = f(a). \tag{1.1}$$

In the preceding lines, $\delta(x-a)$ is just another way to write $\delta_a(x)$.[2] It is a very obvious fact that such a function δ_a does not exist. A function that is zero everywhere except in one point can only have integral zero. The pages of old-fashioned books on quantum mechanics are fraught with such "delta

[1]We will in general note a set of functions whose domain is a set A and who take their values in a set B as $F(A, B)$. Defining the sets A and B is as important for defining a function as giving its "formula" $f(x)$. Consider the catch question to calculate the derivative $Df(x)$ of $f(x) = \ln\ln\sin x$. Using the chain rule one finds $Df(x) = (\cos x)/(|\sin x|\,|\ln\sin x|)$ whose definition domain is a non-empty subset of $\mathbb{R}$. But this is wrong as there is not a single point $x \in \mathbb{R}$ where $\ln\ln\sin x$ is defined. This caveat will play a prominent role throughout Section 3.10 and in Subsection 3.11.3.

[2]We can see here a formal analogy with the definition of the Kronecker delta symbol over a set $\mathcal{S} = \{1, 2, 3 \cdots n\} \subset \mathbb{N}$, which is defined by the properties: $\forall (j,k) \in \mathcal{S}^2$: $(j \neq k \Rightarrow \delta_{jk} = 0)\,\&\,(j = k \Rightarrow \delta_{jk} = 1)$. The Kronecker symbol has the property $(\forall j \in \mathcal{S})\,(\sum_{k=1}^{n} \delta_{jk} a_k = a_j)$. In Dirac's "definition" $x \in \mathbb{R}$ and $a \in \mathbb{R}$ play the same role as the indices $j \in \mathcal{S}$ and $k \in \mathcal{S}$ in the Kronecker symbol, while the integral plays the same role as the sum.

functions", without any mention of the fact that this is a problem. One may think then that the whole of quantum mechanics is based on mathematical nonsense. Fortunately, there exists a mathematically exact way to save the procedure. This was outlined in the fifties by the French mathematician Laurent Schwartz in his book about the theory of distributions [Schwartz (1973)]. One can then settle the problems for physics by postulating that (1.1) is only a symbolic shorthand for the mathematically correct treatment.

Physicists very often take the liberty of using mathematics in their own casual strides as described above, and most of the time they get away with it. But in quantum mechanics they do not, because it backfires on physical issues. In quantum mechanics physicists ignore the correct geometrical meaning of the spinor formalism they use. They invent one of their own, following their intuition, and in doing so they treat spinors like vectors as identified by Cartan, and even vectors like scalars.[3] With these wrong interpretations of the mathematics they obtain physically right answers, because the algebra remains correct. This time, however, the wrong interpretations come at a price. The parallel approach to the mathematics leads to revolutionary *ad hoc* interpretations that are highly counterintuitive and deeply shake our vision of the world that surrounds us. The naturally inborn resistance against these counterintuitive new paradigms is just waved aside by arguing that the ensuing absurdity is a quantum mystery. But a mathematically rigorous approach shows that not all of this is absolute compelling necessity. Hence, for once mathematical rigour reveals itself here as a tool that is of interest to physicists permitting one to investigate physical interpretations, rather than a tedious time-consuming goal in its own right.

In adopting the rigorous approach, we cannot evade the unpleasant necessity of spotting the errors we will come across. But one ought not to do this by just pinpointing the errors and leaving one's colleagues behind to pick up the pieces. We must behave far more responsibly and try to resolve the problems when they arise. This book aims to bring an outright positive message that might sow some seeds of optimism, *viz.* that there might exist a way to better enlightenment in quantum mechanics. I apologize for the times when this may become hidden by the personal style when we oppose the traditional textbooks treatments to the new approach for comparison.

The fundamental concern is to gain more clarity by correcting confusing errors and eliminating unnecessary *ad hoc* interpretations. It can then be

[3]We refer the reader to Footnote 11 of Chapter 3 and to Section 5.7, to discover how very grave some of the mistakes can be.

confidently proclaimed that quantum mechanics gives the right answers and that we can continue to use it, while feeling more comfortable in doing so. The aim is thus to show that there are alternative, mathematically correct derivations for the physics, and that in a rigorous approach some of the quantum mysteries disappear. This justifies *a posteriori* the initial motivation for the quest to understand spinors.

1.2 Paradox: Does quantum mechanics tacitly imply that 0 = 1?

To give the reader an inkling of the kind of mathematical paradoxes he is in for, I will give just this example of what I will call the $0 = 1$ paradox. As described in Section 3.4, to define a spinor in the rotation group one uses isotropic vectors $(X, Y, Z) \in \mathbb{C}^3$, for which $X^2+Y^2+Z^2 = 0$. These spinors code rotations, i.e. group elements. Of course, these complex quantities cannot be particle coordinates $(x, y, z) \in \mathbb{R}^3$, since for particle coordinates one has $x^2 + y^2 + z^2 = r^2 \neq 0, \forall(x, y, z) \neq (0, 0, 0)$. A position vector of a particle is something very different from a rotation or a group element. To follow a particle under rotations we could take $r = 1$. Following a particle with $r = 1$ this way can be done with 3×3 rotation matrices in $\mathbb{R}^3$. But in SU(2) things are not that way; the paradigm is completely different. When Cartan says physicists use spinors like vectors, he pinpoints the fact that they act as though $(X, Y, Z) = (x, y, z)$. Now $|(X, Y, Z)| = 0$, while $|(x, y, z)| = 1$. The Dirac equation can be derived from a reasoning based on the isotropic vector (X, Y, Z), but afterwards, in the calculations, one identifies this isotropic vector with a position vector (x, y, z). This happens in the solution of the wave equation for the hydrogen atom, when one introduces spherical coordinates and harmonic polynomials. In other words, physicists are acting all the time as though $0 = 1$, which is, admittedly, a well-known fact of elementary mathematics. This is only *one* example of a mathematical paradox; there are scores of others.

Such contradictions will not scare those of course, who are unaware of them. They may then proclaim that everything about group theory as used in physics is well known and that there is thus no need to re-explain what spinors are and how group theory is used in quantum mechanics. I may have appeared extremely sarcastic in bringing this to the reader's attention. May he or she forgive me. I am doing it only as a *captatio benevolentiae*, in order to ask him not to hastily make up his mind about this book. Why should I still derive the group theory if it is well known? Is this really suitable as subject matter for a whole book? Still, I do have my reasons to insist on expressing myself and making my points in this way. Even if the group theory is well known, for the aims of the present book, it is just not good enough. The reader will have to learn to see group theory through the eyes of the author. And to get this vision across it will not suffice to merely make reference to an existing textbook.

There is also really nothing to be sarcastic about. Because, miraculously, these calculations, despite all the confusion about the $1 = 0$ issue and other paradoxes, turn out the right physical answers with amazing precision.

1.3 Guiding the reader through this book

There are two proposed paths through the book. The fast track should allow the reader to get an idea of the general structure. Therefore, on a number of occasions the parts of it that the reader might skip on a first reading are flagged. Chapter 2 introduces the reader to some aspects of group theory. The goal here is just to make the readers acquainted with some basic notions that will allow them to read this book without having a prior background in group theory. For most of the readers this kind of introduction might not be necessary, in which case it can be skipped. But on reading it they will probably discover angles of approach they are not familiar with. Chapter 3, about the rotation group, is essential reading as it contains information new even to the reader who is thoroughly acquainted with the subject matter. The fast track-reader can then jump to Chapter 5 to see how the Dirac equation just describes a rotating frame. In fact, Chapter 5 contains a mathematically rigorous proof that the Dirac equation just describes the electron as a spinning top using the language of spinors as the natural tool to describe such dynamical rotations. That spinors are indeed the natural tool to describe a spinning top will have become clear from a reading of Chapter 3. The whole set of Chapters 2–5 constitute a mathematically rigorous, completely self-contained derivation of the Dirac equation from the intuitive idea of a spinning top, that should be accessible to any reader, even if he does not have any prior group-theoretical background. The reader will then know what the Dirac equation *means* and have a clear visual picture for this meaning. This presentation stands thus in marked contrast with the traditional presentation where the Dirac equation is presented as a kind of God-given. Dirac just guessed it, it was then validated by comparison with experiment and we had no further theoretical foundation for it other than the fact that it works. Instead of this *inductive* approach we have now a *deductive* approach for the Dirac equation.

Extremely important is the part about the meaning of spin and the link of this with the isomorphism introduced in Chapter 3. In Chapter 5 a few passages have been flagged to indicate that they could be skipped on a first reading.

Chapter 6 contains a very interesting part of the book, *viz.* a derivation of standard quantum mechanics that allows the reader to understand some of its paradoxes much better. This chapter just builds further on Chapter 5. In Chapter 7 the Bell inequalities are discussed and it is shown that their derivation contains an loophole. This is necessary to validate the approach, as it can be considered as a hidden-variables approach. Chapter 9 is an extension of Chapter 6 for the special and more difficult case of magnetism, while in Chapter 10 the double-slit experiment is discussed.

The reader may then take, on the second reading, the longer path to fill in the gaps.

Chapter 2

Introduction to Groups

(This chapter can be skipped by readers who are familiar with group theory.)

2.1 Definition

A group $(G, \circ)$ is defined as a set G with a composition law $\circ$ for elements $g_j \in G$ with the following properties:

- The set is *closed* under the composition law, which means:

$$\forall g_j \in G, \forall g_k \in G : g_j \circ g_k \in G. \tag{2.1}$$

- The composition law is *associative*, which means:

$$\forall g_j \in G, \forall g_k \in G, \forall g_l \in G : g_j \circ (g_k \circ g_l) = (g_j \circ g_k) \circ g_l. \tag{2.2}$$

- There is an *identity element* $e \in G$ characterized by:

$$\exists e \in G, \forall g_j \in G : g_j \circ e = e \circ g_j = g_j. \tag{2.3}$$

- Each group element $g_j \in G$ has an *inverse element* noted as $g_j^{-1} \in G$ and characterized by:

$$\forall g_j \in G, \exists g_j^{-1} \in G : g_j \circ g_j^{-1} = g_j^{-1} \circ g_j = e. \tag{2.4}$$

It is not necessary that the composition law $\circ$ be *commutative*. The group law is *commutative* when:

$$\forall g_j \in G, \forall g_k \in G : g_j \circ g_k = g_k \circ g_j. \tag{2.5}$$

When the latter axiom is also satisfied, then the group is called *commutative* or *Abelian*. In the reverse case, the group is called *non-commutative* or *non-Abelian*.

Axiom (2.3) is sufficient to show that e must also be unique. Imagine two people who claim they have found an identity element for the group. The first person calls his find e_1, the second person calls his find e_2. As e_1 is an identity element, there must be $e_2 \circ e_1 = e_2$. As e_2 is an identity element, we must have $e_2 \circ e_1 = e_1$. From this it follows that $e_1 = e_2$.

2.2 Remarks on axioms

When we have a set of axioms, we can worry whether the axioms are *independent*, *consistent*, and *complete.*

Independence of a set S of axioms A_j is proved by giving for each axiom $A \in S$ an example of a case whereby all other axioms $A_j \in S \backslash \{A\}$ are satisfied, but A itself is not satisfied. The example of hyperbolic geometry shows that the parallels postulate of Euclidean geometry is independent. Hyperbolic geometry shares all its axioms and postulates with Euclidean geometry, except the parallel postulate. It was discovered (by Bolyai, Lobachevsky and Gauss) when many people had the intuition that the parallels postulate would not be independent from the other postulates and tried to prove it, by a reduction *ad absurdum.* That is, they started from the negation of the parallel postulate and they tried to derive a contradiction from it. That would prove the assumption was wrong and this way prove the parallel postulate. They assumed, thus, that there would be no straight line l' parallel to a given straight line l through a point $P \notin l$, or they assumed that there would be more than just one such straight line. They derived consequences from that assumption and the other axioms of Euclidean geometry, obtaining a whole body of would-be incorrect theorems meant to lead to a contradiction, but the contradiction they were searching for so eagerly refused to materialize. From this, the idea grew that perhaps the parallel postulate was independent after all. The problem was entirely settled by Poincaré [Dieudonnè (1987, p. 220)] who made a one-to-one mapping between the axioms of hyperbolic and Euclidean geometry by building a model of hyperbolic geometry inside Euclidean geometry. This made sure that if there were a contradiction in hyperbolic geometry, there would be one in Euclidean geometry, and *vice versa.*

Getting back to our axioms of a group, we can prove by that same approach of finding a counter-example that the axiom of commutativity for Abelian groups is independent, as non-Abelian groups also exist. Counter-examples for the other axioms exist as well. The one for the associative law is certainly the most difficult one to find.

A set of axioms is self-consistent if one cannot derive a contradiction from them. The simplest way to convince oneself of the self-consistence is to find a model from the physical world that satisfies all the axioms. This is not a rigorous mathematical proof, but at least a good indication that we could call experimental evidence. In this respect, one can believe that Euclidean geometry is self-consistent, and therefore hyperbolic geometry also.

Completeness implies that the whole theory we want to develop can be completely derived from the axioms. According to Gödels theorem a set of axioms is almost always incomplete. We can then add a new independent axiom to the set to render the theory more complete. But as the new axiom is independent, we can also take its negation as a new axiom, such as the example of Euclidean and hyperbolic geometry shows.

2.3 Examples

What the story about hyperbolic geometry illustrates very clearly is that intuition can be dangerous and can lead one astray. It has been explained in Chapter 1 that this is the reason why in mathematics there is so much emphasis on abstraction. We have also explained that the counterpart of this is a kind of lack of intuition. To learn hyperbolic geometry one has to build up a new kind of intuition that runs contrary to the daily-life intuition of Euclidean geometry that feels so comfortable. We cannot say that physicists have not taken the point about the danger of relying too much on "common-sense" intuition to heart. We are confronted with counterintuitive notions in physics on a routine basis. Relativity was considered as counterintuitive when it was introduced and quantum mechanics continues to feel highly counterintuitive. Confronted with that, physicists have built up quantum intuition.

But the tension between rigour and intuition is permanent. The danger of relying too much on one's intuition is constantly lurking around the corner, even within the context of a renewed intuition. Worse, in physics we do not even know what the axioms are supposed to be. They have to be guessed by induction, relying on one's intuition in interpreting experimental evidence. But in the process one should of course not transgress the demarcation line between physics and mathematics and start also interpreting purely mathematical results that might intervene in the calculations. It will be demonstrated in this book that physicists have the tendency to be over-confident about their intuition regarding the meaning of the mathematics they are using. Dirac's delta function discussed in Chapter 1 is a good example of this.

The set of axioms for a group is a choice example of a mathematical presentation that is of an abstraction rendering it *a priori* devoid of any intuition. For the sake of mathematical rigour, this is perfect. For one thing, we will not run head over heels into taking something erroneously for granted based on an unproved intuition, because for the moment, we have not got one. This is what mathematical rigour is about.

But working without intuition is very unpleasant; one has the impression of not understanding what is going on behind the scenes. It is also more difficult to learn things when we have no intuition for what they mean. Examples below are given, which will allow the reader to build up intuition for what groups are and to show him that he is in reality very familiar with them.

The reader who is still unfamiliar with group theory may want to verify in the examples that the four axioms for a group are satisfied and if the group is Abelian or otherwise. He should be able to check what the identity element and the inverse elements are.

• A first example is the symmetry group D_5 of the regular pentagon (see Figure 2.1). The name D_5 reflects that this is a so-called dihedral group.

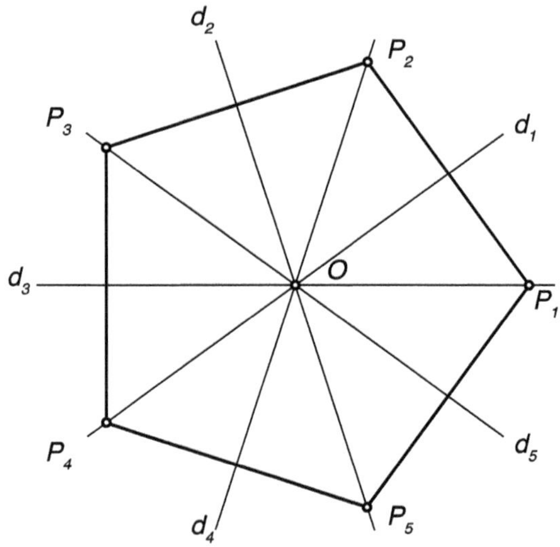

Fig. 2.1 A regular pentagon $P_1P_2P_3P_4P_5$ and the lines d_n, with $n \in [1,5] \cap \mathbb{N}$ that serve to define the reflections A_n in the text.

By this is meant the set of operations that map the pentagon to itself. The five vertices P_j of the pentagon can be represented by the complex numbers: $z_j = e^{\imath 2\pi(j-1)/5} \in \mathbb{C}$. There are five rotations $R_k \in F(\mathbb{C},\mathbb{C})$, where $\forall k \in [0,4] \cap \mathbb{Z}$, $\forall z \in \mathbb{C} : R_k(z) = z\, e^{\imath 2\pi k/5}$, that map the pentagon onto itself. As summarized by the notation $R_k \in F(\mathbb{C},\mathbb{C})$, these rotations are functions. The set $(G_1, \circ)$ of the five functions R_k is a group when the composition law for functions $^\circ$ is taken for $\circ$. This is a *finite* group as it contains a finite number of elements. The identity element is R_0. The inverse element of R_k is R_{5-k}, when $k > 0$ and R_0 if $k = 0$. The group is Abelian as adding up the rotation angles is commutative. But these operations are not the only ones which leave the pentagon invariant. There are also five reflections $A_n \in F(\mathbb{C},\mathbb{C})$ with respect to the lines $d_n = \{z \in \mathbb{C} \parallel \exists r \in \mathbb{R} : z = re^{\imath(n-\frac{1}{2})\frac{2\pi}{5}}\}$. These lines are illustrated in Figure 2.1. The reflections A_n are defined by: $\forall n \in [1,5] \cap \mathbb{N}, \forall z \in \mathbb{C} : A_n(z) = z^{-1}\, e^{\imath 2\pi(2n-1)/5}$. The inverse element of A_n is A_n itself. However, the set G_2 of the five reflections A_n is not a group as the product of two reflections is a rotation, not a reflection, such that the set G_2 is not closed under the composition law of functions. But the set $G = G_1 \cup G_2$ that contains both the reflections and the rotations is closed under the composition law and makes up a group $(G, \circ)$, which is the symmetry group of the pentagon. The "multiplication table" for this group is written in Table 2.1. Such a "multiplication table" is called a Cayley table. The table has been written using the convention that the element on line k and column j corresponds to the group element $g_k \circ g_j$, where by definition operation g_k is executed after operation g_j. This is really important as the group is not Abelian, such that the order of the operations does matter. The reader can check this in the table: for example, $A_3 \circ R_1 = A_5$ and $R_1 \circ A_3 = A_1$, such that $A_3 \circ R_1 \neq R_1 \circ A_3$. The nomenclature "Cayley table" is in principle only used for finite groups, but I will take the liberty to use it also for infinite groups in the next chapter, even if one cannot really write down an infinite table. The Cayley table of the non-Abelian group $(G, \circ)$ contains the Cayley table of the Abelian group $(G_1, \circ)$ in its top left corner. As $G_1 \subset G$, the group $(G_1, \circ)$ is an example of what one calls a *subgroup* of the group $(G, \circ)$. This concept will be defined in Section 2.4. In the bottom right corner of the table one can also check that the composition of two reflections is not a reflection but a rotation. As all elements of the group can be written as products of reflections, it is said that the reflections alone do not constitute a group, but that they *generate* a group.

Table 2.1 Cayley table for the symmetry group of the regular pentagon

$\circ$	R_0	R_1	R_2	R_3	R_4	A_1	A_2	A_3	A_4	A_5
R_0	R_0	R_1	R_2	R_3	R_4	A_1	A_2	A_3	A_4	A_5
R_1	R_1	R_2	R_3	R_4	R_0	A_4	A_5	A_1	A_2	A_3
R_2	R_2	R_3	R_4	R_0	R_1	A_2	A_3	A_4	A_5	A_1
R_3	R_3	R_4	R_0	R_1	R_2	A_5	A_1	A_2	A_3	A_4
R_4	R_4	R_0	R_1	R_2	R_3	A_3	A_4	A_5	A_1	A_2
A_1	A_1	A_3	A_5	A_2	A_4	R_0	R_2	R_4	R_1	R_3
A_2	A_2	A_4	A_1	A_3	A_5	R_3	R_0	R_2	R_4	R_1
A_3	A_3	A_5	A_2	A_4	A_5	R_1	R_3	R_0	R_2	R_4
A_4	A_4	A_1	A_3	A_5	A_1	R_4	R_1	R_3	R_0	R_2
A_5	A_5	A_2	A_4	A_1	A_3	R_2	R_4	R_1	R_3	R_0

• The set $\mathbb{Z}$ of integer numbers forms an Abelian group $(\mathbb{Z}, +)$ under the operation of addition. This is an *infinite* group because it contains an infinite number of elements. The identity element is 0. The inverse element of $j \in \mathbb{Z}$ is $-j$. This group is Abelian. However, $(\mathbb{Z}, \times)$ is not a group. The first three axioms are satisfied, the identity element being the number 1, but (2.4) is not satisfied. This example shows the independence of the axiom expressed in (2.4).

• Of course also $(\mathbb{R}, +)$, where $\mathbb{R}$ is the set of real numbers under addition is an infinite Abelian group. It is a *continuous* group in contrast with $(\mathbb{Z}, +)$ which is *discrete*. Also the vector spaces $(\mathbb{R}^n, \mathbb{R}, +)$ of $\mathbb{R}^n$ over the field $\mathbb{R}$ are groups with respect to vector addition.

• Figure 2.2 illustrates a two-dimensional crystal lattice and its unit cell. It illustrates the specific example of a square lattice. In the most general case, a two-dimensional lattice is defined by a basis of two vectors $\mathbf{a}_1$ and $\mathbf{a}_2$ that are linearly independent, but not necessarily of unit length and orthogonal as in the example of Figure 2.2, where $\mathbf{a}_1 = \mathbf{e}_x$ and $\mathbf{a}_2 = \mathbf{e}_y$. Using this basis, one then constructs the set L of all linear combinations $\mathbf{OP} = \sum_{j=1}^{2} c_j \mathbf{a}_j$ with $c_j \in \mathbb{Z}$. Such a set is sometimes called a $\mathbb{Z}$-module. This set L is a group $(L, +)$ under vector addition and called the Bravais lattice of the two-dimensional crystal. It is the group generated by the two translations with vectors $\mathbf{a}_1$ and $\mathbf{a}_2$ (and their inverses). The translation and position vectors of the Bravais lattice are of the type $\mathbf{OP}$ illustrated in Figure 2.2. The identity element is the null translation $\mathbf{OO}$. The inverse element of $\mathbf{OP}$ is the vector $-\mathbf{OP}$. Like all translation groups, $(L, +)$ is

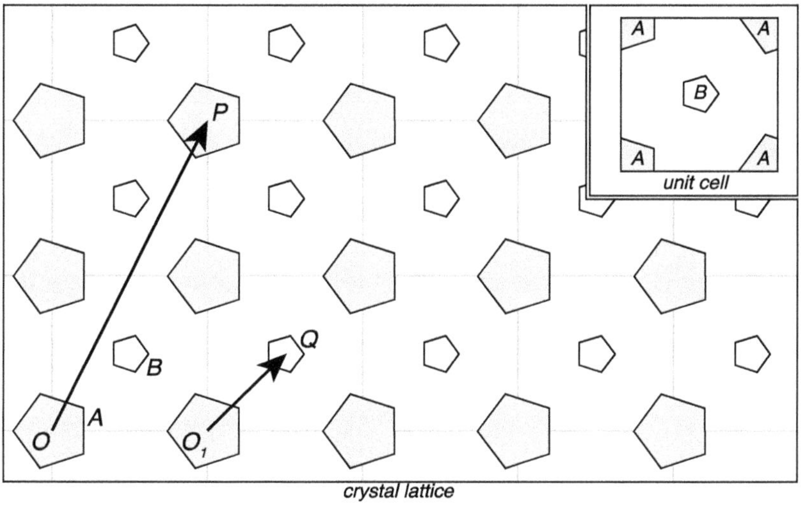

Fig. 2.2 A (metaphorical) two-dimensional crystal lattice with two different atoms (the pentagons A and B) in the unit cell (shown in the inset). The position vectors of the atoms B (of the type $\mathbf{O_1Q}$) with respect to the nodes of the Bravais lattice do not belong to the Bravais lattice and the symmetry group. They belong to the unit cell U. Those of the atoms A belong in this example accidentally to the symmetry group, because the atoms A are situated at nodes of the Bravais lattice. The lattice vector $\mathbf{OP}$ belongs to the Bravais lattice.

Abelian. However the position vectors of the type $\mathbf{O_1Q}$ within a unit cell U do not belong to the symmetry group. In fact, the crystal lattice is the convolution $L * U$ of the Bravais lattice L and the unit cell U. In an analogous way one can construct the Bravais lattice of a three-dimensional crystal by using three linearly independent vectors $\mathbf{a}_1$, $\mathbf{a}_2$ and $\mathbf{a}_3$ to generate the three-dimensional translation group of the lattice. Group theory is very important for the classification of crystal lattices.

• The set $\mathbb{R}$ is not a group for multiplication as 0 has no inverse. However, $(\mathbb{R}\backslash\{0\}, \times)$ is an Abelian group, with 1 as identity element. The same can be said about the set $\mathbb{C}$ and the Abelian group $(\mathbb{C}\backslash\{0\}, \times)$.

• The permutations of the first n integer numbers $j \in [1, n] \cap \mathbb{N}$ form a group under the composition of permutations (understood as the composition $\circ$ of functions) called the symmetric group $(S_n, \circ)$. This is a really nice group to play with. For $n = 5$ the sequence 31452 is a permutation of 12345. It is generally noted as:

$$\begin{pmatrix} 1 & 2 & 3 & 4 & 5 \\ 3 & 1 & 4 & 5 & 2 \end{pmatrix}. \tag{2.6}$$

The identity element is:

$$\begin{pmatrix} 1 & 2 & 3 & 4 & 5 \\ 1 & 2 & 3 & 4 & 5 \end{pmatrix}, \tag{2.7}$$

and the inverse element is:

$$\begin{aligned} \begin{pmatrix} 1 & 2 & 3 & 4 & 5 \\ 3 & 1 & 4 & 5 & 2 \end{pmatrix}^{-1} &= \begin{pmatrix} 3 & 1 & 4 & 5 & 2 \\ 1 & 2 & 3 & 4 & 5 \end{pmatrix} \\ &= \begin{pmatrix} 1 & 2 & 3 & 4 & 5 \\ 2 & 5 & 1 & 3 & 4 \end{pmatrix}. \end{aligned} \tag{2.8}$$

It is thus obtained by reversing the order of the lines and then restoring the order of the columns. Permutation groups for $n > 2$ are non-Abelian, e.g.:

$$\begin{aligned} &\begin{pmatrix} 1 & 2 & 3 \\ 2 & 3 & 1 \end{pmatrix} \begin{pmatrix} 1 & 2 & 3 \\ 1 & 3 & 2 \end{pmatrix} = \begin{pmatrix} 1 & 2 & 3 \\ 2 & 1 & 3 \end{pmatrix}, \qquad \text{while} \\ &\begin{pmatrix} 1 & 2 & 3 \\ 1 & 3 & 2 \end{pmatrix} \begin{pmatrix} 1 & 2 & 3 \\ 2 & 3 & 1 \end{pmatrix} = \begin{pmatrix} 1 & 2 & 3 \\ 3 & 2 & 1 \end{pmatrix}. \end{aligned} \tag{2.9}$$

We see in this example that we omit the symbol $\circ$ for the composition law of functions and just proceed by juxtaposition. We thereby keep the convention for the composition of functions that the operation on the left always comes after the operation on the right. In other words, when group elements are functions then $g_j g_k$ stands for $g_j \circ g_k$. A transposition is a special permutation that permutes two adjacent numbers, e.g.:

$$\begin{pmatrix} 1 & 2 & 3 & 4 & 5 \\ 1 & 3 & 2 & 4 & 5 \end{pmatrix} = (23). \tag{2.10}$$

The transpositions $(j, j+1)$ can be used to generate the whole group. This implies that every permutation can be written as a number of transpositions. A permutation is called odd if it is generated by an odd number of transpositions, it is called even if it is generated by an even number of transpositions. Of course, to be meaningful this definition must be independent of the way one decomposes the permutation in terms of transpositions; the reader might try to construct a proof of this independence. The even permutations form themselves a smaller group, the alternating group A_n. The alternating group A_n is a subgroup of S_n, such that it is a further illustration of the concept of subgroups to be introduced in Section 2.4.

• The special orthogonal group SO(2) of 2×2 matrices $\mathbf{M}$ of the type:

$$\mathrm{SO}(2) : \mathbf{M} = \begin{pmatrix} \cos\alpha & -\sin\alpha \\ \sin\alpha & \cos\alpha \end{pmatrix}, \quad \text{where } \alpha \in \mathbb{R} \tag{2.11}$$

forms a group under matrix multiplication. This is a so-called *matrix group*, and is also an Abelian group. It is called "special" because $\det(\mathbf{M}) = 1$ and "orthogonal" as the matrices are orthogonal. The number 2 in the notation indicates that the matrices are 2×2. The reader will recognize here the matrices for rotations in the plane. The identity element is the unit matrix and the inverse element is obtained by the substitution $\alpha| - \alpha$. The set of rotation matrices is not closed under the operation of addition of matrices.

• The set of numbers $e^{\imath\alpha}, \alpha \in \mathbb{R}$ forms an Abelian group. The identity element corresponds to $\alpha = 0$, and the inverse element is obtained by the substitution $\alpha| - \alpha$. This group is often noted as U(1). It will be shown that U(1) and SO(2) are *isomorphic*. Isomorphism is a very important concept that will be introduced below. Both groups are isomorphic to a third group, *viz.* the group of the rotations around the origin O of a plane.

• The set of three-dimensional rotations R around a fixed point O of $\mathbb{R}^3$ are a group, under the composition of rotations. For the two rotations R_1 and R_2, R_2R_1 is the operation involved in applying rotation R_2 after R_1, i.e. $\forall \mathbf{r} \in \mathbb{R}^3 : [\,R_2R_1\,](\mathbf{r}) = R_2(R_1(\mathbf{r}))$. An elegant proof that the composition of two rotations is another rotation is given in Section 3.5 of Chapter 3, by describing a rotation as the product of two reflections. But it can also be proved in another way, by defining a rotation as an operation that leaves the length r of a vector $\mathbf{r} = \mathbf{OP}$, the point O and the handedness of a reference frame invariant. The inverse rotation is the rotation about the same axis but with the opposite angle. The identity element is the operation that leaves all points of $\mathbb{R}^3$ fixed. The group of three-dimensional rotations is non-Abelian.

• The set of 3×3 matrices used to calculate with rotations is a matrix group, called the *special orthogonal group* SO(3). It is isomorphic with the group of the rotations itself. The terms "special" and "orthogonal" have the same meaning as for SO(2), while 3 stands for the dimension of $\mathbb{R}^3$. The orthogonality of the matrices refers to the property that the columns of these matrices define vectors that are orthogonal (when a normalized orthogonal basis for $\mathbb{R}^3$ is used). The set of rotation matrices is not closed under the operation of addition of matrices. We will see later in this book that this is the reason why we cannot use spinors like vectors (see Subsections 5.1.3 and 5.1.5, and Section 5.2). This fact will become extremely important in the discussion of the superposition principle in quantum mechanics. The structure obtained by making all linear combinations $\mathbf{M} = \sum_j c_j \mathbf{M}_j$ of elements $\mathbf{M}_j \in G$ of a matrix group G, whereby these linear combinations themselves are not necessarily group elements, is called the *group ring*.

An even number of reflections in $\mathbb{R}^3$ defines a rotation, while an odd number of reflections defines a reversal. The determinants of the 3×3 matrices that can be used to make the calculations on rotations, reflections, and reversals are 1 or -1. Reflections are not a group because the product of two reflections is not a rotation. However, reflections can be used to generate the group of rotations and reversals.

• The special unitarian matrix group SU(2) consists of all 2×2 matrices $\mathbf{M}$ of the form:

$$\mathrm{SU}(2) : \mathbf{M} = \begin{pmatrix} a & -b^* \\ b & a^* \end{pmatrix}, \quad \text{where } (a,b) \in \mathbb{C}^2 \ \&\ aa^* + bb^* = 1. \tag{2.12}$$

The group is called "special" because $\det(\mathbf{M}) = 1$, and "unitarian" because $\mathbf{M}^\dagger = \mathbf{M}^{-1}$. The matrices show some structural similarity with those of SO(2). In fact, by making the restriction $a \in \mathbb{R}$, $b \in \mathbb{R}$, we fall back onto SO(2). This is another example of the notion of a subgroup, to be developed in Section 2.4.

• The homogeneous Lorentz transformations of special relativity form a group under the operation of composition. They are generated by the boosts but they also contain rotations. A general transformation is the composition of a boost and a rotation. The group is non-Abelian as it contains the group of three-dimensional rotations.

• The special linear complex matrix group SL(2,$\mathbb{C}$) consists of all 2×2 matrices $\mathbf{M}$ of the form:

$$\mathrm{SL}(2,\mathbb{C}) : \mathbf{M} = \begin{pmatrix} a & b \\ c & d \end{pmatrix}, \quad \text{where } (a,b,c,d) \in \mathbb{C}^4 \ \&\ ad - bc = 1. \tag{2.13}$$

Once again "special" means that the determinants of the matrices are 1. Linear means that there are no further special constraints on the matrices, as matrices are a notation for linear transformations. The symbol $\mathbb{C}$ indicates that the entries of the matrices are complex. We will see this used in Chapter 4 to represent the homogeneous Lorentz group.

• The icosahedral group Y is the group of rotations that leave the icosahedron invariant [Duneau (1994)]. The icosahedron is one of the five regular Platonic solids, illustrated in Figure 2.3. The icosahedral group is isomorphic to the alternating group $(A_5, ^\circ)$, and contains 60 elements. It can be abstractly defined by two generators, R_a and R_b, such that $R_a^5 = R_b^2 = (R_aR_b)^3 = e$, where e is the identity. This abstract definition expresses that R_a is a rotation with a five-fold axis, R_b a rotation with

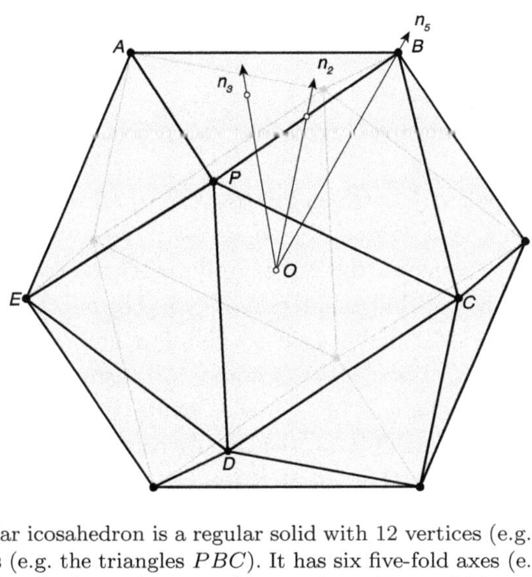

Fig. 2.3 A regular icosahedron is a regular solid with 12 vertices (e.g. P), 30 edges (e.g. PB) and 20 faces (e.g. the triangles PBC). It has six five-fold axes (e.g. n_5), joining the centre O of the icosahedron with one of its vertices, 15 two-fold axes (e.g. n_2) joining O with one of the midpoints of its edges, and ten three-fold axes (e.g. n_3) joining O with one of the centres of gravity of a face. A set of coordinates for the points of the icosahedron is given by: $(0, \pm 1, \pm\alpha)$, $(\pm 1, \pm\alpha, 0)$ and $(\pm\alpha, 0, \pm 1)$ where $\alpha = \frac{1+\sqrt{5}}{2}$ is the golden ratio. The figure was constructed using this algorithm. The following attributions were made: $P(0, 1, \alpha)$, $A(-1, \alpha, 0)$, $B(1, \alpha, 0)$, $C(\alpha, 0, 1)$, $D(0, -1, \alpha)$ and $E(-\alpha, 0, 1)$.

a two-fold axis and $R_a R_b$ a rotation with a three-fold axis in $\mathbb{R}^3$. This is illustrated by a buckyball model in Figure 2.4. The Cayley table for the icosahedral group can be found in [Harter and Weeks (1989)].

2.4 Subgroups

A set $G_1 \subset G$ will be called a subgroup $(G_1, \circ)$ of the group $(G, \circ)$ if it is itself a group. The only thing one has to check for this is if G_1 remains closed under the operation $\circ$, and if the inverse elements of $g \in G_1$ within G also belong to G_1. The other axioms remain satisfied. Some examples of subgroup have already been given, but a few more can be provided, for instance $(\mathbb{R}^2 \times \{0\}, +)$ is a subgroup of $(\mathbb{R}^3, +)$. The group of two-dimensional rotations around O in the Oxy plane is a subgroup of the group of three-dimensional rotations around O in $\mathbb{R}^3$. SO(2) is a subgroup of SU(2), and we will see that SU(2) has indeed also something to do with three-dimensional rotations.

2.5 Homomorphism

A group *homomorphism* f between groups $(G_1, \circ)$ and $(G_2, *)$ is a mapping $f \in F(G_1, G_2)$ such that: $\forall g_j \in G_1, \forall g_k \in G_1 : f(g_j \circ g_k) = f(g_j) * f(g_k)$. Homomorphism is a concept that is not only restricted to its use in groups. The basic idea is always that what one does in G_2 mirrors what one does in G_1, e.g. $e^a e^b = e^{a+b}$ can be used to define a homomorphism between the groups $(\mathbb{R}\backslash\{0\}, \times)$ and $(\mathbb{R}, +)$. A homomorphism does not need to be one-to-one. Not every element of G_2 needs to be an image. It can be that $f(G_1) \subset G_2$, in which case $(f(G_1), *)$ is a subgroup of $(G_2, *)$ as the reader may check. It can also be that an element $h \in G_2$ is the image of several elements $g_j \in G_1$, such that in other words $f^{-1}(h)$ is a subset of G_1, rather than a single element $g \in G_1$. In this respect the reverse image $f^{-1}(e)$ of the identity element $e \in G_2$ is called the *kernel* of f.

When a homomorphism is a one-to-one mapping then it is called an *isomorphism*. When the two sets involved in the mapping are the same, i.e. $G_2 = G_1$, the isomorphism is called a group *automorphism*. More generally, when $G_1 \subset G_2$ it is called a group *endomorphism*. An automorphism is thus an endomorphism for which $G_1 = G_2$. An example of a group isomorphism is the mapping between SO(2) and U(1) provided by the one-to-one correspondence:

$$\begin{pmatrix} \cos\alpha & -\sin\alpha \\ \sin\alpha & \cos\alpha \end{pmatrix} \leftrightarrow e^{\imath\alpha}, \tag{2.14}$$

as the reader may easily check. Both are isomorphic to the group of rotations of the plane. Similarly the group of 3×3 rotation matrices are isomorphic to the group of rotations in $\mathbb{R}^3$.

2.6 Equivalence classes

A relation R between elements of a set S is a subset of $S \times S$. The relation will consist of the ordered couples $(a, b) \in S^2$ between which one draws an arrow from a to b in visualizing the relation. A relation R between elements of S, is an *equivalence relation* if it satisfies the three axioms:

- The relation is *reflexive*, which means that any element of the set is equivalent to itself:

$$\forall a \in S : a \equiv a. \tag{2.15}$$

- The relation is *symmetric*. When a is equivalent to b, then b is equivalent to a:

$$\forall a \in S, \forall b \in S : a \equiv b \Rightarrow b \equiv a. \tag{2.16}$$

- The relation is *transitive*:

$$\forall a \in S, \forall b \in S, \forall c \in S : a \equiv b \,\&\, b \equiv c \Rightarrow a \equiv c. \tag{2.17}$$

Here $a \equiv b$ is a notation for $(a, b) \in R$. For example, in $\mathbb{Z}$, the relation $j \equiv k \Leftrightarrow \exists\ m \in \mathbb{Z} \parallel j - k = 5m$ is an equivalence relation that is noted as $j \equiv k \pmod 5$. When we have an equivalence relation we can build *equivalence classes*: $Cl(a) = \{b \in S \parallel b \equiv a\}$. For the equivalence relation modulo 5 in $\mathbb{Z}$, the classes are $Cl(0)$, $Cl(1)$, $Cl(2)$, $Cl(3)$, $Cl(4)$. The set of these classes is noted as $\mathbb{Z}/5$. Each element of the class can be used to represent the whole class; e.g. both 218 and 3 can be used to represent the class $Cl(3)$. In an *abus de langage* we can say that within the context of $\mathbb{Z}/5$, 218 and 3 are the same thing. On this set one can define an operation $\boxplus$ that renders $(\mathbb{Z}/5, \boxplus)$ a group. One defines $Cl(a) \boxplus Cl(b) = Cl(a+b)$. A different symbol $\boxplus$ is used on the left-hand side, to highlight the fact that it does not have the same meaning as the symbol $+$ on the right-hand side, but in practice the symbol $+$ will be used for both. To prove that the definition of the operation $\boxplus$ makes sense, one must prove that the definition does not depend on the elements a and b that one choses to represent their classes $Cl(a)$ and $Cl(b)$. This is straightforward. In general, the set $\mathbb{Z}/n$ of integer numbers modulo n for a given positive integer number $n \in \mathbb{N}$ forms an Abelian group $(\mathbb{Z}/n, +)$ under the operation of addition. The classes $Cl(j)$ are in general noted as j. The group of equivalence classes modulo $n \in \mathbb{N}$ is one of the simplest examples of a finite group. The identity element is 0. The inverse element of $j \in \mathbb{Z}/n$ is $n - j$ for $j \neq 0$ and 0 for $j = 0$. The reader may try to write the Cayley tables for some groups $(\mathbb{Z}/n, +)$. These groups are isomorphic to the rotation groups of the regular polygons with n vertices.

Isomorphism is an equivalence relation. The concept of equivalence relation permits us also to give a mathematically rigorous formulation of the intuitive notion that isomorphic groups are "the same thing". The matrix group SO(2) is the "same thing" as the group U(1), and both are the "same thing" as the group of two-dimensional rotations in the plane. The groups are isomorphic. We can use one freely to represent the other, because they belong to the same equivalence class. When we use 3×3 rotation matrices to make calculations on three-dimensional rotations, we might as well consider in an *abus de langage* that the rotation matrices *are* rotations.

When physicists speak about group theory they actually refer to a specific field of it, *viz.* group representation theory. We will see that the *abus de langage* which identifies a group with an isomorphic matrix group is the gist of the representation theory used in physics. An isomorphic matrix group is an extremely convenient tool because it turns everything into algebra, permitting us to make all the necessary calculations.

2.7 The assets of abstraction

We have seen that abstraction serves the purpose of mathematical rigour. But there is another reason why abstraction is important in mathematics. We can already see that the definition of a group applies to vastly different sets and vastly different operations. The composition law is sometimes called alternatively a product or a sum, and even then there are different kinds of products and different kinds of sums. But they all satisfy the same four axioms. This is the reason why the definition of the group is rendered abstract. Even the introduction of the unusual symbols $\circ$ and $*$ for the composition laws has been motivated here by the concern to emphasize this abstractness. The specific realization is not specified. The consequence is that when you are able to derive in an abstract way a property of the group from the four abstract axioms, then you will have proved that property for all possible realizations. That is why it is so useful to have abstract structures in mathematics: you save the time of writing out the "same" proof hundreds of times in different guises, i.e. proofs that all have the same structure. The abstraction reflects the desire to lay bare the essence of the common structure and proofs with clinical precision.

For instance, it can be proven that in any group $(G, \circ)$ one can always solve the equation $a \circ x = b$ in x, and that the equation has just one single and unique solution. Here $a \in G$ and $b \in G$ are given group elements, and $x \in G$ is the solution of the equation we want to find. The proof runs in two steps. The first one is heuristic. Suppose first that we have a solution x_0 of this equation, such that truly $a \circ x_0 = b$. That is *a priori* an invalid, gratuitous assumption, but in cheating by making this assumption, it will be possible to obtain some valuable information and eventually cover up for the cheat. Now, a has an inverse element a^{-1}. By multiplying both sides in the identity $a \circ x_0 = b$ to the left with a^{-1} we obtain the new identity: $a^{-1} \circ (a \circ x) = a^{-1} \circ b$. Now we use the associative law to transform this into: $(a^{-1} \circ a) \circ x_0 = a^{-1} \circ b$. Then we use the definition of the inverse to transform this into: $e \circ x_0 = a^{-1} \circ b$. The definition of the identity element transforms this into $x_0 = a^{-1} \circ b$. We know thus that if there is a solution, then it must be $x_0 = a^{-1} \circ b$, such that the solution x_0 is unique if it exists. This is the valuable piece of information we "stole by cheating". Let us now cover up for our foul play. In a further step we will now check that it is indeed a solution. We plug the value of x_0 into $a \circ x = b$ to check if it really satisfies the equation. Now $a \circ x_0 = a \circ (a^{-1} \circ b) = (a \circ a^{-1}) \circ b = e \circ b = b$, where we have used the associative law, the definition of the inverse element and of the identity element. The result $a \circ x_0 = b$ proves then that x_0 is indeed a solution of $a \circ x = b$.

We could have now made this proof for the equation $a + x = b$ in $\mathbb{R}$ using the specific notations in $\mathbb{R}$, or for the equation $R_a R_x = R_b$ for three-dimensional rotations. We would then have written out two proofs. But within the abstract approach, we only need to write up one proof, and it will be valid for all possible realizations. We know thus also automatically that the equation:

$$\begin{pmatrix} 1 & 2 & 3 \\ 2 & 3 & 1 \end{pmatrix} \begin{pmatrix} 1 & 2 & 3 \\ j_1 & j_2 & j_3 \end{pmatrix} = \begin{pmatrix} 1 & 2 & 3 \\ 2 & 1 & 3 \end{pmatrix} \tag{2.18}$$

will be solvable, that it will have a unique solution, and that we have an algorithm to compute this solution. The solution has actually been given in (2.9) above. The point is that the various proofs in the various realizations have the same mathematical structure, just like the axioms we use have the same mathematical structure. This is why abstract structures are very important and useful in mathematics.

Establishing that two different problems have the same structure in mathematics can lead to cross-fertilization between various disciplines. A very successful example of this is the relation between algebra and geometry in analytic geometry. Everything in the algebra can be translated into geometry and *vice versa*. Solutions from one field can be transposed automatically to another field. We see thus that group theory interconnects a large number of realms that at first sight might seem completely unrelated. The point is that the very small set of axioms for a group is already a very powerful set, such that it occupies a very important place in mathematics.

2.8 Intuition and rigour

We can see in the proof that in a group the equation $a \circ x = b$ in x always has a solution, mathematical rigor at work. Every single step is justified by referring to an axiom or a previously established theorem. But we also need examples to be able to make sense of it. The only viable way to learn mathematics is to go backwards and forwards between the two approaches, a rigorous one and a looser and more intuitive one. In good teaching the two approaches must be developed in parallel. Like young lovers, rigour and intuition should walk side by side and hand in hand. That is why the above examples have been given. Intuition can become the poor child of a very rigorous mathematical presentation, while rigour can become the poor child of a very intuitive physical presentation. A treatment of classical mechanics gives one the impression that it is pure mathematics and that one understands everything. This is due to the perfect match between rigour and intuition. When the balance between rigour and intuition is upset, things may start to go wrong in the sense that some things are no longer perfectly understood.

As already described in the Chapter 1, the group theory of the rotation and Lorentz groups is a good example of a situation where this balance has been lost. Because quantum mechanics is formulated in terms of the theory of these groups, it reproduces these problems. The texts are purely algebraic, and the intuition about what the algebra means is missing. It is to this situation one refers by stating: “shut up and calculate”. In many instance physicists may provide some intuition of their own, by stipulating how the algebra in the calculations should be interpreted, but this can be wrong if it is at variance with the interpretation already defined by the mathematics itself. Therefore, acquiring a good insight into the geometrical meaning of the group theory must be the first necessary step of any attempt to understand the physical meaning of quantum mechanics.

2.9 Symmetry

In this section it will be explained that group theory is of utmost importance in physics because it is the prime tool to deal with symmetries. It is often possible to make a drawing that models a group, especially for finite groups. Figure 2.4 gives a drawing for the icosahedral group Y. Figure 2.5 gives a drawing of the group S_4. In both cases, one associates the identity element with a point P with position vector $\mathbf{r} = \mathbf{OP}$ on the map. All other group elements g can then be associated with the points P_g with position vectors $g(\mathbf{r})$. The group elements are thus symbolized by their action on the position vectors $\mathbf{OP}_g = g(\mathbf{r})$ of points. The points P_g serve as images of the group elements g.

It is worth pointing out that the icosahedron I, illustrated in Figure 2.3 and the icosahedral group Y, illustrated in Figure 2.4 are different things, belonging to different worlds. The icosahedron $I \subset \mathbb{R}^3$ in Figure 2.3 can be thought of as representing a real three-dimensional physical object in Euclidean space $\mathbb{R}^3$, that one could hold in one's hand. But Figure 2.4 represents a completely abstract mathematical object that is not tangible and that one cannot hold in one's hand. We can try to imagine this abstract object by visualizing it by a *physical model* in a *model space* $V = \mathbb{R}^3$. Without the model it would be very hard to imagine the abstract mathematical object of the group Y, as it is a set of functions $g \in F(\mathbb{R}^3, \mathbb{R}^3)$, that map vectors $\mathbf{OP} \in \mathbb{R}^3$ onto the vectors $\mathbf{OP}_g \in \mathbb{R}^3$, with the property that $g(I) = I$. We can model this set of functions $G \subset F(\mathbb{R}^3, \mathbb{R}^3)$ by showing how each function works on an arbitrary point P with position vector $\mathbf{OP} \in \mathbb{R}^3 \backslash I$. It is normal to make this choice of model space $V = \mathbb{R}^3$ of points P_g, as we try to model functions $g \in F(V, V)$. The functions g are nothing other than the group elements g that occur in the axioms (2.1)–(2.4). They are thus the objects of study in group theory, not the icosahedron. The icosahedron

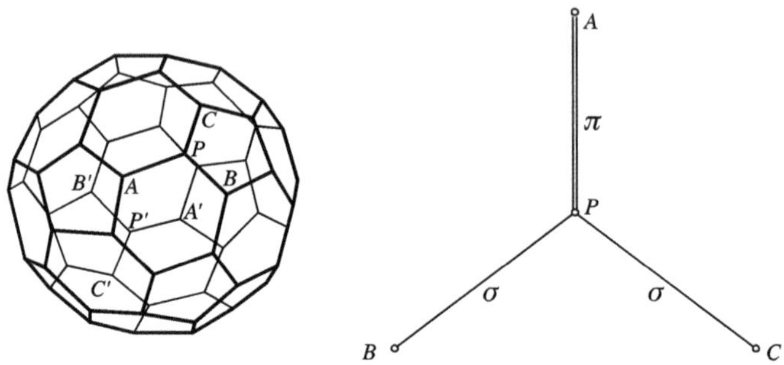

Fig. 2.4 The 60 vertices of the truncated icosahedron in this figure represent the 60 elements of the icosahedral group Y. Well-known physical materializations of this truncated icosahedron are the soccer ball and the C_{60} molecule (often called "buckyball" or buckminsterfullerene). The centre of the buckyball, called O, is not shown. The points P with position vectors **OP** model the icosahedral group. By acting with the icosahedral group on **OP** we obtain the position vectors that model the group elements. As can be seen on the figure by comparing the configuration $PABC$ with the configuration $P'A'B'C'$, each group element has the same environment (reproduced on the right for the first neighbours), such that it is impossible to find out from an inspection of the environment of a group element which group element it is. In the C_{60} molecule, each carbon atom has three first neighbours: one along a double (π-) bond (shared by two hexagons with three-fold symmetry), and two along a simple (σ-) bond (shared by a hexagon and a pentagon (with five-fold symmetry)). The distance along the π-bonds is shorter than along the σ-bonds, such that the truncated icosahedron has no six-fold axes, and contains only three-fold axes. The icosahedral group contains two-fold, three-fold, and five-fold axes, as illustrated in Figure 2.3. If P corresponds to the identity element, A to $g_A \in Y$ and P' to $g \in Y$, then A' corresponds to $g^\circ g_A^\circ g^{-1}$.

can be used to establish the Cayley table of the group, but is not under discussion in the four axioms. In group theory we are also not talking about the vectors **OP** that we use to construct the physical model and which belong to the world $V = \mathbb{R}^3$. We are talking about objects of the world $G \subset F(V, V) = F(\mathbb{R}^3, \mathbb{R}^3)$. This idea of denying citizenship to the vectors will come back in Chapter 3 in an even more radical form, and is a part of the abstract nature of group theory.

Imagine now that such a drawing was a map of the world in which you were living, and that you were trying to figure out where you were by using the map. To your great discomfort you find that from a comparison of your environment with the map you are unable to establish your location, because all local environments look the same. This also implies that you could have identified any point on the map with the identity element, the resulting map is always the same. This is the reason why Galois called group theory the *theory of ambiguity*.

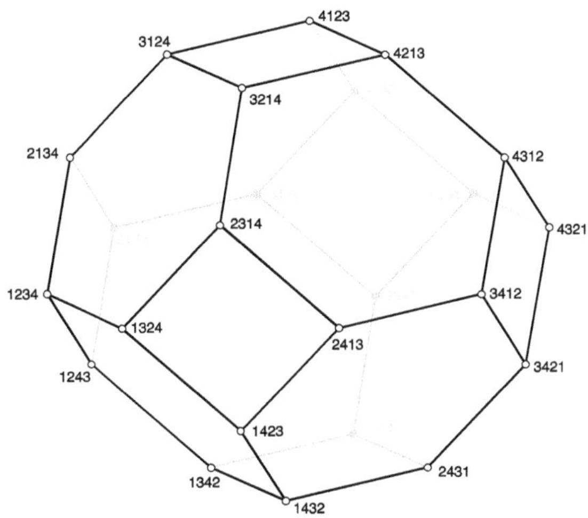

Fig. 2.5 The 24 vertices of the truncated cuboctahedron in this figure represent the 24 elements of the permutation group S_4, whereby the lines represent the transpositions. We see that each vertex has three nearest neighbours. and that each vertex is surrounded by two regular hexagons and a square. Also here one could distinguish two kinds of bonds. Each vertex has two bonds that belong to a square and one bond that does not belong to a square. We see that all points have an identical environment. Each hexagon is isomorphic to S_3. In fact, in each of the hexagons, there is one of the four numbers that remains fixed.

On the C_{60} buckyball, every carbon atom has a double bond and two single bonds. On the cuboctahedron, every vertex has three nearest neighbours, and each vertex is surrounded by a square and a regular hexagon. One can extend this analysis to neighbours of any order. You can also check this in the drawing of the Bravais lattice of a crystal (which visualizes a discrete translation group) in Figure 2.2. The crystal looks the same everywhere.

This is a property of a group. It is due to the existence of group automorphisms C_g defined by: $\forall g \in G : C_g \in F(G,G)$, whereby $\forall a \in G : C_g(a) = g \circ a \circ g^{-1} \in G$. The group automorphisms are themselves a group under the composition law of automorphisms. This looks extremely abstract, but it is extremely important to appreciate that it corresponds with the intuition that the group looks the same everywhere. The automorphism C_g maps each group element a in the environment of the identity element e onto the group element $C_g(a)$ in the environment of the group element g, making the environments look strictly identical. It maps the photograph of one local environment onto the photograph of another local environment, with the effect that all photographs look the same. On the commuting diagram (2.19) we see that applying g to all group elements maps the vertical arrow $e \to a$ to the vertical arrow $g \circ e \to g \circ a$. The first vertical arrow corresponds to a while the second vertical arrow corresponds to $C_g(a)$. We see thus that C_g is an "arrow mapper", mapping arrows onto arrows, as summarized in the commuting diagram (2.19). This diagram shows the action of C_g in function space $F(V, V)$:

$$\begin{array}{ccc} e \in G & \xleftarrow{\;g^{-1}\;} & g \circ e = g \in G \\ \downarrow{\scriptstyle a} & & \downarrow{\scriptstyle C_g(a) = g \circ a \circ g^{-1}} \\ a \in G & \xrightarrow{\;g\;} & g \circ a \in G \end{array} \qquad (2.19)$$

The vertical arrows function as "position vectors" for the elements in the environments. The "arrow mapper" C_g maps position vectors on the photograph of the environment of e to position vectors on the photograph of the environment of g. The action of C_g can also be represented by a diagram in model space. In such a diagram the vectors $\mathbf{OP} = \mathbf{v} \in V$ and $\mathbf{OP}_g = g(\mathbf{v}) \in V$ of the model space V are represented. The maps introduced at the beginning of this section are examples of this approach. The vectors $\mathbf{OP}_g$ here are true "position vectors" associated with group elements. The action of C_g can then be shown on the following commuting diagram for the vectors in model space V, which is used to represent the group elements in function space $F(V, V)$:

$$\begin{array}{ccc} \mathbf{v} \in V & \xleftarrow{\;g^{-1}\;} & g(\mathbf{v}) \in V \\ \downarrow{\scriptstyle a} & & \downarrow{\scriptstyle C_g(a) = g \circ a \circ g^{-1}} \\ \mathbf{v}' = a(\mathbf{v}) \in V & \xrightarrow{\;g\;} & g(\mathbf{v}') \in V \end{array} \qquad (2.20)$$

It can for instance be imagined that G is a rotation group acting on $V = \mathbb{R}^n$. The rotation a rotates the vector $\mathbf{v}$ to $\mathbf{v}' = a(\mathbf{v})$. The rotation $g \circ a \circ g^{-1}$ rotates $g(\mathbf{v})$ to $g(\mathbf{v}')$. It is therefore the "same" rotation as a, after all vectors $\mathbf{v} \in V$ have been bodily rotated to $g(\mathbf{v})$ by g. It can be said that a and $g \circ a \circ g^{-1}$ are *conjugate*. The "arrow mapper" C_g maps a onto its conjugate $g \circ a \circ g^{-1}$. When there is an isomorphic matrix representation for the rotation group, then the matrices $\mathbf{A}$ and $\mathbf{GAG}^{-1}$ will have the same eigenvalues. In the icosahedral group all products RR_aR^{-1} are five-fold axes, all products RR_bR^{-1} are two-fold axes, and all products $RR_aR_bR^{-1} = RR_aR^{-1}RR_bR^{-1}$ are three-fold axes.

It should be noted that in constructing the Figures 2.4 and 2.5, the choice made for the position vector $\mathbf{OP}$ is special with the aim of enhancing the visual appeal, although this is not a necessity and only a matter of aesthetics. Any vector $\mathbf{OP} \in \mathbb{R}^3 \backslash I$ can be taken to generate a diagram that illustrates Y. But to obtain a nice diagram with a buckyball a specific choice must be made. However, the choice $\mathbf{OP} \in I$ is not appropriate as this would make the diagram degenerate, due to the symmetry of I. In other words, a model based on I would not be an isomorphism, as each point of I would represent five different group elements. One says in this respect that the icosahedron I is not a *faithful* model of Y.

The model of the crystal lattice for a translation group is in this respect somewhat special in that the model is based not on a single position vector $\mathbf{OP}$ but a set U of position vectors of model space. This set U is called the *unit cell.* This set consists of the vectors of the type $\mathbf{O_1Q}$ in Figure 2.2, which link a node of a Bravais lattice L to a point in the unit cell U attached to that node. The nodes of the Bravais lattices are position vectors $\mathbf{OP}$ in Figure 2.2, that correspond to elements of the discrete translation group, but the position vectors of atoms in a unit cell, which are not on a node, are not translation vectors of the translation group. Actually, it could be stated that unit cells are used rather than single vectors to model the translation group. A unit cell of a crystal lattice is also a photograph of a local environment, but without group elements.

It is very important that the reader be able to see this intuitive idea of the overall similarity and the reason for it through the forest of abstract notations. The same kind of commuting diagrams as (2.19) and (2.20) relate double bonds to double bonds on the buckyball. So to say, when two persons are living on a same group, then when the first person faxes a photograph of his environment with a church at eight o'clock and a tree at four o'clock to the second person, then the second person will see that it is a perfect photograph of his own environment, with an identical church at eight o'clock and an identical tree at four o'clock. It is the "arrow mapper" C_g that will provide the one-to-one correspondence between the two photographs.

Conjugacy is an equivalence relation, and the corresponding equivalence classes are called *conjugacy classes.* To prove that a is conjugated to itself, we use $g = e$. We obtain then $g \circ a \circ g^{-1} = e \circ a \circ e^{-1} = a$. To prove $a \equiv b \Rightarrow b \equiv a$, we just express $a \equiv b$ by its definition $\exists\, g \in G \parallel b = g \circ a \circ g^{-1}$. The notation implies that the group element that provides for the equivalence $a \equiv b$ has been called g. From this it follows that $a = g^{-1} \circ b \circ g$, showing that the group element that provides the equivalence $b \equiv a$ is g^{-1}. Finally, to prove $a \equiv b \,\&\, b \equiv c \Rightarrow a \equiv c$, we use again the definitions: $(\exists\, g_1 \in G \parallel b = g_1 \circ a \circ g_1^{-1}) \,\&\, (\exists\, g_2 \in G \parallel c = g_2 \circ b \circ g_2^{-1})$. Feeding b from the first equation into the second equation we obtain: $\exists\, g_1 \in G, \exists\, g_2 \in G \parallel c = g_2 \circ g_1 \circ a \circ g_1^{-1} \circ g_2^{-1} = (g_2 \circ g_1) \circ a \circ (g_2 \circ g_1)^{-1}$, which completes the proof.

In the icosahedral group, the map C_g will map a two-fold axis to another two-fold axis, a three-fold axis to another three-fold axis, and a five-fold axis to another five-fold axis. A conjugacy class is thus a set of operators of the same type. This similarity will transpire through the fact that the matrices $\mathbf{A}$ and $\mathbf{GAG}^{-1}$ are linked by a similarity transformation and have the same eigenvalues. It may finally be noted that the mapping: $C \in F(G, F(G, G))$ with $\forall g \in G : C(g) = C_g$ is also a homomorphism. In fact, is is easy to check by using the definitions that $C(g_2 \circ g_1) = C_{g_2 \circ g_1} = C_{g_2} \circ C_{g_1} = C(g_2) \circ C(g_1)$.

One encounters the "arrow mapper" C_g in many texts that contain group theory, be it under the guise of more serious names. For instance the moves you can make on a Rubik's cube are a group, and the mathematical theory [Halberstadt (1980)] that tells you how to "solve" a configuration is based on elementary moves of the type $g \circ a \circ g^{-1}$, because they allow you to translate the instruction that tells you how to apply a particular move (an "arrow") in one environment into an instruction that tells you how to apply a "similar" move in another "similar" environment by conjugation. Consequently one will have to formulate instructions for only a limited set of configuration changes. In matrix form, conjugation corresponds to so-called similarity transformations. Similarity transformations are heavily used in quantum mechanics. In a more general abstract context where one does not necessarily use matrix representations, the terminology "adjoint representation" is being used for the group of transformations C_g as isomorphic images for g.[1]

[1]The arrows are a kind of displacements, and displacements in physical space are described by vectors. In a continuous group one can establish this conceptual link rigorously, by considering infinitesimal arrows. In a limit procedure these infinitesimal arrows will correspond to Lie derivatives, i.e. tangent vectors to the group manifold. The group manifold is here the map that visualizes the continuous group as a curved hyper-surface in some hyper-space. It is in general not a vector space due to the curvature. A precise definition of a Lie group is not given here, because this is a complicated matter

Such an overall similarity (all the photographs look the same) is what one calls *symmetry* and therefore groups embody symmetry. Groups are therefore ideal tools in physics to express invariance and co-variance. In a sense groups are the tools that best embody Einstein's principle of relativity. The equations of physics must be the same in all reference frames. This means that the equations of physics you observe cannot tell you in which reference frame you are, just as you can learn by inspecting the map of a group that you will not be able to tell from your environment where on the group you are. This shows why group theory is so important for physics. With his principle of relativity, in a sense Einstein expressed the fact that the Lorentz transformations form a group.

The all-important symmetry groups needed in physics are the translation groups, rotation groups, and the Lorentz group. They are groups of geometrical transformations. The rotation and translation groups are Euclidean geometry, while the Lorentz group occurs in the geometry of Minkowski space-time: These three groups describe the geometry of physics. Groups are the backbone of geometry. Klein's Erlangen Program aimed at describing all geometry based on groups and invariants. But the group theory used in physics to express the geometry does that most of the time algebraically, by using the matrix representations, although the concepts remain in principle purely geometrical. This is fine, as nothing is more convenient than algebra to make calculations. But without the parallel geometrical track that gives the intuitive meaning for the algebra, it very quickly starts to look impenetrable, even for people like Abdus Salam (see Chapter 1). One must therefore develop the geometrical track, which is why this book claims to be about geometry. It is understood now that this geometrical approach will take us on a journey into the abstract spaces $F(\mathbb{R}^3, \mathbb{R}^3)$ and $F(\mathbb{R}^4, \mathbb{R}^4)$, but that we will be able to model these spaces by using the physical spaces $\mathbb{R}^3$ or $\mathbb{R}^4$. Only the translation groups are Abelian, which is why the other groups are really difficult. The difference between the conceptually simple-looking Abelian groups and the more difficult non-Abelian groups must of course transpire in the algebra of the representation theory. Real or complex numbers are always commuting and can therefore be used to represent Abelian groups, like in the example of U(1). They can of course not be used to represent all details of non-Abelian groups, because they would not be able to reflect the non-commuting character of the operations. To represent

(see[Cornwell (1984)]). The tangent vectors build the Lie algebra associated with a Lie group.

non-Abelian groups we need also non-commuting matrices. There is a rigorous mathematical concept that corresponds to this notion. It is the concept of reducible and irreducible representations, which we will develop below.

2.10 Wave-like eigenfunctions

There is a second reason why group theory is very important. When you move from a place to a similar place within a group you will experience periodicity in your environment. The ideal tools to express such a periodicity are periodic functions, i.e. waves. This can be checked in the simplest case of discrete translation groups, which one uses in crystallography, for example. This can be illustrated with a simple example, based on the model illustrated in Figure 2.6. We start by considering a typical problem of probability calculus, *viz.* a simplified description of the jump diffusion of a single

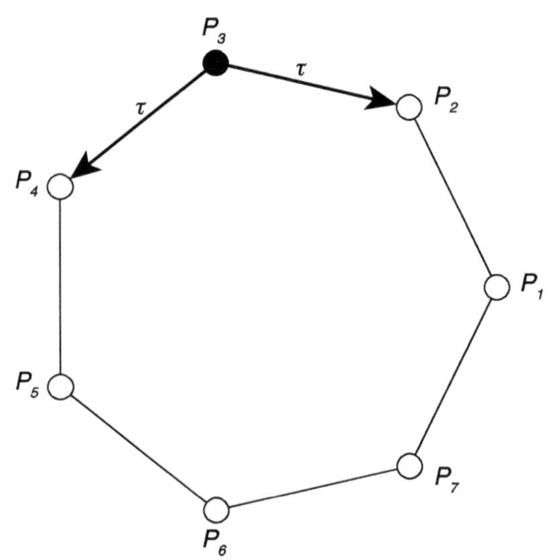

Fig. 2.6 Illustration of the jump model discussed in the text. A particle (indicated by the filled circle) occupies one of the $n = 7$ sites P_j of a regular polygon. The empty sites are indicated by open circles. In the figure the particle is thus situated at site P_3. It has the ability to jump to its first-neighbour sites as indicated by the arrows. The probability of making this jump is expressed in terms of a relaxation time τ. This relaxation time is the same for all first-neighbour jumps, such that the problem has rotational symmetry. This rotational symmetry can be described as translational symmetry with cyclic boundary conditions.

particle on a regular polygon of n vertices. The n vertices are considered to be equivalent and the ability to jump to a nearest-neighbour site is given in terms of a relaxation time τ. The precise jump mechanism itself is ignored and assumed to be infinitely fast. Such models are sometimes used in quasi-elastic neutron scattering studies in solid-state physics [Bée (1988)]. The rate equations can be written in matrix form as:

$$\frac{d}{dt}\,\mathbf{P} = -\frac{1}{\tau}\,\mathbf{MP}. \tag{2.21}$$

Here the $n \times n$ jump matrix $\mathbf{M}$ is defined by:

$$M_{jk} = -\delta_{j,k-1} + 2\delta_{j,k} - \delta_{j,k+1} \tag{2.22}$$

where all indices are to be taken modulo n. The elements $p_j(t)$ of the $n \times 1$ column matrix $\mathbf{P}$ give the probability to find the particle at site j at time t. To solve the set of coupled linear differential equations (2.21) and (2.22) the normal mathematical procedure is to diagonalize $\mathbf{M} = \mathbf{S}^{-1}\mathbf{\Lambda}\mathbf{S}$ such that it can be reduced to n decoupled equations in the new variables $(\mathbf{SP})_j$:

$$\frac{d}{dt}\,\mathbf{SP} = -\frac{1}{\tau}\,\mathbf{\Lambda}\,\mathbf{SP}. \tag{2.23}$$

Let us now consider the problem of the harmonic vibrations (phonons) of a linear chain of n identical atoms of mass μ and linear spacing $\mathbf{a}$, all linked by identical springs with identical spring constants κ. The normal procedure is to introduce cyclic boundary conditions [Ziman (1972)]. This spring model is symbolically illustrated in Figure 2.7. If we note as $\mathbf{u}_j$ the displacement of atom j from its equilibrium position, the dynamical equations can be cast into the matrix form:

$$\frac{d^2}{dt^2}\,\mathbf{U} = -\frac{\kappa}{\mu}\,\mathbf{MU}. \tag{2.24}$$

Here the elements U_j of the $n\times 1$ column matrix $\mathbf{U}$ are the the displacements $\mathbf{u}_j(t)$ of atom j with respect to its equilibrium position. The important point here is that the matrix $\mathbf{M}$ in (2.24) is the same one as in (2.21) and (2.22). Following the mathematics textbook (2.24) can be solved by the same procedure as (2.21) and (2.22), i.e. by diagonalizing $\mathbf{M}$:

$$\frac{d^2}{dt^2}\,\mathbf{SU} = -\frac{\kappa}{\mu}\,\mathbf{\Lambda}\,\mathbf{SU} \tag{2.25}$$

where the quantities $(\mathbf{SU})_j$ are again decoupled. However, in physics textbooks this is at first sight not the solution adopted. Based on intuition one expects the solutions to be wave-like; we are looking for lattice

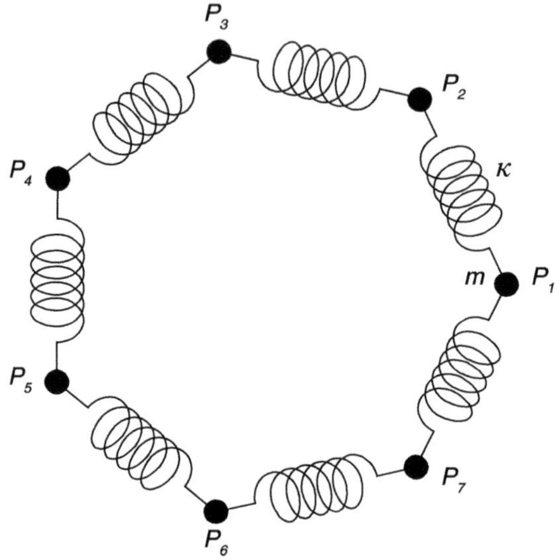

Fig. 2.7 Illustration of the spring model for a linear chain with cyclic boundary conditions discussed in the text. The true geometry of a linear chain of atoms and springs is not a circle, but picturing the $n = 7$ atoms symbolically on a circle at positions P_j reflects the symmetry and the first-neighbour connectivity of the model with its cyclic boundary conditions. The positions on the circle define thus a regular polygon. The atoms all have the same mass m and are linked to their first neighbours by springs of equal strength κ. Translational symmetry with cyclic boundary conditions is thus equivalent to rotational symmetry.

vibrations in the form of phonons. One therefore uses the so-called Bloch *ansatz* to postulate wave-like solutions of the form $\mathbf{u}_j(t) = \mathbf{w}_q(t)\, e^{\imath qj}$ with $q = \mathbf{Q}\cdot\mathbf{a}$ and $\mathbf{w}_q(t) = \mathbf{w}_q(0)\, e^{-\imath\omega t}$, where $\mathbf{Q}$ is a reciprocal lattice vector and $\boldsymbol{a}$ is the basis vector that generates the lattice. This leads to n decoupled equations (labelled q):

$$\frac{d^2}{dt^2}\,\mathbf{w}_q = -\frac{2\kappa}{\mu}\,(1-\cos q)\,\mathbf{w}_q. \tag{2.26}$$

By comparing (2.23) or (2.25) with (2.26) it must now transpire that the Bloch *ansatz* provides us with a very simple means of finding the eigenvectors of $\mathbf{M}$ and thus to diagonalize $\mathbf{M}$. In fact, the eigenvectors of $\mathbf{M}$ are the column matrices $\mathbf{V}^{(q)}$ defined by: $(\mathbf{V}^{(q)})_j = e^{\imath qj}$. This can be seen from:

$$\begin{aligned}(\mathbf{M}\mathbf{V}^{(q)})_j &= \sum_{k=1}^{n} M_{j,k}\, V_k^{(q)} = \sum_{k=1}^{n} (-\delta_{j,k-1} + 2\,\delta_{j,k} - \delta_{j,k+1})\, e^{\imath qk}\\ &= -e^{\imath q\,(j+1)} + 2\, e^{\imath qj} - e^{\imath q(j-1)} = (-e^{\imath q} + 2 - e^{-\imath q})\, e^{\imath qj}\\ &= (2 - 2\,\cos q)\,(\mathbf{V}^{(q)})_j\end{aligned} \tag{2.27}$$

from which we conclude $\mathbf{M}\mathbf{V}^{(q)} = 2\,(1-\cos q)\,\mathbf{V}^{(q)}$. One will recognize that the algebra developed in the derivation of this result is exactly the same as used in deriving (2.26) from (2.24). The Bloch theorem is thus a concealed diagonalization procedure. It was formulated for situations with translational invariance. This is actually our case here since the example embodies the treatment of the problem of a linear chain where one has introduced cyclic boundary conditions at distance n.

There is yet a third angle from which we can approach our jump problem. It permits us to formulate the arguments in a more general context than the two very specific physical models we have used here. We could already invoke the intuitive argument of periodicity described at the beginning of this section, to justify the *ansatz* of a Bloch wave. But the following ideas can also be used.

Consider two polynomials V_a, V_b in one variable x defined by: $(\forall x \in \mathbb{R})\,(V_a(x) = \sum_{k=0}^{K} a_k x^{K-k})$ and $(\forall x \in \mathbb{R})\,(V_b(x) = \sum_{j=0}^{J} b_j x^{J-j})$. The resultant $R(V_a, V_b)$ of these polynomials is defined as the polynomial in $(a_0, a_1 \cdots a_K, b_0, b_1, \ldots, b_J)$ — which we will note as $(\mathbf{a}, \mathbf{b})$ — that vanishes if and only if V_a and V_b have a common root. If the roots of V_a are called $x_1, x_2 \cdots x_K$ and the roots of V_b are called $\xi_1, \xi_2 \cdots \xi_J$ then: $R(V_a, V_b) = a_0^J b_0^K \prod_{k,j}(x_k - \xi_j)$. According to Sylvester $R(V_a, V_b) = \det(\mathbf{M}(\mathbf{a}, \mathbf{b}))$ where:

$$\mathbf{M}(\mathbf{a},\mathbf{b}) = \begin{pmatrix} a_0 & a_1 & a_2 & \cdots & & a_K & 0 & 0 & 0 & \cdots & 0 \\ 0 & a_0 & a_1 & \cdots & & a_{K-1} & a_K & 0 & 0 & \cdots & 0 \\ & & & & \vdots & & & & & & \\ 0 & 0 & 0 & \cdots & & 0 & a_0 & a_1 & \cdots & a_K & 0 \\ 0 & 0 & 0 & \cdots & & 0 & 0 & a_0 & \cdots & a_{K-1} & a_K \\ b_0 & b_1 & b_2 & \cdots & 0 & b_J & 0 & 0 & 0 & \cdots & 0 \\ 0 & b_0 & b_1 & \cdots & 0 & b_{J-1} & b_J & 0 & 0 & \cdots & 0 \\ & & & & \vdots & & & & & & \\ 0 & 0 & 0 & \cdots & & 0 & b_0 & b_1 & \cdots & b_J & 0 \\ 0 & 0 & 0 & \cdots & & 0 & 0 & b_0 & \cdots & b_{J-1} & b_J \end{pmatrix}. \tag{2.28}$$

There are J lines with coefficients from V_a and K lines with coefficients from V_b. This result can be derived by constructing J equations $V_a(x)\,x^j = 0; j = 0, 1, \ldots, J-1$ and K equations $V_b(x)\,x^k = 0; k = 0, 1, \ldots, K-1$. The matrix of this set of $J+K$ equations is exactly $\mathbf{M}(\mathbf{a},\mathbf{b})$. These $J+K$ equations will be satisfied simultaneously if and only if V_a and V_b have a common root x_m in which case we will have:

$$\mathbf{M}(\,\mathbf{a},\mathbf{b}\,)\,(x_m^{K+J-1}, x_m^{K+J-2} \cdots\, x_m^0\,)^\top = (\,0, 0, \ldots, 0)^\top. \tag{2.29}$$

Unless $x_m = 0$ it follows from this that $\det(\mathbf{M}(\,\mathbf{a},\mathbf{b}\,)) = 0$, q.e.d. A superb alternative proof is given in [Gel'fand *et al.* (1994)].

The derivation of the Sylvester determinant and especially the occurrence of (2.29) suggests interpreting the matrix $\mathbf{M}(\,\mathbf{a},\mathbf{b}\,)$ as containing the coefficients of polynomials, its eigenvectors $(x_m^{K+J-1}, x_m^{K+J-2} \cdots\, x_m^0\,)^\top$ being made from powers of a root x_m of these polynomials. The most striking feature of the matrix $\mathbf{M}(\mathbf{a},\mathbf{b})$ is that it contains diagonal stripes, where all lines are repeated identically.

Let us try this interpretation on the $(n \times n)$-matrix $-\mathbf{M}$ defined by (2.22) for the dynamics of a linear chain with cyclic boundary conditions:

$$\begin{pmatrix} -2 & 1 & 0 & \cdots & 0 & 0 & 1 \\ 1 & -2 & 1 & \cdots & 0 & 0 & 0 \\ 0 & 1 & -2 & \cdots & 0 & 0 & 0 \\ & \vdots & & \ddots & & \vdots & \\ 0 & 0 & 0 & \cdots & -2 & 1 & 0 \\ 0 & 0 & 0 & \cdots & 1 & -2 & 1 \\ 1 & 0 & 0 & \cdots & 0 & 1 & -2 \end{pmatrix}. \tag{2.30}$$

This matrix contains in fact a diagonal stripe as described above. There is only one disturbing feature that prevents us from developing the analogy suggested, *viz.* that there are numbers 1 in the top-right and in the bottom-left corners of this matrix. All other lines lead to equations of the type: $(x^2 - (2+\lambda)x + 1)\,x^{n-k-1} = 0$ with k running from 2 to $n-1$. However, it becomes clear, that the lines for $k = 1$ and $k = n$ are no exceptions provided we accept $x^n = 1$ as a compatibility condition. But this compatibility equation now gives us exactly the values of the roots x_m to put into $(x_m^{n-1}, x_m^{n-2} \cdots\, x_m^0\,)^\top$ in order to obtain the eigenvectors of $\mathbf{M}$. (2.29) then gives us the eigenvalues λ_m corresponding to the different eigenvectors built on the roots $x_m = e^{\imath 2\pi m/n}$. Our analogy makes sense and actually constitutes the mathematical proof of the Bloch theorem for a periodic lattice, where one identifies the term $e^{\imath 2\pi jk/n}$ with a wave vector k and a position vector x_j (or x if we ignore the discreteness of the lattice). The diagonal stripes we referred to are in fact the visual expression of the translational symmetry of the matrix $\mathbf{M}$. This translational symmetry corresponds to the fact that (2.22) has the same form for every value of $j \in [1, n] \cap \mathbb{N}$.

From all this it should now transpire that it is possible to diagonalize a huge matrix with translational symmetry analytically by making use of the symmetry. If there is one example that can illustrate the power of group theory in a simple way, and that could convince the reader that it could pay to learn the *Gruppenpest*, then it must be this example of using the symmetry to diagonalize a matrix. Without the symmetry it would most of the time be a hopeless task, as in general the brute-force diagonalization of a $n \times n$ matrix requires the determination of the eigenvalues and the eigenvectors through the solution of the characteristic equation, which is of degree n. In terms of radicals, this is only possible for values of n up to degree 4, such that in general, closed-form analytical expressions for the eigenvalues and eigenvectors cannot be derived.[2] But in our example

---

[2] As a matter of interest, it may be mentioned that Hermite has shown that the quintic equation can be solved in terms of elliptic integrals rather than radicals, but his construction must be considered more as an existence proof than a workable method. Most of the time we will have to make do with numerical solutions. The reader will have probably heard that Galois invented group theory to tackle the problem if the quintic can be solved with radicals. It is noteworthy here that a general method for the equations of degree $n \leq 4$ can be derived by using the method we have used here to derive the Bloch theorem. This permits us to see that a universal strategy underlies all four solutions. One calls a matrix with translational symmetry and cyclic boundary conditions as in (2.30) a *circulant matrix*. The Bloch waves permit us to solve easily the characteristic equation of a circulant matrix as the example of the phonons has shown. The idea is then to find a circulant matrix that has the equation to be solved as its characteristic equation. Expressing the conditions to be fulfilled leads every time to an equation of degree lower than n, such that the situation is simplified. Consider for instance the equation $x^2 + \alpha x + \beta = 0$ and the circulant matrix $\begin{pmatrix} a & b \\ b & a \end{pmatrix}$. The identification with the characteristic equation of the circulant matrix leads to: $-2a = \alpha$ and $a^2 - b^2 = \beta$. This can be solved in a straightforward way. The reader may try to find a solution for the cubic equation based on an analogous approach. It will permit the reader to appreciate how the three numbers 1, $e^{\imath 2\pi/3}$, and $e^{\imath 4\pi/3}$ (the cubic roots of 1) enter the scene. But when we arrive at the quintic, something goes wrong, thwarting the universal strategy. The conditions lead to an equation of degree 6, such that there is no simplification to an equation of lower degree.

the argument of translational invariance, *viz.* the fact that in a sense the situation looks the same everywhere, allows one in a very elegant way to come to terms with the *a priori* monstrous task of finding the eigenvalues and eigenvectors. That the situation looks the same everywhere is embodied by (2.22). Everywhere on the lattice this equation expresses the same relationship between a point with label j and its nearest neighbours labelled $j-1$ and $j+1$.

This is a consequence of the symmetry pointed out in Section 2.9. For an observer living on this lattice it would be impossible to find out from an inspection of his surroundings where on the lattice he was, i.e. what his personal value for j would be. We have seen that this is not a characteristic feature of translational symmetry alone. It also occurs for other symmetry groups. The example with translational invariance was simple, as the underlying translation group is commutative or Abelian.

The eigenvectors of the translation matrices have a wave-like structure, and we have seen that one calls the waves in question *Bloch waves*. The idea is easily generalized to n-dimensional translation groups and lattices. Due to the periodicity of the environments we can expect something analogous for any symmetry group. In daily life, one can go from point A to point B using different roads or different paths. The same applies to the group, and one of the constraints that will tell you which waves you can use is that the periodicities must turn out to be self-consistent when you walk these different paths. The connectivity of the group manifold can therefore be anticipated to play a role. The reader may check that the spherical harmonics for the rotation group, expressed in the spherical coordinates (θ, ϕ) also contain periodic functions in θ and ϕ. In solid-state physics the presence of the underlying Bloch waves is not always manifestly visible. They do come to the fore in the form of true physical waves within a phonon problem, but they do not within the diffusion problem. Nevertheless, they are present in the diffusion problem as well.

Now, in quantum mechanics we have a particle-wave duality that causes significant, conceptual problems. It is relatively hard to imagine how a particle could also be a wave and *vice versa*. It is an interesting speculation one might conceive based on the story with the Bloch waves, that the waves would just be a mathematical tool expressing the symmetry and that the real physics would be just the particles. This is especially tempting because the whole formalism of quantum mechanics seems to be written in group theory. But we will see that group theory and symmetry are not the only culprits for the presence of waves in the formalism. Nevertheless, the question remains interesting, even if the starting point we used to formulate this question is not correct. Let us be *home ludens* and play the game of finding out what kind of mileage one can get in trying to answer the question if the waves could be just a mathematical tool. After going through the whole book, it will be up to the reader to decide for himself what the answer may be.

2.11 Group representations

A (matrix) representation of a group $(G, \circ)$ is an isomorphism between group elements $g \in G$ and $n \times n$ matrices $\mathbf{D}(g)$:

$$\begin{array}{llll} \text{if} & g_1 & \rightarrow & \mathbf{D}(g_1) \\ & g_2 & \rightarrow & \mathbf{D}(g_2), \\ & & & \\ \text{then} & g_2 \circ g_1 & \rightarrow & \mathbf{D}(g_2)\mathbf{D}(g_1). \end{array} \tag{2.31}$$

A well-known example is given by the 3×3 rotation matrices SO(3) as a representation of the rotations in $\mathbb{R}^3$. In this case the matrices operate on 3×1 column vectors which are images of vectors $\mathbf{v} \in \mathbb{R}^3$. The number n is called the dimension of the representation. As already stated, matrix representations permit us to treat the whole group theory algebraically. There also exist infinite-dimensional representations of the rotation and Lorentz groups, but these will not be covered in this book.

2.12 Reducible and irreducible representations

We have seen that C_g maps a group element a to a similar, "conjugate" group element $g \circ a \circ g^{-1}$. In a matrix group it will map matrices $\mathbf{A}$ representing group elements a onto matrices $\mathbf{GAG}^{-1}$ representing group elements $g \circ a \circ g^{-1}$. This kind of mapping is a similarity transformation for matrices. We can now also use matrices $\mathbf{G}$ that do not represent themselves group elements but belong to the group ring to build similarity transformations. The similarity transformations then connect groups that are isomorphic. In fact, when $\mathbf{A} \rightarrow \mathbf{GAG}^{-1}$ and $\mathbf{B} \rightarrow \mathbf{GBG}^{-1}$, then $\mathbf{AB} \rightarrow \mathbf{GABG}^{-1} = \mathbf{GAG}^{-1}\mathbf{GBG}^{-1}$. Similarly, $\mathbf{GA}^{-1}\mathbf{G}^{-1} = (\mathbf{GAG}^{-1})^{-1}$ and $\mathbf{G}\mathbb{1}\mathbf{G}^{-1} = \mathbb{1}$. One can thus obtain equivalent representations of the group by similarity transformations.

A representation is described as *reducible* if a matrix $\mathbf{G}$ of the group ring can be found for a similarity transformation that transforms every matrix in the representation into the same pattern of n diagonal blocks $\mathbf{D}^{(j)}(g), j \in [1, m] \cap \mathbb{N}$, where each of the blocks $\mathbf{D}^{(j)}(g)$ is itself a representation of the group independent of the other blocks. The representation is then said to be decomposed into a direct sum of the $m \in \mathbb{N}$ matrices $\mathbf{D}^{(j)}(g)$:

$$\mathbf{D}(g) = \begin{pmatrix} \mathbf{D}^{(1)}(g) & & & & & \\ & \mathbf{D}^{(2)}(g) & & & & \\ & & \ddots & & & \\ & & & \mathbf{D}^{(j)}(g) & & \\ & & & & \ddots & \\ & & & & & \mathbf{D}^{(m)}(g) \end{pmatrix}. \quad (2.32)$$

In fact, the vector space $(V, K, +)$ of column matrices on which the $\mathbf{D}(g)$ matrices are working is the direct sum of the vector spaces $(V^{(j)}, K, +)$ of column matrices on which the matrices $\mathbf{D}^{(j)}(g)$ are working: $V = V^{(1)} \oplus V^{(2)} \cdots V^{(j)} \cdots \oplus V^{(m)}$. Here K is a notation for a general number field that can be $\mathbb{R}$ or $\mathbb{C}$. If the representation is not reducible, then the representation is called *irreducible*. It is a theorem that Abelian groups have only irreducible representations of dimension 1, while non-Abelian groups can have irreducible representations of dimension greater than 1. A group always has a trivial one-dimensional irreducible representation which maps every group element onto the number 1. The symmetric group $(S_n, ^\circ)$, which is non-Abelian, has another irreducible representation of dimension 1, *viz.* the mapping that attributes to each even permutation the number 1, and to each odd permutation the number -1. Similarly, the non-Abelian group of rotations and reversals in $\mathbb{R}^3$ can be given a one-dimensional "parity" representation in terms of numbers -1 and 1 depending on the question of whether the group element can be obtained from an even or an odd number of reflections. There is a geometrical reason for this similarity. The symmetric group $(S_n, ^\circ)$ can be visualized as the group of rotations and reversals that leave a regular simplex α_{n-1} in $\mathbb{R}^{n-1}$ invariant. A regular simplex α_n is a concept that generalizes the one of an equilateral triangle in $\mathbb{R}^2$, or of a tetrahedron in $\mathbb{R}^3$ to $\mathbb{R}^n$ (see [Coxeter 1963]). It is a regular polytope with $n+1$ vertices in $\mathbb{R}^n$, and is thus a set of $n+1$ points P_j with position vectors $\mathbf{r}_j \in \mathbb{R}^n$ such that $\exists d \in \mathbb{R}, \forall (j,k) \in ([1, n+1] \cap \mathbb{N})^2 : |\mathbf{r}_j - \mathbf{r}_k| = (1 - \delta_{jk})d$. This turns the algebraic concept of a permutation group into a geometrical concept of a symmetry group of a regular polytope, highlighting once more the intimate connection between group theory and symmetries of geometrical objects.[3]

It is a vital task in group theory to find all irreducible representations of a group, because they are the building blocks of all possible representations, reducible or irreducible. That has been done for the Lorentz group and the rotation group, but we will not develop this topic in this book.

There are special group-theoretical techniques to find the matrix $\mathbf{G}$ for the similarity transformation that block-diagonalizes a matrix with a given symmetry to the form given by (2.32), whereby all blocks correspond to irreducible representations. In fact, one can base the heuristics on an idea that is similar to the one we used to find the Bloch waves [Lyonnard (1997)]. But as the group is non-Abelian we will work with eigenvectors that contain non-commuting numbers, and the blocks matrices play the role of such numbers. Block diagonalization of a matrix by exploiting its non-Abelian symmetry is thus analogous to diagonalization of a matrix by exploiting its translational symmetry. To take into account the non-Abelian symmetry we use non-commutative numbers that can be represented by block matrices. This block-diagonalization could be a first step towards the complete diagonalization of the matrix, which should be always feasible provided the dimension of the irreducible representation matrices do not exceed 4. Symmetry can thus be used to block-diagonalize a matrix with a non-Abelian symmetry, and thus also to calculate the exponential of such a matrix, for example. Complete diagonalization of matrices with Abelian symmetry can be obtained by using Bloch waves as described above.

2.13 Eigenvector spaces

The following remark is not really group-theoretical, but it will be of some use later in the book. Figure 2.8 illustrates a jump model on an icosahedron. All local environments on the icosahedron are the same. The figure illustrates the first-neighbour environment of an arbitrary point. We can now imagine that an atom jumps with relaxation time τ between the various vertices of the icosahedron from neighbour to neighbour along the edges

[3] Any choice of $n-1$ points of the simplex α_n defines a $(n-1)$-dimensional hyperplane in $\mathbb{R}^n$, and a permutation of the two remaining points of the simplex corresponds to a reflection with respect to a hyperplane. Therefore, a transposition in the permutation group $(S_{n+1}, ^\circ)$ corresponds to a reflection in $\mathbb{R}^n$ with respect to a $(n-1)$-dimensional hyperplane of $\mathbb{R}^n$. These reflections generate the group of rotations and reversals.

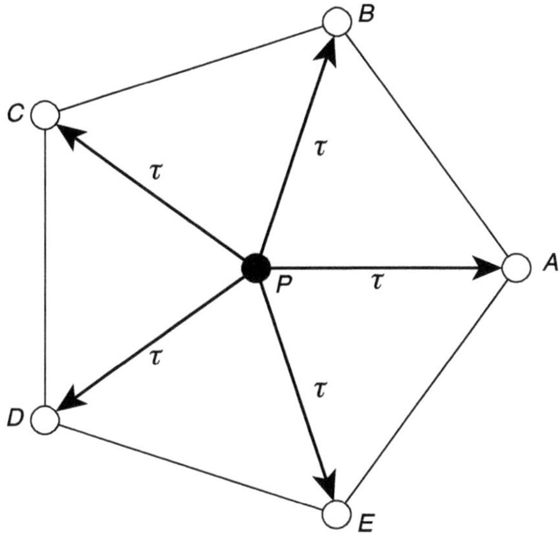

Fig. 2.8 Illustration of a jump model on the icosahedron. Only a pentagonal cap of the icosahedron taken from Figure 2.3 is illustrated, because in each point P of the icosahedron the jump probabilities are the same. The site P is thus situated above the plane defined by the other sites, A, B, C, D, and E. When the atom is at P then it can jump to the five nearest-neighbour sides (A, B, C, D, E) with a relaxation time τ

that join the neighbouring points. The centre of the icosahedron is called O, and the jump matrix $\mathbf{M}$. Just like in the jump model of Figure 2.6 the equations are the "same" everywhere, such that it suffices to draw Figure 2.8 to define the whole jump model and the whole jump matrix. The set of coupled rate equations can be written in matrix form as: $\frac{d}{dt}\mathbf{P} = \frac{1}{\tau}\mathbf{MP}$, where the entries of the matrix $\mathbf{M}$ are given by:

$$M_{jk} = -5\delta_{jk} + s_{jk}, \tag{2.33}$$

where $(\forall k \in [1, 12] \cap \mathbb{N})\,(s_{jk} = 1 \Leftrightarrow k \in S_j \;\&\; s_{jk} = 0 \Leftrightarrow k \notin S_j)$. Here S_j is the set of the five first neighbours of P_j. We will not solve this jump model, but only determine a few eigenvectors of it, using a symmetry argument. Consider for this purpose the position vectors $\mathbf{OP}_j = \mathbf{r}_j = (x_j, y_j, z_j)$ of the points P_j. It is then obvious that:

$$\Big[\sum_{P_k \in S_j} \mathbf{r}_k\Big] \parallel \mathbf{r}_j. \tag{2.34}$$

In fact, the points $P_k \in S_j$ are symmetrically distributed around $\mathbf{OP}_j$ such that their sum must be parallel to $\mathbf{OP}_j$. From this it is easy to see that the 12×1 column vector $[\mathbf{r}_1, \mathbf{r}_2, \ldots, \mathbf{r}_{12}]^\top$ is an "eigenvector" of the jump matrix $\mathbf{M}$. Using the coordinates provided in the caption of Figure 2.3 one can actually check that:

$$\Big[\sum_{P_k \in S_j} \mathbf{r}_k\Big] = (2\alpha - 1)\,\mathbf{r}_j, \tag{2.35}$$

such that the corresponding eigenvalue for the jump problem is $\lambda = (2\alpha - 6)/\tau$. However, we are used to considering eigenvectors strictly as 12×1 column vectors whose entries are scalars. We can remedy this by using an expedient that will often be used in this book: imagine that a $n_2 \times n_3$ matrix $\mathbf{A}$ consists of n_3 column matrices $\mathbf{a}_j$, where $j \in [1, n_3] \cap \mathbb{N}$ and the n_3 matrices $\mathbf{a}_j$ are of the type $n_2 \times 1$. Consider an $n_1 \times n_2$ matrix $\mathbf{B}$. Then $\mathbf{BA}$ is of the type $n_1 \times n_3$ and due to the definition of matrix multiplication, it consists of the n_3 column matrices $\mathbf{Ba}_j$. Working by multiplication on an $n_3 \times n_2$ matrix $\mathbf{A}$ is thus nothing other than working simultaneously by multiplication on n_3 different $n_2 \times 1$ column matrices $\mathbf{a}_j$. Rather than considering $\mathbf{A}$ as a fixed rigid mathematical entity, $\mathbf{A}$ will often be treated as a loose set of column matrices $\mathbf{a}_j$, where the set is assembled by mere juxtaposition, and we can move elements in or out of sets at will, in this way defining various different matrices.

The matrices can be thought of in terms of a kind of Meccano game. The individual pieces of this game are the column matrices. These individual pieces can be put together to make a construction and the constructions can be taken apart again. In this spirit, we can write $\mathbf{r}_j$ as an assembly of three coordinates (x_j, y_j, z_j). The latter implies that $[x_1, x_2, \ldots, x_{12}]^\top$, $[y_1, y_2, \ldots, y_{12}]^\top$, and $[z_1, z_2, \ldots, z_{12}]^\top$ are three linearly independent eigenvectors with the same eigenvalue λ. That is, we can consider $[\mathbf{r}_1, \mathbf{r}_2, \ldots, \mathbf{r}_{12}]^\top$ as a 12×3 matrix, which is the juxtaposition of three 12×1 eigenvectors with eigenvalue λ.

These three eigenvectors build a three-dimensional vector space V of eigenvectors. The eigenvalue λ is degenerate. The three eigenvectors we obtained are expressed with respect to a given choice of basis. By changing the basis they acquire new expressions, and by taking all possible orientations of the basis, the corresponding expressions will run through the whole vector space V. It is however much easier to make physical sense of the 12×3 eigenvector as a whole, than of the separate 12×1 matrices, which are in a sense mere vector projections that depend on an arbitrary choice of basis. In the analogy of the Meccano game, it will be easier to understand what is happing by explaining that the idea is to rotate a crane than by describing one by one the rotations of all the individual Meccano pieces used in the crane. In the latter case one might just miss out on the overall coherence of the global picture and on the general idea. A vector $\mathbf{r}_j$ is a meaningful quantity that does not depend on the choice of a basis. This idea will be used in Chapter 4, where it will be argued that it makes much more physical sense to use two-column quantities as spinors than single-column quantities, as the single-column quantities (which in following Cartan will be called *semi-spinors*) contain only half of the information. This is also the reason why the matrices of SL(2,$\mathbb{C}$) will be used as spinors for the Lorentz group. In a surprising further development it will turn out that the 4×1 matrices of the Dirac theory contain the same information as the SL(2,$\mathbb{C}$) matrices.

2.14 Groups and physics — final remarks

This subsection contains some general remarks about the relationships between physics and group theory, that will permit the reader to see the philosophy behind the methodology and the conceptual links. We have seen that the symmetry of a group corresponds to an idea reminiscent of Einstein's principle of relativity. This can be taken as a first hint of the importance of group theory for physics. There is another angle of approach that illustrates this importance further. In the monumental work of Misner *et al.* [Misner *et al.* (1970)] the authors try to convey the insight to the reader that physics is geometry. But according to Felix Klein's Erlangen Program, geometry is group theory. Combining the two, we can then conclude that physics should be group theory, and this seems to be confirmed in Wigner's work.

An argument developed in Section 2.10 explains why group theory contains waves. This could perhaps be one of the reasons why quantum mechanics is wave mechanics. Quantum mechanics is also matrix mechanics. What we use in quantum mechanics is more specifically group representation theory. The great idea of Descartes was to establish an isomorphism between geometry and algebra. With analytic geometry it becomes possible to check the truth of geometrical theorems mechanically. One could adopt the eccentric viewpoint that one does not need any additional insight to prove the theorems, because all one has to do is process the algebra. One could even go further and make a stand against wanting anything more: intuition kills rigour. (There is another taboo in physics with rather similar discouraging traits: it is generally admitted that quantum mechanics is beyond classical intuition). But for learning the subject matter and gaining insight it is much better to combine both rigour and intuition, such that one can swap constantly from one to the other. In fact, the algebra might prove terribly cumbersome without some clever moves prompted by visual clues on the geometrical side. The reader might try for instance his luck in finding a proof of Morley's theorem that the trisectors of the three angles of a triangle meet at the vertices of an equilateral triangle by purely algebraic methods. He will quickly discover that it is not a trivial task to disentangle the algebra without applying some dedicated identities.

The isomorphism that does the Descartes-like trick for group theory is matrix representation theory. It turns the whole geometry of group theory into matrix algebra just as analytic geometry turns the whole of Euclidean geometry into algebra. The whole subject becomes just a matter of calculating with matrices. Presumably, one may then expect to be able to carry out all the calculations without any insight into the underlying geometry. This is what quantum mechanics is about. In matrix mechanics one can just stick to the algebra without knowing what is going on behind the scenes and blindly follow the leitmotiv summarized in Mermin's witty slogan "Shut up and calculate". Nevertheless, the high level of abstraction and the lack of insight are considered disturbing by many people. In Feynman's words: "I think I can safely say that nobody understands quantum mechanics" [Feynman *et al.* (1964)]. The only way to tackle this problem is finding the meaning of the corresponding geometry, and this is what this book aims to be about.

The reader should be warned, however, that success is not automatically guaranteed. We have seen that the models visualized in Figures 2.6 and 2.7 lead to the same wave functions, even though the physical mechanisms that underly the models are very different. A mechanism for diffusion is all together different from a mechanism for phonon propagation. The wave functions only contain information about the symmetry, not the underlying mechanism. Group theory only deals with the symmetry. In as far as the mere algebra seems to reproduce with great accuracy all we can measure, such that quantum mechanics appears to be a complete theory, it might thus prove impossible to recover the physical mechanism from the theory. Further musings on this theme are given in Chapter 11.

Chapter 3

Spinors in the Rotation Group

No one fully understands spinors. Their algebra is formally understood but their general significance is mysterious. In some sense they describe the "square root" of geometry and, just as understanding the square root of -1 took centuries, the same might be true of spinors.

— Michael Atiyah [Farmelo (2009)]

3.1 Preamble

This section starts with some further remarks about group representation theory, using the group of three-dimensional rotations as a case in point. Some parts of the theory of spinors in the rotation group will be further developed starting from Section 3.2. The reader should remember that it is not the aim to reproduce the full theory of spinors in SO(3) or SU(2). An ample description is already given in many textbooks (for example [Cartan (1981); Chaichian and Hagedorn (1998); Cornwell (1984); Hladik (1996); Inui *et al.* (1990); Jones (1990); Misner *et al.* (1970); Smirnov (1972); Sternberg (1994)]). The aim of this chapter is rather to present the results of the theory in a new light. Wigner identified how important group theory is for quantum mechanics. The approach presented in this book will give a better geometrical insight into this important link. It will focus on a number of important points in the development, that will clarify the spinor idea and by analogy will also make it possible to understand the meaning of spinors in the Lorentz group.

3.2 Tensor products

With two representations $\mathbf{D}^{(1)}$ and $\mathbf{D}^{(2)}$ of dimensions d_1 and d_2 respectively, one can construct a new representation $\mathbf{D}$ of dimension $d_1 d_2$, namely:

$$g \to \mathbf{D}(g) = \mathbf{D}^{(2)}(g) \otimes \mathbf{D}^{(1)}(g), \tag{3.1}$$

which works on vectors $\mathbf{a} = \mathbf{w} \otimes \mathbf{v}$. This is a compact form to state that starting from quantities and transformation laws $w'_j = \sum_m D^{(2)}_{jm}\, w_m$ and $v'_k = \sum_n D^{(1)}_{kn}\, v_n$, one can define quantities $a'_{jk} = w'_j v'_k$, $a_{mn} = w_m v_n$, with a transformation law $a'_{jk} = \sum_{mn} D_{jk,mn}\, a_{mn}$, where $D_{jk,mn} = D^{(2)}_{jm}\, D^{(1)}_{kn}$, as is easily checked. It is also easily checked that these new matrices $\mathbf{D}^{(2)}(g) \otimes \mathbf{D}^{(1)}(g)$ do indeed build a representation. If the eigenvalues corresponding to the eigenvectors $\mathbf{w}_j$ of $\mathbf{D}^{(2)}$ are λ_j, and the eigenvalues corresponding to the eigenvectors $\mathbf{v}_k$ of $\mathbf{D}^{(1)}$ are μ_k, then the eigenvectors of $\mathbf{D}^{(2)}(g)\otimes\mathbf{D}^{(1)}(g)$ will be $\mathbf{w}_j\otimes\mathbf{v}_k$ with eigenvalues $\lambda_j\mu_k$. When the vectors $\mathbf{v}$ and $\mathbf{w}$ are identical, indices (jk) and (kj) must be grouped together, which will reduce the dimension to less than $d_1 d_2$. In the rotation group, this point summarizes the derivation of the whole set of harmonic polynomials (with one additional complication due to the fact that the polynomials are subject to constraints; see below).

3.3 What is a spinor or what kind of thing does a matrix of SU(2) work on?

In applying this idea inversely to the rotation group in $\mathbb{R}^3$, we observe that the eigenvalues for a rotation over an angle φ are $1, e^{\imath\varphi}, e^{-\imath\varphi}$, which is of the type $\lambda_1\mu_1, \lambda_1\mu_2 = \lambda_2\mu_1, \lambda_2\mu_2$, with $\lambda_1 = e^{\imath\varphi/2}$, $\lambda_2 = e^{-\imath\varphi/2}$ and $\mu_1 = e^{\imath\varphi/2}$, $\mu_2 = e^{-\imath\varphi/2}$. This suggests that there might exist a two-dimensional representation $\mathbf{D}$ wherein the eigenvalues are $\lambda_1 = e^{\imath\varphi/2}$ and $\lambda_2 = e^{-\imath\varphi/2}$, and for which the reduction of $\mathbf{D}\otimes\mathbf{D}$ (that is necessary due to $\lambda_1\mu_2 = \lambda_2\mu_1$) corresponds to the three-dimensional representation. The three-dimensional vectors would then in reality be composed quantities of the type $\boldsymbol{\xi}\otimes\boldsymbol{\xi}$ in terms of more basic quantities $\boldsymbol{\xi}$. (Of course, the two-dimensional representation referred to, *viz.* SU(2) is well known. The reader is supposed here to have some knowledge about it in order to be able to follow the argument, but at the end of this chapter, he will perfectly understand this). This finding corresponds to Atiyah's remark [Farmelo (2009)] that the object $\boldsymbol{\xi}$, called a spinor, should be the square root of a vector that is of the type $\boldsymbol{\xi} \otimes \boldsymbol{\xi}$. This surprising possibility has profound consequences: there must exist representations of the rotation group whose matrices do not work on images of vectors, as the two eigenvalues mentioned do not fit into a scheme for vector images. The question arises then on what kind of images these representations might work. The answer is simple: they work on images of rotations. In fact, the group structure of the rotation group exists without any reference to a vector of $\mathbb{R}^3$. All that the group structure really defines is the multiplication table of group elements. Such a multiplication table can be considered as the extrapolation to an infinite group G of what a Cayley table is for a finite group.

Considering the rotations as elements of a group introduces a real paradigm shift: a rotation is considered as a function acting on other rotations rather than on vectors.[1] The action of this function on the other rotations is just a left multiplication with a group element in the abstract group, as illustrated for the group element $g_k \in G$ in the multiplication table for the group G with composition law $\circ$ below:

$$\begin{array}{|c|cccccc|l}
\hline
\circ & g_1 & g_2 & g_3 & \cdots & g_j & \cdots & \\
\hline
g_1 & g_1 \circ g_1 & g_1 \circ g_2 & g_1 \circ g_3 & \cdots & g_1 \circ g_j & \cdots & \\
g_2 & g_2 \circ g_1 & g_2 \circ g_2 & g_2 \circ g_3 & \cdots & g_2 \circ g_j & \cdots & \\
\vdots & \vdots & \vdots & \vdots & & \vdots & & \\
\hline
g_k & g_k \circ g_1 & g_k \circ g_2 & g_k \circ g_3 & \cdots & g_k \circ g_j & \cdots & \leftarrow g_k, \\
\hline
\vdots & \vdots & \vdots & \vdots & & \vdots & & \\
\hline
\end{array} \tag{3.2}$$

where the function $g_k : G \to G;\ g_j \to g_k(g_j) = g_k \circ g_j$. The function g_k is no longer defined by all its function values $g_k(\mathbf{r}), \forall \mathbf{r} \in \mathbb{R}^3$, but all its values $g_k(g_j), \forall g_j \in G$. More rigorously, an arbitrary group element $g \in G$

[1] The reader will recognize here the ideas developed in the discussion about the difference between the icosahedron I and the icosahedral group Y in Section 2.9. The true objects of study in group theory are not the vectors but the elements of $F(\mathbb{R}^3, \mathbb{R}^3)$, as the abstract axioms of a group deny citizenship to the vectors by completely ignoring them and excluding them from the formulation. The composition law that is the subject of study is the composition of functions.

is identified with the function $f_g \in F(G,G)$ that maps G to G according to $f_g : g_j \in G \to f_g(g_j) = g \circ g_j$. The identification is introduced by noting for the sake of simplicity f_g as g. It implies that $f_g \in F(G,G)$ represents $g \in G$. Let this representation f_g of g be called the *automorphism representation.* The simplification of notation that identifies f_g with g is an *abus de langage* but permits us to write $g : g_j \in G \to g(g_j) = g \circ g_j$ as we did in (3.2) and permits grasping more easily the idea of interpreting a rotation as a function that works on other rotations rather than on vectors.[2]

The rotations are no longer considered as functions $f \in F(\mathbb{R}^3, \mathbb{R}^3)$ acting on vectors coded by their coordinates. There is no longer any mention of any vector $\mathbf{r} \in \mathbb{R}^3$ in such an abstract description in terms of rotations as functions acting on other rotations, nor of any length r of such a vector; all the attention is focused on the abstract structure of the group. This results in a minimal description, from which everything that is not essential has been stripped away; for example, any reference to vectors is removed, because while it might be necessary to consider vectors for carrying out the task of creating the group multiplication table, once its structure is in place, these vectors are no longer an indispensable part of the formulation. Perhaps we could reason then as follows: The most fundamental representation of a group should be devoid of any reference to such quantities like vectors. In this minimal approach, all a rotation can work on is another rotation, as this is all the group is about. The most basic representation *must* therefore be one that works on images of group elements. Therefore, the 2×1 column vectors the 2×2 matrices of the representation work on *must* code rotations. It will be shown that this reasoning holds for the rotation group, and that it can be developed by analogy into a guiding principle for the Lorentz group. This paradigm shift in the quantities that are represented by the column matrices can be summarized in the following diagram:

[2]By noting $f_g \in F(G,G)$ as $g \in G$ the group element g is identified with its automorphism representation f_g. This is analogous to identifying a group element with its representation matrix $\mathbf{D}(g)$. Eventually, the group element and a representation of it become intuitively the same thing and one can be substituted for the other. The automorphism representation is based on a homomorphism, e.g. $f_{h \circ g} = f_h \circ f_g$, as is easily proved. The symbol $\circ$ on the right-hand side stands here for the composition law for functions from $F(G,G)$. The proof follows then from: $\forall g_j \in G : (f_h \circ f_g)(g_j) = f_h(f_g(g_j)) = f_h(g \circ g_j) = h \circ (g \circ g_j) = (h \circ g) \circ g_j = f_{h \circ g}(g_j)$. One can also check that $f_{h^{-1}} = (f_h)^{-1}$ and $f_e = 1$, where e notes the unit element of the group, and 1 the identity mapping.

$$
\boxed{
\begin{array}{c}
\begin{array}{ccc}
\mathbf{r}_i \in \mathbb{R}^3 & \xrightarrow{g\in G} & g(\mathbf{r}_i) = \mathbf{r}'_i \in \mathbb{R}^3 \\
\downarrow & & \downarrow \\
\begin{bmatrix} x \\ y \\ z \end{bmatrix} & \xrightarrow{g\in G} & [\,\mathbf{R}_g\,] \begin{bmatrix} x \\ y \\ z \end{bmatrix} = \begin{bmatrix} x' \\ y' \\ z' \end{bmatrix}
\end{array} \\
\underbrace{\qquad\qquad\qquad\qquad\qquad\qquad\qquad\qquad}_{} \\
3 \times 3 \text{ rotation matrices } \mathbf{R}_g \text{ working on } \textit{vectors } \mathbf{r}. \\
\Downarrow \\
\begin{array}{ccc}
g_i \in G & \xrightarrow{g\in G} & g(g_i) = g \circ g_i = g'_i \in G \\
\downarrow & & \downarrow \\
\begin{bmatrix} \xi_0 \\ \xi_1 \end{bmatrix} & \xrightarrow{g\in G} & [\mathbf{D}_g] \begin{bmatrix} \xi_0 \\ \xi_1 \end{bmatrix} = \begin{bmatrix} \xi'_0 \\ \xi'_1 \end{bmatrix}
\end{array} \\
\underbrace{\qquad\qquad\qquad\qquad\qquad\qquad\qquad\qquad}_{} \\
2 \times 2 \text{ rotation matrices } \mathbf{D}_g \text{ working on } \textit{rotations } \boldsymbol{\xi}.
\end{array}
}
\tag{3.3}
$$

In this diagram a group element g is identified with the function $f_g : g_j \in G \to f_g(g_j) = g \circ g_j$. Such functions f_g are called *group automorphisms.*[3] This way, the definition of a rotation has been changed from an element $g \in F(\mathbb{R}^3, \mathbb{R}^3)$ to an element $f_g \in F(G, G)$. The 2×1 column matrices that code the rotations $g_j \in G$ and $f_g(g_j) \in G$, are the so-called *spinors.* The 2×2 SU(2) matrices code $f_g \in F(G, G)$. The representation is thus based on group automorphisms, i.e. the 2×2 representation matrices correspond to group automorphisms, while the spinors correspond to group elements. Whereas $\mathbf{r} = (x, y, z) \in \mathbb{R}^3$ serve as coordinates for vectors, the spinors $\boldsymbol{\xi}$, with $\boldsymbol{\xi}^\top = (\xi_0, \xi_1) \in \mathbb{C}^2$, fulfill the rôle of complex coordinates for rotations. At this point, the reader can already get an inkling that the point we want

[3]This rigorous treatment defines first the functions f_g and then the homomorphism $f \in F(G, F(G, G))$, $f : g \to f_g$, that maps group elements on corresponding group automorphisms. But such scientific rigour tends to hide the true motivation. To get the initial intuitive idea across it is therefore better to use the *abus de langage* discussed in the text.

to make for the rotation group must be true. In fact, as identified in (2.12), and as will be proved in this chapter (in the discussion preceding (3.14)), a rotation matrix in SU(2) is of the form:

$$\begin{pmatrix} a & -b^* \\ b & a^* \end{pmatrix}, \tag{3.4}$$

where $aa^* + bb^* = 1$. It is obvious that the information contents of the second column are completely defined by those of the first column. Hence, within a 2×1 column vector $[a, b]^\top$ the whole information content of the representation is found, and the formalism wherein 2×2 matrices work on 2×2 matrices can be replaced by one wherein the 2×2 matrices work on such 2×1 column matrices. The development will explain in which way this information about the rotation is coded into such column matrices.

It may also be observed that the simplified notation used for the automorphism representation f_g of the group element g by the group element g itself, shows some analogy with the dual-vector notation $\mathbf{a} \in \mathbb{R}^n$ for linear mappings $f_\mathbf{a} \in F(\mathbb{R}^n, \mathbb{R}) : \mathbf{r} \in \mathbb{R}^n \to f_\mathbf{a}(\mathbf{r}) = \mathbf{a} \cdot \mathbf{r} \in \mathbb{R}$.

3.4 Why we need a "*Vielbein*"

This raises the question of the coding: how can one turn the image of a rotation into a 2×1 column vector? A rotation is a linear mapping, and a linear mapping is entirely defined by its restriction to a basis. Hence, to know a rotation completely one must know how it works on the triad (*Dreibein* in German) of three basis vectors of an orthonormal reference frame.

The three vectors have nine components in total, but they are not all independent: the vectors are normalized (three conditions) and mutually orthogonal (three more conditions). Only three independent variables remain: the direction of a first unit vector (e.g. $\mathbf{e}_1 = (x_1, y_1, z_1)$) can be coded with two independent variables, while the direction of a unit vector that is orthogonal to it (e.g. $\mathbf{e}_2 = (x_2, y_2, z_2)$) can then be coded with just one more independent variable. This fixes the value of the third vector $\mathbf{e}_3 = \mathbf{e}_1 \wedge \mathbf{e}_2$. That a rotation is defined by three independent real parameters is also obvious by its description based on Euler angles (α, β, γ), or its description in terms of a rotation axis (which can be defined by a unit vector $\mathbf{n}$, i.e. two independent parameters) and a rotation angle φ.

A scheme for coding the information contained within a triad has been developed by Cartan [Cartan (1981)], clearly showing the leap from vectors to rotations. Two unit vectors of the triad are combined into a single vector by blending them into a single quantity $\mathbf{e}_1 + \imath \mathbf{e}_2 = (x, y, z)$. That is $x = x_1 + \imath x_2, y = y_1 + \imath y_2, z = z_1 + \imath z_2$, where $\mathbf{e}_1 = (x_1, y_1, z_1)$ and $\mathbf{e}_2 = (x_2, y_2, z_2)$. By doing so all the information about the triad is coded unambiguously. It can always be decoded back again: Whatever rotations are performed on this quantity, it will always be possible to identify the rotated images of the two basis vectors $\mathbf{e}_1'$ and $\mathbf{e}_2'$ afterwards by separating the real and imaginary parts.[4]

---

[4]In Chapter 2 it was argued that group theory does not contain vectors as the group elements g are functions $g \in G \subset F(V, V)$. But as these are rather abstract quantities, the concept of a model space has been introduced in an attempt to visualize them. When this idea is applied to the rotation group, the natural choice for the model space would thus be $V = \mathbb{R}^3$, but it turns out that $\mathbb{R}^3$ cannot function as the model space. The vector space $\mathbb{R}^3$ on which the matrices of SO(3) are acting could be considered as a model space for the rotation group according to a diagram of the type of (2.20), but this betrays the original idea of excluding the physical vectors, and using model spaces only to visualize the group elements g. The vector space $\mathbb{R}^3$ cannot function as the model space, because it does not have a subset S of vectors that would represent the group faithfully. As explained in Figure 3.2, each vector of $\mathbb{R}^3$ corresponds to an infinity of group elements. This is reminiscent of the situation with the icosahedral group Y, where the model of the icosahedron I in model space $V = \mathbb{R}^3$ is not an isomorphism for Y, because each point of the icosahedron represents five group elements, and the model is degenerated. For the icosahedral group the degeneracy can be avoided by restricting the model space to $V \backslash I$, i.e. by excluding the points that lead to the degenerated model from the model space. The buckyball is a subset $S \subset V \backslash I$ of this restricted model space. But the same thing cannot be done for the rotation group as the degenerated model is a sphere, which is already the whole model space wherein one would want to build the model. In other words the restriction would be empty.

Could one nevertheless find a genuine vector model according to a diagram of the type of (2.20) to visualize the abstract concept of a rotation group that represents the group faithfully, just like the buckyball does for the icosahedral group? This may look impossible, but against all odds, the group can be represented faithfully by searching for a better vector model based on a less obvious choice $V' \neq V$ for the model space. This choice can be found by going back to the original idea, i.e. a diagram of the type of (2.19). The first faithfull representation of the rotation group that has been proposed in this book visualizes the rotations in terms of triads. Due to the isomorphism, the triads *are* rotations. A triad is a perfect visual picture for a rotation, but it is a set of three vectors rather than a single vector. This suggests that the model space would instead be something like $\mathbb{R}^9$. Fortunately, the triads are themselves already faithfully represented by two vectors, which reduces the model space to $\mathbb{R}^6$, and these can in turn be represented faithfully by isotropic vectors which are elements of $\mathbb{C}^3 \equiv \mathbb{R}^6$. Due to this second series of isomorphisms, the isotropic vectors *are* triads, and due to the first isomorphism they are thus rotations. In summary, this means that isotropic vectors constitute a faithful vector model for the rotation group in the less obvious model space $V' = \mathbb{C}^3$. As it is faithful, it really can be said that group elements are rotated when these isotropic vectors are rotated. This idea thus permits visualizing the abstract concept of the rotation group by

The extrapolation of the algebra involved in the Euclidean distance function of $\mathbb{R}^3$ to $\mathbb{C}^3$ leads to the finding that the quantity $\mathbf{e}_1 + \imath\mathbf{e}_2$ obtained this way by combining two orthogonal unit vectors is a so-called *isotropic vector*, i.e. a vector of "zero length", but the latter formulation is a misuse of language, since after the extrapolation of the Euclidean distance function towards $\mathbb{C}^3$ it no longer defines a distance function. In fact, $x^2 + y^2 + z^2 = (x_1 + \imath x_2)^2 + (y_1 + \imath y_2)^2 + (z_1 + \imath z_2)^2 = 0$, which is completely at odds with a basic axiom for a distance function, *viz.* that it should be positive-definite. (The distance function to be used in $\mathbb{C}^3$ is $xx^* + yy^* + zz^*$.) The transition from vectors to rotations can be made through the use of isotropic vectors. Note that in reality it is not necessary to work on the unit vectors, but only on their directions: a vector remains isotropic when multiplied with a constant. This is why homogeneous coordinates can be used and why representations in terms of homogeneous coordinates (i.e. the harmonic polynomials) are found in the example of the three-dimensional rotation group.

The generalization to higher dimensions of the idea that one must code the whole *n-bein* is a leading principle for representation theory, although it is *a priori* not obvious how one can, for instance, code the four unit vectors of the tetrad (*Vierbein*) in $\mathbb{R}^4$ into one quantity. One runs out of quantities like $\imath$. There is no commutative number field beyond $\mathbb{C}$ and therefore it looks at first sight as though one can only proceed further by introducing non-commutative algebra. (It will be explained later how to overcome these difficulties for $\mathbb{R}^4$, and especially for the Lorentz group of special relativity.)

Eventually, it will be possible to appreciate that the idea of coding the *n-bein* as an image of the group element is rigorously respected and the guiding principle for the development; spinors are nothing other than the appropriate coding in the form of (a set of) column matrices of the elements of the rotation group in $\mathbb{R}^n$ pictured as an *n-bein*. This is precise and clear, and it is noteworthy that such a clear statement is hard to find in the specialized literature, with the immediate consequence that to many the spinor concept looks impenetrable and shrouded in mystery.[5] But if the argument about

---

a vector model according to a diagram of the type of (2.20). The model is not artificial as it is an isomorphism; the model vectors are the isotropic vectors.

[5]Understanding spinors involves thus two things: (1) Understanding the intuitive idea of considering a rotation as a function working on other rotations rather than on vectors,

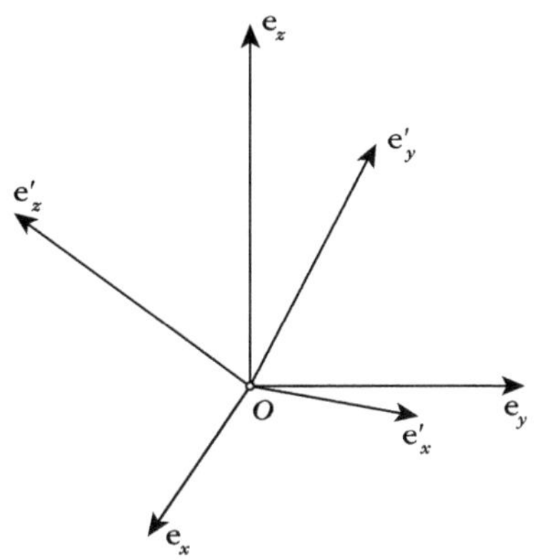

Fig. 3.1 Images of a spinor and a rotated spinor, represented at the origin O of a reference frame. The triad of the three unit vectors $\mathbf{e}_x, \mathbf{e}_y, \mathbf{e}_z$ (along the axes x, y, z of the frame) is the reference spinor that defines the identity element of the rotation group. The rotated spinor is the rotated triad of unit vectors $\mathbf{e}'_x, \mathbf{e}'_y, \mathbf{e}'_z$ (along the rotated axes x', y', z' of the rotated frame).

what is the minimal information which defines a group is combined with the idea that a group element can be visualized as an *n-bein*, the further development looks quite cogent.

To conclude this discussion, we visualize a spinor for the three-dimensional rotation group and its rotation in Figure 3.1. The crucial difference with the rotation of a vector is visualized in Figure 3.2. These two figures summarize one of the major messages of this book. Classical mechanics is very intuitive. The calculations are performed on vectors and using intuition it is possible to visualize such objects. But quantum mechanics looks abstract and devoid of images. This is because the calculations of quantum mechanics are made on spinors, for which a mental picture is not readily available. Figures 3.1 and 3.2 give such an image of a spinor. A spinor is just a set of coordinates for a group element. Any complete set of coordinates will do, and to visualize them the triads of basis vectors that are in one-to-one correspondence with their associated rotations are very convenient. With this image of a spinor we will be able to visualize things again like in classical mechanics.

---

as this is all the structure of a group is about. This corresponds to introducing the automorphism representation. (2) Visualizing the rotations in terms of a triad (as will be further explored in Sections 3.6 and 3.7).

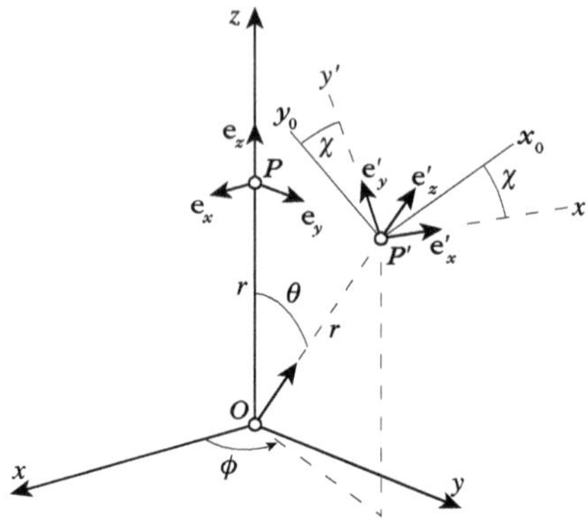

Fig. 3.2 Images of a spinor and a rotated spinor represented on the surface of a sphere (for comparison with the rotation of a vector). The triad of the three unit vectors $\mathbf{e}_x, \mathbf{e}_y, \mathbf{e}_z$ in $P(0,0,r)$ is the reference spinor that defines the identity element of the rotation group. The rotated spinor is the triad in P' of the three rotated unit vectors $\mathbf{e}'_x, \mathbf{e}'_y, \mathbf{e}'_z$. In a description based on vectors, we would represent the rotation by the rotated vector OP' that defines the new z-axis z', with two parameters (θ, ϕ) that define $OP' = (r \sin\theta\cos\phi, r\sin\theta\sin\phi, r\cos\theta)$. The vector $OP'(\theta,\phi)$ does not define the rotation unambiguously, while the spinor does. In fact, after the rotation of OP to OP', the rotated x-axis x_0 and y-axis y_0 may still not be aligned with x' and y'. A further rotation over an angle χ around OP' will make the x and y axes coincide with x' and y'. The supplementary angle χ permits us to define the spinor and the rotation completely. When one must simultaneously visualize several spinors, it is preferable for the sake of clarity to represent them (like we have done here) in different positions P on the surface of a sphere defined by $\mathbf{e}'_z$, rather than all together at the origin.

3.5 Dirac's method to code 3D vectors as 2D complex quantities

For the two-dimensional representation of the rotation group, the two three-dimensional vectors must be combined into an isotropic vector. But there is a problem: how can three-dimensional vectors be coded into a two-dimensional formalism in the first place? The spinors are elements of $\mathbb{C}^2$, while the isotropic vectors are elements of $\mathbb{C}^3$. This is where Dirac's expedient comes in. The reader will recognize in the following heuristics a construction that Dirac used in introducing his famous equation (see Chapter 5). Consider a unit vector $\mathbf{a} = (a_x, a_y, a_z)$. This unit vector will be used to code a reflection with respect to the plane that is normal to this vector.

In fact, as illustrated in Figure 3.3, reflections are the generators of the rotation group and the group can be built starting from these generators.

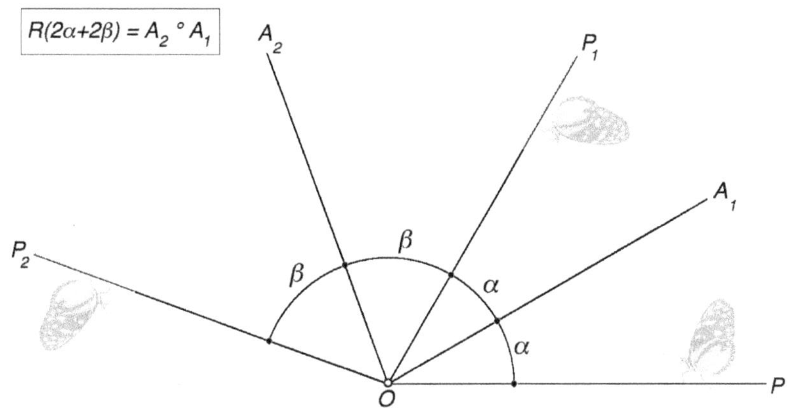

Fig. 3.3 A rotation R in $\mathbb{R}^3$ as the product of two reflections defined by their reflection planes A_1 and A_2. Starting from two arbitrary planes A_1 and A_2 in $\mathbb{R}^3$ that intersect along a straight line n, the plane of the figure is taken perpendicular to the line n and intersects n in the point O. The names A_1 and A_2 of the planes are used to label both their intersections with the plane of the figure and the reflection operations they define. The position vector OP of the point P to be reflected, is at an angle α with respect to A_1, where $A_1(P) = P_1$. The position vector OP_1 is at angle β with respect to A_2. The angle between A_1 and A_2 is then $\alpha + \beta$. As can be seen from their operations on the butterfly, reflections have negative parity, but the product of two reflections conserves the parity. The product of the two reflections is therefore a rotation $R = A_2 \circ A_1$, with axis n and rotation angle $2(\alpha + \beta)$. Only the relative angle $\alpha + \beta$ between A_1 and A_2 appears in the final result, not its decomposition into α and β. Hence, the final result will not be changed when the two planes are turned together as a whole around n keeping $\alpha + \beta$ fixed. This shows that there is an infinite number of ways to decompose a rotation into two reflections. (This is useful for calculating the product of two spatial rotations $R_1 = A_2 \circ A_1$ with axis n_1 and $R_2 = A_4 \circ A_3$ with axis n_2 by using the freedom to choose $A_2 = A_3$ as the plane that contains both n_1 and n_2, such that $R_2 \circ R_1$ reduces then to $A_4 \circ A_1$.) On the other hand, when the plane A_1 is fixed and the plane A_2 allowed to turn, such that $\alpha + \beta$ increases starting from zero, the rotation angle $2(\alpha + \beta)$ runs twice as fast as the angle $\alpha + \beta$ between the reflection planes. When $\alpha + \beta$ reaches the value π, the planes A_1 and A_2 will coincide again (but with opposite orientation normals). The corresponding rotation angle will be 2π and the rotation obtained corresponds to the identity element. To recover the same orientation normal, the plane must turn over 2π, resulting in a 4π rotation.

The first step is to attempt to code the reflections into 2×2 matrices. This leap from vectors to reflections is easier than that from isotropic vectors to rotations. Rotations will follow by combining reflections. The components of the vector $\mathbf{a}$ that defines the reflection A will appear somewhere in the matrix being sought as parameters, but it is not clear how or where. Therefore, we decompose the matrix $\mathbf{A}$ that codes the reflection A defined by $\mathbf{a}$ linearly as $a_x\sigma_x + a_y\sigma_y + a_z\sigma_z$, where $\sigma_x, \sigma_y, \sigma_z$ are unknown matrices, as summarized in the following diagram:

$$\begin{array}{ccc} \mathbf{a} = (a_x, a_y, a_z) \in \mathbb{R}^3 & \xrightarrow[\text{reflection } A]{\textbf{a} \text{ defines a}} & 2 \times 2 \text{ matrix } \mathbf{A} \\ \Big\downarrow \text{definition} & & \Big\downarrow \text{Dirac's heuristics} \\ \mathbf{a} = a_x\mathbf{e}_x + a_y\mathbf{e}_y + a_z\mathbf{e}_z & \xleftarrow[\text{decompositions}]{\text{analogy of}} & \mathbf{A} = \underbrace{a_x\sigma_x + a_y\sigma_y + a_z\sigma_z}_{\text{noted as } \mathbf{a}\cdot\boldsymbol{\sigma}}. \end{array} \tag{3.5}$$

If the matrix σ_x is known, it will indicate where and with which coefficients a_x appears in $\mathbf{A}$. The matrices $\sigma_x, \sigma_y, \sigma_z$, for reflections within $\mathbb{R}^3$, can be found by expressing isomorphically through $\mathbf{AA} = \mathbb{1}$ what defines a reflection, *viz.* that the reflection operator A is idempotent. This will be the case provided the three matrices simultaneously satisfy the six conditions $\sigma_\mu\sigma_\nu + \sigma_\nu\sigma_\mu = 2\delta_{\mu\nu}\mathbb{1}$, i.e. provided one takes the Pauli matrices for $\sigma_x, \sigma_y, \sigma_z$. The matrix $\mathbf{A}$ will be given by:

$$\mathbf{A} = a_x\sigma_x + a_y\sigma_y + a_z\sigma_z = \begin{pmatrix} a_z & a_x - \imath a_y \\ a_x + \imath a_y & -a_z \end{pmatrix}. \tag{3.6}$$

The Pauli matrices are thus reflection operators. For readers used to an approach based on Lie algebra, this may come as a surprise (see Subsection 5.10.1). The diagram in (3.5) shows how within the set of complex 2×2 reflection matrices, the matrices $\sigma_x, \sigma_y, \sigma_z$ play a role that is analogous to that of the basis vectors $\mathbf{e}_x, \mathbf{e}_y, \mathbf{e}_z$ within $\mathbb{R}^3$.

Going beyond the idea that unit vectors identify reflections, $\sigma_x, \sigma_y, \sigma_z$ can be used to code a vector $\mathbf{v}$ of $\mathbb{R}^3$ of any length as a 2×2 matrix $\mathbf{V} = v_x\sigma_x + v_y\sigma_y + v_z\sigma_z$. It is found then that $\mathbf{V}^2 = \mathbf{v}\cdot\mathbf{v}\mathbb{1}$ codes $\mathbf{v}\cdot\mathbf{v}$. More generally, $2\mathbf{w}\cdot\mathbf{v}\,\mathbb{1} = \mathbf{VW} + \mathbf{WV}$. As a reflection defined by a unit vector $\mathbf{a}$ maps a vector $\mathbf{v}$ onto $\mathbf{v} - 2(\mathbf{a}\cdot\mathbf{v})\,\mathbf{a}$, the outcome of operating the reflection A on the vector represented by the matrix $\mathbf{V}$ is given by $-\mathbf{AVA}$. This presents a major problem, because the result is quadratic in the matrix $\mathbf{A}$. In other words, the representation is not linear.[6]

It is obvious that a rotation resulting from two successive reflections A and B respectively defined by unit vectors $\mathbf{a}$ and $\mathbf{b}$, will then be given by $\mathbf{BAVAB}$, which is of the form $\mathbf{SVS}^{-1}$, with $\mathbf{S} = \mathbf{BA}$. To render the

[6] An example of a similar quadratic dependence of a rotation on certain parameters is the expression in $\mathbb{R}^3$ for a rotation with axis $\mathbf{n}$ and angle φ, as given, for example, in equation (3.3.4) in [Chaichian and Hagedorn (1998)].

formalism linear, an attempt will be made to split it by writing $\mathbf{V}$ as $\mathbf{S}_1\mathbf{K}\mathbf{S}_1^{-1}$, such that in the end the group can be represented linearly as matrices $\mathbf{A}$ working on the left on $\mathbf{S}_1$ and/or as matrices $\mathbf{A}^{-1} = \mathbf{A}$ working on the right-hand side of $\mathbf{S}_1^{-1}$. This is self-consistent as $\mathbf{S}_1 \to \mathbf{A}\mathbf{S}_1$ is equivalent to $\mathbf{S}_1^{-1} \to \mathbf{S}_1^{-1}\mathbf{A}^{-1}$.

To obtain more clarity regarding this problem $\mathbf{V}$ can be diagonalized as $\mathbf{S}_1\mathbf{K}\mathbf{S}_1^{-1}$. By doing so, the structure of $\mathbf{S}_1\mathbf{K}\mathbf{S}_1^{-1}$ perfectly matches the structure of $\mathbf{S}\mathbf{V}\mathbf{S}^{-1}$ for a rotation.

3.6 From vectors to spinors: Preliminary description

The representation is not linear because it was applied to vectors rather than to rotations. It should be remembered that the eigenvalues in a representation based on vectors were of the type $\lambda_1^2, \lambda_1\lambda_2, \lambda_2^2$, and that the goal is to find the representation that turns out eigenvalues λ_1, λ_2. The idea of halving the formalism should therefore not come as a surprise, and a vector can be expected to be of the second degree in the more basic quantities. This confirms the fact that $\mathbf{V}$ is a vector, which should in some way be turned into a rotation, by taking a kind of square root (to get from eigenvalues of the type $\lambda_j\lambda_k$ to eigenvalues of the type λ_j).

The full calculations that show how this can be achieved will be given in Section 3.7. The underlying ideas will be described here. As it is the isotropic vectors that must code a rotation, $\mathbf{V}$ must be replaced by a matrix $\mathbf{M}$ that codes an isotropic vector. Suppose as a matter of heuristics that it is also possible to diagonalize the matrix $\mathbf{M}$ as $\mathbf{T}\mathbf{W}\mathbf{T}^{-1}$. Because the structures $\mathbf{T}\mathbf{W}\mathbf{T}^{-1}$ and $\mathbf{S}\mathbf{V}\mathbf{S}^{-1}$ are the same, it will then no longer be possible to tell isotropic vectors and rotations apart if the two diagonalizations are identifiable, provided $\mathbf{W}$ can be identified with the value of $\mathbf{K}$ for some element of the group. This will then be a way to jump logically from a representation in the form of vectors towards a representation in the form of rotations.

But this idea is totally thwarted by the fact that an isotropic vector has "zero length". This implies that $\mathbf{M}^2 = 0$ and therefore that $\mathbf{M}$ cannot be diagonalized because both its eigenvalues are zero. The way out of this impasse is a *re-normalization procedure*. First, diagonalize the matrix $\mathbf{V}$ corresponding to a non-isotropic vector. The eigenvalues turn out to be r and $-r$ where r is the length of the vector. But the amplitude r can be factorized into two terms $\sqrt{r}$, one of which is relegated to the left to combine it with $\mathbf{S}_1$, and one which is moved to the right to combine it with $\mathbf{S}_1^{-1}$. The magic is that when after doing this r is allowed to tend to zero, all quantities remain finite and non-trivial, and this renders the idea of identifying the diagonalization procedures viable again, be it in a modified form.

Simultaneously, a *principle of homogeneity* is recovered that becomes enabled by the prior re-normalization. All remaining expressions become homogeneous in the coordinates (x, y, z). This principle is a kind of surprising mental leap, embodying the leap from vectors to rotations. Without it, all is still in terms of vectors, and it is therefore crucial. The homogeneity principle and the obligation to code the triad become simultaneously enabled by taking the limit $r \to 0$. As mentioned earlier, logically the precise length r of a vector has *a priori* no place in a formulation of the group. Everything should be independent of the precise value of r. It was also stated early on that it is only the directions of the unit vectors of the triad that count. This transpires in the final formalism through the obvious property that the formalism is scale-invariant. This is exactly what the homogeneity principle is about: it leaves scale-invariant homogeneous coordinates. But this is achieved in the most radical fashion: by dismissing the quantity r altogether by making it equal to zero. r is removed from the formalism because r does not appear in the group multiplication table. It has no rôle in the minimal set of parameters that should allow discussing the group table. This reflects the metaphor used in the Introduction that "two chairs define a single table". The mathematical structure should appear from the reasoning without any reference to the underlying meaning that "two points define a straight line". The mathematical structure is provided by the group table, not by interpreting the group elements in terms of how they are operating on vectors. By taking that interpretation too much into account, the fact that a different interpretation exists with a deeper meaning was overlooked, *viz.* one in terms of group elements working on other group elements. Moreover, putting $r = 0$ for a vector renders it easier to express the constraint $x^2 + y^2 + z^2 = r^2$ that defines rotations as isometries, and turns the rotation group into a manifold rather than a vector space. What is left of $\mathbf{W}$ after taking r out can be identified with a reflection matrix, such that the initial goals are fully satisfied.

3.7 From vectors to spinors: Detailed derivation of the expression of the spinors

The ideas applied to the diagonalization of an isotropic vector do indeed reproduce exactly the definition of the spinors for the rotation group as given by Cartan[Cartan (1981)]. It also shows that vectors are second-degree tensor products of spinors. The detail of the calculations described is as follows: the matrix $\mathbf{R}$ that codes the vector $\mathbf{r} = (x, y, z)$ of length $r \neq 0$ is given by:

$$\mathbf{R} = x\sigma_x + y\sigma_y + z\sigma_z = \begin{pmatrix} z & x - \imath y \\ x + \imath y & -z \end{pmatrix}. \tag{3.7}$$

Here the notation (x, y, z) represents a vector that is not of "zero-length", anticipating that eventually it will be replaced by the isotropic vector (x, y, z) defined above. The eigenvalues of this matrix are r and $-r$. Diagonalization leads to:

$$\mathbf{R} = \frac{1}{\sqrt{-2r(x - \imath y)}} \begin{pmatrix} x - \imath y & x - \imath y \\ -z + r & -z - r \end{pmatrix} \begin{pmatrix} r & 0 \\ 0 & -r \end{pmatrix}$$
$$\times \begin{pmatrix} -z - r & -(x - \imath y) \\ z - r & x - \imath y \end{pmatrix} \frac{1}{\sqrt{-2r(x - \imath y)}}, \tag{3.8}$$

and after applying the re-normalization procedure:

$$\mathbf{R} = \frac{1}{\sqrt{-2(x - \imath y)}} \begin{pmatrix} x - \imath y & x - \imath y \\ -z + r & -z - r \end{pmatrix} \begin{pmatrix} 1 & 0 \\ 0 & 1 \end{pmatrix}$$
$$\times \begin{pmatrix} -z - r & -(x - \imath y) \\ z - r & x - \imath y \end{pmatrix} \frac{1}{\sqrt{-2(x - \imath y)}}. \tag{3.9}$$

After taking the limit $r \to 0$ the following is obtained after some algebra for both columns in the left-hand matrix:

$$\begin{pmatrix} \xi_0 \\ \xi_1 \end{pmatrix} = \begin{pmatrix} \pm\sqrt{\frac{x - \imath y}{2}} \\ \pm\sqrt{\frac{-x - \imath y}{2}} \end{pmatrix}, \tag{3.10}$$

in agreement with the result given by Cartan. Note that in the limit $r \to 0$, $(x, y, z) = (x_1 + \imath x_2, y_1 + \imath y_2, z_1 + \imath z_2)$ as discussed above and therefore codes the whole triad of a rotated reference frame. Based on the definitions:

$$x = \xi_0^2 - \xi_1^2, \tag{3.11}$$

$$y = \imath(\xi_0^2 + \xi_1^2), \tag{3.12}$$

$$z = -2\xi_0\xi_1, \tag{3.13}$$

from which it can be seen that (ξ_0, ξ_1) code the whole triad of the rotated reference frame. It must be noted that x, y and z are complex numbers.

For this reason, complex conjugation should not be too routinely used on ξ_0, ξ_1, and other quantities like $x + \imath y$: e.g. the complex conjugate of $x + \imath y$ is not $x - \imath y$ but $x^* - \imath y^*$, etc. ...

There is an alternative way to deduce (3.10–3.13), *viz.* by calculating the representation $\mathbf{GVG}^{-1}$ of the isotropic vector (x, y, z) that is the image of $\mathbf{e}_x + \imath\mathbf{e}_y$ under a general element $\mathbf{G}$ of SU(2) as given by (3.4) (which can now be proved by writing the rotation matrix $\mathbf{G}$ as a product of two reflection matrices $\mathbf{G} = \mathbf{BA}$ following the procedure outlined in Figure 3.3, and deriving from this that $\mathbf{G}^\dagger = \mathbf{G}^{-1}$ and $\det \mathbf{G} = 1$):

$$\begin{aligned} \begin{pmatrix} z & x - \imath y \\ x + \imath y & -z \end{pmatrix} &= \begin{pmatrix} a & -b^* \\ b & a^* \end{pmatrix} \begin{pmatrix} 0 & 2 \\ 0 & 0 \end{pmatrix} \begin{pmatrix} a^* & b^* \\ -b & a \end{pmatrix} \\ &= 2 \begin{pmatrix} -ab & a^2 \\ -b^2 & ab \end{pmatrix} \\ &= 2 \begin{pmatrix} a \\ b \end{pmatrix} \otimes (-b, a). \end{aligned} \tag{3.14}$$

Here $\mathbf{V}$ is the representation of $\mathbf{e}_x + \imath\mathbf{e}_y$. From this we can appreciate that actually $(a, b) = (\xi_0, \xi_1)$, which means that the spinors also occur in the rotation matrices. This also confirms the initial observation that the rotation matrix can be split into two parts.[7] This demonstrates that a vector is not a tensor product $\boldsymbol{\xi} \otimes \boldsymbol{\xi}$ of two spinors $\boldsymbol{\xi}$ as might have been anticipated in Section 3.3, but rather a tensor product of the type $\boldsymbol{\xi}_1 \otimes \boldsymbol{\xi}_2^\dagger$. What was not anticipated at the time is that $\boldsymbol{\xi}_2^\dagger$ does indeed contain the components of $\boldsymbol{\xi}_1$, but in a reshuffled way. Rather than stating that a spinor is the square root of a vector, it should rather be stated that a vector is a bilinear expression of spinors.[8]

[7] In the initial form of (3.7), the isotropic vector $(1, \imath, 0)$ leads to two zeros $[0, 0]^\top$ in the first column, while the corresponding spinor is $[1, 0]^\top$. The re-normalization procedure removes a very inconvenient zero from the formalism. A rotation over an angle φ around the z-axis transforms $[1, 0]^\top$ into $e^{\imath\varphi/2}[1, 0]^\top$, from which it is seen that two quantities that are equal up to a normalization factor nevertheless represent different group elements. This is routinely neglected in quantum mechanics, where spinors are treated like vectors, that one can normalize at will.

[8] The spinor $\boldsymbol{\xi}_2$ corresponds to the second column of the rotation matrix $\mathbf{R}$ that represents the rotation R and whose first column is $\boldsymbol{\xi}_1$. It is the first column of the matrix $\mathbf{R}\,[\,\mathbf{e}_x {\cdot} \boldsymbol{\sigma}\,]$ that represents the reversal $R' = RA_x$. Here A_x is the reflection with respect to the Oyz plane, as its representation matrix is indeed $[\,\mathbf{e}_x {\cdot} \boldsymbol{\sigma}\,] = \sigma_x$. As we always take the first columns of the representation matrices to be our spinors, the spinor $\boldsymbol{\xi}_2$ represents the reversal R'. This spinor is orthogonal to the spinor $\boldsymbol{\xi}_1$ with respect to the Hermitian in-product that is used for complex vector spaces. As the vector space $\mathbb{C}^2$ is

It may finally be noted that when one tries to generalize the presently described method to the Lorentz group, one discovers that the derivation based on the tensor product is more fundamental than the diagonalization procedure (see Footnote 13 in Subsection 3.10.4 and the discussion at the end of Section 4.5). The major issue is to find a way to write a matrix with zero determinant as a tensor product. In the Lorentz group it will not be possible to achieve this by diagonalization.

The basic idea is that a vector $\mathbf{V}$ transforms quadratically according to $\mathbf{V} \to -\mathbf{AVA}$ or $\mathbf{V} \to \mathbf{RVR}^\dagger$, where $\mathbf{R}^\dagger = \mathbf{R}^{-1}$. The isotropic vector which pictures a rotation is a vector and will thus also transform quadratically according to these transformation laws. But this vector is the image of a rotation which transforms linearly. The paradox is solved by showing that the isotropic vector can be written as a tensor product as shown in (3.14) and that the information content of the part of the formalism wherein the column spinor is multiplied to the left is completely equivalent to the information content of the part of the formalism wherein the row spinor is multiplied to the right. In the Lorentz group representation SL(2,$\mathbb{C}$) vectors will still transform according to $\mathbf{V} \to \mathbf{LVL}^\dagger$, but now in general $\mathbf{L}^\dagger \neq \mathbf{L}^{-1}$. It is still possible to code a frame by vectors of "zero length" and to write such vectors as tensor products, but now the information content of the parts that transform by left-hand multiplication and of the parts that transform by right-hand multiplication will no longer be equivalent. This will be discussed in Chapter 4.

The spinors are thus a system of coordinates that define a rotation. As with all systems of coordinates it is defined with respect to a reference frame; in fact, the rotation is represented by a rotated triad. The identification of this rotated triad with a rotation depends on the choice of the reference frame, *viz.* the initial triad one chooses to represent the identity element. This remark will play a role in attempts to define the spin. As the spin is a physical quantity, its definition must be frame-independent.

3.8 A treatment based on the stereographic projection as a source of confusion

There exists an alternative derivation of the two-dimensional representation of the rotation group starting from the stereographic projection of a vector

---

two-dimensional, the relation between $\boldsymbol{\xi}_1$ and $\boldsymbol{\xi}_2$ is bijective, such that the two spinors are isomorphic representations for the group SU(2) of the rotations (without reversals).

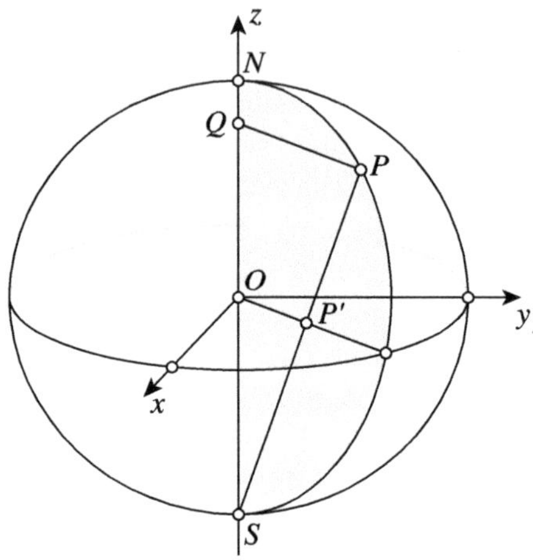

Fig. 3.4 Notations used in the calculations based on the stereographic projection. N and S note the North pole and the South pole of the sphere with centre O and radius $r = 1$. The centre of the stereographic projection is the South pole, such that P is projected onto P'. The point Q is the orthogonal projection of P on the z-axis.

(see [Naimark (1964) and Smirnov (1972)]). A similar derivation for the Lorentz group has been established in [Penrose and Rindler (1984)]. As shown in Figure 3.4, a point $P(x_3, y_3, z_3)$ on the sphere is projected onto the point $P'(x'_3, y'_3, z'_3)$ of the equatorial plane. The sphere has centre O and its radius r is assumed to be $r = 1$, such that $x_3^2 + y_3^2 + z_3^2 = 1$. It may appear pedantic to use the index 3 in the coordinates, but this is done for consistency in the notations of what will follow. Using the theorem of Thales we obtain then:

$$\frac{SP'}{SP} = \frac{OP'}{QP} = \frac{x'_3}{x_3} = \frac{y'_3}{y_3} = \frac{1}{1+z_3} = \frac{x'_3 + \imath y'_3}{x_3 + \imath y_3} = \frac{x'_3 - \imath y'_3}{x_3 - \imath y_3}. \tag{3.15}$$

Introducing $\zeta = x'_3 + \imath y'_3$ one finds:

$$\frac{1}{(1+z_3)^2} = \frac{\zeta\zeta^*}{x_3^2 + y_3^2} = \frac{\zeta\zeta^*}{(1-z_3)(1+z_3)}. \tag{3.16}$$

It can be shown then that a general rotation is a homographic transformation in the variable ζ:

$$\zeta \to \frac{a\zeta - b^*}{b\zeta + a^*}, \tag{3.17}$$

where $aa^* + bb^* = 1$. This represents quite a bit of work, which can be skipped as it is only of secondary importance for the main line of developments. For the interested reader who wishes to check it, the best way to proceed is perhaps to decompose the rotation into the three rotations that define the Euler angles, prove (3.17) for these three Euler rotations separately, and finally use the fact that the homographic transformations form a group. Homographic transformations are the basis of projective geometry, so it should not be a surprise that they appear in an approach based on a stereographic projection.

The next step in the derivation really feels like a rabbit that has been pulled out of a head. One introduces laconically homogeneous coordinates, by putting $\zeta = \xi_1/\xi_0$. This permits the homographic transformation to be presented as a matrix transformation:

$$\begin{pmatrix} \xi_0 \\ \xi_1 \end{pmatrix} \rightarrow \begin{pmatrix} a & -b^* \\ b & a^* \end{pmatrix} \begin{pmatrix} \xi_0 \\ \xi_1 \end{pmatrix}, \tag{3.18}$$

and this way the group structure of SU(2) based on rotation matrices of the form given in (3.4) appears. Simultaneously, one postulates $\xi_0\xi_0^* + \xi_1\xi_1^* = 1$. From this, z_3, $x_3 + \imath y_3$ and $x_3 - \imath y_3$ can be calculated successively such that one obtains finally:

$$x_3 = \xi_0\xi_1^* + \xi_0^*\xi_1, \quad y_3 = \imath(\xi_0\xi_1^* - \xi_0^*\xi_1), \quad z_3 = \xi_0\xi_0^* - \xi_1\xi_1^*. \tag{3.19}$$

Introducing $\zeta = \xi_0/\xi_1$ leaves the reader wondering about the motivation behind it. Moreover, it stirs confusion as it gives the impression that the representation works on column vectors that are images of a *single unit vector* rather than an *isotropic vector* corresponding to a triad, contradicting all the results that have been proved before. As discussed in the preceding lines and in [Coddens (2002)], the substitution $\zeta = \xi_1/\xi_0$ embodies the leap from vectors to group elements, just like the homogeneity principle discussed in the present chapter. In fact, in the substitution $\zeta = \xi_0/\xi_1$, originally $\xi_0 = 1$ is real, but under the action of the group, ξ_0 can acquire complex values. Normalization of the spinors is possible as $aa^* + bb^* = 1$ implies that the norm of a spinor is a group invariant. This normalization reduces the number of independent parameters to three. After the introduction of homogeneous coordinates, $(\alpha\xi_0, \alpha\xi_1)$ is as good a spinor as (ξ_0, ξ_1). The normalization $\xi_0^*\xi_0 + \xi_1^*\xi_1 = 1$ can remove most of the effect of the equivalence $(\xi_0, \xi_1) \equiv (\alpha\xi_0, \alpha\xi_1)$, expressing that both spinors lead to the same ζ-value, but it cannot remove a phase factor $e^{\imath\chi}$. It is this phase factor that stealthily introduces information about the second vector coded into an isotropic vector, as illustrated in Figure 3.2.[9]

All this corresponds to coding a rotation as an isotropic vector, *viz.* coding rotations rather than vectors. This derivation using the stereographic projection is awkward; in Cartan's words, it treats spinors like vectors. The construction defines a unit ray $\{\psi_\chi : \psi_\chi = e^{\imath\chi}\psi\}$ rather than a spinor ψ, keeping the variable $e^{\imath\chi}$ hidden. This tends to prevent discovering a crucial underlying idea rather than helping in discovering it. This is a nice illustration of the point made in the Introduction, *viz.* that it can be extremely difficult to grasp underlying ideas if the presentation is austere or misleading as is the case here.

There is an important consequence of this fact that spinors are not vectors: spinors can, in principle, not be added like vectors. Spinors belong to a manifold, *viz.* the isotropic cone $\mathscr{I}$, not to a vector space. The sum of two isotropic vectors is not necessarily a new isotropic vector. Hence, although the column matrices used in SU(2) look very much like vectors, it has *a priori* no geometrical meaning to add two such column matrices, even though this is perfectly possible algebraically. Similarly, in group representation theory, the only matrices that have a meaningful counterpart in the isomorphism $g \leftrightarrow \mathbf{D}(g)$ are products $\mathbf{D}(g_1)\mathbf{D}(g_2)$ of representation

---

[9]It is perhaps easier to understand this point by remembering that the natural approach consists in deriving (3.17) from (3.18), rather than (3.18) from (3.17). The approach based on the stereographic projection presented here only tries to revert this derivation artificially. To derive (3.17) from (3.18) it suffices to prove that for any spinor $[\,\xi_0, \xi_1\,]^\top$ the stereographic projection of the vector $\mathbf{e}'_z$ of its triad $(\mathbf{e}'_x, \mathbf{e}'_y, \mathbf{e}'_z)$ is $\zeta = \xi_0/\xi_1$. It is then easy to see that ζ contains less information than $[\,\xi_0, \xi_1\,]^\top$. Calculate the matrix $\mathbf{R}(\mathbf{e}'_z, \chi)$ that corresponds to the rotation over an angle χ around $\mathbf{e}'_z$ (this can be done by using a similarity transformation). Consider a rotation with rotation matrix $\mathbf{R}_0$ that aligns $\mathbf{e}'_z$ with $\mathbf{e}_z$. The rotation matrix $\mathbf{R}(\mathbf{e}'_z, \chi)$ is then just given by $\mathbf{R}(\mathbf{e}'_z, \chi) = \mathbf{R}_0^{-1}\,\mathbf{R}(\mathbf{e}_z, \chi)\,\mathbf{R}_0$. This expression for $\mathbf{R}(\mathbf{e}'_z, \chi)$ permits us to check that $\mathbf{R}(\mathbf{e}'_z, \chi)[\,\xi_0, \xi_1\,]^\top$ is just $e^{\imath\chi/2}[\,\xi_0, \xi_1\,]^\top$ (see the derivation of (9.25) in Chapter 9). A rotation by an angle χ around the z'-axis of the triad represented by $[\,\xi_0, \xi_1\,]^\top$ is thus given by the straightforward multiplication: $[\,\xi_0, \xi_1\,]^\top \to e^{\imath\chi/2}[\,\xi_0, \xi_1\,]^\top$. The multiplication by $e^{\imath\chi/2}$ modifies the spinor but it does not affect the ratio $\zeta = \xi_0/\xi_1$, because it is divided out. This shows that the phase factor $e^{\imath\chi/2}$ within a spinor $e^{\imath\chi/2}[\,\xi_0, \xi_1\,]^\top$ codes a rotation angle χ around the z'-axis coded by ζ. The phase factor thus contains a piece of information that is lost when only the stereographic projection ζ is used rather than the spinor $[\,\xi_0, \xi_1\,]^\top$, and the variable $\zeta = \xi_0/\xi_1$ therefore codes less information than the spinor $[\,\xi_0, \xi_1\,]^\top$. To revert this derivation and turn it into a derivation of (3.18) from (3.17), one must recover the information contained within $e^{\imath\chi/2}$ that was lost by putting $\zeta = \xi_0/\xi_1$. This is achieved by "reintroducing" the spinor $[\,\xi_0, \xi_1\,]^\top$ as a set of homogeneous coordinates for ζ. This relationship between ζ and $[\,\xi_0, \xi_1\,]^\top$ plays an important role in Klein's solution of the quintic [Klein (1884)].

matrices. However, one also uses linear combinations $\sum_j c_j\mathbf{D}(g_j)$ in the algebra. These linear combinations define the so-called group ring.

Whereas the approach based on the stereographic projection lacks elegance, it nevertheless cannot be ignored, as it is used in a very important isomorphism between representations based on harmonic polynomials. This isomorphism will be defined in Section 3.10, and its importance for quantum mechanics illustrated in Subsection 3.11.3. This isomorphism is the key to the problem of to what extent one may use spinors as vectors. We give here thus some additional results that we will need at that moment. The calculations based on the stereographic projection work on the real vector $\mathbf{e}_z$ instead of the isotropic vector $\mathbf{e}_x + \imath\mathbf{e}_y$. It is therefore that the coordinates of this vector $\mathbf{e}_z$ have been noted as (x_3, y_3, z_3). It is clear from the structure of these equations in (ξ_0, ξ_1) that $(x_3, y_3, z_3) \in \mathbb{R}^3$. The quantities (x_3, y_3, z_3) used here thus contain less information than the quantities $(x, y, z) \in \mathbb{C}^3$ used in (3.10)–(3.13), which were defined as $(x, y, z) = (x_1 + \imath x_2, y_1 + \imath y_2, z_1 + \imath z_2)$, i.e. the coordinates of $\mathbf{e}_x + \imath\mathbf{e}_y$. From (3.11)–(3.13) we can calculate:

$$\begin{aligned}
&x_1 = \tfrac{1}{2}(\xi_0^2 - \xi_1^2 + \xi_0^{*2} - \xi_1^{*2}), \;\; y_1 = \tfrac{\imath}{2}(\xi_0^2 + \xi_1^2 - \xi_0^{*2} - \xi_1^{*2}),\\
&z_1 = -(\xi_0\xi_1 + \xi_0^*\xi_1^*),\\
&x_2 = \tfrac{\imath}{2}(-\xi_0^2 + \xi_1^2 + \xi_0^{*2} - \xi_1^{*2}), \;\; y_2 = \tfrac{1}{2}(\xi_0^2 + \xi_1^2 + \xi_0^{*2} + \xi_1^{*2}),\\
&z_2 = (\xi_0\xi_1 - \xi_0^*\xi_1^*).
\end{aligned} \tag{3.20}$$

From this the expression of $\mathbf{e}_z = \mathbf{e}_x \wedge \mathbf{e}_y$ can be calculated in terms of the spinors ξ_0 and ξ_1. Using $\xi_0\xi_0^* + \xi_1\xi_1^* = 1$ it can be seen that this expression of $\mathbf{e}_z$ corresponds exactly to (x_3, y_3, z_3) as given in (3.19). This shows that the definition of (ξ_0, ξ_1) derived from the approach based on the stereographic projection is equivalent to the definitions (3.10)–(3.13) (up to the ambiguity that the exact value of the phase factor χ remains unspecified in this construction). While it is possible to determine this way (x_3, y_3, z_3) from (x, y, z), the converse is of course not true, as (x_3, y_3, z_3) does not contain enough information; it contains only two independent real parameters, and the missing parameter is the phase factor χ.

3.9 Harmonic polynomials

3.9.1 *Harmonic polynomials from tensor products*

As explained in the preamble, it is possible, starting from the SU(2) representation, to construct a whole series of higher-dimensional representations that work on tensor products $(\xi_0, \xi_1) \otimes (\xi_0, \xi_1) \otimes \cdots \otimes (\xi_0, \xi_1)$. This will lead to representations of dimension $n+1$ on nth degree polynomials ξ_0^n, $\xi_0^{n-1}\xi_1$, $\xi_0^{n-2}\,\xi_1^2, \cdots, \xi_1^n$.

Let us now introduce harmonic polynomials $P_{\ell,m} \in F(\mathscr{I}, \mathbb{C})$, where $\mathscr{I} \subset \mathbb{C}^3$ is the set $\{(x,y,z) \in \mathbb{C}^3 \parallel x^2+y^2+z^2 = 0\}$, called the isotropic cone. The values of $P_{\ell,m}(x,y,z)$ are defined by considering the tensor product:

$$\underbrace{(\xi_0, \xi_1) \otimes (\xi_0, \xi_1) \otimes \cdots \otimes (\xi_0, \xi_1)}_{n \text{ times}}. \tag{3.21}$$

This tensor product of power n contains the $n+1$ single-term polynomials $P'_{n,k}$ in ξ_0 and ξ_1:

$$\xi_0^n, \quad \xi_0^{n-1}\xi_1, \quad \cdots \quad \xi_0^{n-k}\xi_1^k, \quad \cdots \quad \xi_0\xi_1^{n-1}, \quad \xi_1^n, \tag{3.22}$$

where $P'_{n,k}(\xi_0, \xi_1) = \xi_0^{n-k}\xi_1^k$, and $k \in [0,n] \cap \mathbb{Z}$. The coefficients of the polynomials are less important here. They can be defined by a normalization procedure later on. The point of interest is their functional dependences. The polynomials $P'_{n,k}$ actually belong to $F(\mathbb{C}^2, \mathbb{C})$. They have all the same total degree n, i.e. the degrees $d_{\xi_0} = n-k$ in ξ_0 and $d_{\xi_1} = k$ in ξ_1 satisfy $d_{\xi_0} + d_{\xi_1} = n$.

For even values of $n = 2\ell$, it is possible to transform, by using (3.10)–(3.13), the $n+1 = 2\ell+1$ polynomials $P'_{n,k} \in F(\mathbb{C}^2, \mathbb{C})$ into harmonic polynomials $P_{\ell,m}$ in the variables (x,y,z), such that $P'_{2\ell,k}(\xi_0, \xi_1) = \xi_0^{2\ell-k}\xi_1^k = P_{\ell,m}(x,y,z)$. By using (3.10)–(3.13), the constraint $x^2+y^2+z^2 = 0$ will be implicitly respected. These polynomials will have all the same total degree ℓ in (x,y,z), i.e. $d_x + d_y + d_z = \ell$.

The idea is to write $\xi_0^{2\ell-k}\xi_1^k = \xi_0^{2\ell-2k}\xi_0^k\xi_1^k$, when $2\ell - k > k$; that is, to take k further terms ξ_0 out of $\xi_0^{2\ell-k}$ with the idea of combining them with the k terms ξ_1. For this to be possible, there must be enough terms ξ_0 within $\xi_0^{2\ell-k}$ to do this. This means that there must be more terms in ξ_0 than in ξ_1 within $\xi_0^{2\ell-k}\xi_1^k$. Next, $\xi_0\xi_1$ is substituted by $-z/2$ in the part $\xi_0^k\xi_1^k$ and ξ_0^2 by $(x - \imath y)/2$ in the part $\xi_0^{2\ell-2k}$. For $k = \ell$ this transforms $\xi_0^{2\ell-k}\xi_1^k = \xi_0^\ell\xi_1^\ell$ into an expression proportional to z^ℓ.

When $2\ell - k < k$ then there are more terms in ξ_1 than in ξ_0, and it is therefore possible to take terms ξ_1 out of ξ_1^k to make combinations $\xi_0\xi_1$. When $2\ell - k < k$ one can put $2\ell - k = \kappa$. It is then possible to rewrite $\xi_0^{2\ell-k}\xi_1^k = \xi_0^\kappa\xi_1^{2\ell-\kappa}$. For $2\ell - \kappa > \kappa$ (which is the same as $2\ell - k < k$) it is possible to rewrite $\xi_0^\kappa\xi_1^{2\ell-\kappa} = \xi_0^\kappa\xi_1^\kappa\xi_1^{2\ell-2\kappa}$, substituting ξ_1^2 by $-(x+\imath y)/2$ in the part $\xi_1^{2\ell-2\kappa}$ and $\xi_0\xi_1$ by $-z/2$ in the part $\xi_0^\kappa\xi_1^\kappa$. This way the $2\ell+1$

polynomials $P_{\ell,m}$ are obtained, where $m \in [-\ell,\ell] \cap \mathbb{Z}$, and $m = \ell - k$. These polynomials are thus characterized by a total degree ℓ and a degree $|m|$ in z.

3.9.2 *A generating polynomial*

The rotation matrices for the tensor representation will also be tensor products of the rotation matrices from SU(2). Here the elements of a rotation matrix in SU(2) will be represented with A, B, C, and D rather than a, a^*, b and, $-b^*$ as in (3.4). A rotation will then yield $\xi_0' = A\xi_0 + B\xi_1$, and in the $n+1$-dimensional representation of degree n it can be shown that $\xi_0'^n = (A\xi_0 + B\xi_1)^n$. Cartan [Cartan (1981)] calls $(A\xi_0 + B\xi_1)^n$ the generating polynomial, as the coefficients of the various monomials in A and B yield the polynomials of the representation. One can imagine in an analogous way a generating polynomial $P: P(x,y,z) = (k_x x + k_y y + k_z z)^\ell$ for these polynomials, but this still does not account for the constraints. By using (3.11)–(3.13) and identifying $(A\xi_0 + B\xi_1)^{2\ell} = (k_x x + k_y y + k_z z)^\ell$ it can be seen that:

$$P(x,y,z) = [\,(A^2 - B^2)x - \imath(A^2 + B^2)y - 2ABz\,]^\ell, \tag{3.23}$$

where $k_x = A^2 - B^2$, $k_y = -\imath(A^2 + B^2)$, $k_z = -2AB$. It is easy to check from this that $k_x^2 + k_y^2 + k_z^2 = (A^2 - B^2)^2 - (A^2 + B^2)^2 + (-2AB)^2 = 0$. From this it follows also that P satisfies the Laplace equation $\Delta P = 0$, since $\Delta P(x,y,z) = \ell(\ell-1)(k_x^2 + k_y^2 + k_z^2)(k_x x + k_y y + k_z z)^{\ell-2}$. This fact often serves as a definition for the harmonic polynomials. Applying the identity $k_x^2 + k_y^2 + k_z^2 = 0$ systematically on the coefficients in the development of $P(x,y,z) = (k_x x + k_y y + k_z z)^\ell$ permits the recovery of the correct expressions for all harmonic polynomials in terms of x, y and z. (It may be noted that $\Delta P = 0$ is the Fourier transform of $(k_x^2 + k_y^2 + k_z^2)P = 0$.) We have merely worked out here in more detail a method that has been described by Cartan. We have nevertheless introduced it because it will permit us to touch upon a very important point in Subsection 3.11.3.[10]

---

[10]It is perhaps helpful to check the relationship between harmonic polynomials and spinors on an example. Those with $\ell = 2$ are $Y_{2,-2} = \frac{1}{4}\sqrt{\frac{15}{2\pi}}\,(x - \imath y)^2/r^2 \propto \xi_0^4$, $Y_{2,-1} = \frac{1}{2}\sqrt{\frac{15}{2\pi}}\,(x - \imath y)z/r^2 \propto \xi_0^3\xi_1$, $Y_{2,0} = \frac{1}{4}\sqrt{\frac{5}{\pi}}\,(2z^2 - x^2 - y^2)/r^2 = \frac{1}{4}\sqrt{\frac{5}{\pi}}\,3z^2/r^2 \propto \xi_0^2\xi_1^2$, $Y_{2,1} = -\frac{1}{2}\sqrt{\frac{15}{2\pi}}\,(x + \imath y)z/r^2 \propto \xi_0\xi_1^3$, $Y_{2,2} = \frac{1}{4}\sqrt{\frac{15}{2\pi}}\,(x + \imath y)^2/r^2 \propto \xi_1^4$, where one uses

3.9.3 *Completing the definition*

The harmonic polynomials as described thus far are not yet entirely defined by their construction based on ξ_0 and ξ_1. To complete the definition, it can be stipulated that every harmonic polynomial $P_{\ell m}$ must satisfy the Laplace equation $\Delta P_{\ell m} = 0$, just like the generating function. With a polynomial $P(x,y,z) = (x - \imath y)^{n_1} z^{n_2}$, for $n_2 = 1$ or $n_2 = 0$, it is immediately apparent that $\Delta P = 0$. But for $n_2 \geq 2$ the result is that $(\frac{\partial^2}{\partial x^2} + \frac{\partial^2}{\partial y^2})P(x,y,z) = 0$, such that $\Delta P = \frac{\partial^2}{\partial z^z} P(x,y,z) = 2(x - \imath y)^{n_1}$, which is obviously not zero. The fact that $x^2 + y^2 + z^2 = 0$ can now be used to repair this situation. This will be illustrated on the example $n_2 = 2$. Consider $Q(x,y,z) = Az^2(x - \imath y)^{n_1} - B(x^2 + y^2)(x - \imath y)^{n_1}$. Using $x^2 + y^2 + z^2 = 0$ it can be shown that this is equivalent to $(A + B)z^2(x - \imath y)^{n_1} = (A+B)P(x,y,z)$. For the calculation of $\frac{\partial^2}{\partial x^2} Q(x,y,z)$ only the term $Q_1(x,y,z) = -B(x^2 + y^2)(x - \imath y)^{n_1}$ of $Q(x,y,z)$ needs to be considered, as it is known that $(\frac{\partial^2}{\partial x^2} + \frac{\partial^2}{\partial y^2})(Az^2(x - \imath y)^{n_1}) = 0$. Straightforward calculation shows that $\frac{\partial^2}{\partial x^2} Q_1(x,y,z) = -B(x - \imath y)^{n_1 - 2} [\, n_1(n_1 - 1)(x^2 + y^2) + 4n_1 x \, (x - \imath y) + 2(x - \imath y)^2 \,]$, then $\frac{\partial^2}{\partial y^2} Q_1(x,y,z) = -B(x - \imath y)^{n_1 - 2} [\, -n_1(n_1 - 1)\,(x^2 + y^2) - 4\imath n_1 \, y \,(x - \imath y) + 2(x - \imath y)^2 \,]$, and finally $\frac{\partial^2}{\partial z^2} Q(x,y,z) = 2A(x - \imath y)^{n_1}$. Summing it all up yields then $\Delta Q(x,y,z) = (x - \imath y)^{n_1} [\, 2A - 4B(n_1 + 1) \,]$. By choosing $A = 2(n_1 + 1)B$, one can then satisfy the condition that $\Delta Q(x,y,z) = 0$. For the example of the polynomial $P(x,y,z) = z^2$, $n_1 = 0$. One must then take $A = 2B$, such that $Q(x,y,z) = B\,(2z^2 - x^2 - y^2)$.

---

implicitly $x^2 + y^2 + z^2 = 0$, and which clearly shows how these polynomials are defined by taking a tensor power of (ξ_0, ξ_1). Here $r^2 = x^2 + y^2 + z^2 \neq 0$, which seems to contradict $x^2 + y^2 + z^2 = 0$ but this is due to an isomorphism that allows the replacement of $(x,y,z) \in \mathscr{I}$ with $(x,y,z) \in \mathbb{R}^3$ as will be explained in Section 3.10. The polynomials $Y_{\ell,m}$ are of degree $|m|$ in z (and – after introducing spherical coordinates (r,θ,ϕ) — of degree m in $e^{\imath\phi}$, if one considers them as functions of $F(\mathbb{R}^3,\mathbb{C})$ rather than of $F(\mathbb{C}^3,\mathbb{C})$. It is only for the functions of $F(\mathbb{R}^3,\mathbb{C})$ that one can introduce the spherical coordinates (r,θ,ϕ) needed to define $e^{\imath\phi}$, as discussed in Section 3.10). The normalization factors are defined in such a way that the polynomials become an orthonormal set with respect to the scalar product $\int Y^*_{\ell',m'}(\theta,\phi) Y_{\ell,m}(\theta,\phi) \sin\theta d\theta d\phi = \delta_{\ell\ell'}\delta_{mm'}$, if one considers them as functions of $F(\mathbb{R}^3,\mathbb{C})$ rather than of $F(\mathbb{C}^3,\mathbb{C})$. As functions of $F(\mathbb{C}^3,\mathbb{C})$, they can be normalized using hyper-spherical coordinates $(r,\theta_0,\theta_1,\phi)$. By cyclic permutation of x,y,z one obtains a different set of harmonic polynomials. These sets correspond to a different choice of basis for the spinors. Finally, the fact that the different representations are just based on different tensor powers indicates how Clebsch-Gordon coefficients must be calculated.

3.10 A very important isomorphism within representations based on harmonic polynomials

3.10.1 *First approach*

The method to derive SU(2) that starts from the stereographic projection, while lacking elegance, is needed to establish the existence of an isomorphism used in quantum mechanics. From (3.19) it can be seen that the representation using the stereographic projection is based on the same spinors (ξ_0, ξ_1) as the representation constructed starting from an isotropic vector, but that the (vector) images it uses are different. It is thus actually a different representation, but it is nevertheless isomorphic to the one based on the images of $\mathbf{e}_x + \imath\mathbf{e}_y$. The rotation matrices of the type (3.4) operating on (ξ_0, ξ_1) as an image of $\mathbf{e}_x + \imath\mathbf{e}_y$, are identical to those obtained from (3.17) and operating on the image of $\mathbf{e}_z$ under stereographic projection. To prove this, it will first be necessary to find a result that can be derived from (3.7) and (3.11)–(3.13) and that is actually equivalent to (3.14).

As already explained, the three-dimensional representation with the homogeneous (harmonic) polynomials x, y, z (that are the coordinates of the isotropic vector $\mathbf{e}_x + \imath\mathbf{e}_y$ with representation matrix $\mathbf{V}$) can be derived from the two-dimensional one using the tensor product:

$$\begin{aligned}\mathbf{V} &= \begin{pmatrix} z & x - \imath y \\ x + \imath y & -z \end{pmatrix} \\ &= 2\begin{pmatrix} -\xi_0\,\xi_1 & \xi_0\,\xi_0 \\ -\xi_1\,\xi_1 & \xi_0\,\xi_1 \end{pmatrix} \\ &= \sqrt{2}\begin{pmatrix} \xi_0 \\ \xi_1 \end{pmatrix} \otimes \sqrt{2}\,(-\xi_1, \xi_0). \end{aligned} \tag{3.24}$$

There is some consistency checking to be done here, for instance that the determinant of the 2×2 matrix with the quantities ξ_0, ξ_1 is indeed zero, and also that $z/(x + \imath y) = -(x - \imath y)/z = \xi_0/\xi_1$, etc. Now a rotation with matrix $\mathbf{R}$ will operate on the vector representation $\mathbf{V}$ as $\mathbf{V} \to \mathbf{R}\mathbf{V}\mathbf{R}^\dagger$, where $\mathbf{R}^\dagger = \mathbf{R}^{-1}$. Hence:

$$\begin{aligned}\mathbf{V} &= \sqrt{2}\begin{pmatrix} \xi_0 \\ \xi_1 \end{pmatrix} \otimes \sqrt{2}\,(-\xi_1, \xi_0) \to \\ \mathbf{R}\mathbf{V}\mathbf{R}^\dagger &= \begin{pmatrix} a & -b^* \\ b & a^* \end{pmatrix} \sqrt{2}\begin{pmatrix} \xi_0 \\ \xi_1 \end{pmatrix} \otimes \sqrt{2}\,(-\xi_1, \xi_0) \begin{pmatrix} a^* & b^* \\ -b & a \end{pmatrix}. \end{aligned} \tag{3.25}$$

From the transformation properties of (ξ_0, ξ_1), those of (ξ_0^*, ξ_1^*) can be derived:

$$\begin{pmatrix} \xi_0 \\ \xi_1 \end{pmatrix} \rightarrow \begin{pmatrix} a & -b^* \\ b & a^* \end{pmatrix} \begin{pmatrix} \xi_0 \\ \xi_1 \end{pmatrix}$$

$$\Big\Downarrow \; \textit{Calculating } (\xi_0^*, \xi_1^*) \textit{ from } (\xi_0, \xi_1) \tag{3.26}$$

$$\begin{pmatrix} \xi_1^* \\ -\xi_0^* \end{pmatrix} \rightarrow \begin{pmatrix} a & -b^* \\ b & a^* \end{pmatrix} \begin{pmatrix} \xi_1^* \\ -\xi_0^* \end{pmatrix},$$

from which it can be seen that ξ_1^* transforms as ξ_0 while ξ_0^* transforms as $-\xi_1$. This can also be appreciated from the fact that $\mathbf{R}[\xi_0, \xi_1]^\top$ is equivalent to $[\xi_0^*, \xi_1^*]\mathbf{R}^\dagger$ and from a comparison with (3.25), where $[-\xi_1, \xi_0]$ is also transformed with $\mathbf{R}^\dagger$. Hence $\frac{1}{\sqrt{2}}(x_3 - \imath y_3) = \sqrt{2}\,\xi_0\xi_1^*$ transforms as $\sqrt{2}\,\xi_0^2$, $-z_3 = -(\xi_0\xi_0^* - \xi_1\xi_1^*)$ transforms as $2\xi_0\xi_1$, and $\frac{1}{\sqrt{2}}(x_3 + \imath y_3) = \sqrt{2}\,\xi_0^*\xi_1$ transforms as $-\sqrt{2}\,\xi_1^2$. These transformation laws for the vector (x_3, y_3, z_3) are thus isomorphic to those of the isotropic vector (x, y, z) for which from (3.11)–(3.13) it follows that $x - \imath y = 2\xi_0^2$, $z = -2\xi_0\xi_1$ and $x + \imath y = -2\xi_1^2$.

3.10.2 *An important remark on notation*

Before continuing the development it is necessary to introduce a more compact vector notation $\mathbf{a}{\cdot}\boldsymbol{\sigma}$ for the representation matrix $a_x\sigma_x + a_y\sigma_y + a_z\sigma_z$ that codes a vector $\mathbf{a} \in \mathbb{C}^3$ within SU(2). The reader may notice that this notation has actually already been introduced in (3.5). In order to avoid confusion, it is very important to stress that the notation is purely conventional, especially as it will be further used throughout the rest of the book. It is based on the convention to write the set of the three Pauli matrices $\sigma_x, \sigma_y, \sigma_z$ symbolically together as $\boldsymbol{\sigma} = (\sigma_x, \sigma_y, \sigma_z)$ as though they would constitute a vector of $\mathbb{C}^3$. Note that despite the (potentially misleading) notation, $\boldsymbol{\sigma}$ is not a vector, such that $\mathbf{a}{\cdot}\boldsymbol{\sigma}$ does not represent a scalar quantity, but a vector. Hence, $\mathbf{B}{\cdot}\boldsymbol{\sigma}$ should not be confused with the scalar product of the vector $\mathbf{B}$ with a vector $\boldsymbol{\sigma}$; it is merely the way the vector $\mathbf{B}$ is represented within the group theory of SU(2). Similarly, $\mathbf{L}{\cdot}\boldsymbol{\sigma}$ is not the scalar product of the vector $\mathbf{L}$ with a vector $\boldsymbol{\sigma}$ but just the notation for the vector $\mathbf{L}$ used within the group theory of SU(2).[11]

3.10.3 *Conjugated spinors*

The spinor formalism has thus far been developed on the left-hand column-type spinors in (3.24) but could also have been done on right-hand side line-type spinors in that equation, by multiplication with $\mathbf{R}^\dagger$ on $(-\xi_1, \xi_0)$. By taking Hermitian conjugates, this is equivalent to the column spinor formalism:

$$(-\xi_1, \xi_0) \rightarrow (-\xi_1, \xi_0) \begin{pmatrix} a^* & b^* \\ -b & a \end{pmatrix}$$

$$\Big\Downarrow \quad \textit{Hermitian conjugation} \qquad (3.27)$$

$$\begin{pmatrix} -\xi_1^* \\ \xi_0^* \end{pmatrix} \rightarrow \begin{pmatrix} a & -b^* \\ b & a^* \end{pmatrix} \begin{pmatrix} -\xi_1^* \\ \xi_0^* \end{pmatrix},$$

as this is equivalent to (3.26). The Hermitian conjugates of the line spinors can be called the *conjugate spinors.* They are operated on by $\mathbf{R}$. It has already been noted that the column spinors correspond to the first column of the rotation matrix in (3.4), but now it can be seen that the conjugated spinors correspond to the second column in this rotation matrix. A representation based on the conjugated spinors, or on the line spinors from which they are obtained by taking the Hermitian conjugates, is completely equivalent to the one based on the column spinors.

---

[11]In quantum mechanics, the expression $\frac{\hbar q}{2m_0c}\mathbf{B}\cdot\boldsymbol{\sigma}$ has been interpreted as a true scalar product between vectors $g\mu\mathbf{B}$ and $\frac{\hbar}{2}\boldsymbol{\sigma}$ in the treatment of the anomalous g-factor of the electron (see Section 5.7), and the expression $\mathbf{L}\cdot\boldsymbol{\sigma}$ has been interpreted as a scalar product between true vectors $\mathbf{L}$ and $\boldsymbol{\sigma}$ in the Pauli and Dirac equations, where it has been dubbed "spin-orbit coupling". Finally, $\mathbf{u}\cdot\boldsymbol{\sigma}$, with $\mathbf{u} = \mathbf{p}/p$ has been over-interpreted in terms of a helicity, related to the projection of the "spin" $\frac{\hbar}{2}\boldsymbol{\sigma}$ on the direction of motion $\mathbf{u}$. The over-interpretations lead to the introduction of unadapted concepts and imagery. The occurrence of the terms $\mathbf{B}\cdot\boldsymbol{\sigma}$ and $\mathbf{L}\cdot\boldsymbol{\sigma}$ in the Dirac equation is due to the introduction of a perfectly analogous notation $\mathbf{a}\cdot\boldsymbol{\gamma} = \sum_{\mu=0}^{3} a_\mu\gamma_\mu$ for a four-vector $\mathbf{a} = (a_{ct}, a_x, a_y, a_z)$, where the gamma matrices play the role of a generalization of the Pauli matrices. There is thus no loophole of escape from the inconvenient truth contained in this critical remark.

3.10.4 *The explicit construction of the isomorphism*

More insight into the isomorphism can be obtained from the following elegant explicit construction. The representation matrix of $\mathbf{e}'_z = (x_3, y_3, z_3)$ is $\mathbf{e}'_z{\cdot}\boldsymbol{\sigma}$. This matrix has determinant $-x_3^2 - y_3^2 - z_3^2 = -1$ and eigenvalues -1 and 1 (as it is related by similarity transformation to $\mathbf{e}_z{\cdot}\boldsymbol{\sigma} = \sigma_z$). In fact, any representation matrix of a vector has two opposite eigenvalues r and $-r$, where r is its length. The matrix $\mathbb{1} + \mathbf{e}'_z{\cdot}\boldsymbol{\sigma}$ will therefore have one eigenvalue 0 and one eigenvalue 2. Hence, it will have determinant 0, just like an isotropic vector. It will, however, not correspond to an isotropic vector, nor any vector at all as its two eigenvalues are not opposite. The determinant of the representation matrix $\mathbf{V}$ of an isotropic vector (x, y, z) is zero, while both its eigenvalues are also zero as it is a "zero-length" vector. Using the results of (3.19) and $\mathbb{1} = (\xi_0\xi_0^* + \xi_1\xi_1^*)\mathbb{1}$ we obtain then:

$$\mathbb{1} + \mathbf{e}'_z{\cdot}\boldsymbol{\sigma} = \begin{pmatrix} 2\xi_0\xi_0^* & 2\xi_0\xi_1^* \\ 2\xi_0^*\xi_1 & 2\xi_1\xi_1^* \end{pmatrix} = \sqrt{2}\begin{pmatrix} \xi_0 \\ \xi_1 \end{pmatrix} \otimes \sqrt{2}\,(\xi_0^*, \xi_1^*), \tag{3.28}$$

which fits isomorphically into the scheme of (3.25) and (3.26) for the isotropic vector (x, y, z), because $\mathbf{R}\mathbb{1}\mathbf{R}^\dagger = \mathbb{1}$, as $\mathbf{R}^\dagger = \mathbf{R}^{-1}$. Hence, the incomplete real vector representation based on the coding of the coordinates $x_3 - \imath y_3$, z_3 and $x_3 + \imath y_3$ of the vector $\mathbf{e}'_z = (x_3, y_3, z_3)$ is isomorphic to the complete complex vector representation based on the coding of the triad in terms of the coordinates of the isotropic vector $\mathbf{e}_x + \imath\mathbf{e}_y$. In the incomplete representation the coordinates $(x_3, y_3, z_3) \in \mathbb{R}^3$ are real and satisfy $x_3^2 + y_3^2 + z_3^2 = 1$ (as expressed by $-\det(\mathbf{e}'_z{\cdot}\boldsymbol{\sigma}) = 1$), while in the complete representation the coordinates $(x, y, z) \in \mathbb{C}^3$ are complex and satisfy $x^2 + y^2 + z^2 = 0$ (as expressed by $-\det((x, y, z){\cdot}\boldsymbol{\sigma}) = 0$).

The fact that $x_3^2 + y_3^2 + z_3^2 = 1 \neq x^2 + y^2 + z^2 = 0$ does not negate the existence of such an isomorphism and definitely can be part of it can be understood by decomposing the isomorphism in two steps. The first step is the isomorphism between $\mathbf{e}'_z{\cdot}\boldsymbol{\sigma}$ and $\mathbb{1} + \mathbf{e}'_z{\cdot}\boldsymbol{\sigma}$. The presence of the matrix $\mathbb{1}$ is benign as $\mathbf{R}^{-1} = \mathbf{R}^\dagger$, such that it can be taken in and out of the calculations at any time.[12] This shows that an isomorphism can

[12] Its presence would no longer be benign within SL(2,$\mathbb{C}$) since $\mathbf{L}^{-1} \neq \mathbf{L}^\dagger$, as discussed in Subsection 5.5.2.1. In SL(2,$\mathbb{C}$), the matrix $\mathbb{1} + \mathbf{e}'_z{\cdot}\boldsymbol{\sigma}$ does correspond to the coding of some vector, *viz.* the zero-length vector $\mathbf{e}_{ct} + \mathbf{e}'_z$. After a general Lorentz transformation $\mathbf{L}$ on this vector, the unit matrix will have been will transformed to $\mathbf{L}\,\mathbb{1}\,\mathbf{L}^\dagger \neq \mathbb{1}$ as $\mathbf{L}^{-1} \neq \mathbf{L}^\dagger$. But for those Lorentz transformations $\mathbf{L}$ that are mere rotations, such that $\mathbf{R}^{-1} = \mathbf{R}^\dagger$, the transformations $\mathbb{1} \to \mathbf{R}\,\mathbb{1}\,\mathbf{R}^\dagger$ will preserve the unit matrix. In a sense, $\mathbb{1}+\mathbf{e}'_z{\cdot}\boldsymbol{\sigma}$ corresponds to a vector after all in the extension SU(2) $\to$ SL(2,$\mathbb{C}$). The fact that

change the determinant from -1 to 0. The second step is the isomorphism between the quantity $\mathbb{1} + \mathbf{e}'_z{\cdot}\boldsymbol{\sigma}$ (which does not code a vector) and the quantity $(x, y, z){\cdot}\boldsymbol{\sigma}$ (which codes the isotropic vector (x, y, z)). This is an isomorphism between the spinor (ξ_0^*, ξ_1^*) (occurring on the right-hand side in the tensor product decomposition of $\mathbb{1} + \mathbf{e}'_z{\cdot}\boldsymbol{\sigma}$) and the spinor $(-\xi_1, \xi_0)$ (occurring on the right-hand side in the tensor product decomposition of $(x, y, z){\cdot}\boldsymbol{\sigma}$) and is expressed in (3.26).

The existence of the isomorphism is in some way compulsory: (x_3, y_3, z_3) and (x, y, z) are vector quantities. It is thus natural that they transform the same way in a representation based on vector quantities. Of course (x, y, z) is complex, while (x_3, y_3, z_3) is real, but the construction of the group representation has been developed from the argument that it would always be possible to separate (x_1, y_1, z_1) and (x_2, y_2, z_2) out again.[13] However, it should not be forgetten that the stereographic projection entails in reality a homographic transformation, such that this result is all but trivial.[14]

3.10.5 *Degrees of the harmonic polynomials and degree operators*

3.10.5.1 *The degree of a harmonic polynomial as an eigenvalue of an operator*

The harmonic polynomials $P_{\ell,m} \in F(\mathscr{I}, \mathbb{C})$ are of total degree ℓ and degree $|m|$ in z. After applying the isomorphism they can be considered

---

$\mathbb{1}+\mathbf{e}'_z{\cdot}\boldsymbol{\sigma} = \sqrt{2}\,[\xi_0, \xi_1]^\top \otimes \sqrt{2}\,[\xi_0^*, \xi_1^*]$ and $(x, y, z){\cdot}\boldsymbol{\sigma} = \sqrt{2}\,[\xi_0, \xi_1]^\top \otimes \sqrt{2}\,[-\xi_1, \xi_0]$ share the left-hand side spinor $[\xi_0, \xi_1]^\top$ but differ in the right-hand side spinors $[\xi_0^*, \xi_1^*] \neq [-\xi_1, \xi_0]$ confirms that $\mathbb{1} + \mathbf{e}'_z{\cdot}\boldsymbol{\sigma}$ is not an isotropic vector in SU(2). For an isotropic vector in SU(2), the right-hand side spinor is unambiguously defined by the left-hand side spinor to be $[-\xi_1, \xi_0]$ and nothing else. But again, this ceases to be true in SL(2,$\mathbb{C}$) where $\mathbf{e}_{ct} + \mathbf{e}'_z$ is indeed of the form $\sqrt{2}\,[a, c]^\top \otimes \sqrt{2}\,[a^*, c^*]$ (see (4.12)).

[13]For any real unit vector $\mathbf{n} \in \mathbb{R}^3$, the matrix $\mathbb{1} + \mathbf{n}{\cdot}\boldsymbol{\sigma}$ can be cast in the form $[\varepsilon_0, \varepsilon_1]^\top \otimes [\varepsilon_0^*, \varepsilon_1^*]$. The determinant of the matrix $\mathbb{1} + \mathbf{n}{\cdot}\boldsymbol{\sigma}$ is zero, therefore its two lines are proportional, as well as its two columns. It can thus be written in the form $[\varepsilon_0, \varepsilon_1]^\top \otimes [\varepsilon_2, \varepsilon_3]$. The matrix $\mathbb{1} + \mathbf{n}{\cdot}\boldsymbol{\sigma}$ is Hermitian as the Pauli matrices are Hermitian and the components of $\mathbf{n}$ are real. This proves then that $\varepsilon_2 = \varepsilon_0^*$ and $\varepsilon_3 = \varepsilon_1^*$, which corresponds to the result of (3.28). We see thus that $(\xi_0^*, \xi_1^*) \neq (-\xi_1, \xi_0)$ (as it would imply $\xi_0 = \xi_1 = 0$) in agreement with what was claimed in the previous footnote.

[14]There is a link here with projective geometry where it is proved that there is a perspective mapping between sets of four points on a circle and sets of four straight lines in a beam of lines. Perspective mapping preserves the harmonic ratios involved and is governed by homographic transformations. Projective geometry uses also homogeneous coordinates.

as polynomials $P_{\ell,m} \in F(\mathbb{R}^3, \mathbb{C})$ in real variables $(x, y, z) \in \mathbb{R}^3$, and spherical coordinates (r, θ, ϕ) for (x, y, z) can then be introduced. When they are expressed in spherical coordinates, the polynomials will be noted as $Y_{\ell,m}$. In what follows scalar coefficients in the polynomials will be ignored, as only their functional dependences are here of interest. Expressed in spherical coordinates, the resulting polynomials are of "degree" m in $e^{\imath\phi}$ (whereby the "degree" m can be positive or negative), which permits distinguishing the two polynomials of degree $|m|$ in z just on the basis of their "degree". The polynomials are then of the form $Y_{\ell,m}(\theta, \phi) \propto e^{\imath m\phi} f(\sin\theta, \cos\theta)$ because $\xi_0^{2\ell-k}\xi_1^k \propto e^{-\imath(\ell-k)\phi} \sin^{\ell-k}(\theta) z^k$ (where $z \propto \cos\theta$). This was not possible beforehand when $|m| \neq 0$, as $|m|$ corresponded both to m and $-m$ at the point.

In the following, operators will be introduced that project out the degrees ℓ and m from these polynomials. A representation of degree ℓ is characterized by the fact that for all its polynomials we have $d_x + d_y + d_z = \ell$. It is easy to see that $x\frac{\partial}{\partial x}$ will transform a term $x^{d_x}\, y^{d_y}\, z^{d_z}$ into $d_x x^{d_x}\, y^{d_y}\, z^{d_z}$. But, as within a polynomial there are several terms with different values of d_x, the operator $x\frac{\partial}{\partial x}$ will not have the polynomial as an eigenfunction. However, as $d_x + d_y + d_z = \ell$, the operator $\hat{\mathrm{D}} = \mathbf{r}\cdot\boldsymbol{\nabla} = x\frac{\partial}{\partial x} + y\frac{\partial}{\partial y} + z\frac{\partial}{\partial z}$ will have all the polynomials of a given representation as eigenfunctions with eigenvalue ℓ, therefore $\hat{\mathrm{D}} = \mathbf{r}\cdot\boldsymbol{\nabla}$ is a total-degree operator. This is true for the general case $P_{\ell,m} \in F(\mathscr{I}, \mathbb{C})$, and by isomorphism for $P_{\ell,m} \in F(\mathbb{R}^3, \mathbb{C})$.

The degree operator $\hat{\mathrm{D}}$ is a derivation operator. Suppose we have a derivation operator D working on a product $\prod_{j=1}^n u_j = u_1 u_2 \cdots u_j \cdots u_n$. The derivation yields $D(\prod_{j=1}^n u_j) = \sum_{j=1}^n u_1 u_2 \cdots u_{j-1}(Du_j)\, u_{j+1} \cdots u_n$. The point of interest here in the rotation group is the analogous counterpart $\hat{\mathrm{D}}(\bigotimes_{j=1}^n \boldsymbol{\xi}_j)$ of $D(\prod_{j=1}^n u_j)$, where there is a tensor product $\bigotimes_{j=1}^n \boldsymbol{\xi}_j$ expressed in (3.21), instead of a normal product $\prod_{j=1}^n u_j$, and where each term of the tensor product takes the same value $\boldsymbol{\xi}_j = (\xi_0, \xi_1)$. This way a total degree $n/2$ is obtained, which corresponds to ℓ for $n = 2\ell$.

Using the spherical coordinates within $Y_{\ell,m} \in F(\mathbb{R}^3, \mathbb{C})$, $\pm m$ can be projected out by the operator $\hat{\mathrm{L}}_z = -\imath(x\frac{\partial}{\partial y} - y\frac{\partial}{\partial x})$, which in spherical coordinates is equivalent to $-\imath\frac{\partial}{\partial\phi}$. But the mathematical operator $\hat{\mathrm{L}}_z$ can also be considered as a more general dimensionless operator that can be defined on functions that belong to $F(\mathbb{C}^3, \mathbb{C})$. It is then also defined on the restrictions $F(\mathscr{I}, \mathbb{C})$ and $F(\mathbb{R}^3, \mathbb{C})$. It is only in the restriction $F(\mathbb{R}^3, \mathbb{C})$ that it makes sense to replace it with the expression $-\imath\frac{\partial}{\partial\phi}$. In quantum mechanics we use also the physical operator $-\imath\hbar(x\frac{\partial}{\partial y} - y\frac{\partial}{\partial x})$, which differs from $\hat{\mathrm{L}}_z$ only through the factor $\hbar$. (New notation to distinguish the mathematical and physical operators will not be introduced.) The physical operator is defined only on the restriction $F(\mathbb{R}^3, \mathbb{C})$ and thus a special realization of the more general mathematical operator $\hat{\mathrm{L}}_z$. The same applies for the companion operators $\hat{\mathrm{L}}_x$ and $\hat{\mathrm{L}}_y$. The rotation matrices transform the tensors of the ℓ-dimensional representation, and the degree m will only be projected out as an eigenvalue for the operator $\hat{\mathrm{L}}_z$ when the polynomials have been transformed according to a rotation around the z-axis. Similar statements apply to $\hat{\mathrm{L}}_x$ and $\hat{\mathrm{L}}_y$.

Let us now check what the true meaning of $\hat{\mathrm{L}}_z$ is when defined for $P_{\ell,m} \in F(\mathbb{C}^3, \mathbb{C})$. It will initially be defined for $P_{\ell,m} \in F(\mathscr{I}, \mathbb{C})$, starting from the expression $\xi_0^{d_{\xi_0}} \xi_1^{d_{\xi_1}}$ that defines the harmonic polynomials. Using the definitions, this reduces to: $[x - \imath y]^{d_{\xi_0}/2} [-x - \imath y]^{d_{\xi_1}/2}\, 2^{-(d_{\xi_0} + d_{\xi_1})/2}$. Applying $\hat{\mathrm{L}}_z$ to this yields $\frac{1}{2}\,(d_{\xi_1} - d_{\xi_0})\, \xi_0^{d_{\xi_0}} \xi_1^{d_{\xi_1}}$, where the term $\frac{1}{2}\,(d_{\xi_1} - d_{\xi_0})$ is invariant under the substitution $\xi_0\xi_1 = -z/2$ that one uses in the definitions. The polynomials are thus labelled $(\ell, m) = \frac{1}{2}(d_{\xi_0} + d_{\xi_1}, d_{\xi_1} - d_{\xi_0})$. It is easy to see that $\frac{1}{2}\,(d_{\xi_1} - d_{\xi_0})$ coincides exactly with m in spherical coordinates, as ξ_0 contains $e^{-\imath\phi/2}$ and ξ_1 contains $e^{+\imath\phi/2}$. To define $\hat{\mathrm{L}}_x$ and $\hat{\mathrm{L}}_y$, one must define the spinors differently, for instance by cyclic permutation. The required change of definition corresponds to a change of basis (that can be obtained by a rotation of the triad). For $P_{\ell,m} \in F(\mathbb{C}^3, \mathbb{C})$, $\hat{\mathrm{L}}_z$ is not necessarily a degree operator. In fact, $\hat{\mathrm{L}}_z$ will only be a degree operator for polynomials that represent a rotation around the z-axis. In general, $\mathbf{m}{\cdot}\hat{\mathbf{L}}$ will only be a degree operator for polynomials that represent rotations that orient the triad such that $\mathbf{e}'_z = \mathbf{m}$.

3.10.5.2 *Intermezzo: The link with angular momentum*

At the present stage of development, it is not yet possible to define the physical operator in an intelligible way. This will become only possible in a later stage after introducing the four-vector $(E, c\mathbf{p})$ into the formalism through the phase $(Et - \mathbf{p}\cdot\mathbf{r})/\hbar$. That will only become feasible in Chapter 5. At that point the space-time coordinates $(ct, \mathbf{r})$ will also be introduced through a different approach than the isomorphism of Section 3.10, and the operators can then be defined through the prescriptions $\hat{\mathrm{E}} \to -\frac{\hbar}{\imath}\frac{\partial}{\partial t}$ and $\hat{\mathbf{p}} \to \frac{\hbar}{\imath}\boldsymbol{\nabla}$. This is due to the fact that a Lorentz transformation in free space introduces the coordinates $(\mathbf{r}, t)$ into the phase $e^{-\imath(Et - \mathbf{p}\cdot\mathbf{r})/\hbar}$ of the wave function.

The interpretation of $\hat{\mathrm{L}}_\mu$ as an angular momentum operator is a third guise of the operator initially introduced for $F(\mathbb{C}^3,\mathbb{C})$. It will become mathematically equivalent to the one defined by restricting $F(\mathbb{C}^3,\mathbb{C})$ to $F(\mathbb{R}^3,\mathbb{C})$.

It will be necessary to reconsider the problem of the definition of the operators $\hat{\mathrm{L}}_x, \hat{\mathrm{L}}_y, \hat{\mathrm{L}}_z$ several times during this book in order to discuss all their different facets. We cannot discuss all these different aspects at once at some well-chosen strategic point in the presentation. We will have to build up the whole realm of angular-momentum and degree operators gradually, and what we derive at a given step will be necessary for intermediate developments until we introduce the next step. There will be several stages and at each stage we will be able to complete the picture a bit more and discover new faces for these angular-momentum operators. But their most general definition will remain one based on degree operators. There is thus a profound link between angular momentum and the degrees of harmonic polynomials, and the notion of a degree operator is more general and more fundamental.

Here it is already possible to discuss the angular-momentum operators as degree operators defined on polynomials $P_{\ell,m} \in F(\mathbb{C}^3,\mathbb{C})$, and their restrictions $P_{\ell,m} \in F(\mathscr{I},\mathbb{C})$ and $Y_{\ell,m} \in F(\mathbb{R}^3,\mathbb{C})$. They will become physical operators following the introduction of $Et - \mathbf{p}\cdot\mathbf{r}$. Their relationship with the spin operators in SU(2) will be considered in terms of how they operate on the original spinor (ξ_0,ξ_1). To give this a meaning, it will be necessary to return to the initial definition of $\hat{\mathrm{L}}_z$ as operating on functions $P_{\ell,m} \in F(\mathbb{C}^3,\mathbb{C})$. With reference to the Dirac equation, it will be necessary to consider SU(2) as a subgroup of SL(2,$\mathbb{C}$) where the single spinor (ξ_0,ξ_1) will give rise to two "semi-spinors" (a,c) and (b,d). Their application to the solution of the wave equations for the hydrogen atom will necessitate a discussion of what they become in the case of planar motion. This corresponds to the restriction of the rotation group to the Abelian subgroup of the rotations in a plane, when the essential parts of the polynomials then become similar to polynomials encountered in solid-sate physics for problems with translational invariance with cyclic boundary conditions.

3.10.5.3 *The operator* $\hat{\mathbf{L}}^2$

In order to allow us to present the relation between spin and statistics in Subsection 3.11.4 one further consideration is needed. $\hat{\mathbf{L}}^2$ operates on functions $P_{\ell,m} \in F(\mathbb{C}^3, \mathbb{C})$ as an operator that yields information about the total degree ℓ. In fact, on the set of harmonic polynomials $P \in F(\mathbb{C}^3, \mathbb{C})$, characterized by $\Delta P = 0$, the operator $\hat{\mathbf{L}}^2 = \hat{\mathrm{L}}_x^2 + \hat{\mathrm{L}}_y^2 + \hat{\mathrm{L}}_z^2$ reduces to $\hat{\mathrm{D}}^2 + \hat{\mathrm{D}}$, such that it automatically projects out $\ell(\ell+1)$. (This will be discussed in more detail in Section (12.2)). The fact that the polynomials should satisfy the Laplace equation $\Delta P = 0$ (already described in Section 3.9) is essential here. $\hat{\mathbf{L}}^2$ can thus be used to obtain information about the degree ℓ, which is just a fact of Euclidean geometry. There is thus no such thing as the square of the length of an angular-momentum vector, that would take a strange expectation value $\hbar^2\ell(\ell+1)$ and would not be a true square due to some mysterious quantum effect. The term $\ell(\ell+1)$ already exists within the mathematics before any application of it to physics, as the operator $\hat{\mathbf{L}}^2$ corresponds to the much more general operator $\hat{\mathrm{D}}^2 + \hat{\mathrm{D}}$, where $\hat{\mathrm{D}}$ can be defined without any reference to angular momentum. (Also $\hat{\mathrm{L}}_z$ can be defined on $F(\mathbb{R}^3, \mathbb{C})$ without any reference to angular momentum.)

3.10.5.4 *The importance of the interpretation of* $\hat{\mathbf{L}}^2$ *in terms of a degree operator*

It is apparent that the operator $\hat{\mathbf{L}}^2$ does not project out the square of the value projected out by the operator $\hat{\mathbf{L}}$, *even not in mathematics*. The issue here is that the correspondence between physical quantities and operators as introduced in physics textbooks is wrong. This will be discussed further in Section 12.2. The fact that the operators are not correctly defined in textbooks is proved beyond any appeal by the fact that they yield $\ell(\ell+1)$ rather than ℓ^2 for the eigenvalue of the operator $\hat{\mathbf{L}}^2$, which is supposed to correspond to the quantity ℓ^2. It is for this reason that the alternative interpretation of the operators in terms of degree operators will also be pursued. Such a change of perspective will not upset the traditional calculations, such as those for the energy levels of the hydrogen atom. Of course, degrees of polynomials are quantized quantities.

3.10.5.5 *The importance of the interpretation of* $\hat{\mathbf{L}}_z$ *in terms of a degree operator*

While it has been proved above that $\hat{\mathrm{L}}_z$ is a degree operator, there is nothing that supports the idea that one can interpret $\hat{\mathrm{L}}_z$ as an operator that would correspond to an angular momentum component, according to some rule. The rule fails for $\hat{\mathrm{L}}^2$ as it yields a nonsensical eigenvalue $\ell(\ell+1)$. It is admittedly tempting to use the rule as it looks as though physical operators can be derived by just considering their action on $e^{-\imath(Et-\mathbf{p}\cdot\mathbf{r})/\hbar}$, but this is not part of group theory.

The way physical operators are defined is an extrapolation that can only be introduced using three ingredients: restricting $F(\mathbb{C}^3,\mathbb{C})$ to $F(\mathbb{R}^3,\mathbb{C})$, then invoking the isomorphism of Section 3.10 (as further developed in Subsection 3.11.3), *and most importantly, by not asking what happens with the definition of the angular momentum after a minimal substitution is introduced to account for the presence of a potential*, especially if that potential does not have rotational symmetry.

The third ingredient is highly questionable as it just corresponds to a lack of rigor. After the minimal substitution (to be justified in Section 5.6), the operators will in general no longer act on a wave function that has the structure $e^{-\imath(Et-\mathbf{p}\cdot\mathbf{r})/\hbar}$. There are several problems that arise here:

- First, imagine a motion characterized by $k_x = x^2$ and $k_y = y^2$. We have then $\psi = e^{\imath(k_x x+k_y y-\omega t)} = e^{\imath(x^3+y^3-\omega t)}$. There is then an ambiguity in the definition of the operator $-\imath\frac{\partial}{\partial x}$. Used on $\psi(x,y,z,t) = e^{\imath(k_x x+k_y y-\omega t)}$, the operator yields $-\imath\frac{\partial\psi}{\partial x} = k_x\psi$. Used on $\psi(x,y,z,t) = e^{\imath(x^3+y^3-\omega t)}$ it yields $-\imath\frac{\partial\psi}{\partial x} = 3x^5\psi$. The true meaning of $-\imath\frac{\partial}{\partial x}$ would certainly be $-\imath\frac{\partial\psi}{\partial x} = 3x^5\psi$ as ψ is considered as a function of the type $\psi\in F(\mathbb{R}^4,\mathbb{C}) : (x,y,z,t) \to \psi(x,y,z,t)$. By introducing a second meaning for $-\imath\frac{\partial}{\partial x}$, the procedure that should enable us to recover k_x can then be saved.[15] The rule is thus that k_x and k_y should be replaced by their values only after the derivation is performed. This means that k_x and k_y are treated as numerical values without taking into account how they depend on (x,y,z,t).
- In Section 5.6, it will be demonstrated that the minimal substitution serves to define the purely kinetic part of $\hat{\mathbf{p}}$ and $\hat{\mathrm{E}}$ in the presence of a potential. In fact, the prescriptions $\hat{\mathrm{E}} \to -\frac{\hbar}{\imath}\frac{\partial}{\partial t}$ and $\hat{\mathbf{p}} \to \frac{\hbar}{\imath}\boldsymbol{\nabla}$ yield the values for the total energy-momentum, not for the purely kinetic part of the energy-momentum. These purely kinetic parts are needed to define the instantaneous Lorentz transformation. It is the precise meaning of the minimal substitution that will indicate that p_x and p_y

[15] The true meaning of $e^{-\imath(Et-\mathbf{p}\cdot\mathbf{r})/\hbar}$ will only survive in the case of *uniform motion*. For uniform circular motion it will be possible to derive an expression $e^{-\imath(\omega t-k_\varsigma\varsigma)}$ where ς is a path length along the circle.

must be replaced by their values only after the partial derivations have been performed. In a sense, the minimal substitution lifts an ambiguity between the total energy-momentum and the kinetic energy-momentum that exists when the particle is in free space, by showing that the quantities needed in a Lorentz transformation are the kinetic part of the energy-momentum.

- After establishing this rule it would seem then that the generalization is still valid. But the definitions $\hat{\mathrm{E}} \rightarrow -\frac{\hbar}{\imath}\frac{\partial}{\partial t}$ and $\hat{\mathbf{p}} \rightarrow \frac{\hbar}{\imath}\boldsymbol{\nabla}$ have been derived for a *scalar* wave function within the context of the Schrödinger equation. It will be shown (in Section 9.5 and more particularly in Subsection 9.5.2) that the definitions must be reconsidered when we use a *multi-component spinor* wave function.
- Even with a scalar wave function, $\hat{\mathrm{L}}_j$ are not the operators for the x_j-components of the angular momentum. This follows from an analysis of the action of the operators $\hat{\mathrm{L}}_j$ on the harmonic polynomials (see Subsection 6.2.9.3). The operator $\hat{\mathrm{L}}_j$ expresses the total angular momentum when this angular momentum is aligned with the x_j-axis. In general, when the angular momentum is aligned with the unit vector $\mathbf{m}$, the angular momentum operator will be $\mathbf{m}\cdot\hat{\mathbf{L}}$, where $\hat{\mathbf{L}} = (\hat{\mathrm{L}}_x, \hat{\mathrm{L}}_y, \hat{\mathrm{L}}_z)$. This is consistent with the result obtained in Subsection 3.10.5.1.

A large number of paradoxes about angular momentum in quantum mechanics can be avoided by rejecting the procedure by which physical operators are defined using these three ingredients. And on the basis of our findings with $\hat{\mathrm{L}}^2$ and our criticism of the third ingredient of the procedure, we have good grounds for doing so. We are not forced to accept at face value that $\hat{\mathrm{L}}_x$, $\hat{\mathrm{L}}_y$, and $\hat{\mathrm{L}}_z$ are operators that would correspond to the x-, y- and z-components of the angular momentum for a particle in orbital motion around an arbitrary axis $\mathbf{m}$. If one rejects the procedure, then $\hat{\mathrm{L}}_x$, $\hat{\mathrm{L}}_y$, and $\hat{\mathrm{L}}_z$ will only keep the meaning of degree operators. But $\hat{\mathrm{L}}_z$ has a physical role, as it intervenes in the calculation of the Zeeman splitting in a magnetic field. It will take a long journey before we will be able to sort this out. We will have to clarify the meaning of spin before we can address this problem. The reader is not asked at this stage to reject the definition of the angular-momentum operators in physics. For the time being the reader is only asked to accept that a certain caution will be observed in the text. This caution will consist in interpreting angular momentum operators only in terms of degree operators.

3.11 Important consequences for quantum mechanics

3.11.1 *Why the probability densities are expressed as $\psi^*\psi$ or $\Psi^\dagger\Psi$*

From the quadratic expressions in equations (3.11, 3.12, 3.13), we see already transpire here an essential feature that we find back in quantum mechanics. Probabilities behave like charges; they are conserved quantities. In relativity, they become parts of a more general probability "charge-current" four-vector, subject to a continuity equation. The continuity equation expresses that probability is a conserved quantity. As also in the Lorentz group four-vectors will be second-degree tensors of spinors, this will explain why probability densities are expressed through a quadratic expression $\Psi^\dagger\Psi$, which is part of a probability charge-current four-vector $(\Psi^\dagger\Psi, \Psi^\dagger\boldsymbol{\alpha}\,\Psi)$ subject to a continuity equation, whereby Ψ is a spinor. Here, $\boldsymbol{\alpha}$ stands for the triplet of Dirac matrices $(\alpha_x, \alpha_y, \alpha_z)$. This will be rendered more precise later on.

3.11.2 *Topology and 4π turns*

It may be noted that rotations operate on reflections in a different way than they operate on vectors: there is a factor of two in the respective angles involved. This only becomes paradoxical if these different representations based on reflections and based on vectors become confused. In fact, the rotations around an axis can be considered as the product of two reflections, as illustrated in Figure 3.3. Consider the first reflection as fixed and the second reflection allowed to turn. When the second reflection plane has turned over an angle φ, the rotation will have an angle 2φ. When the reflection plane of the second reflection (or its normal) is turned over an angle of π, in terms of the ensuing rotation a full turn will have been made, but the normal vector will have been multiplied by -1. Hence, a 4π turn must be made to recover the initial normal vector. This is of course algebraically related to the fact that vectors are second-rank tensors in terms of spinors. This is a property of the wave function in quantum mechanics that is considered as counterintuitive, but it follows from the mathematics of the rotation group. This fact has often been mapped onto a topological argument that describes the same concept but creates more mystery rather than clarifying it. In fact, as two opposite vectors define the same reflection, it is pertinent to identify them within the rotation group. Through the arguments permitting the removal r from the formalism and the introduction of homogeneous coordinates it transpired that the relevant parameters needed to define reflections within the context of the rotation group are not really the unit vectors but the directions of these vectors. This set of directions is $\mathbb{R}P^2$. These directions must be represented in one-to-one correspondence with a hemisphere rather than with a sphere. There is thus a necessity to identify opposite points on the sphere as both representing the same element of $\mathbb{R}P^2$, and to glue the hemisphere to itself along its edge. But an inspection of two nearby directions OP and OQ, where P and Q are both on the edge of the hemisphere, to see how this gluing must be done, will reveal that gluing simultaneously both P to its counterpart P' and Q to its counterpart Q' can only be achieved by doing it in a Möbius-like fashion. This is illustrated in Figure 3.5.

The analogy is nice. When a full loop is made on a Möbius ring, the surface normal becomes inverted. When a full turn is made in the rotation group, the normal to the second reflection plane that defines a family of rotations is also inverted. This is a topological argument and a number of topological constructions have been developed to illustrate this, *viz.* in Figure 41.6 of [Misner *et al.* (1970)] and in [Feynman and Weinberg (1987)]. In Figure 41.6 of [Misner *et al.* (1970)] it is possible to follow the fate of four threads that can be considered as materializing four elements of $\mathbb{R}P^2$.

3.11.3 *Two movies for the price of one*

Attention must also be drawn to a further particularity, an important remark, that will allow us to discover also something quite beautiful about the formalism of quantum mechanics. It has been shown that the harmonic polynomials build representations. This is all fine if we think about the polynomials in (x, y, z) as abstract quantities. But as physicists, we are used to thinking of (x, y, z) as the coordinates of a particle in $\mathbb{R}^3$, while on the other hand, in the spinor formalism, (x, y, z) are not the coordinates of a particle, but of an isotropic vector that codes a rotation, and the numbers (x, y, z) belong to the isotropic cone $\mathscr{I}$ of $\mathbb{C}^3$. Let us note for this subsection the coordinates of the isotropic vector as (X, Y, Z) to make the distinction.

We could hit here the same kind of confusion between vectors and group elements as with the stereographic projection. It is obvious from the structure of the harmonic polynomials that they are deduced from the constraint $\forall (X, Y, Z) \in \mathscr{I}$, $X^2 + Y^2 + Z^2 = 0$, while for particle coordinates

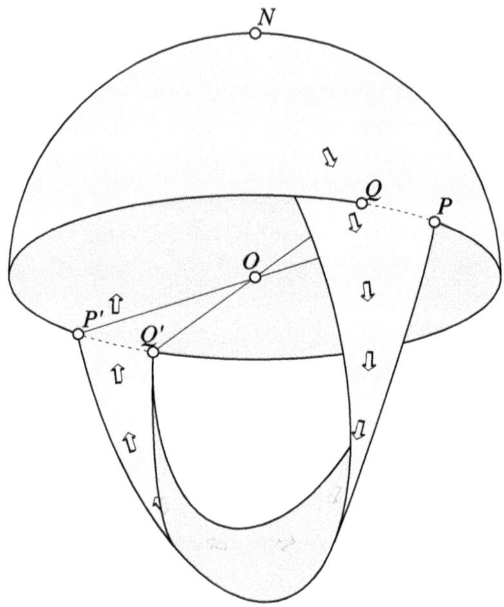

Fig. 3.5 The hemisphere with centre O and radius ON represents the set of all reflection normals OP. The two antipodal points P and P' on the edge of the hemisphere correspond to an identical reflection, as do the two antipodal points Q and Q'. When P leaves the hemisphere, P' enters it, such that P seems to "jump" to P'. There is no real discontinuity in this; it is only like jumping from 12 to zero on the reading of a clock at midnight. To visualize the true connectivity, the hemisphere must be glued to itself along its edge, such that in the end a surface without boundaries is obtained. For instance P must be glued to P' and Q to Q'. This can only be achieved by doing it in a Möbius fashion, gluing the outside of the hemisphere to its inside, resulting in a surface that has only one side. This is illustrated by the arrows on the twisted ribbon introduced to connect P to P' and Q to Q'. The ribbon is entirely metaphorical and only serves to illustrate the connectivity. It shows parts that must be "glued" together because they are connected and indicates how they are connected by illustrating how they must be "glued" together. The ribbon must be considered as attached to the whole equator, rather than just to the segments PQ and $P'Q'$. Its length PP' can be considered as zero as the topological argument is independent from the choice of a metric.

$(x, y, z) \in \mathbb{R}^3$ it must follow that $(x, y, z) \neq (0, 0, 0) \Rightarrow x^2+y^2+z^2 = r^2 \not\equiv 0$. Quantum mechanics describes the rotation of the electron by coding it as a spinor based on an isotropic vector (X, Y, Z) with $X^2 + Y^2 + Z^2 = 0$. But when it comes to solving the Dirac or Schrödinger equation one substitutes quite happily the particle coordinates (x, y, z) into the variables (X, Y, Z) that occur in the harmonic polynomials $P(X, Y, Z)$. This is the

$0 = 1$ paradox mentioned in Section 1.2. This can be summarized in the following schema:

$$
\boxed{
\begin{array}{l}
\mathbf{r} = (x,y,z) \in \mathbb{R}^3 \\
\quad \Big\downarrow \text{spinor field of triads } (\xi_0,\xi_1) \\
(\xi_0(\mathbf{r}), \xi_1(\mathbf{r})) \\
\quad \Big\downarrow \underbrace{(\xi_0,\xi_1)\otimes(\xi_0,\xi_1)}_{\text{2 spinors}} \quad \rightarrow \quad \underset{\text{isotropic vector}}{(X,Y,Z) \in \mathscr{I} \subset \mathbb{C}^3} \\
(X(\mathbf{r}), Y(\mathbf{r}), Z(\mathbf{r})) \\
\quad \Big\downarrow \underbrace{(X,Y,Z)\otimes(X,Y,Z)\cdots\otimes(X,Y,Z)}_{\ell \text{ terms in } (X,Y,Z)} \quad \rightarrow \quad \underset{\text{polynomials } P_{\ell,m}}{P_{\ell,m}(X,Y,Z)} \\
P_{\ell,m}(X(\mathbf{r}), Y(\mathbf{r}), Z(\mathbf{r})) \\
\quad \Big\downarrow \begin{array}{l}\text{Quantum mechanics:} \\ \text{Isomorphism } \mathscr{I} \equiv \mathbb{R}^3\end{array} \quad \boxed{(X,Y,Z) \equiv (x,y,z)} \\
\psi(\mathbf{r}) = P_{\ell,m}(\mathbf{r}) \\
\quad \Big\downarrow \text{Spherical coordinates } \mathbf{r} \rightarrow (r,\theta,\phi) \\
\psi(\mathbf{r}) = F(r) Y_{\ell,m}(\theta,\phi).
\end{array}
}
\tag{3.29}
$$

This schema is valid for the Schrödinger equation for the hydrogen atom and has to be modified for the Dirac equation, which uses a different type of wave function. Comprehensive understanding of what a spinor is, is essential to discern that the apparent contradiction, which consists in identifying $(X, Y, Z) \equiv (x, y, z)$ in this schema, constitutes a serious conceptual difficulty. The quantities (θ, ϕ) in $Y_{\ell,m}(\theta, \phi)$ contain less information than (X, Y, Z) in the original harmonic polynomials $P_{\ell,m}(X, Y, Z)$. From Section 3.10, the reader will already be able to guess the solution. It is the power of analyticity in $\mathbb{C}$ that accomplishes a miracle here. The algebra from the isotropic cone $\mathscr{I}$ can be extended to the whole of $\mathbb{C}^3$, and then restricted again to $\mathbb{R}^3$. The original and final domains only have the trivial $\mathbf{0}$ vector in common but the whole algebraic group structure remains the same and carries through by isomorphism. The vector $\mathbf{0}$ cannot be used to represent a triad of basis vectors whose length would not be normalized to 1. It also does not correspond to a spinor that is normalized to 1. The concept that spinor space would contain just one point $\mathbf{0}$ of physical space $\mathbb{R}^3$ is thus incorrect (see Subsection 5.3.3).

Due to this isomorphism, it will be possible to identify the particle coordinates with the spinor parameters, remembering that the meaning of the symbols (x, y, z) and (X, Y, Z) is totally different. *A priori* one might fear that the structure of the equation $X^2 + Y^2 + Z^2 = 0$ will be preserved in the isomorphism such that it can never correspond to $x^2 + y^2 + z^2 = r^2 \not\equiv 0$. But as shown in Section 3.10, there is an isomorphism that maps spinor coordinates (X, Y, Z) onto vector coordinates (x, y, z), in conformity with the two equations. It is therefore that we really needed it to define this isomorphism, as it is crucial for the validity of the calculations of quantum mechanics. This may be summarized by stating that besides the isomorphism $\mathscr{I} \equiv SU(2)$ between isotropic vectors and rotations (coded by spinors), quantum mechanics uses an isomorphism $\mathscr{I} \equiv \mathbb{R}^3$ between vectors and spinors.

Construction of the spinor began by choosing $(1, \imath, 0)$ as the isotropic vector. It therefore looked as though the coordinates (x, y, z) would be those of $\mathbf{e}_z$, i.e. of the normal to the plane of motion if this motion were in the Oxy plane. But it would have been equally possible to take the isotropic vector $(X, Y, Z) = (0, 1, \imath) = \mathbf{e}_y + \imath\mathbf{e}_z$ rather than the isotropic vector $(X, Y, Z) = (1, \imath, 0) = \mathbf{e}_x + \imath\mathbf{e}_y$ as the starting point for the development. The point is that (3.10)–(3.13) would have remained the same, and only the values of (X, Y, Z) in these equations change. With this choice of basis, in a right-handed frame, the starting value (x, y, z) that is isomorphic to the starting value $(X, Y, Z) = (0, 1, \imath)$ is $(1, 0, 0)$.[16] Thus, (x, y, z) can really be made to correspond to particle coordinates.

The isomorphism carries through in the different representations based on harmonic polynomials of degree ℓ. In the original harmonic polynomials $P(X, Y, Z)$, $P \in F(\mathbb{C}^3, \mathbb{C})$ is a function of complex variables that takes complex values, while in the isomorphic image, $P(x, y, z)$, $P \in F(\mathbb{R}^3, \mathbb{R}) \vee P \in F(\mathbb{R}^3, \mathbb{C})$ is a function of real variables that takes real or complex values. In this respect, a change to spherical coordinates (θ, ϕ) based on the sequence of substitutions $(x, y, z) \to (x', y', z') = (x/r, y/r, z/r)$, $(x', y', z') = (\sin\theta\cos\phi, \sin\theta\sin\phi, \cos\theta)$, and finally $(x' + \imath y', x' - \imath y', z') = (e^{\imath\phi}\sin\theta, e^{-\imath\phi}\sin\theta, \cos\theta)$ to go from real spherical harmonics $P \in F(\mathbb{R}^3, \mathbb{R})$ to complex spherical harmonics $P \in F(\mathbb{R}^3, \mathbb{C})$, can only be defined

[16]It follows then that $\mathbb{1} + \mathbf{e}_x{\cdot}\boldsymbol{\sigma} = \begin{pmatrix} 1 & 1 \\ 1 & 1 \end{pmatrix} = \sqrt{2}\begin{pmatrix} \xi_0 \\ \xi_1 \end{pmatrix} \otimes (\xi_0^*, \xi_1^*)\sqrt{2}$, where $\xi_0 = \xi_1 = \sqrt{-\imath/2}$, which can be calculated from the values of (X, Y, Z). A rotation over an angle φ around the z-axis in the spinor formalism will then correspond exactly to a rotation over the same angle φ around the z-axis of the vector $\mathbf{e}_x$.

within the isomorphic representation with $P \in F(\mathbb{R}^3, \mathbb{R})$.[17] This representation is incomplete, as transpires from the fact that it is based on only two independent variables (θ, ϕ) rather than three. It is not possible to define an analogous set of angular variables (Θ, Φ) for (X, Y, Z), not even with $(\Theta, \Phi) \in \mathbb{C}^2$, as it is impossible to satisfy an equation $\cos^2\Theta + \sin^2\Theta\cos^2\Phi + \sin^2\Theta\sin^2\Phi = 0$, since $(\forall\alpha \in \mathbb{C})(\cos^2\alpha + \sin^2\alpha = 1)$. Certainly, sines and cosines of imaginary arguments introduce hyperbolic functions, but the relation $\cosh^2 x - \sinh^2 x = 1$ still has 1 and not a zero on its right-hand side.[18]

The fact that the isomorphism is not trivial, and perhaps even accidental, is evident from the unusual nature of the relationship between $\mathbb{1} + \mathbf{e}'_z\cdot\boldsymbol{\sigma}$ and $(\mathbf{e}'_x + \imath\mathbf{e}'_y)\cdot\boldsymbol{\sigma}$ discussed in Section 3.10. This suggests that the textbook solutions of the Schrödinger and Dirac equations for the hydrogen atom are *ad hoc*. It is not the introduction of spherical coordinates that results in the rotational symmetry, but the use of harmonic polynomials. These should preferably be taken from $F(\mathbb{C}^3, \mathbb{C})$, as the ones from $F(\mathbb{R}^3, \mathbb{C})$ based on real spherical coordinates are not able to reflect the spinor symmetry. A much more relevant set of parameters would be the rotation angle and the

[17] It is worth noting in this respect that in textbooks the normalization of the spherical harmonics, expressed in the coordinates (x, y, z) and the ensuing Clebsch-Gordon coefficients are (somewhat aberrantly) defined on the basis of the polynomials $P \in F(\mathbb{R}^3, \mathbb{R}) \vee P \in F(\mathbb{R}^3, \mathbb{C})$ rather than $P \in F(\mathbb{C}^3, \mathbb{C})$, since they are explicitly based on the introduction of spherical coordinates. One could also define a normalization procedure for $P \in F(\mathbb{C}^3, \mathbb{C})$, based on the fact that a rotation is defined by three independent real parameters. The normalized volume element $\frac{1}{4\pi^2}\sin^2(\varphi/2)\sin\theta\, d\varphi\, d\theta\, d\phi$ of the Haar integral for the rotation group (see for example Appendix C of the book of H.F. Jones [Jones (1990)]) can in this respect be considered as derived from the volume element $r^3\sin^2\theta_0\sin\theta_1\, dr\, d\theta_0\, d\theta_1\, d\phi$, that occurs when one introduces the four-dimensional hyper-spherical coordinates $(r, \theta_0, \theta_1, \phi)$ discussed in Footnote 10 of Subsection 3.9.2. One just has to make the correspondence $(\theta_0, \theta_1, \phi) \leftrightarrow (\varphi/2, \theta, \phi)$ with $r = 1$. In the Haar integral the coordinate φ of a rotation defines its rotation angle, and the coordinates (θ, ϕ) its rotation axis $\mathbf{n}$. These are the coordinates to be used in the generalization towards $P \in F(\mathbb{C}^3, \mathbb{C})$, as generalizing $(\theta, \phi) \in \mathbb{R}^2 \to (\Theta, \Phi) \in \mathbb{C}^2$ is not viable. This could lead to more complete hyper-spherical harmonics $Y(\varphi, \theta, \phi)$. This study has not investigated this possibility, nor how they could be put into correspondence with the traditional spherical harmonics in terms of functional dependence, normalization factors and Clebsch-Gordon coefficients. The fact that these more general functions are of no use in for example, the solutions of the Schrödinger or Dirac equations for the hydrogen atom, is due to the fact that they use the isomorphism defined in Section 3.10.

[18] Equations of the type $(\Delta + V(r))\psi = 0$ are of course related to equations $\Delta\psi = 0$ that define both types of harmonic polynomials $P(X, Y, Z)$ and $P(x, y, z)$. The solutions of $\Delta\psi = 0$ in terms of (X, Y, Z) or (x, y, z) do not depend on the real or complex character of the arguments.

spherical coordinates of the rotation axis. It will be necessary to refine this statement in Subsection 5.4.3.

It is quite beautiful that due to this analytic extension the algebra plays us two movies at the same time. On the isotropic cone $\mathscr{I}$ the algebra with the variables $(X, Y, Z) \in \mathscr{I}$ plays us the movie of the rotations, while on $\mathbb{R}^3$ the same algebra plays us the movie of particle motion.

In a perfectly analogous way, the algebra on the light cone $\mathscr{C}$ in $\mathbb{C}^4$ in the Lorentz group will describe the orientation of the axes of a reference frame, while on $\mathbb{R}^4$ the very same algebra will describe its translational motion, and vectors such as the probability current-density four-vector $(\Psi^\dagger\Psi, \Psi^\dagger\boldsymbol{\alpha}\Psi)$. It is important to remember that the variables (x, y, z, t) in $\Psi(x, y, z, t)$ used to describe the probability density will have a completely different meaning and range of values when $\Psi(X, Y, Z, T)$ is used to describe the orientation of the basic tetrad. Not making the distinction between the variables (X, Y, Z, T) on the light cone $\mathscr{C}$ that serve to code a tetrad or a part of a tetrad, and the coordinates $(x, y, z, t) \in \mathbb{R}^4$ of a particle, will suggest the wrong notion of an electron that would travel at the speed of light although it does not have zero rest mass.[19]

In fact, a physically clear picture of a wave function $\Psi(x, y, z, t)$ can be given: it defines in each point $(x, y, z, t) \in \mathbb{R}^4$ of space-time an orientation of the tetrad coded by the four complex parameters (X, Y, Z, T) that will occur in the 4×1 matrix $\Psi \in \mathscr{C}$. These components are traditionally expressed as functions of $(x, y, z, ct) \in \mathbb{R}^4$ in a way that one can easily confuse the variables $(X, Y, Z, cT) \in \mathscr{C}$ with the coordinates $(x, y, z, ct) \in \mathbb{R}^4$ themselves.

3.11.4 *The relation between spin and statistics*

All the finite-dimensional representations of the rotation group are tensor powers $(\xi_0, \xi_1) \otimes (\xi_0, \xi_1) \otimes \cdots \otimes (\xi_0, \xi_1)$. The tensor power can be even, such that the tensor components can than be identified with $Y_{\ell,m}$. The tensor power can also be odd and can then be identified with the components of a tensor product $Y_{\ell,m} \otimes (\xi_0, \xi_1)$. As can be seen from the Rodrigues formula to be discussed in Subsection 5.1.2, and as already discussed in Subsection 3.11.2, when we make a rotation of 2π on a spinor (ξ_0, ξ_1), it changes sign. This is often paraphrased by stating that SU(2) is a double covering of

[19] One can find this confusion in textbooks, where it is discussed in terms of a so-called *Zitterbewegung*, an alleged quantum mystery.

SO(3). It is now possible to see what happens in the tensor representations with these changes of sign. In the even-power tensor representations, they will mutually cancel each other out two by two, as there is an even number of spinors (ξ_0, ξ_1) in the tensor product. But in the odd-power tensor representations the change of sign within one of the terms will not be cancelled out. As discussed in Subsection 3.10.5 and will be further discussed in Subsection 5.10.1.6 and Section 12.2, the quantum number ℓ in the expectation value $\ell(\ell+1)\hbar^2$ of the physical operator $\hat{\mathbf{L}}^2$ corresponds to the degree ℓ of the representation, such that odd and even powers correspond to half-integer and integer angular momenta respectively. This relationship between the mathematical operator $\hat{\mathbf{L}}^2$ and $\hat{\mathrm{D}}^2 + \hat{\mathrm{D}}$ was discussed in Subsection 3.10.5.

There is an isomorphism between the isotropic cone $\mathscr{I}$ (as a manifold to visualize SU(2)) and $\mathbb{R}^3$ that is used in quantum mechanics. From a radical viewpoint, this means that $\mathbb{R}^3$ can be used to visualize SU(2). Imagine now a circular orbit for a particle (centred at the origin). On this circular orbit, the particle can be considered as moving on SU(2), due to the isomorphism that was introduced. There is then a problem due to the fact that when a full turn 2π is made on the circular orbit in $\mathbb{R}^2$, a rotation of 2π is performed in SU(2). But this means that the wave function will have changed sign, while we will have come back to the initial position $\mathbf{r}$. The wave function is then not a true function.

An attempt could be made to avoid this by only choosing representations with an even power in the tensor product. But such an even power leads to a representation with an odd number of spherical harmonics. There will be physical considerations that constrain the choice of representation. One of these is that the experimentally observed Zeeman effect is explained by assuming that electron spin represents a magnetic moment. The current induced by the orbital motion of the electron also represents a magnetic moment.

When an external magnetic field is activated both the intrinsic and orbital magnetic dipole moments will have different energies depending on the question of whether they are parallel or anti-parallel to the external magnetic field. This means that there must always be an even number of magnetic sub-states within the degenerated state obtained when there is no magnetic field, because they manifest themselves when the degeneracy is lifted as momentum-up and momentum-down states. It is therefore necessary to use an odd-power tensor product representation in order to obtain a description that accounts successfully and consistently for the experimentally observed Zeeman splitting. Even before adopting the physical explanation in terms of some magnetic dipole moments, it is necessary to take a representation that contains an even number of substates in order to be in agreement with the observed Zeeman splitting. It is clear from this that an electron should always be described with an SU(2) representation in order to account for the observed Zeeman splitting. Other physical problems could however be treated by a harmonic-polynomial representation. It will be discussed in Subsection 6.2.6 how it is possible to make sure that the SU(2) wave function remains a true function, by replacing $\mathbb{R}^3$ with a Riemann surface. However, difficulties remain in describing the combined state of several electrons.

Consider a problem with two electrons. These two electrons with their spin could also be described by a tensor product. Imagine these two electrons are on a circular orbit. When both electrons are rotated by π, each electron will have taken up the previous position of its counterpart. The combined wave function will then be multiplied by -1 while the state cannot be distinguished from the initial one. As one rotation corresponds to two spinors, $\psi_1 = (\xi_0, \xi_1)$ could be chosen for one electron and $-(\xi_0, \xi_1)$ for the other electron right from the start. But the problem will return in various different guises when there are more electrons. There is however a simple and convenient general rule to settle this problem once and for all: replace the simple tensor products with anti-symmetrical tensor products in the spirit of Slater determinants.[20] The explanation described here for the so-called Pauli exclusion principle was discovered by Feynman [Feynman and Weinberg (1987)], even if he did not discuss it really within the context of group theory.

Confusing $(x, y, z) \in \mathbb{R}^3$ and $(X, Y, Z) \in \mathscr{I}$, or more generally $(x, y, z, ct) \in \mathbb{R}^4$ and $(X, Y, Z, cT) \in \mathscr{C}$ on the light cone in $\mathbb{C}^4$, leads to the notion that the wave function changes sign when the positions of two fermions are exchanged [Feynman and Weinberg (1987); Gross *et al.* (1991)]. The canonical notion, however, is that the wave function changes sign when the orientations of two tetrads is exchanged. In the French translation of Feynman's work there is a footnote from the French physicist Lévy-Leblond who points out the confusion between $\mathbb{R}^4$ and $\mathscr{C}$ in this argument. He was right to do so, but the isomorphism $\mathscr{I} \equiv \mathbb{R}^3$ introduced in Section 3.10 provides the missing link in this puzzle.

---

[20]The analogue of a Slater determinant for spinors has actually to be defined. If $\boldsymbol{\xi} = [\xi_0, \xi_1]^\top \in F(\mathbb{R}^3, \mathscr{I})$ and $\boldsymbol{\eta} = [\eta_0, \eta_1]^\top \in F(\mathbb{R}^3, \mathscr{I})$ are two spinor-valued functions, this could be $\Psi \in F(\mathbb{R}^3 \times \mathbb{R}^3, \mathbb{C}^2 \times \mathbb{C}^2) : (\mathbf{r}_1, \mathbf{r}_2) \to \Psi(\mathbf{r}_1, \mathbf{r}_2) = \boldsymbol{\xi}(\mathbf{r}_1) \otimes \boldsymbol{\eta}(\mathbf{r}_2) - \boldsymbol{\xi}(\mathbf{r}_2) \otimes \boldsymbol{\eta}(\mathbf{r}_1)$.

It must be clear now that there are two different types of representations of the rotation group that one uses in quantum mechanics. To characterize the position of an electron on its orbit, one can use the incomplete vector-type representations based on spherical harmonics $Y_{\ell,m}$. For the spin of the electron we will need a two-component spinor (ξ_0, ξ_1) of SU(2).

It may be finally noted that in his treatment of the spinors in the rotation group Cartan[Cartan (1981)] also jumps backwards and forwards between $(X, Y, Z) \in \mathbb{C}^3$ when they are on the isotropic cone $\mathscr{I}$ and $(x, y, z) \in \mathbb{R}^3$ when they describe the coordinates of a real vector without any warning or acknowledgement. This is a real distraction for the reader and a further illustration of the point made in the Introduction about the difficulty of understanding the underlying ideas from an austere presentation.

Chapter 4

Spinors in the Homogeneous Lorentz Group

(This chapter can be skipped on a first reading.)

4.1 It takes two different zero-length vectors to code the whole tetrad

In this chapter it will be attempted to build a representation of the group of Lorentz transformations in $\mathbb{R}^4$ based on the same principle that the column matrices the representation matrices work upon must be images of Lorentz transformations through the coding of a tetrad. However, this turns out to be more difficult than one would expect on the basis of experience gained in developing SU(2). Therefore, even though the chapters of this book are placed in a logical order, some readers may find it helpful on a first reading to jump to Chapter 5, where we derive the Dirac equation, and then come back to Chapter 4. This will give a good idea about the motivation for the construction of the Lorentz spinors in Chapter 4.

The *Vielbein* needed will now be a tetrad as illustrated in Figure 4.1. One of the difficulties is that in SL(2,$\mathbb{C}$) it is not possible to code the whole tetrad into a single column vector along the same lines as in the rotation group. In fact, it will be necessary to code the tetrad into two column vectors. The origin of this difference with the rotation group resides in the fact that there are no further quantities like $\imath$ to combine the components of the tetrad into one single expression in a way that would still allow retrieval of these components from the linear combination. The whole problem is thus rooted in the fact that there is no commutative field of numbers beyond $\mathbb{C}$. Due to this, coding the information content of four vectors into a single vector such that after applying Lorentz transformations on it, this information could be retrieved again unambiguously (like in SU(2)), is just impossible.

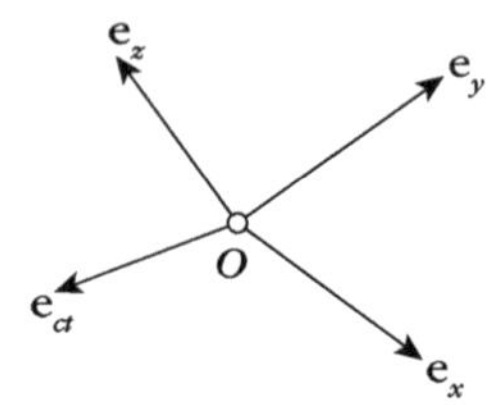

Fig. 4.1 Just like in the rotation group, the Lorentz transformations are completely defined by the unit vectors of a normalized orthogonal basis $\mathbf{e}_x$, $\mathbf{e}_y$, $\mathbf{e}_z$, $\mathbf{e}_{ct}$. Such a tetrad can be considered as a spinor. Combinations like $\mathbf{e}_x \pm \imath\mathbf{e}_y$, $\mathbf{e}_{ct} \pm \mathbf{e}_z$ were called semi-spinors by Cartan.

Nevertheless, the leading principle behind group representation theory remains finding a way to code the *Vielbein*. This idea will also be true for spinors in the Lorentz group: a spinor is the image of a group element L and codes the image $(\mathbf{e}'_{ct}, \mathbf{e}'_x, \mathbf{e}'_y, \mathbf{e}'_z) = (L(\mathbf{e}_{ct}), L(\mathbf{e}_x),\ L(\mathbf{e}_y),\ L(\mathbf{e}_z))$ of a complete tetrad or *Vierbein* $(\mathbf{e}_{ct}, \mathbf{e}_x, \mathbf{e}_y, \mathbf{e}_z)$ of mutually orthogonal unit vectors of an orthonormal basis of $\mathbb{R}^4$ under the action of this group element L.[1] This means that it should be possible to code the whole tetrad of basis vectors into a spinor.

First, a remark about the special nature of the Minkowski pseudo-metric used in the Lorentz group should be made. Let us assume that the signature of the pseudo-metric is $(1, -1, -1, -1)$ such that the pseudo-norm of $(ct, x, y, z,)$ is $c^2t^2 - x^2 - y^2 - z^2$. There are then two very distinct ways of obtaining a vector of "zero length". In fact, one possibility is to take e.g. $\mathbf{e}_x + \imath\mathbf{e}_y$, which is an isotropic vector built according to the same idea as in the rotation group. But it is also possible to take $\mathbf{e}_{ct} + \mathbf{e}_z$, which is a null vector or a point on the light cone $\mathscr{C}$ in Minkowski space-time. As then $(\forall\alpha \in \mathbb{R})(\alpha(\mathbf{e}_{ct} + \mathbf{e}_z) \in \mathscr{C})$, $\mathbf{e}_{ct} + \mathbf{e}_z$ defines the light ray to which it belongs. One can imagine an analogy of the latter quantity with an isotropic vector, by assuming that the quantities are not combined as $\mathbf{e}_z + \imath\mathbf{e}_{ct}$, due to the fact that the signature is not $(1, 1, 1, 1)$. With an isotropic vector like $\mathbf{e}_x + \imath\mathbf{e}_y$, both parts $\mathbf{e}_x$ and $\mathbf{e}_y$ can be reconstructed by separating the real and imaginary parts. Is it also possible to reconstruct $\mathbf{e}_z$ and $\mathbf{e}_{ct}$ from the null vector $\mathbf{e}_{ct} + \mathbf{e}_z$?

It will be shown how the whole information content of the tetrad is coded in an element of SL(2,$\mathbb{C}$), such that in a sense these 2×2 matrices

[1]The notation $\mathbf{L}$ is used to note a representation of an element L of the Lorentz group, just as $\mathbf{R}$ and $\mathbf{B}$ are used to note matrices that represent rotations and boosts. In general this should not lead to confusion with the angular-momentum operator $\hat{\mathbf{L}}$.

can be seen as the spinors of the Lorentz group. In conformity with a terminology used by Cartan, each column of such a 2×2 matrix can be considered as a semi-spinor. Together these two semi-spinors are coding two zero-length vectors, $\mathbf{e}_x + \imath \mathbf{e}_y$ and $\mathbf{e}_{ct} + \mathbf{e}_z$.[2] From the knowledge of these two zero-length vectors, the whole tetrad can be reconstructed. In fact from $\mathbf{e}_x + \imath \mathbf{e}_y$ the vectors $\mathbf{e}_x$ and $\mathbf{e}_y$ can be reconstructed by separating the real and imaginary parts. The vector $\mathbf{e}_{ct} - \mathbf{e}_z$ can then also be reconstructed as the unique vector with the proper norm that is simultaneously orthogonal to $\mathbf{e}_x$, $\mathbf{e}_y$ and $\mathbf{e}_{ct} + \mathbf{e}_z$. And this allows us to reconstruct $\mathbf{e}_{ct}$ and $\mathbf{e}_z$ from $\mathbf{e}_{ct} + \mathbf{e}_z$ and $\mathbf{e}_{ct} - \mathbf{e}_z$. This gives a clear and neat geometrical meaning to the concept of a spinor: spinors code group elements, and this is achieved by coding the complete image of an orthonormal basis.

[2]This will be discussed in Section 4.6. Some authors consider the 2×1 column matrices as the spinors. This is of course a matter of definition. But the information content of the semi-spinors is always larger than that of the 2×1 column matrices that can be "derived" from the "zero-length" vectors. The introduction of a "zero-length" vector looks arcane and has no further natural justification if the basic idea that we want to define a tetrad is abandoned. It is the definition of the tetrad that necessitates the combination of two orthogonal unit vectors into a pair, and it is this orthogonality of unit vectors that implies then that the pair forms a "zero-length" vector. Furthermore, two complementary "zero-length" vectors must be chosen. Two "zero-length" vectors will be considered as truly complementary only when they permit reconstruction of the whole tetrad (see Section 4.6). It is for these reasons that it is more suitable to consider the 2×2-matrices of SL(2,$\mathbb{C}$) as the true spinors. This choice of definition permits eluding Atiyah's verdict cited at the beginning of Chapter 3. The true spinors contain then the complete information about the tetrad, but scattered over the two semi-spinors. One could try to consider the two semi-spinors as separate entities, but the necessity of the constraint $ad - bc = 1$ (see (4.9)) imposed on these two semi-spinors is difficult to interpret geometrically. There is thus little hope that one could make sense of the semi-spinors separately as truly meaningful mathematical objects, that one could call then spinors. The constraint $ad - bc = 1$ rather seems to emphasize that one must consider the 2×2 matrices of SL(2,$\mathbb{C}$) as indivisible wholes. The image of the Meccano game introduced in Section 2.13 can be used here. A group element corresponds to a tetrad, and the tetrad is a Meccano construction of four unit vectors. Each *Vielbein* introduced in Section 3.4 as an image of a group element is such a Meccano construction of unit vectors. In the formalism, the *Vielbein* can be expressed as a set of complex column vectors which are themselves constructions containing two unit vectors. Rather than describing the transformation of the individual pieces of the construction, it is much easier to make sense of the construction when described as a whole. It is the global structure that counts, not the individual building blocks. In SU(2) isotropic vectors were used as an equivalent construction for two mutually orthogonal unit vectors, allowing the pair to be treated as a whole and describing the rotations of the triads as a whole. The whole issue is thus to find equivalent constructions.

4.2 The representations SL(2,ℂ)

At this stage it may be noted that it is possible to code a four-dimensional vector $\mathbf{v} = (x, y, z, ct)$ in a two-dimensional formalism, by taking:

$$\begin{aligned}\mathbf{v} = (x, y, z, ct) \to \mathbf{V} &= ct\mathbb{1} + (\sigma_x x + \sigma_y y + \sigma_z z)\\ &= \begin{pmatrix} ct + z & x - \imath y \\ x + \imath y & ct - z \end{pmatrix}.\end{aligned} \tag{4.1}$$

There is an alternative vector representation given by:

$$\begin{aligned}\mathbf{v} = (x, y, z, ct) \to \mathbf{V}^{\star} &= ct\mathbb{1} - (\sigma_x x + \sigma_y y + \sigma_z z)\\ &= \begin{pmatrix} ct - z & -(x - \imath y) \\ -(x + \imath y) & ct + z \end{pmatrix}.\end{aligned} \tag{4.2}$$

Note that the symbol $\star$ is different from the symbol $*$ used for complex conjugation. The $\star$ operation is also different from Hermitian conjugation for which the symbol $\dagger$ will be used. The elements of $\mathbf{V}^{\star}$ are the minors of the matrix $\mathbf{V}^{\top}$, therefore in these representations $\mathbf{V}^{\star\star} = \mathbf{V}$ and $\mathbf{V}^{\star}\mathbf{V} = (\det \mathbf{V})\,\mathbb{1} = (c^2t^2 - x^2 - y^2 - z^2)\,\mathbb{1}$. For unit vectors $\mathbf{a}$, we thus have $\mathbf{A}^{\star} = \mathbf{A}^{-1}$. Two representations $\mathbf{V}$ and $\mathbf{V}^{\star}$ are linked by a parity transformation, as $\mathbf{V} = ct\mathbb{1} + \mathbf{r}{\cdot}\boldsymbol{\sigma}$ is related to $\mathbf{V}^{\star} = ct\mathbb{1} - \mathbf{r}{\cdot}\boldsymbol{\sigma}$ through a parity transformation $\mathbf{r}| - \mathbf{r}$. The two SL(2,ℂ) representations can thus be called left-handed and right-handed representations.

The scalar product of two four-vectors $\mathbf{v} * \mathbf{w}$ is now given by $\frac{1}{2}\,(\mathbf{V}^{\star}\mathbf{W} + \mathbf{W}^{\star}\mathbf{V})$. The notation $*$ is used for the Minkowski scalar product in $\mathbb{R}^4$ in order to distinguish it from the dot product in $\mathbb{R}^3$. In this formalism, the reflection defined by a unit vector $\mathbf{a}$ whose matrix is $\mathbf{A}$ operates on a vector $\mathbf{v}$ whose matrix is $\mathbf{V}$ as $\mathbf{V} \to -\mathbf{A}\mathbf{V}^{\star}\mathbf{A}$. A rotation leads to $\mathbf{V} \to \mathbf{B}\mathbf{A}^{\star}\mathbf{V}\mathbf{A}^{\star}\mathbf{B}$, showing it is necessary to leap backwards and forwards between two different vector representations in order to obtain a complete formalism for the Lorentz group, including the reflection operators that generate it. The formalism is, however, a true representation for the group of the Lorentz transformations that does not include these reflections.

4.3 Dirac's expedient again

It is possible to return to a complete and pure reflection formalism as was present in the rotation group by combining the two vector representations into a single one as follows:

$$\mathbf{v} = (x, y, z, ct) \rightarrow \begin{pmatrix} \mathbf{0} & \mathbf{V} \\ \mathbf{V}^{\star} & \mathbf{0} \end{pmatrix} = ct\gamma_{ct} + x\gamma_x + y\gamma_y + z\gamma_z, \tag{4.3}$$

since the reflection then works according to:

$$\begin{aligned} \begin{pmatrix} \mathbf{0} & \mathbf{V} \\ \mathbf{V}^{\star} & \mathbf{0} \end{pmatrix} &\rightarrow -\begin{pmatrix} \mathbf{0} & \mathbf{A} \\ \mathbf{A}^{\star} & \mathbf{0} \end{pmatrix} \begin{pmatrix} \mathbf{0} & \mathbf{V} \\ \mathbf{V}^{\star} & \mathbf{0} \end{pmatrix} \begin{pmatrix} \mathbf{0} & \mathbf{A} \\ \mathbf{A}^{\star} & \mathbf{0} \end{pmatrix} \\ &= -\begin{pmatrix} \mathbf{0} & \mathbf{A}\mathbf{V}^{\star}\mathbf{A} \\ \mathbf{A}^{\star}\mathbf{V}\mathbf{A}^{\star} & \mathbf{0} \end{pmatrix}, \end{aligned} \tag{4.4}$$

with the reflection operator:

$$a_{ct}\gamma_{ct} + a_x\gamma_x + a_y\gamma_y + a_z\gamma_z = \begin{pmatrix} \mathbf{0} & \mathbf{A} \\ \mathbf{A}^{\star} & \mathbf{0} \end{pmatrix} = \underset{\sim}{\mathbf{A}}, \tag{4.5}$$

squaring to unity:

$$\begin{pmatrix} \mathbf{0} & \mathbf{A} \\ \mathbf{A}^{\star} & \mathbf{0} \end{pmatrix} \begin{pmatrix} \mathbf{0} & \mathbf{A} \\ \mathbf{A}^{\star} & \mathbf{0} \end{pmatrix} = \begin{pmatrix} \mathbb{1} & \mathbf{0} \\ \mathbf{0} & \mathbb{1} \end{pmatrix}, \tag{4.6}$$

and $\mathbf{A}^{\star}\mathbf{V}\mathbf{A}^{\star} = (\mathbf{A}\mathbf{V}^{\star}\mathbf{A})^{\star}$. Here, γ_{ct}, γ_x,γ_y,γ_z, are the Dirac gamma matrices with the usual commutation relations: $\gamma_\mu\gamma_\nu + \gamma_\nu\gamma_\mu = 2g_{\mu\nu}\mathbb{1}$. This becomes then equivalent to the four-dimensional description that can be derived from Dirac's expedient, and actually it would be more logical to start the development from (4.6). The notation $\underset{\sim}{\mathbf{A}}$ will be used for the 4×4 matrices to distinguish them from the 2×2 matrices $\mathbf{A}$ and $\mathbf{A}^{\star}$, as they will often be used all within the same context.

The difficulty resides in choosing the correct form of (4.6) that is needed to fall back onto (4.1) and (4.2). The form of (4.6) is a standard form used by Cartan. It may not be the standard form a physicist is used to, but it can be argued that this is not important as Pauli has shown that all representations are equivalent up to a similarity transformation (see also Footnote 3 below). But with these other equivalent representations it would not be as easy to discover the representations defined by (4.1) and (4.2).

The two-dimensional matrices can only yield a representation of the proper Lorentz transformations. The Lorentz reflections cannot be accounted for in a single two-dimensional representation, due to the fact that there is not a fourth 2×2 matrix that anti-commutes with $\sigma_x, \sigma_y, \sigma_z$.

It is thus not possible to build up the representation SL(2,$\mathbb{C}$) from reflections as with SU(2); a different kind of heuristics is needed. The need for four anti-commuting γ-matrices requires the use of a four-dimensional representation, which is the Dirac representation.

In deriving (4.6) by using the Dirac expedient, there is some liberty in the choice of the matrix γ_{ct}, even after having fixed γ_x, γ_y and γ_z. In fact, one can find five mutually anti-commuting γ-matrices. The fifth one in Cartan's choice is:

$$\gamma_5 = \begin{pmatrix} \mathbb{1} & \mathbf{0} \\ \mathbf{0} & -\mathbb{1} \end{pmatrix}, \tag{4.7}$$

which is thus a matrix that anti-commutes with all of the matrices γ_{ct}, γ_x, γ_y and γ_z. In Dirac's choice the roles of γ_{ct} and γ_5 have been inverted. In his choice,[3] the matrix corresponding to (x, y, z, ct), is then:

$$\mathbf{v} = (x, y, z, ct) \to \begin{pmatrix} ct\mathbb{1} & \mathbf{r}\cdot\boldsymbol{\sigma} \\ -\mathbf{r}\cdot\boldsymbol{\sigma} & -ct\mathbb{1} \end{pmatrix} = ct\gamma_5 + x\gamma_x + y\gamma_y + z\gamma_z. \tag{4.8}$$

Cartan's choice appears much simpler, because it permits the use of only matrices that have a block form, where the blocks on one diagonal (the main or the secondary one) are always zero. Appendix A contains a discussion on how one can discover the representation SL(2,$\mathbb{C}$) from the Cartan representation.

The situation in SL(2,$\mathbb{C}$) will be analogous to the one in the rotation group SU(2). Considering the representation as working on group elements,

[3]Intuitively it is obvious that it should be immaterial if γ_{ct} or γ_5 is chosen to build the representation. In fact, the five matrices γ_{ct}, γ_x, γ_y, γ_z, γ_5 can be used to build a representation of the transformation group SO(3,2) of $\mathbb{R}^5$ that preserves the metric with signature $(+ + - - -)$. Replacing γ_{ct} by γ_5 then just corresponds to changing the way Minkowski space-time is embedded within this space $\mathbb{R}^5$. The way a space is embedded into a higher-dimensional space can of course not affect its geometry. Just imagine that Euclidean geometry would depend on the way we orient the plane within $\mathbb{R}^n$, with $n \geq 3$. The plane geometry would then contain information about these other dimensions! Therefore the representations must be equivalent. This can be also checked algebraically by calculating the eigenvalue equations in both representations for the matrices that code a four-vector, e.g. $(E, c\mathbf{p})$. One finds $(\lambda^2 - (E^2 - c^2p^2))^2 = 0$ in both representations, such that the two matrices have the same eigenvalues $-m_0c^2$ (twice), and m_0c^2 (twice). The fact that they have the same eigenvalues implies that these two matrices can be obtained from one another by a similarity transformation. As γ_{ct} and γ_5 anti-commute, any linear combination $\gamma_4 = (\cos\alpha)\gamma_{ct} + (\sin\alpha)\gamma_5$ will also satisfy the anti-commutation relations $\gamma_\mu\gamma_\nu + \gamma_\nu\gamma_\mu = 2g_{\mu\nu}\mathbb{1}$. These ideas of equivalence between different four-dimensional subspaces of $\mathbb{R}^5$ with the ones of Subsection 5.4.2.3 (see also the discussion of the analogy with the buckyball model for the icosahedral group from Section 2.9) may give a geometrical insight into Pauli's theorem. The ideas of Subsection 5.4.2.3 are needed here to cover any further *internal* similarity transformations that may occur *within* a given four-dimensional subspace of $\mathbb{R}^5$.

the representation is already linear. But if it is considered as working on four-vectors, then it acts quadratically. The four-vectors are associated with reflection matrices. These cannot be coded within SL(2,ℂ). However, the four-dimensional Dirac representation, given by (4.4), is generated by such reflections and it acts quadratically on reflection matrices as defined by (4.5). It will be possible to derive the linear formalism for the group elements from the quadratic formalism for four-vectors by considering two special zero-length four-vectors that together will code the whole tetrad. This will be shown in Section 4.6 for SL(2,ℂ).

4.4 Coding a tetrad in SL(2,ℂ) using tensor products that involve the two semi-spinors

The problem with attempting to code the tetrad within SL(2,ℂ) is that there is no linear combination that would allow one to code the whole information content of a tetrad into a single zero-length vector. It is possible to code $\mathbf{e}_{ct} + \mathbf{e}_z$ and $\mathbf{e}_x + \imath\mathbf{e}_y$ separately, but combining them linearly into a single vector within the 2×2 matrix formalism in such a way that they can be separated out again unambiguously fails due to the non-existence of another commuting number α that could be used like $\imath$ to code the whole tetrad as $\mathbf{e}_x + \imath\mathbf{e}_y + \alpha(\mathbf{e}_{ct} + \mathbf{e}_z)$. One way would be to try to use a quaternion for α but this would imply immediately that the dimension of the representation has to be increased from 2 to 4. In order to obtain unambiguous coding and decoding *using a commutative number field*, it would be necessary to consider α as a *variable*, such that the coding would become a linear polynomial in α. Finally, also combining the two zero-length vectors by using the tensor product leads to a 4×4 matrix formalism rather than a 2×2 one. Despite these problems a matrix with zero determinant in the 2×2 matrix formalism of SL(2,ℂ) seems to have exactly the right amount of independent real parameters, *viz.* six, that are required to code a general Lorentz transformation, and it does not require the presence of a variable α for that. As the determinant of a matrix that codes a vector corresponds to the square of the length of that vector, it is tempting to identify a matrix with zero determinant with a zero-length vector. This will have to be rendered more precise later on when it will be shown that one must also keep track of a phase factor. In fact, in general one will construct the zero-length vectors using two orthogonal unit vectors, and this procedure defines only five independent parameters. The lacking sixth parameter is this phase factor.

The coding can be achieved by exploring another, much less obvious possibility. The idea is to develop two separate "spinor"-like quantities (in the sense of the construction developed within the rotation group) which will acquire the status of semi-spinors within the Lorentz group, e.g. $\boldsymbol{\eta} = (\eta_0, \eta_1)$ for $\mathbf{e}_{ct} + \mathbf{e}_z$, and $\boldsymbol{\xi} = (\xi_0, \xi_1)$ for $\mathbf{e}_x + \imath\mathbf{e}_y$, and to construct (several) tensor products from them, such as $\boldsymbol{\xi}^\top \otimes \boldsymbol{\eta}$, which then in principle should be able to code the whole tetrad. (It will be revealed later that the correct expression should actually be $\boldsymbol{\xi}^\top \otimes \boldsymbol{\eta}^*$.)

The idea is thus that in order to avoid a situation in which a tensor product would raise the dimension of the representation matrices from 2×2 to 4×4, the tensor product is coded at "spinor" level rather than at vector level. It is the fact that a "spinor" behaves like a "square root of a vector" which enables one to construct a tensor product whereby the matrix formalism remains at the 2×2 level. The development will show that it will be necessary to slightly modify this idea to make things work.

4.5 A very important difference between SL(2,ℂ) and SU(2)

In the 4×4 representation, a general Lorentz group reflection is represented by the transformation: $\underset{\sim}{\mathbf{V}} \rightarrow -\underset{\sim}{\mathbf{A}}\underset{\sim}{\mathbf{V}}\underset{\sim}{\mathbf{A}}$. Therefore, in $\mathrm{SL}(2,\mathbb{C})$ a general Lorentz group reflection is represented by the transformations: $\mathbf{V} \rightarrow -\mathbf{A}\mathbf{V}^\star\mathbf{A}$ and $\mathbf{V}^\star \rightarrow -\mathbf{A}^\star\mathbf{V}\mathbf{A}^\star$. Here, $\mathbf{A}$ and $\underset{\sim}{\mathbf{A}}$ are the matrices that correspond to the four-vector $\mathbf{a}$ that is normal to the reflection hyperplane. As the reflections can be used to generate the Lorentz group, a general rotation, boost, or Lorentz transformation are thus of the form: $\underset{\sim}{\mathbf{V}} \rightarrow \underset{\sim}{\mathbf{B}}\underset{\sim}{\mathbf{A}}\underset{\sim}{\mathbf{V}}\underset{\sim}{\mathbf{A}}\underset{\sim}{\mathbf{B}}$. With $\underset{\sim}{\mathbf{L}} = \underset{\sim}{\mathbf{B}}\underset{\sim}{\mathbf{A}}$, it follows that: $\underset{\sim}{\mathbf{V}} \rightarrow \underset{\sim}{\mathbf{L}}\underset{\sim}{\mathbf{V}}\underset{\sim}{\mathbf{L}}^{-1}$. This leads to $\mathbf{V} \rightarrow \mathbf{B}\mathbf{A}^\star\mathbf{V}\mathbf{A}^\star\mathbf{B}$ and $\mathbf{V}^\star \rightarrow \mathbf{B}^\star\mathbf{A}\mathbf{V}^\star\mathbf{A}\mathbf{B}^\star$. Now the matrices $\mathbf{A}$ and $\mathbf{B}$ are Hermitian, such that $\mathbf{A}^\star\mathbf{B} = (\mathbf{B}\mathbf{A}^\star)^\dagger$. The general form of a rotation, a boost, or a Lorentz transformation is thus: $\mathbf{V} \rightarrow \mathbf{L}\mathbf{V}\mathbf{L}^\dagger$.[4] Note that in general $\mathbf{L} \neq \mathbf{L}^\dagger$, as $\mathbf{L}$ must code six real parameters rather than three. A general Lorentz transformation matrix $\mathbf{L}$ is given by:

$$\mathbf{L}(a,b,c,d) = \begin{pmatrix} a & b \\ c & d \end{pmatrix}, \tag{4.9}$$

[4]With $\mathbf{L} = \mathbf{B}\mathbf{A}^\star$, and using the Hermitian properties of the matrices $\mathbf{A}$ and $\mathbf{B}$ in SL(2,ℂ) it follows then also that: $\mathbf{A}^\star\mathbf{B} = \mathbf{L}^\dagger$, $\mathbf{A}\mathbf{B}^\star = \mathbf{L}^{-1}$ and $\mathbf{B}^\star\mathbf{A} = \mathbf{L}^{\dagger-1}$, and we also have: $\mathbf{V}^\star \rightarrow \mathbf{L}^{\dagger-1}\mathbf{V}^\star\mathbf{L}^{-1}$. This shows that the 4×4 Lorentz matrix $\underset{\sim}{\mathbf{L}}$ must contain the blocks $\mathbf{L}$ and $\mathbf{L}^{\dagger-1}$ while its inverse $\underset{\sim}{\mathbf{L}}^{-1}$ must contain $\mathbf{L}^{-1}$ and $\mathbf{L}^\dagger$.

with $ad - bc = 1$. Based on this definition it follows that:

$$\mathbf{L} = \begin{pmatrix} a & b \\ c & d \end{pmatrix}, \qquad \mathbf{L}^{\dagger} = \begin{pmatrix} a^* & c^* \\ b^* & d^* \end{pmatrix},$$
$$\mathbf{L}^{-1} = \begin{pmatrix} d & -b \\ -c & a \end{pmatrix}, \qquad (\mathbf{L}^{-1})^{\dagger} = \begin{pmatrix} d^* & -c^* \\ -b^* & a^* \end{pmatrix}. \tag{4.10}$$

This is quite natural; it is obvious that an alternative representation can be built through the isomorphism of group elements $g \rightarrow g^{-1}$. If $g \rightarrow \mathbf{D}(g)$ is a representation built on left multiplication, then $g \rightarrow \mathbf{D}^{-1}(g)$ is a representation built on right multiplication, and vice versa. Similarly, if $g \rightarrow \mathbf{D}(g)$ is a representation built on left multiplication, then $g \rightarrow \mathbf{D}^{\dagger}(g)$ is a representation built on right multiplication, and vice versa.

This is then achieved by isomorphisms of representation matrices $\mathbf{D}(g) \leftrightarrow \mathbf{D}^{\dagger}(g)$ and $\mathbf{D}(g) \leftrightarrow \mathbf{D}^{-1}(g)$. The 4×4 Lorentz matrix $\underset{\sim}{\mathbf{L}}$ is given by:

$$\underset{\sim}{\mathbf{L}} = \begin{pmatrix} \mathbf{L} & \\ & \mathbf{L}^{\dagger -1} \end{pmatrix}, \quad \underset{\sim}{\mathbf{L}}^{-1} = \begin{pmatrix} \mathbf{L}^{-1} & \\ & \mathbf{L}^{\dagger} \end{pmatrix}. \tag{4.11}$$

This is easy to check by using the identities derived in Footnote 4. (See Subsection 5.5.2.1.) Due to the way the rotation matrices $\mathbf{R}$ act on vectors in SU(2), *viz.* $\mathbf{V} \rightarrow \mathbf{RVR}^{\dagger}$ with $\mathbf{R}^{\dagger} = \mathbf{R}^{-1}$, the composition of isomorphisms $\mathbf{D}(g) \leftrightarrow \mathbf{D}^{\dagger -1}(g)$ does not lead to a new, different matrix formalism in the rotation group. But in the Lorentz group the situation is more complicated than in the rotation group because $\mathbf{L}^{\dagger}$ and $\mathbf{L}^{-1}$ are different matrices. If we use the transposition of matrices rather than Hermitian conjugation then even more different representations can be derived. One of these corresponds to punctuated spinors (see below).

The discussion of some subtle points has at this stage been relegated to Appendix B. From this discussion it emerges that the codings of the two zero-length vectors must contain contributions of both types $(\eta_0, \eta_1)^{\top}$ and (η_0^*, η_1^*). The circumstance that $\mathbf{L}^{-1} \neq \mathbf{L}^{\dagger}$, is responsible for the fact that it is no longer possible to use within the Lorentz group the same diagonalization procedure to jump from vectors to group elements, as has been used in Section 3.7 for the rotation group. In the Lorentz group it is no longer possible to identify something that has a structure $\mathbf{LVL}^{\dagger}$ with something that has a structure $\mathbf{SVS}^{-1}$. It is here indeed no longer possible to "halve" the formalism by a diagonalization procedure to render it linear, as the operations on the left hand side are no longer identical to the operations on the right hand side, as was the case in the rotation group. In other words, the two halves would not be identical. Within the Lorentz group a procedure other than diagonalization is thus required to "halve" the formalism into two parts that are identical. To obtain two halves the vector $\mathbf{V}$ must be written in the form $\boldsymbol{\xi}^\top \otimes \boldsymbol{\eta}^*$, where $\boldsymbol{\eta}^*$ is identical to $\boldsymbol{\eta}^{\top\dagger}$.

4.6 The two types of zero-length vectors in the Lorentz group that define a tetrad

Defining the tetrad requires the following codings:

$$\begin{aligned} \mathbf{e}_{ct} + \mathbf{e}_z &= (1,0,0,1) \rightarrow \mathbf{V}_1 = \begin{pmatrix} 2 & 0 \\ 0 & 0 \end{pmatrix}, \\ \mathbf{e}_x + \imath\mathbf{e}_y &= (0,1,\imath,0) \rightarrow \mathbf{V}_2 = \begin{pmatrix} 0 & 2 \\ 0 & 0 \end{pmatrix}. \end{aligned} \tag{4.12}$$

A general Lorentz transformation with matrix $\mathbf{L}$ is given by (4.9), with $ad - bc = 1$, and transforms the two vectors $\mathbf{V}_j$ into $\mathbf{V}_j' = \mathbf{L}\mathbf{V}_j\mathbf{L}^\dagger$, given by:

$$\begin{aligned} \mathbf{V}_1' &= \begin{pmatrix} 2aa^* & 2ac^* \\ 2ca^* & 2cc^* \end{pmatrix} = 2\begin{pmatrix} a \\ c \end{pmatrix} \otimes (a^*, c^*), \\ \mathbf{V}_2' &= \begin{pmatrix} 2ab^* & 2ad^* \\ 2cb^* & 2cd^* \end{pmatrix} = 2\begin{pmatrix} a \\ c \end{pmatrix} \otimes (b^*, d^*), \end{aligned}$$

from which it can be seen that the tetrad is coded by two spinor-like quantities (ξ_0, ξ_1) and (η_0, η_1), that combine into a quantity:

$$\mathbf{V}_2' = 2\begin{pmatrix} \xi_0 \\ \xi_1 \end{pmatrix} \otimes (\eta_0^*, \eta_1^*), \tag{4.13}$$

with $\xi_0\eta_1 - \xi_1\eta_0 = 1$ and whereby

$$\mathbf{V}_1' = 2\begin{pmatrix} \xi_0 \\ \xi_1 \end{pmatrix} \otimes (\xi_0^*, \xi_1^*) \tag{4.14}$$

automatically codes $\mathbf{V}_1'$. SU(2) and SL(2,$\mathbb{C}$) can then be compared according to the following diagram for the isotropic vector $\mathbf{e}_x + \imath\mathbf{e}_y \in \mathscr{I}$:

$$\begin{array}{ccc} \mathbf{V} \equiv \mathbf{e}_x + \imath\mathbf{e}_y \in \mathscr{I} & \underline{\underline{\qquad}} & \mathbf{V} \equiv \mathbf{e}_x + \imath\mathbf{e}_y \in \mathscr{I} \\ \Big\downarrow \begin{array}{l} \mathbf{V}' = \mathbf{R}\mathbf{V}\mathbf{R}^\dagger \\ \mathbf{R} \in \mathrm{SU}(2) \end{array} & & \Big\downarrow \begin{array}{l} \mathbf{V}' = \mathbf{L}\mathbf{V}\mathbf{L}^\dagger, \\ \mathbf{L} \in \mathrm{SL}(2,\mathbb{C}) \end{array} \\ \underbrace{\mathbf{V}' = \sqrt{2}\begin{pmatrix} \xi_0 \\ \xi_1 \end{pmatrix} \otimes \sqrt{2}(-\xi_1 \;\; \xi_0)}_{\mathbf{V}' \equiv \mathbf{e}_x' + \imath\mathbf{e}_y' \in \mathscr{I}} & & \underbrace{\mathbf{V}' = \sqrt{2}\begin{pmatrix} a \\ c \end{pmatrix} \otimes \sqrt{2}(b^* \;\; d^*)}_{\mathbf{V}' \equiv \mathbf{e}_x' + \imath\mathbf{e}_y' \in \mathscr{C}} \end{array} \tag{4.15}$$

In SU(2) a single spinor $[\,\xi_0 \;\; \xi_1\,]^\top$ codes a whole rotation because $[\,\xi_0 \;\; \xi_1\,]^\top$ and $[\,-\xi_1 \;\; \xi_0\,]$ can be derived one from another. In SL(2,ℂ), the two quantities $[\,a \;\; c\,]^\top$ and $[\,b^* \;\; d^*\,]$ cannot be derived from one another. They are only related by the constraint $ad - bc = 1$. One of them alone is thus not capable of coding the full information about the tetrad. This is the reason why the name "semi-spinors" is preferred. The two semi-spinors must thus be subjected to Lorentz transformations to obtain a complete representation of the homogeneous Lorentz group, *viz.* $[\,a \;\; c\,]^\top$ acted upon by left multiplication by a Lorentz matrix $\mathbf{L}$, and $[\,b^* \;\; c^*\,]$ acted upon by right multiplication by a Lorentz matrix $\mathbf{L}^\dagger$. The latter is equivalent to a semi-spinor $[\,b \; d\,]^\top$ being acted upon by left multiplication by a Lorentz matrix $\mathbf{L}$. The matrix in (4.9) can thus be considered as a full spinor. It is in fact obtained as the juxtaposition of the two semi-spinors. The reader will recognize that this way of writing a spinor as the juxtaposition of two semi-spinors corresponds to the derivation of the linear formalism for the SL(2,ℂ) representation matrices when they work on group elements, from the quadratic action of these representation matrices when they work on four-vectors. As for SU(2), it was necessary to start from a special "zero-length" vector to obtain this derivation, a result anticipated in Section 4.3. In (4.15) $\mathbf{e}'_x + \imath\mathbf{e}'_y \in \mathscr{C}$ has been noted for SL(2,ℂ) because the original isotropic vector $\mathbf{e}_x + \imath\mathbf{e}_y \in \mathscr{I} \subset \mathscr{C}$ can acquire a time-component under the action of Lorentz transformations. This does not happen under the action of rotations in SU(2).

The fact that $\mathbf{V}'_1$ and $\mathbf{V}'_2$ are written as tensor products automatically reflects that they are zero-length vectors. The canonical starting values are $(\xi_0, \xi_1) = (1, 0)$ and $(\eta_0, \eta_1) = (0, 1)$. From the knowledge of the spinors, the images of $\mathbf{e}_{ct} + \mathbf{e}_z$ and $\mathbf{e}_x + \imath\mathbf{e}_y$ can be reconstructed unambiguously.

The converse is *not* true. The quantity (ξ_0, ξ_1) is only determined by $\mathbf{e}'_{ct} + \mathbf{e}'_z$ up to a "phase factor". It is easy to see that $[a, c]^\top \to e^{\imath\chi}[a, c]^\top$, $[a^*, c^*] \to e^{-\imath\chi}[a^*, c^*]$ does not modify $\mathbf{V}'_1$. Similarly, (ξ_0, ξ_1) and (η_0, η_1) are not completely determined by $\mathbf{e}'_x + \imath\mathbf{e}'_y$. Even the combined knowledge of $\mathbf{e}'_{ct} + \mathbf{e}'_z$ and $\mathbf{e}'_{ct} - \mathbf{e}'_z$, is not sufficient to reconstruct (ξ_0, ξ_1) and (η_0, η_1) completely. In fact, a rotation around the z-axis, with rotation matrix $\mathbf{R}_z$, will leave the two zero-length vectors $\mathbf{V}_1 \leftrightarrow \mathbf{e}_{ct} + \mathbf{e}_z$ and $\mathbf{V}_4 \leftrightarrow \mathbf{e}_{ct} - \mathbf{e}_z$ unchanged (i.e. $\mathbf{R}_z\mathbf{V}_1\mathbf{R}_z^\dagger = \mathbf{V}_1$, $\mathbf{R}_z\mathbf{V}_4\mathbf{R}_z^\dagger = \mathbf{V}_4$), such that $\mathbf{L}\mathbf{R}_z\mathbf{V}_1\mathbf{R}_z^\dagger\mathbf{L}^\dagger = \mathbf{L}\mathbf{V}_1\mathbf{L}^\dagger$, and $\mathbf{L}\mathbf{R}_z\mathbf{V}_4\mathbf{R}_z^\dagger\mathbf{L}^\dagger = \mathbf{L}\mathbf{V}_4\mathbf{L}^\dagger$. The resulting quantities will thus not contain information about $\mathbf{R}_z$ as this information has been "squared" out. The quantities (ξ_0, ξ_1) and (η_0, η_1) (which do contain the information about $\mathbf{R}_z$) are thus not completely determined. Similarly, the combined knowledge of $\mathbf{V}_2' \leftrightarrow \mathbf{e}_x' - \imath\mathbf{e}_y'$ and $\mathbf{V}_3' \leftrightarrow \mathbf{e}_x' + \imath\mathbf{e}_y'$ will not be sufficient to reconstruct (ξ_0, ξ_1) and (η_0, η_1) completely. Here, it is a boost along the z-axis with boost matrix $\mathbf{B}_z$ that will leave $\mathbf{V}_2$ and $\mathbf{V}_3$ unchanged. The combined knowledge of $\mathbf{V}_1'$ and $\mathbf{V}_2'$ does, however, permit reconstruction of (ξ_0, ξ_1) and (η_0, η_1). The same applies for the combinations $(\mathbf{V}_1', \mathbf{V}_3')$, $(\mathbf{V}_4', \mathbf{V}_2')$, and $(\mathbf{V}_4', \mathbf{V}_3')$.

These results can be obtained after some tedious algebra, by just attempting to reconstruct (ξ_0, ξ_1) and (η_0, η_1) from the zero-length vectors one assumes to be given. But the point can also be understood (in a less detailed way) geometrically. Once the starting values for the spinors have been fixed, e.g. by adopting the canonical values $(\xi_0, \xi_1) = (1, 0)$ and $(\eta_0, \eta_1) = (0, 1)$ for them, the values of the four zero-length vectors $\mathbf{V}_j$ will be unambiguously defined for any subsequent Lorentz transformation with matrix $\mathbf{L}(a, b, c, d)$. Imagine now that both $\mathbf{V}_1' \leftrightarrow \mathbf{e}_{ct}' + \mathbf{e}_z'$ and $\mathbf{V}_3' \leftrightarrow \mathbf{e}_x' + \imath\mathbf{e}_y'$ are known. By separating the real and imaginary parts in $\mathbf{e}_x' + \imath\mathbf{e}_y'$, $\mathbf{e}_x'$ and $\mathbf{e}_y'$ can then be reconstructed unambiguously. The combined knowledge of $\mathbf{e}_{ct}' + \mathbf{e}_z'$, $\mathbf{e}_x'$ and $\mathbf{e}_y'$ permits then reconstructing also $\mathbf{e}_{ct}' - \mathbf{e}_z'$ unambiguously. Finally, combining $\mathbf{e}_{ct}' - \mathbf{e}_z'$ and $\mathbf{e}_{ct}' + \mathbf{e}_z'$ will permit recovering the whole tetrad. As each Lorentz transformation with matrix $\mathbf{L}(a, b, c, d)$ is in one-to-one correspondence with the tetrad it generates by operating on the canonical basis, knowing the whole tetrad means that the whole Lorentz transformation is known. Hence, in order to know if it is possible to reconstruct (ξ_0, ξ_1) and (η_0, η_1) completely from the knowledge of a combination of zero-length vectors, it suffices to check if one can reconstruct the other zero-length vectors of the tetrad from the combination given. It takes the presence of three of the four unit vectors $\mathbf{e}_\mu'$ within the combination $(\mathbf{V}_j', \mathbf{V}_k')$ to have a complete description.

4.7 Dotted spinors

In addition to the four matrices defined in (4.10), one can introduce the 2×2 matrices: $\dot{\mathbf{L}} = \mathbf{L}^{\dagger\top}$. We then have $\mathbf{L} = \mathbf{L}_2\mathbf{L}_1$, such that $\mathbf{L}^\dagger = \mathbf{L}_1^\dagger\mathbf{L}_2^\dagger$, and $\mathbf{L}^{\dagger\top} = \mathbf{L}_2^{\dagger\top}\mathbf{L}_1^{\dagger\top}$, which again respects the same order of operations as in $\mathbf{L} = \mathbf{L}_2\mathbf{L}_1$. The entries in the matrix $\mathbf{L}^{\dagger\top}$ will thus all be calculated correctly. The only thing that has changed is that the places of the elements with indices 12 and 21 have been swapped. It is possible to make an error of interpretation about the meaning of the elements of the matrices, by basing oneself on the places where these elements occur in their matrix. If the operation is performed twice, the entries 12 and 21 will be in the same place again. The important point is however the self-consistency of the calculations, not on which place in a matrix formalism the elements are positioned. The matrices $\dot{\mathbf{L}}$ are used in the representation theory. Their advantage is that they can be used in left multiplication. The point is that (4.13) and (4.14) show that two representations of the SL(2,$\mathbb{C}$)-type out of the four given in (4.10) are needed to perform all the necessary calculations, one with the elements of $\mathbf{L}$ in (4.10), and one based on the elements of $\mathbf{L}^\dagger$. The disadvantage of $\mathbf{L}^\dagger$ is that it requires right multiplication. By pulling back $\mathbf{L}^\dagger$ to $\dot{\mathbf{L}}$ it will be possible to perform all the calculations by left multiplication. The indices of the elements involved in the representation based on $\mathbf{L}^{\dagger\top}$ are then written with punctuated indices. This pull-back from $\mathbf{L}^\dagger$ to $\dot{\mathbf{L}}$ changes the equations according to the substitution:

$$\begin{aligned} \begin{pmatrix} \xi_0' \\ \xi_1' \end{pmatrix} \otimes (\eta_0', \eta_1') &= \mathbf{L} \begin{pmatrix} \xi_0 \\ \xi_1 \end{pmatrix} \otimes (\eta_0, \eta_1)\mathbf{L}^\dagger \qquad \rightarrow \\ \begin{pmatrix} \xi_0' \\ \xi_1' \end{pmatrix} \otimes \begin{pmatrix} \eta_0' \\ \eta_1' \end{pmatrix} &= \mathbf{L} \otimes \dot{\mathbf{L}} \begin{pmatrix} \xi_0 \\ \xi_1 \end{pmatrix} \otimes \begin{pmatrix} \eta_0 \\ \eta_1 \end{pmatrix}, \end{aligned} \tag{4.16}$$

which is a way to write the formalism in terms of transformations of 4×1 matrices rather than 2×2 matrices. The components η_0, η_1 correspond thus to the punctuated indices, as they transform according to $\dot{\mathbf{L}}$ rather than according to $\mathbf{L}$. They are therefore written as $\dot{\eta}_0, \dot{\eta}_1$ in order to keep track of the transformation matrix required. This shows that it is possible to use a representation in terms of quantities $(\xi_0\dot{\eta}_0, \xi_0\dot{\eta}_1, \xi_1\dot{\eta}_0, \xi_1\dot{\eta}_1)^\top$ and obtain this way a representation working on 4×1 matrices. While this pull-back expedient permits the calculations to be transcribed much more mechanically by carrying them all out in left multiplication, the disadvantage is that the intricate technicalities of the algebraic moves involved are beyond reckoning, such that the formalism will appear inscrutable if these moves are not described in minute detail. If the Dirac representation is described properly in terms of matrices of SL(2,$\mathbb{C}$), and these 2×2 matrices used as spinors, it will be apparent that the natural combination to be used is $\mathbf{L}$, $\mathbf{L}^{-1\dagger}$ rather than $\mathbf{L}$, $\dot{\mathbf{L}}$ (see for instance (5.33) below) such that the punctuated spinors can then be avoided. The dotted spinors are also essential when one wishes to use differential operators in tensor representations. Perhaps it is possible to define a formalism that would allow for a distinction between left and right differentiation, but it would certainly be quite complicated.

4.8 Missing phase factors or boost parameters

To know if a given combination of zero-length vectors is sufficient to reconstruct the whole tetrad, it suffices to count the total number of independent parameters. Only if this number is six will the information be complete; for instance, $\mathbf{e}_x + \imath\mathbf{e}_y$ does not contain the complete information as it only contains five independent real parameters. A boost parameter is still needed to make the description of the tetrad complete. (Such a boost parameter will also be refered to as a phase factor in an *abus de language*.) This boost parameter will also be present in the quantities (ξ_0, ξ_1) and (η_0, η_1) after applying subsequent Lorentz transformations on the canonical starting values $(1, 0)$ and $(0, 1)$. Similarly, a and c contain four real parameters. They contain the information about the three parameters contained in $\mathbf{e}_{ct} + \mathbf{e}_z$ and the phase factor. That $\mathbf{e}_{ct} + \mathbf{e}_z$ contains three parameters can be seen from the fact that $\mathbf{V}_1$ is Hermitian. This can also be appreciated from the fact that from knowing $\mathbf{e}_{ct} + \mathbf{e}_z$ alone, it is impossible to reconstruct $\mathbf{e}_{ct}$ and $\mathbf{e}_z$ separately.

This situation of a phase factor that is easily overlooked is quite analogous to the one with the phase factor that is missed in describing the spinors of the rotation group starting from the stereographic projection of a unit vector, as discussed in Section 3.8. There, the information about a second unit vector that is needed to make the description of the triad complete, is also missing, and this information must also be coded as a phase factor, inadvertently introduced at a later stage when one makes a transition from Cartesian to homogeneous coordinates, and from a homographic transformation to a linear mapping embodied by a 2×2 matrix.

4.9 The coding of the tetrad in the Dirac representation

A nice result can be obtained by considering also the "zero-length" vectors:

$$\begin{aligned} \mathbf{e}_x - \imath\mathbf{e}_y = (0, 1, -\imath, 0) \rightarrow \mathbf{V}_3 &= \begin{pmatrix} 0 & 0 \\ 2 & 0 \end{pmatrix}, \\ \mathbf{e}_{ct} - \mathbf{e}_z = (1, 0, 0, -1) \rightarrow \mathbf{V}_4 &= \begin{pmatrix} 0 & 0 \\ 0 & 2 \end{pmatrix}. \end{aligned} \tag{4.17}$$

The Lorentz transformation with matrix $\mathbf{L}$ transforms these two vectors $\mathbf{V}_j$ into:

$$\begin{aligned} \mathbf{V}'_3 &= \begin{pmatrix} 2ba^* & 2bc^* \\ 2da^* & 2dc^* \end{pmatrix} = 2\begin{pmatrix} b \\ d \end{pmatrix} \otimes (a^*, c^*), \\ \mathbf{V}'_4 &= \begin{pmatrix} 2bb^* & 2bd^* \\ 2db^* & 2dd^* \end{pmatrix} = 2\begin{pmatrix} b \\ d \end{pmatrix} \otimes (b^*, d^*). \end{aligned} \tag{4.18}$$

The 4×4 matrix built on the tensor product:

$$\begin{aligned} \begin{pmatrix} a & b \\ c & d \end{pmatrix} \otimes \begin{pmatrix} a^* & c^* \\ b^* & d^* \end{pmatrix} &= \frac{1}{2}\begin{pmatrix} \mathbf{V}'_1 & \mathbf{V}'_3 \\ \\ \mathbf{V}'_2 & \mathbf{V}'_4 \end{pmatrix} \\ &\leftrightarrow \frac{1}{2}\begin{pmatrix} \mathbf{e}'_{ct} + \mathbf{e}'_z & \mathbf{e}'_x - \imath\mathbf{e}'_y \\ \\ \mathbf{e}'_x + \imath\mathbf{e}'_y & \mathbf{e}'_{ct} - \mathbf{e}'_z \end{pmatrix}, \end{aligned} \tag{4.19}$$

mimics then the structure $ct\mathbb{1} + x\sigma_x + y\sigma_y + z\sigma_z \leftrightarrow \mathbf{e}'_{ct}\mathbb{1} + \mathbf{e}'_x\sigma_x + \mathbf{e}'_y\sigma_y + \mathbf{e}'_z\sigma_z$ of the 2×2 representation matrices on a larger scale. (4.19) depends on the choice of in which order the components of the tensor product are noted. Here, the "spinor-preserving" order is used:

$$\begin{aligned} \mathbf{L} \otimes \mathbf{L}^\dagger &= \begin{pmatrix} a & b \\ c & d \end{pmatrix} \otimes \begin{pmatrix} a^* & c^* \\ b^* & d^* \end{pmatrix} \\ &= \begin{pmatrix} aa^* & ac^* & ba^* & bc^* \\ ca^* & cc^* & da^* & dc^* \\ ab^* & ad^* & bb^* & bd^* \\ cb^* & cd^* & db^* & dd^* \end{pmatrix}, \end{aligned} \tag{4.20}$$

which is perhaps less obvious than the order:

$$\begin{aligned} \mathbf{L} \otimes \mathbf{L}^\dagger &= \begin{pmatrix} a & b \\ c & d \end{pmatrix} \otimes \begin{pmatrix} a^* & c^* \\ b^* & d^* \end{pmatrix} \\ &= \begin{pmatrix} aa^* & ab^* & ba^* & bb^* \\ ac^* & ad^* & bc^* & bd^* \\ ca^* & cb^* & da^* & db^* \\ cc^* & cd^* & dc^* & dd^* \end{pmatrix}, \end{aligned} \tag{4.21}$$

which might look more logical but scatters the components of spinors $\boldsymbol{\xi} = (a, b)$ and $\boldsymbol{\eta} = (b, d)$ over different blocks, thereby rendering the block structure in terms of SL(2,$\mathbb{C}$) less evident. Of course these choices are linked by a similarity transformation. The whole four-dimensional coding can then be written as $\frac{1}{4}\sum \mathbf{V}'_\mu \tilde{\mathbf{V}}_\mu = \frac{1}{4}\sum \mathbf{V}'_\mu \otimes \mathbf{V}_\mu$, where each 4×4 matrix $\tilde{\mathbf{V}}_\mu$ has a block structure $\mathbb{1} \otimes \mathbf{V}_\mu$ that is obtained from the corresponding 2×2 matrix $\mathbf{V}_\mu$ by replacing its elements $v_{ij}^{(\mu)}$ by $v_{ij}^{(\mu)}\mathbb{1}$. (In other words, three blocks of $\frac{1}{2}\tilde{\mathbf{V}}_\mu$ are zero matrices and one block is a unit matrix. The "coefficients" $\mathbf{V}'_\mu$ are supposed to work on the unit matrix, thereby transforming it into the matrix $\mathbf{V}'_\mu$ that codes $\mathbf{V}_\mu$ in the new frame.)

This shows that the most simple and symmetrical coding of a Lorentz transformation is done with two light rays travelling in opposite directions, by using $\mathbf{V}_1$ and $\mathbf{V}_4$. The matrix $\mathbf{V}_4$ can now be considered as the representation of $\mathbf{e}'_{ct} + \mathbf{e}'_z$ in the alternative representation, i.e. $\mathbf{V}_4 = \mathbf{V}_{\bar{1}}^\star$. In other words, an isomorphism can be built thus:

$$\begin{pmatrix} \mathbf{V}_1 & \mathbf{0} \\ \mathbf{0} & \mathbf{V}_{\bar{1}}^\star \end{pmatrix} \leftrightarrow \begin{pmatrix} \mathbf{V}_1 & \mathbf{0} \\ \mathbf{0} & \mathbf{V}_4 \end{pmatrix}. \tag{4.22}$$

This leads to the conclusion that in the representation based on the Dirac matrices, the operator on the left-hand side (which is defined by the single "zero-length" vector $\mathbf{e}_{ct} + \mathbf{e}_z$), or alternatively the reflection operator:

$$\begin{pmatrix} \mathbf{0} & \mathbf{V}_{\bar{2}}^\star \\ \mathbf{V}_2 & \mathbf{0} \end{pmatrix} \leftrightarrow \begin{pmatrix} \mathbf{0} & \mathbf{V}_3 \\ \mathbf{V}_2 & \mathbf{0} \end{pmatrix}, \tag{4.23}$$

(which is defined by a single isotropic vector $\mathbf{e}_x + \imath\mathbf{e}_y$), is able to code the whole tetrad, provided the phase factors in the matrices $\mathbf{V}_1$ and $\mathbf{V}_4$ (or $\mathbf{V}_2$ and $\mathbf{V}_3$) are monitored when following the formalism. The phase factor is essential as the couples of vectors $(\mathbf{e}_{ct}, \mathbf{e}_z)$ or $(\mathbf{e}_x, \mathbf{e}_y)$ that can be reconstructed from the "zero-length" vectors coded in $(\mathbf{V}_1, \mathbf{V}_4)$ and $(\mathbf{V}_2, \mathbf{V}_3)$ respectively only contain five independent real parameters. The physical meaning of the isomorphism $\mathbf{V}_{\bar{1}}^\star \leftrightarrow \mathbf{V}_4$ is clear. The second quantity codes the light ray in the opposite direction of the light ray coded by $\mathbf{V}_1$ in one two-dimensional representation. The first quantity codes the principal light ray in the representation based on the inverse matrices. But it is obvious that in the Lorentz group, the inverse transformation is given by making the substitution $\mathbf{v}| - \mathbf{v}$. All this becomes possible by the fact that the Dirac representation contains the two alternative two-dimensional representations combined with the fact that the coding $\mathbf{V}_{\bar{1}}^\star$ of a light ray in the alternative two-dimensional representation is equivalent to the coding $\mathbf{V}_4$ of the light ray that travels in the opposite direction of $\mathbf{V}_1$ in the original two-dimensional representation.

4.10 The last parameter in any group SO(n_1,n_2) is always a "phase factor"

In summary, the coding of a general Lorentz transformation can be seen conceptually in terms of an isotropic vector plus a phase factor, or two light rays travelling oppositely plus a "phase factor": The situation will be discussed first from the viewpoint that the Lorentz transformation is coded with the aid of an isotropic vector, using an analogy with SO(4) and assuming the fourth coordinate is called u. The idea is to try to imagine what a four-dimensional rotation could be. Let us start from the Oxy plane. Both the z and the u axis are then orthogonal to this plane. By definition a rotation in the Oxy plane will leave both these axes fixed, as it only affects the coordinates x and y in (x, y, z, u). But under an exchange of rôles this also means that a rotation in the Ozu plane will not have any effect in the Oxy plane. A phase factor must therefore be used to keep track of this "twist" between the two couples of axes x, y and z, u. The phase factor codes the direction of the actual z axis within the one-dimensional continuum of directions in the Ozu plane that are orthogonal to the Oxy plane. The fact that the fourth dimension cannot be seen makes this phase factor easily overlooked. In the Lorentz group making a rotation about an axis in the Ozu plane that is different from the z-axis implies that there has been previously a boost. This implies that in the frame wherein this rotation takes place, the clocks will not be ticking at the same rate as they would have ticked in a rest frame, if a rotation had been carried out in that rest frame. When the Lorentz transformation is coded by two light rays travelling in opposite directions along the z-axis, the presence of a phase factor indicates that the Lorentz transformation is not a pure boost along the z-axis, but also contains a rotation in the Oxy-plane.

It has already been noted that an analogous situation with a phase factor that is easily overlooked exists within the rotation group. In fact, in $\mathbb{R}^n$ the *Vielbein* is completely coded by the $n-1$ unit vectors $\mathbf{e}_1, \mathbf{e}_2, \ldots \mathbf{e}_{n-1}$ of an orthonormal basis. The first unit vector $\mathbf{e}_1$ brings in $n-1$ free parameters, as it has n coordinates satisfying a normalization condition. The second unit vector $\mathbf{e}_2$ brings in $n-2$ free parameters since it is not only normalized but also is orthogonal to $\mathbf{e}_1$. The last vector $\mathbf{e}_{n-1}$ must only bring in one number that defines the orientation of both $\mathbf{e}_{n-1}$ and $\mathbf{e}_n$ with respect to all other vectors. Both in the rotation and in the Lorentz group, this last free parameter occurs as a phase factor. Without this phase factor, the vectors $\mathbf{e}_2$ and (in the Lorentz group) $\mathbf{e}_3$ would still be able to twist with respect to the other unit vectors.

4.11 Expressing the tetrad in terms of more physical parameters

From all this it can be appreciated that the coding in terms of an isotropic vector plus a phase factor is closer to the geometrical viewpoint of rotations than the coding in terms of two opposite light rays plus a phase factor, which is more physical. The whole tetrad becomes unambiguously defined by coding the space-like triad $(\mathbf{e}_x, \mathbf{e}_y, \mathbf{e}_z)$ through the expression $e^{\imath\chi}(\mathbf{e}_x + \imath\mathbf{e}_y)$, whereby the phase factor $e^{\imath\chi}$ indirectly defines the third element $\mathbf{e}_z$ of this triad. The fourth element of the tetrad $\mathbf{e}_{ct}$ is uniquely defined by the triad due to the orthonormality of the basis.

In the rotation group a general rotation can also be expressed in terms of other parameters like the three Euler angles, or the rotation axis and the rotation angle. These expressions are derived in many textbooks. Similar expressions for a general homogeneous Lorentz transformation in terms of the more physical parameters like the boost vector $\mathbf{v}$ and rotation axis and angle $(\mathbf{n}, \varphi)$ have been derived in Appendix C. It is shown there how the tetrad parameters (a, b, c, d) can be derived from the parameters $(\mathbf{v}, \mathbf{n}, \varphi)$ and *vice versa*. In conclusion, it is thus possible to find the coding of a general Lorentz transformation in two different sets of six parameters, *viz.* the ones that define the tetrad and those that define the intrinsic physical parameters.

4.12 Final considerations

The problem of the physical contents of a spinor has not always been coped with satisfactorily in the literature. For example, in [Misner *et al.* (1970)] a description in terms of flags and flag poles has been developed based on the idea that the whole information content of the six parameters would not be contained in the formalism. This is due to the bias introduced by wanting to represent the spinors on a single sphere. A very nice description is then introduced in terms of the appearance of the night sky. But this represents only one semi-spinor, while the whole information content is scattered over two complementary semi-spinors. The complete information content should be represented in terms of two spheres.

It is in this respect perhaps worthwhile to point out that other authors have coded the tetrad of the Lorentz group through more elaborate formalisms. For example, Newman and Penrose [Newman and Penrose (1962)] have coded the four vectors $\mathbf{e}_x \pm \imath\mathbf{e}_y$, $\mathbf{e}_{ct} \pm \mathbf{e}_z$ (which make up the so-called null tetrad) separately. The coding through SL(2,$\mathbb{C}$) as developed here is, however, the most concise one possible, and it is not necessary to layer more structure upon the formalism to be able to work with null tetrads, as they are already perfectly accounted for in SL(2,$\mathbb{C}$).

The Lorentz group has a six-dimensional real-parameter set. The two columns in a matrix of SL(2,$\mathbb{C}$) belong both to $\mathbb{C}^2$. Together, they would be equivalent to $\mathbb{C}^4$, i.e. eight real parameters, but the constraint that the metric should be conserved reduces the number of free real parameters to six. It must be stressed that after allowing for the constraints due to the metric, the six remaining real parameters are now completely free and not subject to any further constraints. The construction detailed here squeezes the whole null tetrad with its six independent real parameters into a two-dimensional formalism, and this might help in keeping the calculations as simple as possible.

It may be noted that by definition:

$$\begin{aligned} \begin{pmatrix} \xi_0 \\ \xi_1 \end{pmatrix} &= \mathbf{L} \begin{pmatrix} 1 \\ 0 \end{pmatrix} = \begin{pmatrix} a \\ c \end{pmatrix}, \\ \begin{pmatrix} \eta_0 \\ \eta_1 \end{pmatrix} &= \mathbf{L} \begin{pmatrix} 0 \\ 1 \end{pmatrix} = \begin{pmatrix} b \\ d \end{pmatrix}. \end{aligned} \tag{4.24}$$

It can be seen from this that in a sense the spinors of the Lorentz group are the elements of SL(2,$\mathbb{C}$) themselves, as given by (4.9). This is consistent with the fact that SL(2,$\mathbb{C}$) is already linear in its own right.

From a single representation, one can build a set of eight equivalent representations by taking various combinations of the inverse, the Hermitian conjugate, or the transposed of the matrices. Four of the combinations are given in (4.10). The other combinations are the dotted counterparts of these. For each of these eight types of matrices one can derive the corresponding type of semi-spinors. In most texts this plethora of different types of spinors are introduced *ex cathedra*, together with a terminology of normal and conjugated, punctuated and non-punctuated, covariant and contra-variant spinors, without referring to the fundamental reason for their existence. This can be a source of awe and puzzlement, especially since the notations become freighted with additional marks in order to make the distinction between the various types. These notations become necessary if one really wishes to carry out detailed calculations in practical applications, but the reader may appreciate from the present chapter that for the needs of identifying the underlying ideas a notation based on the quantities a, b, c, d and complex conjugation is sufficient.

With some perspicacity it could have been anticipated from the very form of (4.9) that it should not be possible to code the whole tetrad into one column vector and that two were required. In fact, from the knowledge of one column and the condition $ad - bc = 1$ it is not yet possible to determine the elements from the other column as in SU(2) on the basis of (3.4).

Chapter 5

The Dirac Equation from Scratch

Philosophy is written in this grand book — I mean the universe — which stands continually open to our gaze, but it cannot be understood unless one first learns to comprehend the language and interpret the characters in which it is written. It is written in the language of mathematics, and its characters are triangles, circles, and other geometrical figures, without which it is humanly impossible to understand a single word of it; without these, one is wandering around in a dark labyrinth.

— Galileo Galilei [Galilei (1623)]

5.1 The Dirac equation in free space — first approach

5.1.1 *Introduction*

In this chapter it is shown that the Dirac equation can be derived from the Rodrigues equation. In a first approach the underlying ideas will be described, but in a more critical examination of the results obtained this way two worrisome features will gradually emerge. The first is the absence of a one-to-one equivalence between the Rodrigues formula and the Dirac equation. The second is the fact that the definition of spin that one can introduce based on this approach is not covariant. By the time we start realizing this it will become compulsory to revise our copy and start all over again. We could have skipped this first "wrong" heuristic approach to obtain a more compact presentation, but it would then be difficult for the reader to understand how we obtained the definitive version.

A useful identity that will be used throughout the rest of the book, and that is obtained by straightforward calculation, is the following:

$$[\,\mathbf{a}\cdot\boldsymbol{\sigma}]\,[\,\mathbf{b}\cdot\boldsymbol{\sigma}] = \mathbf{a}\cdot\mathbf{b}\,\mathbb{1} + \imath\,(\mathbf{a}\wedge\mathbf{b})\cdot\boldsymbol{\sigma}. \tag{5.1}$$

The reader is advised to learn it by heart or to keep it close at hand, in order to avoid performing the same calculation over and over gain.

5.1.2 *The Rodrigues equation for a rotating frame*

The contents of this chapter can be no better summarized then by paraphrasing Galilei's quotation at the beginning of it.

Quantum mechanics cannot be understood unless one first learns to comprehend the language and interpret the characters in which it is written. It is written in the language of spinors. Without understanding spinors it is humanly impossible to understand a single word of it. Without understanding spinors, one is wandering around in a dark labyrinth.

In this chapter it will be proved that the Dirac equation describes spinning particles in the language of spinors. A spinor codes a rotation or a Lorentz transformation by coding the complete orientation of the basic triad or tetrad. Up to now, only purely mathematical results have been derived, but from now on the demarcation line between the mathematics and the physics will be crossed. As soon as this happens it becomes possible to question the assumptions introduced. Therefore, the transgressions should be limited to a strict minimum.

Let us turn back to rotations. They are a subgroup of the Lorentz group. Using the precise meaning of a spinor it can now be checked if the spin as described in the Dirac equation corresponds to a rotating particle. The following example will be instrumental in illustrating how the spinor formalism works. The Rodrigues formula:

$$\cos\frac{\varphi}{2}\mathbb{1} - \imath\,\mathbf{n}\cdot\boldsymbol{\sigma}\sin\frac{\varphi}{2} \tag{5.2}$$

for a rotation $R(\mathbf{n},\varphi)$ over an angle φ around an axis with unit vector $\mathbf{n}$, can be derived by considering two reflections A and B defined by planes which have unit normals $\mathbf{a}$ and $\mathbf{b}$ and intersect along $\mathbf{n}$ at an angle $\varphi/2$, such that $\mathbf{a}\cdot\mathbf{b} = \cos\frac{\varphi}{2}$. (On (5.2) it is easy to check the properties $\mathbf{R}^\dagger = \mathbf{R}^{-1}$ and $\det\mathbf{R} = 1$ needed to prove (3.4).) This construction was introduced in Figure 3.3. The rotation will then have an angle φ and an axis, which is the intersection of the two planes and coded by a unit vector $\mathbf{n}$ defined by $\sin\frac{\varphi}{2}\,\mathbf{n} = \mathbf{a}\wedge\mathbf{b}$. Here $\mathbf{n}$ will give the sense of the rotation according to the right-hand rule if the reflection B comes after the reflection A. In representation theory (with the conventions of Eq. (3.7)) and using (5.1) this leads to $\mathbf{R}(\mathbf{n},\varphi) = [\,\mathbf{b}\cdot\boldsymbol{\sigma}\,]\,[\,\mathbf{a}\cdot\boldsymbol{\sigma}\,] = (\mathbf{a}\cdot\mathbf{b})\,\mathbb{1} + \imath\,[\,(\mathbf{b}\wedge\mathbf{a})\cdot\boldsymbol{\sigma}\,]$, which is the result announced.[1]

Let us check (5.2) for a rotation around the z-axis. The reference frame is given by $\mathbf{e}_x = (1,0,0)$, $\mathbf{e}_y = (0,1,0)$, $\mathbf{n} = \mathbf{e}_z = (0,0,1)$, such that $\mathbf{e}_x + \imath\mathbf{e}_y = (1,\imath,0)$. Hence, we have $x = 1$, $y = \imath$. According to (3.10) the spinor is given by:

$$\begin{pmatrix}\xi_0\\ \\ \xi_1\end{pmatrix} = \begin{pmatrix}\sqrt{(x-\imath y)/2}\\ \\ \sqrt{(-x-\imath y)/2}\end{pmatrix} = \begin{pmatrix}1\\ \\ 0\end{pmatrix}. \tag{5.3}$$

Now $\cos\frac{\varphi}{2}\mathbb{1} - \imath\sigma_z\sin\frac{\varphi}{2}$ becomes:

$$\begin{pmatrix}e^{-\imath\varphi/2} & 0\\ \\ 0 & e^{+\imath\varphi/2}\end{pmatrix}. \tag{5.4}$$

Operating on the spinor, this yields:

$$\begin{pmatrix}\sqrt{(x'-\imath y')/2}\\ \\ \sqrt{(-x'-\imath y')/2}\end{pmatrix} = \begin{pmatrix}\xi_0'\\ \\ \xi_1'\end{pmatrix} = \begin{pmatrix}e^{-\imath\varphi/2}\\ \\ 0\end{pmatrix}, \tag{5.5}$$

such that $(x'-\imath y')/2 = e^{-\imath\varphi}$ and $(-x'-\imath y')/2 = 0$. From this $x' = e^{-\imath\varphi}$ and $y' = \imath e^{-\imath\varphi}$can be derived, and $x_1' = \cos\varphi$, $x_2' = -\sin\varphi$, $y_1' = \sin\varphi$, $y_2' = \cos\varphi$, such that $\mathbf{e}_x' = (\cos\varphi, \sin\varphi)$, and $\mathbf{e}_y' = (-\sin\varphi, \cos\varphi)$, which corresponds indeed to a rotation over an angle φ around the z-axis.

---

[1]Note that the conventions used in (3.7) for the rotation group are consistent with those used in (4.1) for the Lorentz group, such that the rotation group is embedded correctly within SL(2,$\mathbb{C}$). Alternative conventions exist for (3.7) and (4.1), which can be obtained by replacing $\imath\,|-\imath$ in both equations. Such substitutions lead to isomorphic representations. It suffices to make the substitution $\imath\,|-\imath$ throughout to swap from one representation to the other. With these conventions, (3.7) and (4.1) are compatible, and we have $\sigma_x\sigma_y = \imath\sigma_z$, $\sigma_y\sigma_z = \imath\sigma_x$, $\sigma_z\sigma_x = \imath\sigma_y$, which may be noted as $\sigma_x\sigma_y = \imath\sigma_z(cycl.)$. With the alternative choice we have $\sigma_x\sigma_y = -\imath\sigma_z(cycl.)$ and the Rodrigues equation would have been $\cos\frac{\varphi}{2}\mathbb{1} + \imath\mathbf{n}\cdot\boldsymbol{\sigma}\sin\frac{\varphi}{2}$.

Let us now code a *dynamical* rotation with angular velocity ω around the axis with unit vector $\mathbf{n}$. According to the Rodrigues formula this must be:

$$\cos\frac{\omega t}{2}\mathbb{1} - \imath\,[\,\mathbf{n}\cdot\boldsymbol{\sigma}\,]\,\sin\frac{\omega t}{2}. \tag{5.6}$$

For the example of a spinning motion around the z-axis, (5.4) becomes:

$$\begin{pmatrix} e^{-\imath\omega t/2} & 0 \\ 0 & e^{+\imath\omega t/2} \end{pmatrix}, \quad \psi(t) = e^{-\imath\omega t/2}\begin{pmatrix} 1 \\ 0 \end{pmatrix}. \tag{5.7}$$

Here, $\psi = [\,\xi_0, \xi_1\,]^\top$ is the corresponding spinor, taken from (5.5). In the general case of (5.6), the spinor will thus evolve with time according to:

$$\psi(t) = \left(\cos\frac{\omega t}{2}\mathbb{1} - \imath\,[\,\mathbf{n}\cdot\boldsymbol{\sigma}\,]\sin\frac{\omega t}{2}\right)\psi(0). \tag{5.8}$$

Of course the Rodrigues formula is not directly intuitive, but the reader can replace φ by ωt in the example worked out in (5.3)–(5.5) to convince himself of the fact that this represents a rotating frame and to "see" in these equations the frame turning "with his own eyes". In Figure 5.1 the Rodrigues formula is illustrated for a fixed axis $\mathbf{n}$ and varying $\varphi = \omega t$. Derivation of (5.8) produces a differential equation for $\psi(t)$. From:

$$\frac{d}{dt}\psi(t) = \frac{\omega}{2}\left(-\sin\frac{\omega t}{2}\mathbb{1} - \imath\,[\,\mathbf{n}\cdot\boldsymbol{\sigma}\,]\,\cos\frac{\omega t}{2}\right)\psi(0), \tag{5.9}$$

and using $[\mathbf{n}\cdot\boldsymbol{\sigma}]^2 = \mathbb{1}$ we obtain:

$$\frac{d}{dt}\psi(t) = -\imath\,[\,\mathbf{n}\cdot\boldsymbol{\sigma}\,]\,\frac{\omega}{2}\psi(t). \tag{5.10}$$

By applying the equation to itself one can also derive:

$$\frac{d^2}{dt^2}\,\psi = -\frac{\omega^2}{4}\,\psi. \tag{5.11}$$

5.1.3 *Lifting the equation from the SU(2) representation to the Dirac representation*

From now on the variable τ will represent the time in a frame where the electron is translationally at rest. The four-vector $(c\tau, 0, 0, 0)$ in a moving frame will be noted as (ct, x, y, z). By putting $\hbar\omega/2 = m_0c^2$ (which could

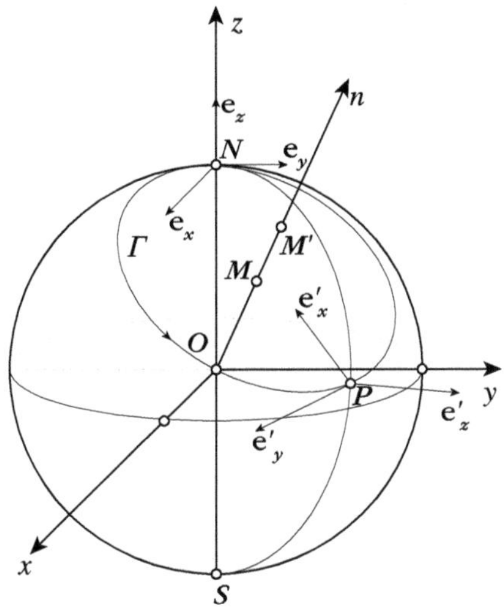

Fig. 5.1 The image $\mathbf{e}'_x, \mathbf{e}'_y, \mathbf{e}'_z$ of the spinor that corresponds to a rotation over an angle φ around the fixed axis $\mathbf{n} \parallel OM'$. The identity element corresponds to the triad $\mathbf{e}_x, \mathbf{e}_y, \mathbf{e}_z$. The vector OP gives the direction of $\mathbf{e}'_z$. When φ varies, the point P describes the circle Γ with centre $M \in OM'$. The point P that corresponds to the identity element is $N \in \Gamma$. The point N belongs to Γ as the identity element corresponds to the rotation with angle $\varphi = 0$, whatever the value of $\mathbf{n}$. For the sake of clarity, the triads are represented at different positions P on the surface of a sphere rather than all together in a big tangle at the origin. This convention was introduced in Figure 3.2.

be interpreted loosely by stating that the whole rest energy of the electron is due to its rotation with angular momentum $\hbar/2)^2$ we obtain:

$$\frac{1}{c^2}\frac{d^2}{d\tau^2}\psi \equiv \left[\frac{1}{c^2}\frac{\partial^2}{\partial t^2} - \frac{\partial^2}{\partial x^2} - \frac{\partial^2}{\partial y^2} - \frac{\partial^2}{\partial z^2}\right]\psi = -\frac{m_0^2c^2}{\hbar^2}\psi, \tag{5.12}$$

which is the Klein-Gordon equation. From this it becomes obvious that the Dirac operator:

$$\frac{1}{c}\frac{d}{d\tau} \to \gamma_{ct}\frac{1}{c}\frac{d}{d\tau} = \sum \gamma_\mu \frac{\partial}{\partial x_\mu}, \tag{5.13}$$

where the gamma matrices are defined by $\gamma_\mu\gamma_\nu + \gamma_\nu\gamma_\mu = 2g_{\mu\nu}\mathbb{1}$ and $g_{\mu\nu}$ is the metric tensor, expresses the derivation with respect to proper time in the representation of the Lorentz group spanned by the Dirac matrices.

[2]Of course, introducing $\hbar\omega/2 = m_0c^2$ is completely *ad hoc*. In a sense it is a cheat as it is done only to obtain the same results as in quantum mechanics. However, an *a posteriori* motivation for it will be found unexpectedly in Subsection 6.2.10.

When this and $\hbar\omega/2 = m_0c^2$ is used on (5.10) we obtain:

$$\sum \gamma_\mu \frac{\partial}{\partial x_\mu}\psi \simeq -\imath\,[\,\mathbf{n}\cdot\boldsymbol{\sigma}\,]\,\frac{m_0 c}{\hbar}\psi, \tag{5.14}$$

which is, except for the factor $\mathbf{n}\cdot\boldsymbol{\sigma}$, the Dirac equation for a free electron. The symbol $\simeq$ has been used here because the matrices on the two sides of this would-be equation are not of the same dimensions. On the left-hand side there are quantities from a four-dimensional formalism, while on the right-hand side there still remain quantities from a two-dimensional formalism, thus requiring the right hand side to be lifted to a four-dimensional formalism.

In the example of the rotation around the z-axis it is possible to check that $[\,\mathbf{n}\cdot\boldsymbol{\sigma}\,]\,\psi(t) = \psi(t)$, because for $\mathbf{n} = \mathbf{e}_z$ the matrix $\mathbf{n}\cdot\boldsymbol{\sigma}$ becomes diagonal (with entries 1 and -1) and the element on the second row of a spinor that codes a rotation around the z-axis is zero. But this is not general. Without the assumption $\mathbf{n} = \mathbf{e}_z$, it is not obvious that the Dirac equation can be derived from the Rodrigues equation. Hence, the derivation that would apply for $\mathbf{n} = \mathbf{e}_z$ is not generally valid.[3]

However, let us decompose the spinor $\psi(t)$ in the *vector basis* of the eigenvectors of $\mathbf{n}\cdot\boldsymbol{\sigma}$. It must be emphasized that this is a strange and hybrid construction, which has no directly obvious geometrical meaning, as spinors do not belong to a vector space.[4] It is only a calculation expedient. As

---

[3]There exists a brute-force shortcut to this problem. It consists in deriving the Dirac equation for the special case $\mathbf{n} = \mathbf{e}_z$, where $[\,\mathbf{n}\cdot\boldsymbol{\sigma}\,]\,\psi$ can be replaced by ψ and the Lorentz covariance of the resulting simplified equation used to claim that it is generally valid. This leads to the desired result, and could be considered as a derivation of the Dirac equation from scratch. But it is a logically flawed proof for a correct result. The derivation hides two important difficulties, that will necessitate reconsidering how one can derive the Dirac equation in a logically correct way:

(1) The Dirac equation derived this way is not equivalent to the Rodrigues formula. This issue will be addressed in the lines of the main text that follow immediately after this footnote, where an alternative derivation will be given that does not rely on the shortcut outlined here. This non-equivalence could be summerized symbolically by: (Rodrigues $\Rightarrow$ Dirac), but $\neg$ (Dirac $\Rightarrow$ Rodrigues). This non-equivalence transpires in the fact that the Dirac equation does not contain the explicit mention of the rotation axis $\mathbf{n}$, while the Rodrigues formula does.

(2) The simplification $[\,\mathbf{e}_z\cdot\boldsymbol{\sigma}\,]\,\psi = \psi$ is not Lorentz covariant.

The lack of equivalence mentioned under point (1) implies that the equation could have unwanted solutions that are physically meaningless. This problem will lead to a whole discussion of the definition of spin in Section 5.4.

[4]This point has been discussed on several occasions in Chapter 3, e.g. in Section 3.8, based on preliminary remarks made during the discussion of the groups SO(2) and SO(3)

the matrix $\mathbf{n}\cdot\boldsymbol{\sigma}$ is Hermitian, its eigenvalues (1 and -1) are real, and its eigenvectors are orthogonal. In other words, ψ splits into two parts, $c_+\psi_+$ and $c_-\psi_-$. These two quantities are not spinors, but vector projections of spinors.[5] But ψ_+ is a solution of $\frac{d}{dt}\psi_+(t) = -\imath\frac{\omega}{2}\psi_+(t)$ and therefore of the Dirac equation:

$$\boxed{\sum \gamma_\mu \frac{\partial}{\partial x_\mu}\psi_+ = -\imath\,\frac{m_0 c}{\hbar}\psi_+,} \tag{5.15}$$

while ψ_- is a solution of $\frac{d}{dt}\psi_-(t) = +\imath\frac{\omega}{2}\psi_-(t)$, and therefore of another Dirac-like equation with a reversed sign:

$$\boxed{\sum \gamma_\mu \frac{\partial}{\partial x_\mu}\psi_- = +\imath\,\frac{m_0 c}{\hbar}\psi_-.} \tag{5.16}$$

As the two equations (5.15) and (5.16) have the same solutions, it follows ultimately that ψ as a whole is a solution of the Dirac equation.[6] But this derivation does not run both ways, and only allows for a special linear combination of the type $c_+\,\psi_+\,e^{-\imath\frac{\omega t}{2}} + c_-\,\psi_-\,e^{+\imath\frac{\omega t}{2}}$. In fact, in the two-dimensional formulation given by (5.10), the rotation matrix

---

in Section 2.3. This implies that not every linear combination of two spinors is a new spinor, nor does the vector projection of a spinor need to be a spinor. Note that conceptually, a linear combination of two isotropic vectors in the context of the rotation group is not necessarily an isotropic vector. Also, the linear combination of two rotation matrices is not necessarily a new rotation matrix. Such a linear combination belongs to the *group ring* rather than to the group itself. Summing spinors is thus only defined on the group ring. Of course this may look as a nitpicking remark from a purely algebraic viewpoint, if one lacks the geometrical insight about the true meaning of a spinor. In Section 5.2 it will be demonstrated that in quantum mechanics this remark is conceptually very important.

[5]For $\mathbf{n} = \mathbf{e}_z$ we have $\psi = \psi_+$ as only $\psi_+ = [1,0]^\top e^{-\imath\varphi/2}$ is non-zero. This can be considered as corresponding to a clockwise rotation with matrix $\mathbf{R}(\mathbf{e}_z,\varphi)$ operating on $[1,0]^\top$. The combination of "spin up" with a negative frequency in the argument of the exponential corresponds then to a clockwise rotation of a right-handed frame. The "spin down" spinor $[0,1]^\top = \sigma_x[1,0]^\top$ corresponds then to a left-handed frame. Operating $\mathbf{R}(\mathbf{e}_z,\varphi)$ on this spinor yields $[0,1]^\top e^{\imath\varphi/2}$. The combination of spin down with a positive frequency in the argument of the exponential corresponds thus to a clockwise rotation of a left-handed frame. These two solutions $\mathbf{R}(\mathbf{e}_z,\varphi)[1,0]^\top$ and $\mathbf{R}(\mathbf{e}_z,\varphi)\sigma_x[1,0]^\top$ are a vector basis for the two-dimensional vector space that contains the spinors, but also non-spinor elements. To span a four-dimensional vector space, a left-handed representation of SL(2,$\mathbb{C}$) is added. This will become very clear in Subsection 5.5.2.2.

[6]The initial motivation for the statement that the two equations have the same solutions was the idea that the four eigenvectors proposed in physics textbooks are claimed to correspond to "spin up" and "spin down" combined with "positive" and "negative" energies. But this is a physical argument and not mathematically rigorous. The four solutions can be obtained from one another by substitutions $\mathbf{n}| -\mathbf{n}$ and/or $\omega| -\omega$. The substitution

$\cos\frac{\omega t}{2}\,\mathbb{1} - \imath\,[\,\mathbf{n}\cdot\boldsymbol{\sigma}\,]\,\sin\frac{\omega t}{2}$ corresponds to the spinor $\psi = (\cos\frac{\omega t}{2} - \imath n_z \sin\frac{\omega t}{2}, -\imath(n_x + \imath n_y)\sin\frac{\omega t}{2})^\top$ as can be seen by operating the matrix on the reference starting spinor $(1,0)^\top$. The eigenvectors of the matrix $\mathbf{n}\cdot\boldsymbol{\sigma}$ in (5.10) are $\psi_+ = (n_z + 1, n_x + \imath n_y)^\top$ for the eigenvalue $+1$ and $\psi_- = (n_z - 1, n_x + \imath n_y)^\top$ for the eigenvalue -1. The meaningful linear combination $c_+\,\psi_+\,e^{-\imath\frac{\omega t}{2}} + c_-\,\psi_-\,e^{+\imath\frac{\omega t}{2}}$ that permits recovery of ψ corresponds to the choice $c_+ = 1/2$, $c_- = -1/2$. By identifying the pre-factors of $\cos\frac{\omega t}{2}$ and $\sin\frac{\omega t}{2}$ in $c_+\,\psi_+\,e^{-\imath\frac{\omega t}{2}} + c_-\,\psi_-\,e^{+\imath\frac{\omega t}{2}}$ it is easy to check that the only linear combination that leads to a meaningful spinor structure $\psi = (\cos\frac{\omega t}{2} - \imath\tilde{n}_z\sin\frac{\omega t}{2}, -\imath(\tilde{n}_x + \imath\tilde{n}_y)\sin\frac{\omega t}{2})^\top$ is the one given by $c_+ = 1/2$, $c_- = -1/2$. Here, $(\tilde{n}_x, \tilde{n}_y, \tilde{n}_z)$ is the general solution for the rotation axis that one would be searching for, assuming that the solution (n_x, n_y, n_z) is not unique and an attempt is made to find the other solutions. These other solutions would be alien to the original problem that brought us to (5.10). But such other solutions do not exist, and therefore other linear combinations are not meaningful. It may also be noted that both eigenvalues of the Dirac equation have a two-dimensional vector space of eigenvalues. The liberty to choose a basis corresponds to a choice of $\mathbf{n}$. In textbooks, the choice $\mathbf{n} = \mathbf{e}_z$ is made.

$\mathbf{n} | -\mathbf{n}$ corresponds to a change between a right-handed and a left-handed representation SL(2,$\mathbb{C}$). As discussed before, the four-dimensional representation contains both, and has, of course, four eigenvectors. Each of the two SL(2,$\mathbb{C}$) representations contains only two eigenvectors, with opposite associated frequencies. It would thus be better to state that we try to combine the four possibilities into one equation, whereby one pair of solutions corresponds to a right-handed representation SL(2,$\mathbb{C}$) and the other pair to a left-handed representation SL(2,$\mathbb{C}$). In a certain presentation of this approach one may even reshuffle the contents of the pairs, but this is not important.

Another motivation is that there is no way in mathematics to tell $\imath$ and $-\imath$ apart. Both quantities are just defined simultaneously as the two solutions of the equation $x^2 = -1$. There is nothing in their definition that differentiates them. There is thus ambiguity between them in the sense given by Galois (discussed in Section 2.9). The substitution $\imath | -\imath$ is an isomorphism. This is only a heuristic argument based on a mathematical intuition. The isomorphism is seen at work for the first time in Footnote 1. Extending these two SU(2) representations leads to the left- and right-handed SL(2,$\mathbb{C}$) representations, which are combined together in a single Dirac equation according to Footnote 5, rendering the Dirac equation *isomorphic to itself.* But the counter-examples of the SU(2) representations and the SL(2,$\mathbb{C}$) representations, which are not self-isomorphic shows that the heuristic argument alone is not sufficient to justify the statement. Without noting the self-isomorphism, the heuristic argument is only a logically flawed proof for a correct result.

We may note that this derivation is very different from the traditional scheme proposed by Dirac. Historically, Dirac proposed to write:

$$\boxed{E\,\mathbb{1} = \boldsymbol{\alpha}\cdot c\mathbf{p} + \beta m_0 c^2.} \tag{5.17}$$

Here, $\boldsymbol{\alpha} = (\alpha_x, \alpha_y, \alpha_z)$ and β are so-called *Dirac matrices*. They are defined by the request that $(\boldsymbol{\alpha}\cdot c\mathbf{p} + \beta m_0 c^2)^2 = (c^2\mathbf{p}^2 + m_0^2 c^4)\,\mathbb{1}$, such that squaring (5.17) leads to the relativistic energy-momentum conservation law:

$$\boxed{E^2 = c^2\mathbf{p}^2 + (m_0 c^2)^2.} \tag{5.18}$$

The request leads to the conditions:

$$\alpha_j\alpha_k + \alpha_k\alpha_j = 2\delta_{jk}\mathbb{1}, \quad \beta^2 = \mathbb{1}, \quad \beta\alpha_j + \alpha_j\beta = 0\mathbb{1}. \tag{5.19}$$

Finally, Dirac made the substitutions:

$$\boxed{\hat{\mathrm{E}} \to -\frac{\hbar}{\imath}\frac{\partial}{\partial t}, \quad \hat{\mathbf{p}} \to \frac{\hbar}{\imath}\boldsymbol{\nabla}.} \tag{5.20}$$

These substitutions had already been used in the "derivation" of the Schrödinger equation. Dirac merely guessed his equation. The ultimate justification for what Dirac proposed is that the equation passed the test of comparison with experiment with flying colours. But it is completely unsatisfactory not to have any other justification for an equation other than the fact that it works without knowing what is going on behind the scenes. In the present approach, we make *tabula rasa* with the derivations and the rules of traditional quantum mechanics that are so difficult to understand. The Dirac equation is derived from the picture of a spinning electron, and the rules then derived from it. For example, the rules of (5.20) will be a simple consequence of the Dirac equation, rather than axiomatic elements needed to derive the equation.[7]

[7]Inspection of (5.38) reveals that it is the left-handed representation $-\frac{\hbar}{\imath}\,[\,\frac{\partial}{\partial ct} - \boldsymbol{\nabla}\cdot\boldsymbol{\sigma}\,]$ that works on the spinor Ψ. This is due to the fact that in products of SL(2,$\mathbb{C}$) matrices one must always alternate between left-handed and right-handed representation matrices. It is then from the expression $-\frac{\hbar}{\imath}\,[\,\frac{\partial}{\partial ct} - \boldsymbol{\nabla}\cdot\boldsymbol{\sigma}\,]\,\Psi$ that one can naturally derive the definitions $\hat{\mathrm{E}} = -\frac{\hbar}{\imath}\frac{\partial}{\partial ct}$ and $\hat{\mathbf{p}} = \frac{\hbar}{\imath}\boldsymbol{\nabla}\cdot\boldsymbol{\sigma}$. The reason why the energy-momentum four-vector and the four-gradient can be substituted for one another in the schematic of (5.17) is that they are both four-vectors and that the schematic is just the way one has to code four-vectors in the Dirac representation of the Lorentz group.

Let us also introduce a notation. The operator on the left-hand side of (5.38) can be written as $-\frac{\hbar}{\imath}\gamma_{ct}\frac{\partial}{\partial ct} - \frac{\hbar}{\imath}\gamma_x\frac{\partial}{\partial x} - \frac{\hbar}{\imath}\gamma_y\frac{\partial}{\partial y} - \frac{\hbar}{\imath}\gamma_z\frac{\partial}{\partial z}$. All signs in it are of the same type. Using the operator definitions this becomes: $\frac{1}{c}\hat{\mathrm{E}}\gamma_{ct} - \hat{\mathrm{p}}_x\gamma_x - \hat{\mathrm{p}}_y\gamma_y - \hat{\mathrm{p}}_z\gamma_z$, where the signs are no longer all of the same type. Therefore, one introduces the conventions $\gamma^{ct} = \gamma_{ct}$,

5.1.4 *Embedding of SU(2) within SL(2,ℂ)*

The notation $c_+\,\psi_+\,e^{-\imath\frac{\omega t}{2}} + c_-\,\psi_-\,e^{+\imath\frac{\omega t}{2}}$ shows that negative frequencies already occur within the rotation group. This has no relationship with anti-particles, as the context is purely geometrical. In general, it suffices to change the sense of the rotation to change the sign of the frequencies. As stated in Footnote 1 the representation SU(2) has been embedded within SL(2,ℂ) self-consistently. The rotation matrix $\cos\frac{\omega t}{2}\,\mathbb{1} - \imath\,\mathbf{n}{\cdot}\boldsymbol{\sigma}\,\sin\frac{\omega t}{2}$ can then be used also in SL(2,ℂ) and written as $\frac{1}{2}\,\mathbf{N}\,e^{-\imath\omega t/2} + \frac{1}{2}\,\mathbf{N}^\star\,e^{+\imath\omega t/2}$. Here $\mathbf{N}$ and $\mathbf{N}^\star$ correspond to $\mathbb{1} + \mathbf{n}{\cdot}\boldsymbol{\sigma}$ and $\mathbb{1} - \mathbf{n}{\cdot}\boldsymbol{\sigma}$.[8] The coding of a vector $(x, y, z) = \mathbf{r}$ as a 2×2-matrix $\mathbf{R}$ in the representation SU(2) with signature $+{+}+$ is given by (3.24). As $\det\mathbf{R} = -\mathbf{r}^2$, this coding is compatible with the embedding of $\mathbf{r}$ as a four-vector $(ct, x, y, z) = (0, \mathbf{r})$ in SL(2,ℂ) where the signature is $+ - --$ (corresponding to the metric $c^2t^2 - x^2 - y^2 - z^2$). The change of signs in the signatures from $+{+}+$ for $\mathbf{r}^2 = x^2+y^2+z^2$ in SU(2) to $- - -$ for $-\mathbf{r}^2$ in SL(2,ℂ) is automatically introduced by the change of rule from $\mathbf{R}^2 = \mathbf{r}^2\mathbb{1}$ to $\det\mathbf{R} = -\mathbf{r}^2$.

5.1.5 *Caveat: This is not yet quantum mechanics*

This way the Dirac equation has been derived and it has been found that it describes a particle by attaching a reference frame to it and by treating this co-rotating frame with the aid of group representation theory. The equation just expresses (5.6) using the derivative $\gamma_{ct}\,\frac{\partial}{\partial c\tau} = \sum_\mu \gamma_\mu \frac{\partial}{\partial x_\mu}$ with respect to the proper time τ of the co-moving frame. As the derivation operator $\frac{1}{c}\frac{d}{d\tau}$ is part of a four-vector $(\frac{1}{c}\,\frac{d}{dt}, \boldsymbol{\nabla})$, it is treated as a reflection operator. Thus, the Dirac equation treats a spinning particle subjected to Lorentz transformations. The spinning particle is first treated in its rest frame, and then Lorentz invariance is used to express this in any other moving frame.

$\gamma^{\,x} = -\gamma_x$, $\gamma^{\,y} = -\gamma_y$, $\gamma^{\,z} = -\gamma_z$, which can be summarized as $\gamma^\mu = \sum_\nu g^{\mu\nu}\gamma_\nu$. These conventions permit $\frac{1}{c}\hat{\mathrm{E}}\gamma_{ct} - \hat{\mathrm{p}}_x\gamma_{\mathrm{x}} - \hat{\mathrm{p}}_y\gamma_y - \hat{\mathrm{p}}_z\gamma_z$ to be rewritten as $\sum_\mu \gamma^\mu\hat{\mathrm{p}}_\mu$ wherein again all signs are of the same type. Dirac defined the gamma matrices starting from $(\gamma_{ct}E + \gamma_x cp_x + \gamma_y cp_y + \gamma_z cp_z)^2 = (m_0c^2)^2\mathbb{1}$ rather than $(\gamma^{\,ct}E + \gamma^{\,x}cp_x + \gamma^{\,y}cp_y + \gamma^{\,z}cp_z)^2 = (m_0c^2)^2\mathbb{1}$, such that the text book definitions are the exact opposite of those described in this book. But in both cases it must be remembered that in the expressions that contain only one type of sign, $(\hat{\mathrm{E}}, c\hat{\mathbf{p}})$ and $(\frac{\partial}{\partial ct}, \boldsymbol{\nabla})$ are combined with gamma matrices of opposite types.

[8]It is tempting to interpret here $\mathbf{N}$ and $\mathbf{N}^\star$ as the "zero-length" vectors $\mathbf{e}_{ct} + \mathbf{n}$ and $\mathbf{e}_{ct} - \mathbf{n}$, which would give a beautiful interpretation of the formalism in terms of light rays. But this is not correct. It will be explained later that the fact that the unit matrix seems to be associated with $\mathbf{e}_{ct}$ in SL(2,ℂ) is a very subtle point.

The pure-state solutions of the Dirac equation and the original Rodrigues equation are not equivalent, as the pure states of the Dirac equation are vector projections of the Rodrigues spinors, which are only identical to the true spinors when $\mathbf{n} \parallel \mathbf{e}_z$. In all other cases, the vector projections no longer keep their clear initial geometrical meaning. As the Dirac equation has simply been guessed, it may be necessary to call upon the Rodrigues equation for matters of deciding what is a pure state.

The substitutions that are needed to introduce the electromagnetic potentials into the Dirac equation will be discussed in Section 5.6. As the Schrödinger equation can be derived from the Dirac equation with an electromagnetic four-potential, it is evident that a lot can be derived about two prominent equations that are used in quantum mechanics from a simple *ansatz* of a rotating particle. The only unjustified issue is the gimmick of replacing ω according to $\hbar\omega/2 = m_0c^2$, which is inspired by the relations of Planck and Einstein. Hence it seems that the demarcation line between physics and mathematics has not been transgressed too much. A possible justification for introducing the Planck-Einstein relations will be discussed in Subsection 6.2.10.

Of course, *it cannot possibly be claimed at the present stage of the development in the book that a derivation of quantum mechanics would have been given.* What the reader has been able to discover up to now, are only the geometrical contents of a *mathematical language*, upon which quantum mechanics is built. There are two reasons for this:

(1) Hitherto, all the ingredients used in the approach have been classical, while quantum mechanics contains aspects (like the superposition principle) that appear totally counterintuitive from a classical viewpoint. Therefore, these aspects have certainly not been touched upon in this derivation.

(2) Another aspect that has not yet been touched upon in the preceding pages is the probabilitistic character of quantum mechanics, as everything that has been derived is conceptually deterministic. This is not necessarily counterintuitive, as classically there is no impediment to swapping from a deterministic formulation to a probabilistic one by averaging over certain variables. However, there is no such free choice between two alternatives in the standard interpretation of quantum mechanics, where a description in terms of probabilities is considered an *absolutely necessity.* These two aspects are thus beyond what has been achieved up to now by pure mathematics.

5.2 A warning about the superposition principle and the negative energies

Physicists use spinors like vectors. — Elie Cartan

It may come as a surprise that the structure of the Dirac equation and the Schrödinger equation rather lies on the mathematical side of the demarcation line between the mathematics and the physics. In fact, it is very often stated that the superposition principle in quantum mechanics is a direct consequence of the linearity of the Dirac equation or the Schrödinger equation. The superposition principle intervenes for instance in the double-slit experiment, which Feynman called "the only mystery of quantum mechanics". The superposition principle corresponds thus to something that is beyond classical intuition. One would therefore expect that the Dirac equation cannot be derived from purely geometrical arguments as in this book.

• *Superposition principle.* The superposition principle is not as obvious a consequence of the linearity of the Dirac equation as is commonly stated, due to the fact that the wave functions are spinors. As it is not meaningful to add spinors like vectors, the superposition principle is not granted by the linearity of the equation. Only one special linear combination will lead to a true spinor. The superposition principle implies that spinors can be treated as vectors. From this it must be obvious that the superposition principle is definitely not part of the entirely classical derivation that has been presented in the preceding pages. This difficulty will be returned to later on.

• *Negative energies.* In the present first approach, the correlated notion that the negative frequency solution $c_+ \, \psi_+ \, e^{-\imath \frac{\omega t}{2}}$ alone makes physical sense in its own right, rather than merely being a further meaningless vector projection of a true spinor $c_+ \, \psi_+ \, e^{-\imath \frac{\omega t}{2}} + c_- \, \psi_- \, e^{+\imath \frac{\omega t}{2}}$, is alien to the conceptually classical derivation given above. Furthermore, there is absolutely nothing in this derivation of the Dirac equation that implies that negative frequencies would correspond to advanced waves or negative energies. It could be asserted also that negative frequencies correspond to clockwise rather than anticlockwise rotations, because changing the sign of the frequency corresponds in essence to changing the sense of the rotation. A spinor corresponds to a *single rotation* of a *single* particle.

• *Conclusion.* With the proviso that the present derivation of the Dirac equation is still flawed (see Section 5.3), it can already be stated that

> the Dirac equation belongs to the realm of (relativistic) classical physics. It is only the way in which it will be used which will turn it into quantum mechanics.

There are several reasons for this. Spinors are not vectors, such that the superposition principle is not an immediate consequence of the linearity of the Dirac equation. The use of a vector formalism based on spinors will be discussed in the Section 5.4 in terms of the group ring. Further reasons supporting this assertion are the issues of the negative frequencies and the probabilities.[9]

5.3 Guidance through the rest of this chapter

5.3.1 *Free-space Dirac equation*

The preceding sections aimed to show the reader that the free-space Dirac equation can be derived by just expressing that an electron spins. The approach adopted so far will reveal itself only as a first approximation. The rest of this chapter will be devoted to ironing out the difficulties arising from the fact that the derivation so far proposed in this book is not as neat as it could be. Eventually it will be possible to give an exact and rigorous mathematical proof that

> the Dirac equation describes in a relativistically covariant way and by using the language of spinors that the electron spins like a top.

This will not be easy to prove. The major steps are summarized in Table 5.2 and briefly described below.

---

[9] At least in the present first approach, a true spinor contains both positive and negative frequencies, and the two contributions must have equal weights $c_+ = \frac{1}{2}$ and $c_- = -\frac{1}{2}$. It is known that Feynman searched for an explanation for the fact that the wave functions he had to use contained both "advanced" and "retarded" waves on a same footing. In the Rodrigues equation something similar is obtained in a very simple way, such that this seems to offer an explanation for Feynman's result. But this is superficial and incorrect. In fact, this "property" will be dismissed in Section 5.4.

That the actual approach is only a first heuristic approximation will start to transpire at the time it will be attempted to understand the formalism that is used to describe the spin of the electron within quantum mechanics. In Footnote 3 a brute-force derivation of the Dirac equation was proposed. Taking $\mathbf{n} = \mathbf{e}_z$ in the Rodrigues equation yields the Dirac equation, but the derivation is only valid for the special case $\mathbf{n} = \mathbf{e}_z$. One can then try to show that the Dirac equation is generally valid by transforming the equation obtained for $\mathbf{n} = \mathbf{e}_z$ covariantly to other frames where $\mathbf{n} \neq \mathbf{e}_z$. As it seems obvious that $\mathbf{n}$ is a vector, one may expect that in changing reference frames the identity $\mathbf{n} = \mathbf{e}_z$ will be transformed into $\mathbf{n}' = \mathbf{e}'_z$, but this is not true. This will reveal that the definition of spin must be based on the quantity $\mathbf{e}'_z$ rather than on the quantity $\mathbf{n}'$, and that the derivation of the Dirac equation cannot be based on the Rodrigues formula when $\mathbf{n} \neq \mathbf{e}_z$. This is the subject of Section 5.4. This section is lengthy, as it is necessary that the reader understands that at the onset there are two plausible alternative ways ($\mathbf{n}'$ and $\mathbf{e}'_z$) to build up the theory. Only one choice is viable. If this choice were introduced without discussing why the alternative is wrong, then it may look to the reader as coming out of the blue. It is also because the development of this point is rather lengthy that we outline here how it fits into the larger picture, such that the reader does not lose sight of what we eventually are aiming at.

In developing Section 5.4, a new difficulty will be encountered, and it is a major one. It will look as though it is impossible to derive the Dirac equation without making an approximation, such that the Dirac equation would not be rigorously exact and general. Therefore in Section 5.5 an exact Dirac-like equation will be derived that does not contain this approximation.

Surprisingly, it will be possible to derive the Dirac equation from this Dirac-like equation, without introducing any approximations at all. It will require a non-trivial move: it consists in adopting solutions that are different from the ones we naively had in mind at the outset, but by the context this tricky move will just present itself quite naturally. This will also allow one to understand the nature of the solutions of the equation and to see how eventually all the pieces fall into place.

5.3.2 *Dirac equation with potential*

A further concern is justifying the minimal substitution, because this should in principle open the door to a complete understanding of full-fledged quantum mechanics. Indeed, the whole of quantum mechanics is derived from two master equations: the Dirac equation and the Schrödinger equation (which can be derived from the Dirac equation). Full understanding of the origin of this minimal substitution enables full understanding of the whole formalism of quantum mechanics. But in traditional treatments, the minimal substitution is introduced by analogy with classical mechanics, without any further discussion or justification as though it were trivial and self-evident. Ironically enough, we are simultaneously being told that quantum mechanics is radically different from classical mechanics. Why then should it be taken for granted that this substitution can yield the correct equation? Has this not just been a serendipitous guess? This problem will be solved in Section 5.6. The solution will lead to a good understanding of the meaning of the minimal substitution in quantum mechanics.

5.3.3 *Strategy*

The goal of this chapter will then have been reached, *viz.* deriving the two wave equations on which all further quantum mechanical calculations are based. This part of quantum mechanics will have then been derived *deductively* from the *ansatz* that the electron spins, in marked contrast with the historical evolution where it was derived *inductively* from experimental observations by Heisenberg or guessed by Dirac and experimentally validated. The inductive derivation can only be presented as a set of rules. Not knowing where these rules come from makes it hard to analyse difficult theoretical problems in depth.[10] The deductive derivation puts us in a much more comfortable position to investigate paradoxes and the meaning of the results. In reading this chapter, the reader will perhaps draw strength from the perspective that the ultimate aim is to obtain such a deductive derivation based on a clear visual picture.

The approach in this book is at variance with a viewpoint that spinors are not rotating in physical space, in a way similar to the one encountered with isospin. This viewpoint was summarized by Cartan who stated in [Cartan (1981)]:

[10] Of course one can try to justify *a posteriori* the rules obtained inductively by presenting them as deductively derived from a set of axioms. This is then what could be called an interpretation of quantum mechanics. While the interpretation of classical mechanics is very intuitive, the set of axioms in traditional quantum mechanics is much less so. It really presents a number of arcane traits and traditional quantum mechanics even contains some errors. In the approach of this book the interpretation must follow naturally from the mathematical meaning of the spinor quantities used.

> Certain physicists regard spinors as entities which are, in a sense, unaffected by the rotations which classical geometrical entities (vectors etc.) can undergo, and of which the components in a given reference frame are susceptible to undergo linear transformations which are in a sense autonomous.

Cartan qualified this viewpoint as "startling". In fact, the claim that spinors do not turn in physical space could be attributed to a lack of understanding of the geometrical meaning of spinors. As they are based on coordinates $(x, y, z) \parallel x^2 + y^2 + z^2 = 0$, one may quickly come to the conclusion that spinors cannot have anything to do with real Euclidean space (see for example [Biedenharn and Louck (1985)]). Too quickly, as a matter of fact, since it has been explained that (x, y, z) are used to represent a triad of basis vectors, which indeed turns in physical space. Therefore, this isospin-inspired viewpoint may be *ad hoc* and neither compelling nor unique. It is certainly justifiable to question this traditional viewpoint on the basis of the exact derivation of the Dirac equation that will be presented here. The viewpoint based on the analogy with isospin contains even more exceptional assumptions than the isospin model itself, as in the spin operator $\hat{S}_z$, the index z refers to physical space despite the denial that the spinor would turn in physical space. In the isospin operator $\hat{I}_z$ the index z does not refer to physical space, such that the postulates of this formalism are less demanding.

The traditional approach also applies the Dirac equation to neutrinos for which a zero rest mass is assumed. It could be held against the present approach that it cannot describe neutrinos, as it assumes a non-zero rest mass for the particle described. But the existence of neutrino oscillations indicates that neutrinos do not have zero rest mass. We could thus turn this argument around and use it against the traditional approach, because it "predicts" neutrinos of zero rest mass.

It may finally be noted that a rotating frame is a problem in a relativistic context. At a large enough distance from the origin of the frame it would imply motion faster than light. There is also the problem of Lorentz contraction as identified by Einstein in his example of a rotating disk. However, the spinors and the frame are only used here as a useful set of coordinates to describe the spinning electron, not to describe the space-time as it would be observed in a macroscopic frame that would co-rotate with the electron.

In the next chapters the Schrödinger and Dirac equations will be used to tackle quantum mechanical problems. It will then become apparent that alternative derivations exist for the theoretical results that are quite different from those presented in traditional treatments. These alternative derivations have a clear geometrical meaning and there is no need to invoke mysterious quantum effects to explain or interpret them. The whole becomes then more clear and physical. This is really what the analogy with the problem of Dirac's "delta function" introduced in (1.1) is all about. The main aim of this book is actually to investigate to what extent we can solve the so-called quantum mysteries by using a purely deductive approach.

5.4 Spin and the group ring

5.4.1 *Spin as a set of spinors*

5.4.1.1 *Something puzzling*

The meaning of the eigenvector $\psi_+ = (n_z + 1, n_x + \imath n_y)^\top$ of the matrix $\mathbf{n}{\cdot}\boldsymbol{\sigma}$ (that codes the vector $\mathbf{n}$) in Subsection 5.1.3 can be understood as follows. It is not a true spinor, as it is obtained from cutting the true spinor into two parts ψ_+ and ψ_-. The reflection operator $\mathbf{n}{\cdot}\boldsymbol{\sigma}$ does not work on images of vectors from $\mathbb{R}^3$ because it is working quadratically on vectors: vectors behave as tensors of rank 2 within SU(2). The reflection operator $\mathbf{n}{\cdot}\boldsymbol{\sigma}$ must thus work on images of rotations, reflections, and reversals, which can be represented by true spinors. It follows that a reflection operator cannot have another group element as an eigenvector under the form of a true spinor, whereby the corresponding eigenvalue would be 1. In fact, the reflection operator $\mathbf{n}{\cdot}\boldsymbol{\sigma}$ will transform a reflection into a rotation, a rotation into a reversal, and a reversal into a rotation. In other words, it changes the nature of the group element. If the eigenvector ψ_+ were a true spinor corresponding to the eigenvalue 1, this would imply that it corresponds to a group element that remains invariant under the reflection, and thus has not changed its nature. Hence, the eigenvector of a reflection operator cannot possibly be a spinor that corresponds to a group element. Nevertheless, these eigenvectors are definitely used in physics. This is puzzling and leaves one wondering what the meaning of the eigenvector could be.

5.4.1.2 *The eigenvector is a set*

The answer is that it is a hybrid quantity that corresponds to an element of the group ring. In fact, the algebra does not work only on the group but on the entire group ring. If the eigenvector has a meaning at all, then this meaning can only be searched for within the calculus of the group ring. There is thus a need to find a linear combination of several group elements, for example (in the simplest possible approach) of two operators $\hat{\mathrm{O}}_1$ and $\hat{\mathrm{O}}_2$ such that $[\,\mathbf{n}{\cdot}\boldsymbol{\sigma}\,]\,\hat{\mathrm{O}}_1 = \hat{\mathrm{O}}_2$ and $[\,\mathbf{n}{\cdot}\boldsymbol{\sigma}\,]\,\hat{\mathrm{O}}_2 = \hat{\mathrm{O}}_1$. These two identities summarize how a reflection operator is supposed to work on a group element.

Let us establish what kind of operator could be associated this way with a reflection defined by its reflection normal $\mathbf{n}$. The effect of $\mathbf{n}{\cdot}\boldsymbol{\sigma}$ on another reflection is a rotation, unless we take $\mathbf{n}{\cdot}\boldsymbol{\sigma}$ itself, and then it yields the identity operator. It thus seems indicated to take $\mathbb{1}$ and $\mathbf{n}{\cdot}\boldsymbol{\sigma}$ together. According to the same logic as has been used for spinors, the eigenvector with eigenvalue 1 codes $\mathbb{1} + \mathbf{n}{\cdot}\boldsymbol{\sigma}$ (as it corresponds to the first column of this quantity). In fact, the spinor of a group element was obtained by taking the first column of the 2×2 matrix that corresponds to the group element, and here ψ_+ is obtained by taking the first column of $\mathbb{1} + \mathbf{n}{\cdot}\boldsymbol{\sigma}$. The quantity ψ_+ actually codes sets $e^{\imath\chi}\,\{\mathbb{1}, \mathbf{n}{\cdot}\boldsymbol{\sigma}\}$, where $e^{\imath\chi}$ is a phase factor (which, as will be demonstrated, is not immaterial). The idea is thus to code the set $\{\mathbb{1}, \mathbf{n}{\cdot}\boldsymbol{\sigma}\}$ by the sum of its two elements. We have in fact $(\mathbf{n}{\cdot}\boldsymbol{\sigma})\,(\mathbb{1} + \mathbf{n}{\cdot}\boldsymbol{\sigma}) = \mathbb{1} + \mathbf{n}{\cdot}\boldsymbol{\sigma}$. This would correspond to $\mathbf{n}{\cdot}\boldsymbol{\sigma}\,\{\mathbb{1}, \mathbf{n}{\cdot}\boldsymbol{\sigma}\} = \{\mathbb{1}, \mathbf{n}{\cdot}\boldsymbol{\sigma}\}$, such that the set is an eigenvector. The transition from the 2×2 matrix $\mathbb{1} + \mathbf{n}{\cdot}\boldsymbol{\sigma}$ to its 2×1 counterpart $\psi_+ = (n_z + 1, n_x + \imath n_y)^\top$ can be made by operating with it on $[1, 0]^\top$, which comes down to just taking its first column.

5.4.1.3 *Rotations*

Let us now show that the sets $e^{\imath\chi}\,\{\mathbb{1}, \mathbf{n}{\cdot}\boldsymbol{\sigma}\}$, with $\chi \neq 0$ can code rotations. A general rotation over an angle φ around an axis $\mathbf{n}$ will be given by: $\hat{\mathrm{O}}_1 = \cos\frac{\varphi}{2}\,\mathbb{1} - \imath\,[\,\mathbf{n}{\cdot}\boldsymbol{\sigma}\,]\,\sin\frac{\varphi}{2}$. The group element $\hat{\mathrm{O}}_2 = [\,\mathbf{n}{\cdot}\boldsymbol{\sigma}\,]\,\hat{\mathrm{O}}_1$ must then be taken to complete the set that contains $\hat{\mathrm{O}}_1$. One can verify that this leads to $\hat{\mathrm{O}}_1 + \hat{\mathrm{O}}_2 = e^{-\imath\varphi/2}(\,\mathbb{1} + \mathbf{n}{\cdot}\boldsymbol{\sigma}\,)$. From this form it is easy to see that $[\,\mathbf{n}{\cdot}\boldsymbol{\sigma}\,]\,(\hat{\mathrm{O}}_1 + \hat{\mathrm{O}}_2) = \hat{\mathrm{O}}_1 + \hat{\mathrm{O}}_2$. It can be rewritten as:

$$(\,\mathbb{1} + \mathbf{n}{\cdot}\boldsymbol{\sigma})\hat{\mathrm{O}}_1 = e^{-\imath\varphi/2}(\,\mathbb{1} + \mathbf{n}{\cdot}\boldsymbol{\sigma}). \tag{5.21}$$

This shows that the eigenvector does not code a state of the system, but merely a set $e^{-\imath\varphi/2}\,\{\mathbb{1}, \mathbf{n}{\cdot}\boldsymbol{\sigma}\} \rightsquigarrow \{\hat{\mathrm{O}}_1, \hat{\mathrm{O}}_2\}$. The symbol $\rightsquigarrow$ is used here to indicate that summing the elements of the sets yields the same result. Of course, only the decomposition into $\hat{\mathrm{O}}_1$ and $\hat{\mathrm{O}}_2$ makes sense in terms of group elements; the decomposition into $e^{-\imath\varphi/2}\mathbb{1}$ and $e^{-\imath\varphi/2}\ \mathbf{n}{\cdot}\boldsymbol{\sigma}$ is just algebra. The ray $\{\psi_\varphi(\mathbf{n}) : \psi_\varphi(\mathbf{n}) = e^{-\imath\varphi/2}\,\psi(\mathbf{n})\}$ based on $\mathbb{1} + \mathbf{n}{\cdot}\boldsymbol{\sigma}$, coding $\psi(\mathbf{n}) = \{\mathbb{1}, \mathbf{n}{\cdot}\boldsymbol{\sigma}\}$ corresponds then to the infinite set of all the rotations and reversals which have $\mathbf{n}$ as the rotation axis. It can therefore be stated that it is the ray based on $\mathbf{n}$. Each member $e^{-\imath\varphi/2}\psi(\mathbf{n})$ of the ray corresponds to the two-element subset of the ray that contains the rotation and the reversal whose rotation angle is φ.

5.4.1.4 *Reversals*

The "spin down" eigenvector ψ_- is also a set, *viz.* $e^{-\imath\varphi/2}\,\{\mathbb{1}, -\mathbf{n}{\cdot}\boldsymbol{\sigma}\}$. In fact, $(\mathbf{n}{\cdot}\boldsymbol{\sigma})\,(\mathbb{1} - \mathbf{n}{\cdot}\boldsymbol{\sigma}) = -(\mathbb{1} - \mathbf{n}{\cdot}\boldsymbol{\sigma})$. For this case it is wise to start from a rotation around $-\mathbf{n}$, e.g. $\hat{\mathrm{O}}'_1 = \cos\frac{\varphi}{2} + \imath\mathbf{n}{\cdot}\boldsymbol{\sigma}\sin\frac{\varphi}{2}$. Now $\hat{\mathrm{O}}'_2$ must be defined as $-[\,\mathbf{n}{\cdot}\boldsymbol{\sigma}\,]\,\hat{\mathrm{O}}'_1$, to make sure that an eigenvalue of -1 is obtained, such that $\hat{\mathrm{O}}'_2 = \imath\,(\cos\frac{(\varphi+\pi)}{2} + \imath\,[\,\mathbf{n}{\cdot}\boldsymbol{\sigma}\,]\sin\frac{(\varphi+\pi)}{2})$. Then $[\,\mathbf{n}{\cdot}\boldsymbol{\sigma}\,]\,\hat{\mathrm{O}}'_1 = -\hat{\mathrm{O}}'_2$ and $[\,\mathbf{n}{\cdot}\boldsymbol{\sigma}\,]\,\hat{\mathrm{O}}'_2 = -\hat{\mathrm{O}}'_1$. The eigenvector $(\hat{\mathrm{O}}'_1 + \hat{\mathrm{O}}'_2)$ is equal to $e^{-\imath\varphi/2}\,(\mathbb{1} - \mathbf{n}{\cdot}\boldsymbol{\sigma})$, and thus codes the set $e^{-\imath\varphi/2}\,\{\mathbb{1}, -\mathbf{n}{\cdot}\boldsymbol{\sigma}\} \leadsto \{\hat{\mathrm{O}}'_1, \hat{\mathrm{O}}'_2\}$.

5.4.1.5 *Conceptual remarks*

The phase factor $e^{-\imath\varphi/2}$ is not scrutable in the eigenvector calculations for the operator $\mathbf{n}{\cdot}\boldsymbol{\sigma}$. In classical vector calculus, two eigenvectors with a different pre-factor are considered to be equivalent, and therefore can be normalized. However, for spinors the phase in the pre-factor is important, and two two-element subsets with different phase vectors code different sets of rotations. The reflection operator $\mathbf{n}{\cdot}\boldsymbol{\sigma}$ has an infinity of two-element sets as eigenvectors, that differ only by a phase vector in their coding, but are very distinct sets. (One can also consider that the reflection operator $\mathbf{n}{\cdot}\boldsymbol{\sigma}$ has the infinite set that is the union of all two-element sets as an eigenvector.)

The eigenvectors are eigenvectors of the reflection operator $\mathbf{n}{\cdot}\boldsymbol{\sigma}$, not of the rotation matrix that describes the motion of the electron, and they code sets. For rotations, the eigenvalue for a set is $+1$ for a rotation around $\mathbf{n}$ and -1 for a rotation around $-\mathbf{n}$. If one assumes that $\mathbf{n}$ is only allowed to take values in one hemisphere, then this formalism serves to distinguish clockwise and anticlockwise rotations. In the exceptional case that $\mathbf{n} = \mathbf{e}_z$ the eigenvectors correspond to true spinors. We can then interpret them as clockwise and anticlockwise rotations. But in all other cases, an eigenvector is only a vector projection of a spinor, such that it does not correspond to a group element. The case with $\mathbf{n} = \mathbf{e}_z$ can be considered as a numerical coincidence, due to the fact that one of the two spinor components is zero.[11]

5.4.2 *Alas, the definition of spin introduced is not covariant*

5.4.2.1 *The sets defined do not survive rotations*

A paradox ...

There is a problem with the transformation properties of (5.21). If both sides of this equation are multiplied to the left by a general rotation matrix $\mathbf{R}$, the left-hand side can be rewritten as $\mathbf{R}(\mathbb{1}+\mathbf{n}{\cdot}\boldsymbol{\sigma})\mathbf{R}^{-1}\mathbf{R}\hat{\mathrm{O}}_1$. Here $\mathbf{R}(\mathbb{1}+\mathbf{n}{\cdot}\boldsymbol{\sigma})\mathbf{R}^{-1}$ makes then sense as the transformation of the operator $\mathbb{1}+\mathbf{n}{\cdot}\boldsymbol{\sigma}$ under a rotation R (in the sense given by (2.19)). The concept of this operator is adding the identity element $\mathbb{1}$ to a reflection operator $\mathbf{n}{\cdot}\boldsymbol{\sigma}$ defined by a vector $\mathbf{n}$, which is also coded as $\mathbf{n}{\cdot}\boldsymbol{\sigma}$. Within SU(2), vectors $\mathbf{a}{\cdot}\boldsymbol{\sigma}$ are indeed transformed according to $\mathbf{a}{\cdot}\boldsymbol{\sigma} \to \mathbf{R}\,[\,\mathbf{a}{\cdot}\boldsymbol{\sigma}\,]\,\mathbf{R}^{-1}$. Following this concept to the letter, $\mathbf{R}(\mathbb{1}+\mathbf{n}{\cdot}\boldsymbol{\sigma})\mathbf{R}^{-1}$ will be the value of the operator after a rotation R. One might be tempted to think of interpreting this as proof that adding $\mathbb{1}$ to the coding $\mathbf{n}{\cdot}\boldsymbol{\sigma}$ of a vector transforms like a vector. But the quantity $\mathbb{1}+\mathbf{n}{\cdot}\boldsymbol{\sigma}$ occurs also on the right-hand side of this equation, such that there seems to be a contradiction, as the right-hand side is only transformed by left multiplication with $\mathbf{R}$, instead of being transformed also by a similarity transformation as on the left-hand side.

... and its solution:

The origin of this paradox is that contrary to what might be inferred from the way it is noted, the quantity $\mathbf{n}$ *is not a vector*. It cannot be considered as a vector as it does not transform like a true vector. This is a really surprising fact, which must be explained carefully. It should be noted that $\mathbf{n}$ occurs with two different meanings in (5.21). It is not a true vector when considered as a quantity that defines the axis of a rotation. On the other hand, when $\mathbf{n}$ is considered as an instantaneous quantity used to define an "instantaneous" reflection operator $\mathbf{n}{\cdot}\boldsymbol{\sigma}$, then it is a true vector. It is necessary to introduce two different notations to clearly distinguish the

[11]The coincidence resides in the fact that the general expression of the eigenvector is $\psi_+ = [1+n_z, n_x+\imath n_y]^\top$, which reduces to $[2,0]^\top$ for $\mathbf{n} = \mathbf{e}_z$. This is then blindly normalized to $[1,0]^\top$, creating the illusion of the presence of a true spinor by hiding the true structure $e^{\imath\chi}(\mathbb{1}+\mathbf{n}{\cdot}\boldsymbol{\sigma})$.

true vector and the rotation axis, by noting the reflection operator rather as $\mathbf{s}\cdot\boldsymbol{\sigma}$, in order to analyse the problem with (5.21). That the quantity $\mathbf{n}$ that defines the rotation axis of a rotation R_1 is not a true vector, but a vector-valued function, follows from the fact that for an arbitrary rotation R, the quantity $R(\mathbf{n})$ will not be the rotation axis of RR_1.[12] On the other hand, after an arbitrary rotation R, the new value of a true vector like $\mathbf{e}_z$ will definitely become $R(\mathbf{e}_z)$.

The construction of the left-hand side of the equation starts from the rotation axis $\mathbf{n}$ of the rotation $\hat{\text{O}}_1$. Howere, $\mathbf{n}$ is not a true vector; it corresponds to a rotation axis and as such does not transform as a vector. But an instantaneous real vector $\mathbf{s}$ can be associated with it, which could be transformed like a vector. The real vector $\mathbf{s}$ instantaneously takes the same value as $\mathbf{n}$. Perhaps a good analogy would be to consider $\mathbf{n} \in F(G, \mathbb{R}^3)$ as a function defined on the group G of the rotations, while $\mathbf{s}$ would be the instantaneous value $\mathbf{s} = \mathbf{n}(R_1) \in \mathbb{R}^3$ that this function takes at the point $R_1 \in G$. After an arbitrary rotation R, we have that $R(\mathbf{s}) \neq \mathbf{n}(RR_1)$. The quantities $\mathbf{s}$ and $\mathbf{n}$ are thus different quantities, but they momentarily take the same value. This existence of an instantaneous equality could be called a local illusion. The reflection operator $\mathbf{s}\cdot\boldsymbol{\sigma}$ defined by the true vector $\mathbf{s}$ transforms under a rotation R as $\mathbf{s}\cdot\boldsymbol{\sigma} \to \mathbf{R}\,[\,\mathbf{s}\cdot\boldsymbol{\sigma}\,]\,\mathbf{R}^{-1}$. From the way the quantities have been defined on the left-hand side of (5.21) it is thus obvious that the quantity that intervenes here is $\mathbb{1} + \mathbf{s}\cdot\boldsymbol{\sigma}$, which truly transforms as $\mathbb{1} + \mathbf{s}\cdot\boldsymbol{\sigma} \to \mathbf{R}(\mathbb{1} + \mathbf{s}\cdot\boldsymbol{\sigma})\mathbf{R}^{-1}$. In fact, the whole discussion started from the question of what the meaning of an eigenvector of the reflection operator $\mathbf{s}$ could be, and then the eigenvector was constructed from $\hat{\text{O}}_1$ by operating $\mathbb{1} + \mathbf{s}\cdot\boldsymbol{\sigma}$ on $\hat{\text{O}}_1$.

The right-hand side of the equation is, however, derived from the calculation of the eigenvector $\hat{\text{O}}_1 + \hat{\text{O}}_2$ constructed this way from $\hat{\text{O}}_1$. In the

---

[12]This can be easily proved in two steps:

(1) Prove that the rotation axis $\mathbf{n}$ of an arbitrary rotation R is not $\mathbf{e}'_z = R(\mathbf{e}_z)$, unless $\mathbf{n} = \mathbf{e}_z$. When $\mathbf{n} \neq \mathbf{e}_z$, there will be an angle $\chi \neq 0$ between $\mathbf{n}$ and $\mathbf{e}_z$. R is now applied to the identity element. Under the rotation R the vector $\mathbf{e}_z$ will turn around $\mathbf{n}$ to $\mathbf{e}'_z$ preserving the angle $\chi \neq 0$ with $\mathbf{n}$. This implies that $\mathbf{e}'_z \neq \mathbf{n}$.

(2) Consider the case that R_1 is a rotation around the z-axis, whereby the rotation axis $\mathbf{n}'$ of R is not the z-axis. For R_1, $\mathbf{n} = \mathbf{e}'_z = \mathbf{e}_z$. For the rotation $\mathbf{RR}_1$, $R(\mathbf{n}) = R(\mathbf{e}'_z) = R(\mathbf{e}_z) = \mathbf{e}''_z$. Let the rotation axis of $\mathbf{RR}_1$ be noted as $\mathbf{n}''$. As for the rotation $\mathbf{RR}_1$ the value of $\mathbf{e}''_z = \mathbf{RR}_1(\mathbf{e}_z)$ will be different from the value of $\mathbf{n}''$, owing to the first step, it follows that $\mathbf{n}'' \neq R(\mathbf{n})$, which completes the proof. A rotation about $\mathbf{n}$ is illustrated in Figure 5.1.

term $\hat{\mathrm{O}}_1 = [\,(\,\mathbb{1} + \mathbf{n}{\cdot}\boldsymbol{\sigma}\,)\,e^{-\imath\varphi/2\tau} + (\,\mathbb{1} - \mathbf{n}{\cdot}\boldsymbol{\sigma}\,)\,e^{\imath\varphi/2}\,]/2$, the quantity $\mathbf{n}$ has the meaning of a rotation axis. In the term $\hat{\mathrm{O}}_2 = [\,\mathbf{s}{\cdot}\boldsymbol{\sigma}\,]\,\hat{\mathrm{O}}_1$, however, $\mathbf{s}$ is a vector that defines a reflection operator in the part $\mathbf{s}{\cdot}\boldsymbol{\sigma}$, while $\mathbf{n}$ defines again a rotation axis within the explicit expression for the part $\hat{\mathrm{O}}_1$. The part $[\,\mathbf{s}{\cdot}\boldsymbol{\sigma}\,]\,[\,\mathbf{n}{\cdot}\boldsymbol{\sigma}\,]$ in the detailed further calculation of the right-hand side yields $\mathbb{1}$, while the part $[\,\mathbf{s}{\cdot}\boldsymbol{\sigma}\,]\,\mathbb{1}$ yields $\mathbf{s}{\cdot}\boldsymbol{\sigma}$. But the equality of $[\,\mathbf{s}{\cdot}\boldsymbol{\sigma}\,]\,[\,\mathbf{n}{\cdot}\boldsymbol{\sigma}\,] = \mathbb{1}$ will not be conserved under a similarity transformation, clearly revealing that the local illusion is just an illusion. The quantity $\mathbf{s}{\cdot}\boldsymbol{\sigma}$ corresponds to a true vector and will thus transform according to a similarity transformation under a rotation, but the quantity $\mathbf{n}{\cdot}\boldsymbol{\sigma}$ does not code a true vector. The transformation of $[\,\mathbf{s}{\cdot}\boldsymbol{\sigma}\,]\,[\,\mathbf{n}{\cdot}\boldsymbol{\sigma}\,] + [\,\mathbf{s}{\cdot}\boldsymbol{\sigma}\,]$ will then become: $\mathbf{R}\,[\,\mathbf{s}{\cdot}\boldsymbol{\sigma}\,]\mathbf{R}^{-1}\,\mathbf{R}\,[\,\mathbf{n}{\cdot}\boldsymbol{\sigma}\,] + \mathbf{R}[\,\mathbf{s}{\cdot}\boldsymbol{\sigma}\,] = [\,\mathbf{s}'{\cdot}\boldsymbol{\sigma}\,]\,\mathbf{R}\,[\,\mathbf{n}{\cdot}\boldsymbol{\sigma}\,] + \mathbf{R}\,[\,\mathbf{s}{\cdot}\boldsymbol{\sigma}\,]$. The right-hand side should thus in principle not transform as a vector. But even if it did, then it should contain a multiplication by $\mathbf{R}^{-1}$ to the right, which would have to be done on both sides. This would leave $\mathbf{R}\hat{\mathrm{O}}_1\mathbf{R}^{-1}$ on the left-hand side, which would not agree with the transformation properties of $\hat{\mathrm{O}}_1$ which must transform according to $\hat{\mathrm{O}}_1 \to \mathbf{R}\hat{\mathrm{O}}_1$.

This way it can be seen that under rotation the original meaning of the equation gets completely lost, because the definition is based on a local illusion. The meaning of the equation is thus not rotationally invariant. In fact, if an arbitrary rotation R is applied to two rotations $R_1(\mathbf{n}, \varphi_1)$ and $R_2(\mathbf{n}, \varphi_2)$ which have the same rotation axis but different rotation angles, the corresponding products $RR_1(\mathbf{n}, \varphi_1)$ and $RR_2(\mathbf{n}, \varphi_2)$ will no longer have the same rotation axis; the rotation axis does not transform as a vector. Therefore, the set of all rotations that share the same rotation axis $\mathbf{n}$ and is defined by the spin breaks apart under a subsequent rotation R if R has a rotation axis that is different from $\mathbf{n}$. The spin defined this way is thus an ephemeral phenomenon.[13] The definition and the interpretation of the spin in terms of the set $\{\hat{\mathrm{O}}_1, \hat{\mathrm{O}}_2\}$ by means of (5.21) are not rotationally invariant. On the left-hand side of (5.21), the rotated vector $\mathbf{s}' = R(\mathbf{s})$ calculated from $\mathbf{s}'{\cdot}\boldsymbol{\sigma} = \mathbf{R}\,[\,\mathbf{s}{\cdot}\boldsymbol{\sigma}\,]\,\mathbf{R}^{-1}$ will not correspond to the rotation axis $\mathbf{n}'$ of $\mathbf{R}\hat{\mathrm{O}}_1$. Let us call $\hat{\mathrm{O}}_1' = \mathbf{R}\hat{\mathrm{O}}_1$. Then the set $\{\hat{\mathrm{O}}_1', \hat{\mathrm{O}}_2'\}$ (where

[13] It is easy to check with the Rodrigues formula that starting from rotations $\cos\frac{\varphi_1}{2}\,\mathbb{1} - \imath\sin\frac{\varphi_1}{2}\,[\,\mathbf{n}_1{\cdot}\boldsymbol{\sigma}\,]$ that share the same rotation axis $\mathbf{n}_1$ but have different rotation angles φ_1, the products $(\cos\frac{\varphi_2}{2}\,\mathbb{1} - \imath\sin\frac{\varphi_2}{2}\,[\,\mathbf{n}_2{\cdot}\boldsymbol{\sigma}\,])\,(\cos\frac{\varphi_1}{2}\,\mathbb{1} - \imath\sin\frac{\varphi_1}{2}\,[\,\mathbf{n}_1{\cdot}\boldsymbol{\sigma}\,])$ where $\mathbf{n}_2$ and φ_2 are fixed, will no longer share the same rotation axis if $\mathbf{n}_1 \neq \mathbf{n}_2$. The geometrical counterpart of this algebraic argument has already been used in a special case, *viz.* in the second part of the proof in Footnote 12.

$\hat{O}'_2 = [\,\mathbf{n}'{\cdot}\boldsymbol{\sigma}\,]\,\hat{O}'_1$), that will define the new spin will not be $\{\mathbf{R}\hat{O}_1, \mathbf{R}\hat{O}_2\}$, as in general $\hat{O}'_2 \neq \mathbf{R}\hat{O}_2$.

5.4.2.2 *Intermezzo: What now?*

Eventually, the fact that the definition of the spin based on the "fake" vector $\mathbf{n}$ is not rotationally invariant and based on a local illusion will necessitate a reconsideration of the definition of spin and the true meaning of the Dirac equation. It will be shown why that conclusion is compelling.

Dirac just guessed his equation. The first time one comes across it, it really looks arcane, and one does not know what it means. Discovering the true meaning of a spinor has made it possible to obtain a nice visual interpretation of the Dirac equation. The idea is very natural as it is just based on the true mathematical meaning of the spinor quantities. There is no need to invent this meaning using a very ingenious mental construction. It is already there, just by the fact that the mathematical language that we use describes group elements. We have therefore become convinced that this must be the way to give meaning to this equation. If this does not work, then it will be very hard to find another way to give meaning to it. We have also discovered a very appealing interpretation of the difficult concept of spin.

But for the time being, all these kindergarten dreams have been smashed to smithereens. As announced in Subsection 5.1.1, two problems remain. The relationship between the Dirac equation and the Rodrigues formula does not seem to be a one-to-one correspondence. This raises the fear that the Dirac equation may not keep track of the complete information about the spin, and then the dynamics derived from it might be wrong. It has also been discovered that it was naive to think that the physical spin axis of an electron would be just the mathematical rotation axis $\mathbf{n}{\cdot}\boldsymbol{\sigma}$, because $\mathbf{n}$ does not transform like a vector. An attempt will now be made to try to repair this situation.

5.4.2.3 *Concepts of covariance*

- *Mathematical formalism.* During the preceding developments it has become clear that the idea to base the definition of spin on $\mathbf{n}$ is not viable, because such a definition is not covariant. It is not even covariant in a restriction to the rotation group. Checking the covariance under transformations of a group is a simple mathematical procedure. The discussion of this procedure can be enriched by adding more physically intuitive arguments to it. The wrong definition of spin based on $\mathbf{n}{\cdot}\boldsymbol{\sigma}$ will be referred to as the "spinning-frame model". The new covariant definition of spin will be based on $\mathbf{e}'_z\cdot\boldsymbol{\sigma}$, and corresponds to the description of what will be referred to as the "spinning-top model".

Let us try to render covariant within SU(2) an equation that is of the type:

$$\frac{d}{dc\tau}\psi = [\,\mathbf{w}\cdot\boldsymbol{\sigma}\,]\,\psi, \tag{5.22}$$

where $\mathbf{w}$ is some vector, e.g. $\mathbf{w} = -\imath\frac{m_0 c}{\hbar}\mathbf{s}$. The point is that this equation is equivalent to:

$$\frac{d}{dc\tau}\mathbf{R}\psi = \mathbf{R}\,[\,\mathbf{w}\cdot\boldsymbol{\sigma}\,]\,\mathbf{R}^{-1}\mathbf{R}\psi, \tag{5.23}$$

where the 2×2 matrix $\mathbf{R}$ is an arbitrary fixed element of SU(2). The equation is obtained by multiplying both sides of (5.22) to the left with $\mathbf{R}$ and inserting $\mathbf{R}^{-1}\mathbf{R} = \mathbb{1}$. As in SU(2), true vectors $\mathbf{w}$ transform according to $\mathbf{w}\cdot\boldsymbol{\sigma} \rightarrow \mathbf{w}'\cdot\boldsymbol{\sigma} = \mathbf{R}\,[\,\mathbf{w}\cdot\boldsymbol{\sigma}\,]\,\mathbf{R}^{-1}$, and spinors transform according to: $\psi \rightarrow \psi' = \mathbf{R}\psi$, the equation can be rewritten as:

$$\frac{d}{dc\tau}\psi' = [\,\mathbf{w}'\cdot\boldsymbol{\sigma}\,]\,\psi', \tag{5.24}$$

which proves the covariance of the equation. In this equation $\mathbf{w}$ can be replaced by $-\imath\frac{m_0 c}{\hbar}\mathbf{e}'_z$, because $\mathbf{e}_z$ is a true vector. It cannot be replaced by $-\imath\frac{m_0 c}{\hbar}\mathbf{n}$, because $\mathbf{n}$ is not a true vector. The fact that $\mathbf{w}$ can be replaced by $-\imath\frac{m_0 c}{\hbar}\mathbf{e}'_z$ is important. This covariance of $-\imath\frac{m_0 c}{\hbar}\mathbf{e}'_z$ will be used together with the coincidence that $\mathbf{e}'_z = \mathbf{n}$ for a rotation around the z-axis to derive an alternative, covariant Dirac-like equation from the Rodrigues formula for a rotation around the z-axis.

• *Frame-independent definitions.* This principle of covariance has a mathematical counterpart, *viz.* that the definition of a meaningful mathematical quantity should not depend on the choice of a particular reference frame or of a set of coordinates. It is this requirement that distinguishes $\mathbf{n}$ and $\mathbf{e}'_z$. Given any triad, it is possible to tell what the value of $\mathbf{e}'_z$ is without knowing what the initial triad was that defined the identity element. The decision will not depend on the reference frame. But it is not possible to tell what the value of $\mathbf{n}$ is without knowing what the initial triad that defines the identity element was. The definition of $\mathbf{n}$ is therefore frame-dependent. For this reason, $\mathbf{n}$ is not a meaningful mathematical quantity. In more physical language, postulating that $\mathbf{n}$ has some value in some situation would imply that an absolute reference frame had been selected. There is thus a very logical mathematical requirement that is equivalent with the absence of an absolute reference frame in physics. The value of $\mathbf{e}'_z$ can be defined without any knowledge about the original orientation of the z-axis. There is thus no need to keep track of the original orientation of the reference frame. Completely the opposite is true for the value of $\mathbf{n}$, whose definition is frame-dependent, and depends completely on the original orientation of the reference frame. Due to this memory effect, the simple steady-state situation whereby the triad spins around a fixed axis $\mathbf{a} \neq \mathbf{e}'_z$, will be described as a spinning motion around a varying rotation axis if, following the mathematical tradition, one specifies the rotation axis by the value of $\mathbf{n}$, whose definition depends on the choice of a reference frame.

• *Physically meaningful definition of a spin axis.* There is a difference between the physical and mathematical notions of a rotation axis. The mathematical rotation axis corresponds to the value of $\mathbf{n}$ for one rotation of SU(2) in a given reference frame. The physical rotation axis does not correspond to a single group element; it corresponds to *a rotational or spinning motion.* This physical rotation axis can be defined by instantaneous inspection of the motion. There is no need to know the previous history. It is not necessary to know which frame has historically been used to define the identity element, as is the case for the mathematical rotation axis for a single group element. Such a single group element is not a set containing several spinors that can be used to describe rotational motion. Therefore, it can at the very best only be considered as a snapshot of a spinning motion.

• *Principle of relativity and groups.* The fact that vectors transform according to a similarity transformation is an example of Einstein's principle of relativity, as discussed in Section 2.9. A nice example of this is the C_{60} molecule as a model for the icosahedral group (illustrated in Figure 2.4). Each carbon molecule can be identified with a group element. It suffices to identify one arbitrarily selected carbon atom $C_{\mathbb{1}}$ with the identity element $\mathbb{1}$ and then identify any other carbon atom C_g with the group element g that turns $C_{\mathbb{1}}$ into the position of C_g. Each carbon atom has three first neighbours, one first neighbour at a shorter distance along a double bonding, and two first neighbours at a longer distance along a single bonding. For every carbon atom its local environment looks the same. It is also for this reason that any carbon atom can be selected to be identified with the identity element. To know which neighbour of C_g will be at the other end of the double bonding, it suffices to calculate $g^\circ h^\circ g^{-1}$, where h is the group element that corresponds to the neighbour along the double bonding of the carbon atom that corresponds to the identity element. The group element $g^\circ h^\circ g^{-1}$ corresponds then to the neighbour along the double bonding of C_g. All the local environments are this way equivalent by similarity transformations. This impossibility of establishing location on the group is also evidenced by the fact that the buckyball has the same radius of curvature all over. This principle is also valid for the group of homogeneous Lorentz transformations and for the Poincaré group, where the fact that it is impossible to locate a group element on the basis of information about its environment corresponds to Einstein's principle of relativity. In an infinitesimal environment, the analogue of the the bonds OA, OB and OC for the buckyball in Figure 2.4 can be written as Δh. By choosing an appropriate parameter ϵ one can then calculate $dh/d\epsilon$, which defines a vector of tangent space. These tangent vectors will be discussed in Subsection 5.10.1.2.[14]

[14]It will be possible to appreciate the present footnote fully in Section 5.10 in the discussion on the Lie algebra. To compare local environments, it is necessary to use similarity transformations. The infinitesimal bonds Δh will then transform as $\Delta h \to g^\circ \Delta h^\circ g^{-1}$, such that the tangent vectors will transform according to $dh/d\epsilon \to g^\circ (dh/d\epsilon)^\circ g^{-1}$. In SU(2), this corresponds to exactly the same transformation law as for the vectors of $\mathbb{R}^3$. In this sense the tangent vectors transform quadratically. But the tangent vectors of SU(2) cannot be identified with the vectors of $\mathbb{R}^3$. This becomes very clear for the homogeneous Lorentz group where the tangent space is six-dimensional, while $\mathbb{R}^4$ is four-dimensional. However, the vectors of $\mathbb{R}^3$ or $\mathbb{R}^4$ could be incorporated into the tangent space to the group manifold by considering in both cases the inhomogeneous group that contains also the translations. In a homogeneous group, the transformations are of the type $\mathbf{v} \to \mathbf{M}\mathbf{v} + \mathbf{v}_0$. The matrix calculations then no longer comply with the definition of a representation of a group as given in (2.31), as this definition only allows for transformations of the type $\mathbf{v} \to \mathbf{M}\mathbf{v}$. This problem is analogous to the one of rotations and translations in the plane $\mathbb{R}^2$. By introducing homogeneous coordinates (x, y, z), the inhomogeneous equation $Ax + By + C = 0$ is transformed into a homogeneous equation $Ax + By + Cz = 0$. The projective plane with its points characterized by the coordinates (x, y, z) can be geometrically interpreted as the set of directions of $\mathbb{R}^3$. This is projective geometry (in the projective space $\mathbb{R}P^2$) and it permits the recovery of homogeneous transformation laws. The same thing can be done for the Poincaré group by introducing the projective space $\mathbb{R}P^4$. The vectors of $\mathbb{R}^3$ or $\mathbb{R}^4$ will then also transform quadratically under the transformations of homogeneous groups. Of course, the group elements of such homogeneous groups transform like $h \to g^\circ h$, i.e. linearly. The group elements correspond to the spinors: as explained in Section 3.3, spinors *are* group elements. It is possible to see then that vectors transform "quadratically", while spinors transform linearly. In a more loose approach, one can define spinors as corresponding only to column matrices rather than a set of column matrices. Such spinors will also transform linearly but they no longer have a nice, geometrical meaning, as will be discussed in Footnote 22. This way, the spinors seem to be in Atiyah's word "the square root of geometry" (see the citation at the beginning of Chapter 3). A more rigorous formulation would be

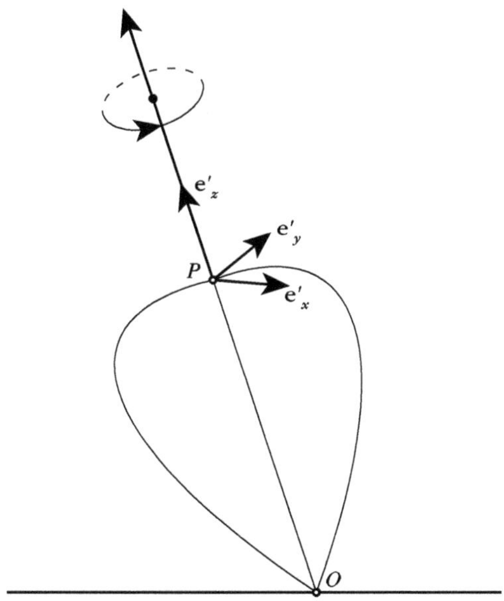

Fig. 5.2 The specificity of the spinning-top model is that the fixed rotation axis is $\mathbf{e}'_z$ rather than $\mathbf{n}$. This way, it becomes possible to identify it with something material that transforms like a vector, such as a symmetry axis of the mass distribution. This can be compared with the spinning-frame model which corresponds to Figure 5.1.

5.4.2.4 *The axis of a spinning top*

The spinning-top model is illustrated in Figure 5.2. It is not equivalent to the Rodrigues formula, like the spinning-frame model, but to a Rodrigues-like formula. This Rodrigues-like formula expresses a spinning motion around $\mathbf{e}'_z$ that can be derived from the Rodrigues formula by considering the special case $\mathbf{e}'_z = \mathbf{n} = \mathbf{e}_z$ where it coincides with the Rodrigues formula. The Rodrigues-like formula can afterwards be generalized to cases where $\mathbf{e}'_z \neq \mathbf{e}_z$. It is an equation with spin axis.

In the image of a spinning top, the rotation axis always corresponds to the physical axis of the mass distribution. It could be identified by putting marks at the two points on the surface of the top defining the axis, and using these marks to define the vector $\mathbf{e}'_z$ of the triad. Observers inspecting the top in different frames will agree that the top is spinning around the

---

that vectors transform as rank-2 tensors in a context where spinors transform like rank-1 tensors.

axis $\mathbf{e}'_z$ of the triad (that some observer may have attached to the top ages ago), even if they will attribute different values to the coordinates of $\mathbf{e}'_z$.

Such a steady state whereby the rotation axis of the spinning motion remains fixed is not the most general possible motion. The spin axis may undergo precession around an axis. If this precession axis is fixed, the situation would also correspond to a steady state, but of a different kind than the steady state without precession. This precession axis could then be called a rotation axis of order 2, as compared with the axis of a rotation without precession, where the rotation axis would be of order 1. The precession axis itself may undergo secondary precession around a fixed axis of order 3. And this can go on to any order. This way, a whole hierarchy of steady states of different types corresponding to rotation axes of different orders can be defined. Each of these steady states will have its own Rodrigues-like formula that could then possibly be rendered Lorentz-covariant in the form of a Dirac-like equation. It is thus possible to conceive a whole series of Dirac-like equations of increasing complexity.

The lesson to be taken from this is that it is not possible to think of $\mathbf{e}'_z$ as a rotation axis, because this would overly restrict the types of motion and steady states that must be considered. For the most general motion it will only be possible to define an instantaneous axis. Only when the top is in some steady state will it become possible to define a rotation axis of some order. Instead it is necessary to think of $\mathbf{e}'_z$ as a physical vector that is not defined by some rotation, but by some other criterion that is more physical, such as the symmetry axis of the mass distribution, and that can be used to keep track of the rotational motion.

5.4.3 *Spinning tops: A definition of spin that survives rotations*

5.4.3.1 *An ambiguity due to the choice of the z-axis as the rotation axis*

In the preceding lines the construction of the spin operator has been based on $\mathbf{n}\cdot\boldsymbol{\sigma}$, where $\mathbf{n}$ is the mathematical rotation axis. However, textbooks always treat the case that $\mathbf{n}$ is aligned with the z-axis. There is thus some ambiguity in that the spin operator could also be based on the vector $\mathbf{e}'_z$ of the triad $\mathbf{e}'_x = R(\mathbf{e}_x)$, $\mathbf{e}'_y = R(\mathbf{e}_y)$, $\mathbf{e}'_z = R(\mathbf{e}_z)$ that characterizes a general rotation R, rather than the rotation axis $\mathbf{n}$ of R. However, that possible ambiguity is never discussed. This is not suprising as it is hard to spot. One must indeed have worked extensively with group theory before one starts wondering about the difference between $\mathbf{n}$ and $\mathbf{e}'_z$.

The author noticed that his colleagues who where discussing about spin, were thinking about it as an object that they could truly rotate like $\mathbf{e}'_z$, and how this was different from the concept of spin used in the first approach to the Dirac equation described in the preceding pages where the spin is identified with $\mathbf{n}$. By trying to figure out the difference between the eigenvectors of $\mathbf{n}\cdot\boldsymbol{\sigma}$ and of $\mathbf{e}'_z\cdot\boldsymbol{\sigma}$ it is possible to discover that the definition of spin based on $\mathbf{n}$ is not covariant while the definition based on $\mathbf{e}'_z$ is. At face value, this finding seems to require the withdrawal of the whole set of assumptions from which the Dirac equation was derived earlier, but it will turn out that the gist of the idea can be preserved.[15,16]

To study the case of the definition of spin for the spinning top, it is necessary to start from the situation where the two possible concepts of spin coincide, *viz.* the case where the rotation axis is the z-axis, since then $\mathbf{n} = \mathbf{e}_z$. This special case can be used as the starting point. The new definition can then be generalized by covariance to the case where $\mathbf{n} \neq \mathbf{e}'_z$. Actually, this has already been done in (5.22)–(5.24), where it was pointed out that the equations are covariant for $\mathbf{w} = \mathbf{e}'_z$ but not for $\mathbf{w} = \mathbf{n}'$.

---

[15]There is thus a clear distinction between the pictures of the spinning top (defined by $\mathbf{e}'_z$) and the spinning frame (defined by $\mathbf{n}$). For a given rotation with its well-defined associated values of $\mathbf{n}$ and $\mathbf{e}'_z$, both reflection operators $\mathbf{n}\cdot\boldsymbol{\sigma}$ and $\mathbf{e}'_z\cdot\boldsymbol{\sigma}$ will define sets, but these sets are very different. The sets based on $\mathbf{n}$ contain rotations whose triads have different values of $\mathbf{e}'_z$ as illustrated in Figure 5.1, while the sets based on $\mathbf{e}'_z$ contain rotations whose axes $\mathbf{n}$ are different (see also Footnote 18).

The sets based on $\mathbf{n}$ cannot be transformed as a whole, as no similarity transformation exists that would transform $\mathbf{n}$ as a vector according to $\mathbf{n}\cdot\boldsymbol{\sigma} \to \mathbf{R}\,[\,\mathbf{n}\cdot\boldsymbol{\sigma}\,]\,\mathbf{R}^{-1}$. (For example, the radius of the circle Γ in Figure 5.1 varies with $\mathbf{n}$ such that it is obvious that the sets of end points P of $\mathbf{e}'_z$ cannot be transformed into one another.) They "evaporate" under rotations.

It will be shown that the sets based on $\mathbf{e}'_z$ are robust against rotations and can be rotated as a whole.

[16]As already noted, there is a difference in the behaviour of $\mathbf{n}$ and $\mathbf{e}'_z$ under rotations. It can already be appreciated that there is a difference in behaviour between $\mathbf{e}'_z$ and $\mathbf{n}$ for the identity element, for which $\mathbf{n}$ can be given any value, while $\mathbf{e}'_z$ must be $\mathbf{e}_z$. One can make the difference even more clear by the following consideration:

When a rotation with a value $\mathbf{e}'_z$ is continually transformed to another rotation with value $\mathbf{e}''_z$, the vector $\mathbf{e}'_z$ will describe a continuous path on the unit sphere between $\mathbf{e}'_z$ and $\mathbf{e}''_z$. But $\mathbf{n}$ does not have to describe such a continuous path if one wishes to go continuously from a rotation $R(\mathbf{n}_1, \varphi_1)$ to a rotation $R(\mathbf{n}_2, \varphi_2)$. Let us start from a rotation with parameters $(\mathbf{n}_1, \varphi_1)$. First, continuously reduce the rotation angle φ from φ_1 to 0, keeping $\mathbf{n}_1$ fixed. But as $\varphi = 0$ corresponds to the identity element it follows that $(\mathbf{n}_1, 0) \equiv (\mathbf{n}_2, 0)$. A discontinuous jump can thus be made in the value of the parameter $\mathbf{n}$ from $\mathbf{n}_1$ to $\mathbf{n}_2$, while in reality nothing happens. Finally, φ can be increased continuously from 0 to φ_2, keeping $\mathbf{n}_2$ fixed.

(The reader might on a first reading consider jumping to Subsection 5.4.4.)

5.4.3.2 *Ideas to preserve about negative energies and the mystery of quantum superposition*

After replacing $\mathbf{n}$ by $\mathbf{e}'_z$ in order to repair the spinning-frame model for the fact that it is not physical, some of the nice properties of the spinning-frame model continue to be valid within the spinning-top model that results. They deal with the two objections raised in Section 5.2, *viz.* the unjustified use of the superposition principle and the fact that the negative frequencies do not correspond to antiparticles. The fact that spin corresponds to a set can resolve these problems. This will be demonstrated here only for the spinning-frame model, because the arguments become somewhat masked by a coincidence in the spinning-top model as will be explained later.

The Rodrigues rotation matrix is $\frac{1}{2}\left[\, e^{-\imath\varphi/2}(\mathbb{1}+\mathbf{n}{\cdot}\boldsymbol{\sigma})+e^{+\imath\varphi/2}(\mathbb{1}-\mathbf{n}{\cdot}\boldsymbol{\sigma})\,\right]$. By taking $\hat{\mathrm{O}}_1+\hat{\mathrm{O}}_2$ with $\hat{\mathrm{O}}_2=\mathbf{n}{\cdot}\boldsymbol{\sigma}\,\hat{\mathrm{O}}_1$ this yields just $\psi_+=e^{-\imath\varphi/2}(\mathbb{1}+\mathbf{n}{\cdot}\boldsymbol{\sigma})$, as though $\psi_-=e^{+\imath\varphi/2}(\mathbb{1}-\mathbf{n}{\cdot}\boldsymbol{\sigma})$ would have been dropped. Hence, the spinor is cut into two parts, which should in principle not be done, as spinors are not vectors. Moreover the parts contain only one frequency.

• *Superposition of states.* On first inspection, the set $\{\hat{\mathrm{O}}_1,\hat{\mathrm{O}}_2\}$ occurs in the calculations as a mysterious quantum superposition of two states. It should be obvious, however, that there is no quantum mystery here; we just use the superposition to define a set that corresponds to a vector state. This way, provided the real normalization factors are ignored, a vector state ψ_+ appears as the superposition of two spinor states ψ and $[\,\mathbf{n}{\cdot}\boldsymbol{\sigma}\,]\,\psi$. Simultaneously, a spinor state ψ appears as the superposition of two vector states ψ_+ and ψ_-. These superposition states are not mysterious but simply a way to treat both spinors and vectors within a single framework and to link them to one another. The superposition of states has been discussed here within a purely geometrical framework, such that there is no mystery. The same will be true in the spinning-top model.[17] The superposition of states is an expedient to treat vectors within a formalism that is not originally designed for them. The spin of an electron has to be a vector as it does not correspond to a single spinor. It corresponds to a movie of rotating spinors; a single spinor is only a still of this movie.

---

[17]The duality between the notions that "two vectors define a spinor" and "two spinors define a vector" is related to the duality between the notions that "two straight lines through the origin define a plane through the origin" and "two planes through the origin define a straight line through the origin".

• *Negative energies.* The solutions ψ_+ and ψ_- now contain only one frequency. The same will be true for the eigenvectors of the spinning-top model. Single negative frequencies can then be interpreted in two ways: viz. as a consequence of inverting the sense of the rotation, or of inverting the charge. The generic SU(2) solution consists of course in inverting the sense of the rotation $\varphi \to -\varphi$. But as $\varphi = \omega t$, a formalism for antiparticles based on negative frequencies $\omega \to -\omega$ (or alternatively a picture $t \to -t$ of travelling backwards in time, as described by Feynman) can be added on. All the fuss about the Dirac sea to explain negative energies ($E = \hbar\omega$) within the choice $\omega \to -\omega$ results from interpreting the negative frequencies too literally. The description given here represents an alternative solution for the paradox of the negative frequencies, and is very different from Majorana's solution. It is based on the observation that negative frequencies already exist within the geometry of SU(2), such that they should not be over-interpreted in terms of negative energies. This does not exclude the possibilities that Majorana fermions could exist.

5.4.3.3 *An alternative definition of spin based on the spinning-top model*

The choice of $\mathbf{e}'_z{\cdot}\boldsymbol{\sigma}$ as the spin operator will also be able to explain why the eigenvector of the spin operator contains only one sign of the frequency. The reason for this is the same as evoked previously, *viz.* that the superposition of states simply serves to treat vectors within the framework of the group theory by considering them as sets of spinors. There is also no quantum mystery here in the definition of spin and in the superposition of states. But accidentally, it will be possible here to consider that the superposition of states, needed to define a vector state, behaves as a single state, due to a coincidence.

In defining the set that defines the axis of the spinning frame, the quantity $\mathbb{1} + \mathbf{n}{\cdot}\boldsymbol{\sigma}$ is used to correspond to $\mathbf{n}$. In the isomorphism discussed in Section 3.10, the spinor $[\xi_0, \xi_1]^\top$ related to the isotropic vector $\mathbf{e}'_x + \imath\mathbf{e}'_y$ is also related to the corresponding axis vector $\mathbf{e}'_z$ by a procedure of adding a unit matrix to its coding so as to obtain $\mathbb{1} + \mathbf{e}'_z{\cdot}\boldsymbol{\sigma}$ from $\mathbf{e}'_z{\cdot}\boldsymbol{\sigma}$. It has been shown that $\mathbf{e}'_z$ contains less information than $\mathbf{e}'_x + \imath\mathbf{e}'_y$. It is due to this incompleteness that $\mathbf{e}'_z$ does not correspond to a single spinor but defines a whole set of them, characterizing a state. This stands in marked contrast with the isotropic vector $\mathbf{e}'_x + \imath\mathbf{e}'_y$ which contains the complete information about a single spinor.

This incompleteness can be used to describe spinning tops. In general, $\mathbf{n} \neq \mathbf{e}'_z$. It is only for a rotation around the z-axis that $\mathbf{n} = \mathbf{e}'_z = \mathbf{e}_z$. In the following description, the images $\mathbf{e}'_z$ of the triad vector $\mathbf{e}_z$ under the rotations will be considered as the axis of a solid top.[18] Rather than starting from spinors and figuring out how a vector can be built from them, the discussion for a spinning top starts from a vector and checks how it can be decomposed it into a set of spinors.

As already mentioned, it will appear in the following as though a single element from a ray is a pure spinor state rather than a vector state constructed as a set containing spinors of opposite handedness. The eigenvectors will in fact have a form that conjures up the illusion that they are true spinors. One should, however, not forget that a pure spinor can never be the eigenvector of a reflection operator, as a reflection changes the handedness of a spinor. But we will first start from this illusion and show afterwards that the true result must nevertheless be interpreted as a sum of two spinors.[19]

The problem of the spin of the top can now be addressed with a similarity transformation. For $\mathbf{n} = \mathbf{e}_z$, we have $\mathbf{e}'_z{\cdot}\boldsymbol{\sigma} = \sigma_z$. The eigenvectors of $\mathbf{e}'_z{\cdot}\boldsymbol{\sigma}$ are then $\psi_+ = e^{\imath\chi}[1,0]^\top$ with eigenvalue $+1$, and $\psi_- = e^{\imath\chi}[0,1]^\top$ with eigenvalue -1. An attempt can now be made to interpret the eigenvector as a pure spinor state using (3.10), despite the fact that a reflection operator can never have a pure spinor state as an eigenvector, because a reflection operator changes the handedness of the spinors. As already intimated, the solution will look just like a pure spinor state, and the same will also be true for ψ_-.

From the definition of a spinor in (3.10), it follows that the value of ψ_+ for $\chi = 0$ corresponds to $(x, y, z) = (1, \imath, 0)$. It corresponds thus to the

---

[18]The set of points N on the unit sphere that are the end points of the axes $\mathbf{n}$ of all rotations that map $\mathbf{e}_z$ (with end point Z) onto some fixed value of $\mathbf{e}'_z$ (with end point Z') is the great circle that bisects the segment ZZ' of the great circle through Z and Z'. The set of end points Z' that image $\mathbf{e}'_z$ for all triads that have $\mathbf{n}$ as rotation axis is the small circle with centre N that contains Z. Now a rotation can transform a great circle into another great circle, but it cannot move a small circle into a small circle of a different radius. The great circles, which are sets defined by vectors $\mathbf{e}'_z$, are therefore robust against rotations, while the small circles, which are sets defined by vectors $\mathbf{n}$, are not. This confirms why the definition of spin based on $\mathbf{e}'_z$ is covariant, while the one based on $\mathbf{n}$ is not.

[19]The fact that we can interpret the spinor as a single state is due to the fact that it contains only the first column of the matrix $\mathbb{1} + \mathbf{e}'_z{\cdot}\boldsymbol{\sigma}$. The second column of that matrix is a null-vector (in contrast with the second column of a true rotation matrix).

starting value of the isotropic vector, which is the identity operation. For $\chi \neq 0$, it corresponds to a rotation around the z-axis (over an angle -2χ).

By using again the definition of the spinor in (3.10), one finds that ψ_- corresponds to $(x, y, z) = (-1, \imath, 0)$ for $\chi = 0$. This corresponds (for a right-handed frame) to $\mathbf{e}'_x = -\mathbf{e}_x$, $\mathbf{e}'_y = \mathbf{e}_y$, and $\mathbf{e}'_z = -\mathbf{e}_z$, i.e. to a rotation over an angle π around the y-axis. For $\chi \neq 0$, this corresponds to a subsequent rotation (over an angle $+2\chi$) around the z-axis, as can be checked by applying this rotation to the spinor. The rotation over an angle π around the y-axis, is coded by the matrix $-\imath\sigma_y$. A rotation over an angle 2χ around the z-axis is coded by a diagonal matrix, with $e^{-\imath\chi}$ and $e^{+\imath\chi}$ on the diagonal. After calculation of the product and identifying it with the general Rodrigues formula, it is possible to see then that this product codes a rotation over an angle π around the axis $(-\sin\chi, \cos\chi, 0)$. The mathematical rotation axes $\mathbf{n}$ form thus a fan that contains all the unit vectors of the Oxy plane (as already anticipated geometrically in Footnote 18). In other words, the various elements of the unit ray $\{\psi_\chi : \psi_\chi = e^{\imath\chi}\psi\}$ do not even share the same mathematical rotation axis $\mathbf{n}$. This is a heteroclitic set of rotations, which only have in common that $\mathbf{e}'_z$ ends up in $-\mathbf{e}_z$.

It is much more simple to interpret these two sets of rotations as a kind of representation of the vector $\mathbf{e}_z$ or the vectors $\pm\mathbf{e}_z$. This should not come as a surprise, as it has been derived within a logic of vector calculus rather than within a logic of spinors. Spinors cannot be eigenvectors of reflection operators. When χ is varying with time, one can only make sense of the set by considereing it as describing a frame rotating around $\mathbf{e}_z$, whereby the previous history (concerning the change of the tilt of the rotation axis (e.g. from $\mathbf{e}_z$ to $-\mathbf{e}_z$)) no longer matters.

What has been described here for $\mathbf{e}_z$ will be valid for any other axis. The vector $\mathbf{e}_z$ is coded by σ_z. After a rotation R it will be transformed to $\mathbf{e}'_z$ coded by $\mathbf{e}'_z{\cdot}\boldsymbol{\sigma} = \mathbf{R}\sigma_z\mathbf{R}^{-1}$. The new eigenvectors will be $\psi'_+ = \mathbf{R}\psi_+$ and $\psi'_- = \mathbf{R}\psi_-$, as is easily checked. Hence, everything obtained for the spin operator $\mathbf{e}_z{\cdot}\boldsymbol{\sigma}$, will be valid for the spin operator $\mathbf{e}'_z{\cdot}\boldsymbol{\sigma}$ by similarity transformation. In such situations, it may even be more difficult to make sense of the sets of rotations in terms of $\mathbf{n}$. Only the fact that all these rotations share $\mathbf{e}'_z$ will characterize them.

5.4.3.4 *Comparison of the spinning-top and the spinning-frame models*

There are a number of features that the models for the spinning top and the spinning frame have in common. The eigenvalue equation for the spin operator does not define a single wave function or spinor, but a whole infinite set of them. This set defines a vector state rather than a spinor state. The spin operator is thus an expedient to treat vector states in a group theory that in principle would not allow for their existence; it describes a vector state as an infinite set of spinor states. Within SU(2), the equation $[\,\mathbf{n}\cdot\boldsymbol{\sigma}\,]\,\psi = \lambda\psi$ contains three unknown real parameters (which define the complete spinor ψ) and only two independent known parameters (contained in $\mathbf{n}\cdot\boldsymbol{\sigma}$). This clearly shows that it defines an infinite one-parameter set of spinors, that can be identified with a vector state corresponding to $\mathbf{n}$. The equation $[\,\mathbf{n}\cdot\boldsymbol{\sigma}\,]\,\psi = \lambda\psi$ defines thus a set of spinors that we call a vector, just as the equation $x^2 + y^2 = r^2$ in analytic geometry defines a set of points that we call a circle, but the routine of calculating eigenvectors and eigenvalues in linear algebra, where no attention needs to be paid to phase factors, is a serious barrier to realizing that it defines a set. The same argument can be applied to the eigenvector equation $[\,\mathbf{e}'_z\cdot\boldsymbol{\sigma}\,]\,\psi = \lambda\psi$.

If the axis $\mathbf{e}_z$ of a spinning top is tilted by a rotation $R(\mathbf{n}_1, \varphi_1)$ such that it becomes $\mathbf{e}'_z$, then in physics there is not so much interest in knowing the complicated values for the axis $\mathbf{n}$ and rotation angles φ of $R(\mathbf{e}'_z, \omega t)^\circ R(\mathbf{n}_1, \varphi_1)$, where $R(\mathbf{e}'_z, \omega t)$ are the further rotations of the tilted top around its tilted axis $\mathbf{e}'_z$. The only part worthy of attention will be $R(\mathbf{e}'_z, \omega t)$.

The difference between the spinning top and the spinning frame is that it is possible to rotate $\mathbf{e}_z$ as a vector, while under a rotation the rotation axis $\mathbf{n}$ does not rotate like a vector, but "evaporates". In other words, there is a difference between the physical rotation axis of a spinning top, which is given by the common axis $\mathbf{e}'_z$ of the whole set of rotations $R(\mathbf{e}'_z, \omega t)$, and the mathematical rotation axes that correspond to all the absolute instantaneous rotation axes $\mathbf{n}(t) \neq \mathbf{e}'_z$ of the rotations $R(\mathbf{e}'_z, \omega t)^\circ R(\mathbf{n}_1, \varphi_1)$. For a spinning top, the true axes $\mathbf{n}(t)$ are not important. The physical parameters of a spinning top that are needed are $\mathbf{e}'_z$ and its angular velocity ω. For the spinning top, the description in terms of this alternative spin matches these needs perfectly, and the unit ray $\{\psi_\chi : \psi_\chi = e^{\imath\chi}\psi\}$ corresponds exactly to the physical rotation axis. We can identify a spinning top with the unit ray $\{\psi_\omega : \psi_\omega = e^{\imath\omega t}\psi\}$, and this unit ray is exactly the eigenvector of the spin operator $\mathbf{e}'_z\cdot\boldsymbol{\sigma}$ defined by the physical rotation axis $\mathbf{e}'_z$ of the top.

The physics of the spinning top requires thus that $\mathbf{e}'_z$ is used as the spin vector rather than $\mathbf{n}$. Here again the quantity $\psi_1 + \psi_2$ that codes the set $\{\hat{\mathrm{O}}_1, \hat{\mathrm{O}}_2\}$, is a superposition of states, that in principle would not seem to correspond to anything that is easy to mentally visualize or make sense of. The solution to this problem of superposition is the same as earlier stated, but there are differences in the details between the spinning top and the spinning frame. When the sets used are defined with respect to $\mathbf{e}_z$ rather than with respect to $\mathbf{n}$, the sum $\psi_1 + \psi_2$ will become algebraically undistinguishable from a pure state, such that it then no longer will appear to be a superposition, despite the fact that it truly is. Let us take for $\hat{\mathrm{O}}_1$ a rotation around the z-axis. The corresponding spinor ψ_1 will then be $[e^{-\imath\varphi/2}, 0]^\top$. The associated group element $\hat{\mathrm{O}}_2 = [\,\mathbf{e}_z{\cdot}\boldsymbol{\sigma}\,]\,\hat{\mathrm{O}}_1$, will have as corresponding spinor $\psi_2 = [\,\mathbf{e}_z{\cdot}\boldsymbol{\sigma}\,]\,\psi_1 = \psi_1$. The eigenvector ψ_+ of the reflection operator $\mathbf{e}_z{\cdot}\boldsymbol{\sigma}$ will be $\psi_+ = \psi_1 + \psi_2 = 2\psi_1$, and after normalizing, the eigenvector is a true spinor due to the coincidence that $\psi_1 = \psi_2$.[20]

5.4.3.5 *A quantum superposition of states that looks like a spinor*

Let us imagine that it has not yet been decided what the meaning of the spin operator $\mathbf{e}_z{\cdot}\boldsymbol{\sigma}$ is, such that it can still be interpreted in two ways: As the value of the rotation axis $\mathbf{n}$ of the rotation $\hat{\mathrm{O}}_1$ or as the value of the vector $\mathbf{e}'_z$ of the triad that defines the rotation $\hat{\mathrm{O}}_1$.

But it is now possible to ask if such an eigenvector equation that leads to a pure spinor state exists for all orientations of the rotation axis. For $\hat{\mathrm{O}}_1$, the eigenvector ψ_+ will be $\psi_1 + \psi_2$ with $\psi_2 = [\,\mathbf{w}{\cdot}\boldsymbol{\sigma}\,]\,\psi_1$. Let us now operate on $\hat{\mathrm{O}}_1$ with an arbitrary rotation R such as to obtain a general rotation. This will change the value of $\mathbf{w}$ to $\mathbf{w}'$ and ψ_1 to $\psi'_1 = \mathbf{R}\psi_1$. To obtain a pure state, it is necessary that $\psi'_2 = [\,\mathbf{w}'{\cdot}\boldsymbol{\sigma}\,]\,\psi'_1$ obtained with $\mathbf{w}'{\cdot}\boldsymbol{\sigma}$ from ψ'_1 is again proportional to ψ'_1. By operating with $\mathbf{R}$ on both sides of $\psi_2 = [\,\mathbf{w}{\cdot}\boldsymbol{\sigma}\,]\,\psi_1$, we obtain $\mathbf{R}\psi_2 = \mathbf{R}[\,\mathbf{w}{\cdot}\boldsymbol{\sigma}\,]\,\psi_1$. But this can be rewritten as: $\mathbf{R}\psi_2 = \mathbf{R}\,[\,\mathbf{w}{\cdot}\boldsymbol{\sigma}\,]\,\mathbf{R}^{-1}\mathbf{R}\psi_1 = \mathbf{R}\,[\,\mathbf{w}{\cdot}\boldsymbol{\sigma}\,]\,\mathbf{R}^{-1}\psi'_1$. Now, if $\mathbf{w} = \mathbf{e}_z$ then $\mathbf{R}[\,\mathbf{w}{\cdot}\boldsymbol{\sigma}\,]\,\mathbf{R}^{-1} = \mathbf{w}'{\cdot}\boldsymbol{\sigma}$, as $\mathbf{e}_z$ is a vector, while if $\mathbf{w} = \mathbf{n}$, $\mathbf{R}\,[\,\mathbf{w}{\cdot}\boldsymbol{\sigma}\,]\,\mathbf{R}^{-1} \neq \mathbf{w}'{\cdot}\boldsymbol{\sigma}$, as $\mathbf{n}$ is not a true vector. Moreover, in the option where $\mathbf{w}$ corresponds to $\mathbf{e}'_z$ (rather than to $\mathbf{n}$), $\mathbf{R}\psi_2$ will be the value of ψ'_2. As $(\exists\chi \in \mathbb{R})(\psi_1 = e^{\imath\chi}\psi_2)$, we will have that $\psi'_1 = e^{\imath\chi}\psi'_2$ for the same value of χ, such that $\psi'_1 + \psi'_2$ will again lead to a pure state after normalization.[21] An analogous reasoning on the conjugate spinors can be made for the spin-down eigenvectors.

---

[20]The necessity to consider things this way follows from the general form of the eigenvector ψ_+ for a general reflection operator $\mathbf{w}{\cdot}\boldsymbol{\sigma}$. Before normalization this eigenvector takes the form $\psi_+ = e^{\imath\chi}[\,1 + w_z, w_x + \imath w_y\,]^\top$. From this it can be seen that the true eigenvector must contain $1 + w_z$. But in the limit $\mathbf{w} \to \mathbf{e}_z$, the quantity $1 + w_z$ can after normalization become confused with w_z or 1.

From this discussion it is possible to see that the definition of the spin operator for a spinning top must be based on the vector $\mathbf{e}_z$ of the triad, that it corresponds to the physical axis of the dynamical rather than the geometrical rotation of the spinning top, and that its eigenvector corresponds to a pure energy state that contains only one sign for $\hbar\omega$, even if a rotation matrix itself in general contains contributions with both ω and $-\omega$. The vector state is also obtained here by describing it as a superposition of spinor states, even if this is masked by the fact that ψ_1 and ψ_2 are proportional. This superposition is just a means to describe vector states. The eigenvector is thus also here not a mysterious quantum superposition of two states that would be difficult to understand on the basis of common sense.

The solution proposed here for the paradox of the quantum superposition of states is not claimed to be general. In the discussion of the double-slit experiment it will be shown that the superposition principle and the mysterious quantum superposition states can be avoided by using a completely different strategy.

5.4.4 *Generalization of the spin concept to space-time*

5.4.4.1 *From rotational covariance to Lorentz covariance*

The previous lengthy discussions show that the definition of spin is a very subtle problem and a Gordian knot where a whole bunch of difficulties come together.

(1) The problem of the "negative energies".
(2) The problem of the meaning of the superposition principle.
(3) The ambiguity between $\mathbf{n}$ and $\mathbf{e}'_z$.
(4) The absence of a parameter specifying the spin axis in the Dirac equation, while such a parameter is present in the Rodrigues equation.
(5) The difference between spinors and vectors.
(6) The difference between physical and mathematical rotations.
(7) The need to invoke Einstein's principle of relativity to choose between the models of the spinning top and the spinning frame, etc. ...

[21]This proportionality between $\psi'_1 = [\,\xi_0, \xi_1\,]^\top$ and $\psi'_2 = [\,\mathbf{e}'_z{\cdot}\boldsymbol{\sigma}\,]\,\psi'_1$ can also be checked by using the expression for $\mathbb{1} + \mathbf{e}'_z{\cdot}\boldsymbol{\sigma}$ given by (3.28) to calculate $[\,\mathbb{1} + \mathbf{e}'_z{\cdot}\boldsymbol{\sigma}\,]\,[\,\xi_0, \xi_1\,]^\top$ using $\xi_0\xi_0^* + \xi_1\xi_1^* = 1$.

Let us give an outline of the path that must be followed to render the spinning-top model Lorentz covariant. The intention is to describe a spinning top whose physical rotation axis is given by $\mathbf{e}'_z$, where the triad of the top is defined by the spinor $\psi = [\xi_0, \xi_1]^\top$. The spinor $[\xi_0, \xi_1]^\top$ thus describes an arbitrary triad of the set a spinning top runs through if its spin axis is aligned with $\mathbf{e}'_z$. The vector $\mathbf{e}'_z$ is then, according to (3.19), coded by:

$$\boxed{\mathbf{e}'_z \cdot \boldsymbol{\sigma} = \begin{pmatrix} \xi_0\xi_0^* - \xi_1\xi_1^* & 2\xi_0\xi_1^* \\ 2\xi_0^*\xi_1 & \xi_1\xi_1^* - \xi_0\xi_0^* \end{pmatrix} = 2\psi \otimes \psi^\dagger - \mathbb{1}.} \tag{5.25}$$

It is easy to check that $[\,\mathbf{e}'_z \cdot \boldsymbol{\sigma}\,]\,\psi = \psi$. In fact, $(2\psi \otimes \psi^\dagger - \mathbb{1})\psi = \psi$. For arbitrary 2×1 matrices A, B, C, one can change the order of the calculations $(A \otimes B^\dagger)C = A(B^\dagger C)$, such that $(\psi \otimes \psi^\dagger)\psi = \psi(\psi^\dagger\psi)$. Using $\psi^\dagger\psi = 1$ leads to the desired result.

The result $[\,\mathbf{e}'_z \cdot \boldsymbol{\sigma}\,]\,\psi = \psi$ has been derived already (towards the end of Subsection 5.4.3.3) by considering a rotation R that maps $\mathbf{e}_z$ onto $\mathbf{e}'_z$. Let the spinor for a rotation around $\mathbf{e}_z$ be called ψ_0. The special case of the equation $[\,\mathbf{e}_z \cdot \boldsymbol{\sigma}\,]\,\psi_0 = \psi_0$ is trivially valid. The general case $[\,\mathbf{e}'_z \cdot \boldsymbol{\sigma}\,]\,\psi = \psi$ follows then from $\mathbf{R}[\,\mathbf{e}_z \cdot \boldsymbol{\sigma}\,]\,\mathbf{R}^{-1}\mathbf{R}\psi_0 = \mathbf{R}\psi_0$. This approach was thus based on rotational covariance. This way $(2\psi \otimes \psi^\dagger - \mathbb{1})\psi = \psi$ can also be derived from $(2\psi_0 \otimes \psi_0^\dagger - \mathbb{1})\psi_0 = \psi_0$ by using rotational covariance. When a rotation matrix $\mathbf{R}$ is applied on the left to ψ_0, a rotation matrix $\mathbf{R}^\dagger$ must also be applied on the right to $\psi_0^\dagger$. This yields $(2\,\mathbf{R}\psi_0 \otimes \psi_0^\dagger\mathbf{R}^\dagger - \mathbb{1})\,\mathbf{R}\psi_0 = \mathbf{R}\psi_0$, as $\mathbf{R}^\dagger = \mathbf{R}^{-1}$. It can thus be considered that $\mathbb{1}$ stands for $\mathbf{R}\mathbb{1}\mathbf{R}^\dagger$ in this calculation. This way, the result for ψ is derived from the result for ψ_0 by rotational covariance. From now on we will use ω_0 for the value of ω in a rest frame, and use ω for its value in a moving frame.

As $\frac{d}{d\tau}\psi_0 = -\imath\frac{\omega_0}{2}\,[\,\mathbf{e}_z \cdot \boldsymbol{\sigma}\,]\;\psi_0$, also $\mathbf{R}\frac{d}{d\tau}\psi_0 = -\imath\frac{\omega_0}{2}\mathbf{R}\,[\,\mathbf{e}_z \cdot \boldsymbol{\sigma}\,]\,\mathbf{R}^{-1}\;\mathbf{R}\psi_0$, as long as the group elements remain restricted to the subgroup of rotations. Hence, the equation $\frac{d}{d\tau}\psi = -\imath\frac{\omega_0}{2}\,[\,\mathbf{e}'_z \cdot \boldsymbol{\sigma}\,]\,\psi$ is also covariant under rotations. As $[\,\mathbf{e}'_z \cdot \boldsymbol{\sigma}\,]\,\psi = \psi$, the equation $\frac{d}{d\tau}\psi = -\imath\frac{\omega_0}{2}\,[\,\mathbf{e}'_z \cdot \boldsymbol{\sigma}\,]\,\psi$ can be simplified to $\frac{d}{d\tau}\psi = -\imath\frac{\omega_0}{2}\psi$, which transforms covariantly under rotations. It even transforms covariantly under all operations of the homogeneous Lorentz group and it has the form of the textbook Dirac equation. At first sight it seems as though the Dirac equation is the perfect generalization to the Lorentz group of the equation for the spinning-top model.

But the simplification $[\,\mathbf{e}'_z \cdot \boldsymbol{\sigma}\,]\,\psi = \psi$ is only covariant under rotations. It is not Lorentz covariant, because the identity $(2\psi \otimes \psi^\dagger - \mathbb{1})\,\psi = \psi$ is not Lorentz covariant. In fact, for a general Lorentz transformation with matrix $\mathbf{L}$ in SL(2,ℂ), $(2\mathbf{L}\psi_0 \otimes \psi_0^\dagger \mathbf{L}^\dagger - \mathbb{1})\mathbf{L}\psi_0 \neq \mathbf{L}\psi_0$, because $\mathbf{L}^\dagger \neq \mathbf{L}^{-1}$, as stressed in Section 4.5. Therefore, the equation cannot be simplified to the Dirac form if it must be possible to transform its information content covariantly. The textbook Dirac equation is itself covariant, but is not the covariant generalization of $\frac{d}{d\tau}\psi = -\imath\frac{\omega_0}{2}\,[\,\mathbf{e}'_z{\cdot}\boldsymbol{\sigma}\,]\,\psi$.

When (5.25) (with $\mathbf{s} = \mathbf{e}'_z$) is used on (5.22):

$$\frac{d}{d\tau}\psi = -\imath\frac{\omega_0}{2}\,[\,\mathbf{s}{\cdot}\boldsymbol{\sigma}\,]\,\psi, \tag{5.26}$$

the equation can be written as

$$\boxed{\frac{d}{d\tau}\psi = -\imath\frac{\omega_0}{2}\,(2\psi \otimes \psi^\dagger - \mathbb{1})\,\psi.} \tag{5.27}$$

The metric of the Lorentz group does not warrant $\psi^\dagger\psi = 1$ within SL(2,ℂ), such that this equation cannot be further simplified. By using the quantity $(2\psi \otimes \psi^\dagger - \mathbb{1})$ to express $\mathbf{e}'_z{\cdot}\boldsymbol{\sigma}$ in the equation it is important that $\mathbf{e}'_z{\cdot}\boldsymbol{\sigma} = (2\psi \otimes \psi^\dagger - \mathbb{1})$ always co-moves with ψ as it should, such that the information about $\mathbf{e}'_z{\cdot}\boldsymbol{\sigma}$ is not lost. This new equation is non-linear; worse, it is not covariant in SL(2,ℂ). But it can be rendered covariant by lifting it to the Dirac representation, because in the Dirac representation vectors transform again by a similarity transformation $\mathbf{e}_z{\cdot}\gamma \rightarrow \underset{\sim}{\mathbf{L}}\,[\,\mathbf{e}_z{\cdot}\gamma\,]\,\underset{\sim}{\mathbf{L}}^{-1}$ rather than by a transformation of the type $\mathbf{e}_z{\cdot}\boldsymbol{\sigma} \rightarrow \mathbf{L}\,[\,\mathbf{e}_z{\cdot}\boldsymbol{\sigma}\,]\,\mathbf{L}^\dagger$ that prevails within SL(2,ℂ). A large effort will be made hereafter to develop a generalization of the equation $\frac{d}{d\tau}\psi = -\imath\frac{\omega_0}{2}\,[\,\mathbf{e}'_z{\cdot}\boldsymbol{\sigma}\,]\,\psi$ that is covariant and allows recovery of the information about the rotation axis at any time. The development will actually result in a generalization that keeps track of the information content of the whole 2×2 spinor of SL(2,ℂ), rather than of the partial information contained in the 2×1 SU(2)-spinor ψ used here.

5.5 Fully deterministic free-space Dirac-like equation: Exact derivation of the Dirac equation

5.5.1 *Possible loss of information about the spin axis in the traditional Dirac equation*

It has been shown that the Dirac equation does not contain an explicit mention of the actual value of the spin axis. It is not immediately obvious if this is physically damaging. It is difficult to obtain a clear insight in this situation. Accepting the standard viewpoint that the rotation of the electron corresponds to a magnetic dipole moment that is parallel to the spin axis, for an electron at rest, this magnetic dipole moment will not interact with, for instance, the electric Coulomb field of the nucleus in the hydrogen problem. But it *will* interact with an external magnetic field. This magnetic field can be postulated to be aligned with the z-axis, in order to keep the treatment of the corresponding Dirac equation close to the textbook treatment. When the electron moves within the Coulomb field of a nucleus, the moving magnetic dipole will give rise to an induced electric dipole, which will interact with the Coulomb field of the nucleus. Non-relativistically this will not change the motion of the centre of mass, but relativistically it will. This can be seen from a reasoning *ex absurdo*. In fact, even if one assumed initially that it would not alter the motion of the centre of mass and only introduced a change in the rotation, just as in the non-relativistic case, the ensuing change of rotational energy would change the relativistic mass. It would then be natural to conclude that the orientation of the electric dipole within the Coulomb field affects the motion of the centre of mass anyway. This could be a tiny loophole in the Dirac equation that does not follow the spin axis.

That the information about the spin axis $\mathbf{n}$ (or $\mathbf{e}'_z$) is not coded explicitly into the equation is very obvious from the fact that the eigenvalues -1 and $+1$ of the free-space equation both have a two-dimensional vector space of eigenvectors wherein any set of basis vectors can be chosen; there is no further information available that would prompt the choice of one basis over another.[22] Now, in a different point of space-time there will be again such a two-dimensional subspace with its possible arbitrary

[22]This can be seen very clearly in the Cartan representation. The argument can be translated to the Dirac representation by a similarity transformation. Within the Cartan representation the matrices of SL(2,$\mathbb{C}$) are used as spinors rather than one-column matrices, because the one-column quantities contain only half of the information, and as such do not have geometrical meaning. This argument was preempted in Section 2.13 by the description of a jump model on an icosahedron. The situation with the Dirac equation is analogous with the situation in this jump model. The eigenvalues of the 4×4 matrix $\sum_\mu \gamma^\mu c p_\mu$ that occurs in the Dirac equation are twofold degenerate. Based on the discussion in Section 2.13 it is preferable to consider two-column eigenvectors rather than one-column eigenvectors, because this allows the preservation of the geometrical meaning of the "eigenvectors" and does not break up the SL(2,$\mathbb{C}$) matrices. (See also Footnote 26 of this chapter.) In the Dirac equation, the information content of a single column should in principle have even less geometrical meaning, because spinors are not vectors. Having established that it is possible to reason on the SL(2,$\mathbb{C}$) matrices rather than on one-column vectors, the following analogy can be drawn: in the jump model presented in Section 2.13 the choice of the basis $\mathbf{e}_x, \mathbf{e}_y, \mathbf{e}_z$ for $\mathbb{R}^3$ is arbitrary. When an

choices. But arbitrary choices in both spinors may not be consistent with the Lorentz transformation that transforms the first spinor into the second spinor. One may hope that the differential equation takes care of this consistently. It is in principle possible to calculate the spin axis from a spinor, but as the simplification made here is not Lorentz covariant, there is a concern that no longer the correct spinor is selected in the second point of space time. In a spinor field, such as for the hydrogen atom, it is possible that the exact spinor values, $\psi(\mathbf{r}_1)$ and $\psi(\mathbf{r}_2)$, in two different points $\mathbf{r}_1$ and $\mathbf{r}_2$, could be incorrectly defined, and that this may lead to values for the spin that are mutually inconsistent. It would therefore be convenient if it were possible to introduce a correct Lorentz-covariant constraint for the choice of the second spinor, by introducing and following explicitly the spin axis $\mathbf{s}$ (or $\mathbf{e}'_z$). Transforming this constraint covariantly would then make sure that the choices are also Lorentz-transformed covariantly.

If the information about the spin axis could be preserved, it might be possible to obtain a derivation that runs mathematically both ways in the sense that the complete solutions ψ that belong to a state could be reconstructed. This complete solution ψ must contain six independent parameters.[23]

As keeping track of the information about the spin axis in a self-consistent way could be necessary to obtain a correct description of the physics, it will be attempted to derive an equation that does so. As a Lorentz transformation contains both a boost and a rotational part, one might thus postulate that it is also important to keep track of the boost

---

eigenvalue is n-fold degenerate, then it has an n-dimensional vector space of eigenvectors, e.g. $\mathbb{C}^n$. Within this vector space any basis can be chosen. This corresponds only to a change of orientation of the reference frame. The analogue for the Dirac equation consists in choosing different 2×2 spinors with a different direction for the spin axis $\mathbf{s}$. The spin axis $\mathbf{s}$ is thus not specified by the equation and any such can be chosen.

[23]It was extremely difficult to decide with complete certainty if the Dirac equation was really wrong in that it would not keep track of the information about the spin axis. In fact, even if one does not specify this information explicitly, it could still implicitly be carried along within the calculations by covariance. In fact, this is the case within the subgroup of rotations due to the identity $\mathbf{e}'_z{\cdot}\boldsymbol{\sigma} = 2\psi \otimes \psi^\dagger - \mathbb{1}$, but not within the homogeneous Lorentz group as has become obvious from the discussion that lead to (5.27). It is this discussion that suggests that the textbook Dirac equation may not be complete, because the simplification $(2\psi \otimes \psi^\dagger - \mathbb{1})\psi = \psi$ is not Lorentz covariant. As already suggested, incompleteness could be a problem. The best way to settle this issue is to keep track of the information explicitly and to compare the solutions of the equation obtained with those of the textbook one.

vector. This would lead to yet another equation than will be derived here. In that equation it would be necessary to replace $\mathbf{s}$ by a six-component complex (tensor) quantity $\mathbf{s}_1 + \imath\mathbf{s}_2$. This would correspond to the classical notion that when all initial conditions are fixed, the whole orbit can be described. But if everything is fixed, so will be the total energy. It is only by not fixing the total energy that it will be possible discover later on that not all values for the total energy are allowed.

5.5.2 *Keeping track of the spin axis*

5.5.2.1 *Preliminaries*

First, the equation will be derived within the rest frame of the electron, then it will be generalized covariantly to an arbitrary frame. Let the 2×2 spinor matrix of SL(2,$\mathbb{C}$) that corresponds to the representation wherein $\mathbf{v}$ corresponds to $\mathbf{V}$ be called Ψ, and the 2×2 spinor that corresponds to the representation wherein $\mathbf{v}$ corresponds to $\mathbf{V}^\star$ be called $\Psi^\star$. It was demonstrated in Subsection 5.4.4.1 that in generalizing the simplification:

$$(2\psi \otimes \psi^\dagger - \mathbb{1})\psi = \psi \tag{5.28}$$

to SL(2,$\mathbb{C}$), the derivation no longer works because for a 2×2 Lorentz transformation matrix $\mathbf{L}$ within SL(2,$\mathbb{C}$), we have in general $\mathbf{L}^\dagger \neq \mathbf{L}^{-1}$. Vectors in SL(2,$\mathbb{C}$) transform according to $\mathbf{V} \to \mathbf{LVL}^\dagger$ rather than according to $\mathbf{V} \to \mathbf{LVL}^{-1}$. The vector $\mathbf{s}{\cdot}\boldsymbol{\sigma} = 2\psi \otimes \psi^\dagger - \mathbb{1}$ transforms within SU(2) according to $\mathbf{V} \to \mathbf{RVR}^{-1}$ only because $\mathbf{R}^\dagger = \mathbf{R}^{-1}$. This seems to suggest that the Dirac equation is only an approximation based on a treatment of the spin as a three-dimensional quantity. When $2\psi \otimes \psi^\dagger - \mathbb{1}$ is considered as an Euclidean vector that is only transformed by rotations according to Galilean invariance, the derivation works. But when it becomes a four-vector following Lorentz invariance, everything breaks down.

With the Dirac representation, however, vectors transform again according to $\underset{\sim}{\mathbf{V}} \to \underset{\sim}{\mathbf{L}}\,\underset{\sim}{\mathbf{V}}\,\underset{\sim}{\mathbf{L}}^{-1}$ (the symbol $\sim$ is used here to flag 4×4 matrices that belong to the Cartan representation, such that they can be distinguished from 2×2 matrices that belong to an SL(2,$\mathbb{C}$) representation). This is confusing as one starts to wonder why the simplification does not then hold sway anyway. It must now be explained why moving from SL(2,$\mathbb{C}$) to the Dirac representation does not change this state of affairs. In fact, it leads to a *different simplification than the one that would lead to the Dirac equation, and thus leads to a different covariant equation.* For a four-vector a_μ of unit length that is coded by $\mathbf{A} = a_{ct}\mathbb{1} + \mathbf{a}\cdot\boldsymbol{\sigma}$ in SL(2,$\mathbb{C}$) the reflection operator:

$$\underset{\sim}{\mathbf{A}} = \begin{pmatrix} & \mathbf{A} \\ \mathbf{A}^\star & \end{pmatrix} \tag{5.29}$$

can be considered. Here, $\mathbf{A}^\star = \mathbf{A}^{-1}$ for a four-vector of unit length. This reflection operator corresponds to the 4×4 Dirac representation of a_μ and satisfies the condition $\underset{\sim}{\mathbf{A}}^2 = \mathbb{1}$. (For spatial reflections one would obtain rather $-\mathbb{1}$. This problem can be avoided by using a different choice for the signature of the metric.) It will operate on another four-vector v_μ according to: $\underset{\sim}{\mathbf{V}} \to \underset{\sim}{\mathbf{A}}\,\underset{\sim}{\mathbf{V}}\,\underset{\sim}{\mathbf{A}}^{-1}$. The effect of a reflection is explicitly:

$$\begin{aligned} \underset{\sim}{\mathbf{V}} = \begin{pmatrix} & \mathbf{V} \\ \mathbf{V}^\star & \end{pmatrix} &\to -\ \underset{\sim}{\mathbf{A}}\,\underset{\sim}{\mathbf{V}}\,\underset{\sim}{\mathbf{A}} \\ &\to -\begin{pmatrix} & \mathbf{A} \\ \mathbf{A}^\star & \end{pmatrix}\begin{pmatrix} & \mathbf{V} \\ \mathbf{V}^\star & \end{pmatrix}\begin{pmatrix} & \mathbf{A} \\ \mathbf{A}^\star & \end{pmatrix}. \end{aligned} \tag{5.30}$$

The result is (see also (4.4)):

$$-\begin{pmatrix} & \mathbf{A}\mathbf{V}^\star\mathbf{A} \\ \mathbf{A}^\star\mathbf{V}\mathbf{A}^\star & \end{pmatrix}. \tag{5.31}$$

This implies that $\mathbf{V} \to -\mathbf{A}\mathbf{V}^\star\mathbf{A}$ and $\mathbf{V}^\star \to -\mathbf{A}^\star\mathbf{V}\mathbf{A}^\star$. This way, reflections cause a jump between the two different SL(2,$\mathbb{C}$) representations. This is due to the fact that SL(2,$\mathbb{C}$) is not able to accommodate for reflections. Treating reflections goes beyond the framework of the SL(2,$\mathbb{C}$) representation. Only even products of reflections, i.e. true Lorentz transformations do not force a change of representation. Those true right-handed Lorentz transformations are composed of an even number of reflections, and are therefore block diagonal within the Dirac representation. Two such reflections will define a Lorentz transformation $\underset{\sim}{\mathbf{L}} = \underset{\sim}{\mathbf{B}}\,\underset{\sim}{\mathbf{A}}$, acting on v_μ according to: $\underset{\sim}{\mathbf{V}} \to \underset{\sim}{\mathbf{B}}\,\underset{\sim}{\mathbf{A}}\,\underset{\sim}{\mathbf{V}}\,\underset{\sim}{\mathbf{A}}^{-1}\underset{\sim}{\mathbf{B}}^{-1} = \underset{\sim}{\mathbf{L}}\,\underset{\sim}{\mathbf{V}}\,\underset{\sim}{\mathbf{L}}^{-1}$. From:

$$\begin{pmatrix} \mathbf{0} & \mathbf{B} \\ \mathbf{B}^\star & \mathbf{0} \end{pmatrix}\begin{pmatrix} \mathbf{0} & \mathbf{A} \\ \mathbf{A}^\star & \mathbf{0} \end{pmatrix}\begin{pmatrix} \mathbf{0} & \mathbf{V} \\ \mathbf{V}^\star & \mathbf{0} \end{pmatrix}\begin{pmatrix} \mathbf{0} & \mathbf{A} \\ \mathbf{A}^\star & \mathbf{0} \end{pmatrix}\begin{pmatrix} \mathbf{0} & \mathbf{B} \\ \mathbf{B}^\star & \mathbf{0} \end{pmatrix} =$$

$$\begin{pmatrix} \mathbf{0} & \mathbf{B}\mathbf{A}^\star\mathbf{V}\mathbf{A}^\star\mathbf{B} \\ \mathbf{B}^\star\mathbf{A}\mathbf{V}^\star\mathbf{A}\mathbf{B}^\star & \mathbf{0} \end{pmatrix}, \tag{5.32}$$

we obtain $\mathbf{L} = \mathbf{B}\mathbf{A}^\star$. From this and using $\mathbf{A}\mathbf{A}^\star = \mathbb{1}$ (which is true by definition), we obtain $\mathbf{A}\mathbf{B}^\star = \mathbf{L}^{-1}$. As the matrices $\mathbf{A}$ and $\mathbf{B}$ are Hermitian, we have $\mathbf{A}^\star\mathbf{B} = \mathbf{L}^\dagger$, and so we obtain:

$$\underset{\sim}{\mathbf{L}} = \begin{pmatrix} \mathbf{L} & \mathbf{0} \\ \mathbf{0} & \mathbf{L}^{-1\dagger} \end{pmatrix} \qquad \& \qquad \underset{\sim}{\mathbf{L}}^{-1} = \begin{pmatrix} \mathbf{L}^{-1} & \mathbf{0} \\ \mathbf{0} & \mathbf{L}^{\dagger} \end{pmatrix}. \tag{5.33}$$

Hence, the four different quantities from (4.10) all play a role. The matrix $\underset{\sim}{\mathbf{L}}$ and its inverse work as follows on a general four-vector in a similarity transformation:

$$\begin{pmatrix} \mathbf{0} & \mathbf{V} \\ \mathbf{V}^\star & \mathbf{0} \end{pmatrix} : \qquad \mathbf{V} \to \mathbf{L}\mathbf{V}\mathbf{L}^\dagger, \qquad \mathbf{V}^\star \to \mathbf{L}^{-1\dagger}\mathbf{V}^\star\mathbf{L}^{-1}. \tag{5.34}$$

From this it can be seen that the Lorentz transformation corresponds to: $\mathbf{V} \to \mathbf{L}\mathbf{V}\mathbf{L}^\dagger$ and $\mathbf{V}^{-1} \to \mathbf{L}^{-1\dagger}\mathbf{V}^{-1}\mathbf{L}^{-1}$. This proves $\mathbf{V} \to \mathbf{L}\mathbf{V}\mathbf{L}^\dagger$ within SL(2,$\mathbb{C}$) as stated, even if $\underset{\sim}{\mathbf{V}} \to \underset{\sim}{\mathbf{L}}\,\underset{\sim}{\mathbf{V}}\,\underset{\sim}{\mathbf{L}}^{-1}$ within the Dirac representation.[24] Now the wave function corresponds to a group element, and the SL(2,$\mathbb{C}$) matrices can be used as spinors containing all the information about the group elements Ψ and $\Psi^{\dagger-1}$, which become this way simply notations for the group elements $\mathbf{L}$ and $\mathbf{L}^{\dagger-1}$. The wave function is thus of the form:

$$\boldsymbol{\Psi} = \begin{pmatrix} \Psi & \\ & \Psi^{-1\dagger} \end{pmatrix}. \tag{5.35}$$

5.5.2.2 *The correct Dirac-like equation*

The wave functions contain the blocks Ψ and $\Psi^{-1\dagger}$ on the diagonal. Here $\Psi^{-1\dagger}$ is the counterpart $\Psi^\star$ of Ψ in the $\star$-representation. As a general Lorentz transformation is obtained from an even number of reflections, the block-diagonal structure is correct. (5.27) becomes then:

$$\begin{aligned} &\begin{pmatrix} & \frac{d}{dc\tau}\mathbb{1} \\ \frac{d}{dc\tau}\mathbb{1} & \end{pmatrix} \begin{pmatrix} \Psi & \\ & \Psi^{-1\dagger} \end{pmatrix} \\ &\qquad = -\imath\frac{m_0 c}{\hbar} \begin{pmatrix} & \mathbf{s}\cdot\boldsymbol{\sigma} \\ \mathbf{s}\cdot\boldsymbol{\sigma} & \end{pmatrix} \begin{pmatrix} \Psi & \\ & \Psi^{-1\dagger} \end{pmatrix}, \end{aligned} \tag{5.36}$$

or:

$$\begin{aligned} &\begin{pmatrix} & \frac{d}{dc\tau}\mathbb{1} \\ \frac{d}{dc\tau}\mathbb{1} & \end{pmatrix} \begin{pmatrix} \Psi & \\ & \Psi^{-1\dagger} \end{pmatrix} \\ &\qquad = -\imath\frac{m_0 c}{\hbar} \begin{pmatrix} & \mathbf{S} \\ -\mathbf{S}^\star & \end{pmatrix} \begin{pmatrix} \Psi & \\ & \Psi^{-1\dagger} \end{pmatrix}. \end{aligned} \tag{5.37}$$

---

[24]For any group element $a \in G$ the $G \to G$ map $h_a : g \to h_a(g) = a \circ g \circ a^{-1}$ is an isomorphism, as $h_a(g_2 \circ g_1) = h_a(g_2) \circ h_a(g_1)$. This is seen at work here and the structure of the equations is preserved due to this isomorphism. This is Einstein's principle of relativity. It expresses the fact that the transformations of physics must build a group.

Here $\mathbf{S} = [\,\mathbf{s}\cdot\boldsymbol{\sigma}\,]$. When Ψ is a rotation, $\Psi^{\dagger-1} = \Psi$, as $\mathbf{R}^{\dagger} = \mathbf{R}^{-1}$. It is due to this fact that there is certainty about the sign in front of the term $\mathbf{s}\cdot\boldsymbol{\sigma}$ that operates on $\Psi^{\dagger-1}$. The fact that $\mathbf{S}^{\star} = -\mathbf{s}\cdot\boldsymbol{\sigma}$ explains then the term $-\mathbf{S}^{\star}$. This way (5.27) is reproduced in the two SL(2,$\mathbb{C}$) representations, and it can be seen that the spin transforms as an axial vector, as it has the same sign in the left-handed representation as in the right-handed representation. For the columns $\psi_j \in \{\psi, \psi_c^{\dagger}\}$ of Ψ we obtain then the simplification $[\,\mathbf{s}\cdot\boldsymbol{\sigma}\,]\psi_j = \pm\psi_j$ that is not covariant, but yields the Dirac equation. The operator $\frac{d}{dc\tau}$ will later become a four vector $(\frac{\partial}{\partial ct}, \boldsymbol{\nabla})$. It is therefore normal to combine it with γ_{ct}. Similarly, $\mathbf{s} = \mathbf{e}'_z$ has been coded as a vector (that will become a four-vector). A truly covariant form of (5.36) could be:

$$\underbrace{\begin{pmatrix} \mathbf{0} & \frac{1}{c}\frac{\partial}{\partial t}\mathbb{1} + \boldsymbol{\nabla}\cdot\boldsymbol{\sigma} \\ \frac{1}{c}\frac{\partial}{\partial t}\mathbb{1} - \boldsymbol{\nabla}\cdot\boldsymbol{\sigma} & \mathbf{0} \end{pmatrix}}_{vector} \underbrace{\begin{pmatrix} \Psi & \mathbf{0} \\ \mathbf{0} & \Psi^{-1\dagger} \end{pmatrix}}_{spinor} = -\imath\frac{m_0 c}{\hbar} \underbrace{\begin{pmatrix} \mathbf{0} & s_{ct}\mathbb{1} + \mathbf{s}\cdot\boldsymbol{\sigma} \\ -(s_{ct}\mathbb{1} - \mathbf{s}\cdot\boldsymbol{\sigma}) & \mathbf{0} \end{pmatrix}}_{axial\ vector} \underbrace{\begin{pmatrix} \Psi & \mathbf{0} \\ \mathbf{0} & \Psi^{-1\dagger} \end{pmatrix}}_{spinor}. \tag{5.38}$$

The symmetry of the various quantities that intervene in this equation has also been indicated in the equation. (5.38) could perhaps be rewritten in a more familiar notation with Einstein summation convention:

$$\boxed{[\,\gamma^{\mu} c\hat{\mathrm{p}}_{\mu}\,]\,\boldsymbol{\Psi} = m_0 c^2\,[\,\gamma_5\gamma_{\mu}s_{\mu}\,]\,\boldsymbol{\Psi},} \tag{5.39}$$

where $\hat{\mathrm{E}} = -\frac{\hbar}{\imath}\frac{\partial}{\partial t}$ and $\hat{\mathbf{p}} = \frac{\hbar}{\imath}\boldsymbol{\nabla}$. It reduces to (5.36) for a rotation around the z'-axis within a frame at rest. We have then $s_t = 0$ and can drop the terms $\boldsymbol{\nabla}\cdot\boldsymbol{\sigma}$. (5.38) "squares" to a Klein-Gordon equation (as will be proved in Subsection 5.5.2.4). But (5.38) does not need "squaring" to decouple Ψ and $\Psi^{\star}$ (as in the traditional Dirac equation) as they are already decoupled.[25]

[25]In the traditional Dirac equation the vector matrix on the right-hand side of (5.36) is replaced by $\mathbb{1}$, such that this results in a set of two coupled equations $[\,\frac{1}{c}\frac{\partial}{\partial t}\mathbb{1} - \boldsymbol{\nabla}\cdot\boldsymbol{\sigma}\,]\,\Psi = -\imath\frac{m_0 c}{\hbar}\Psi^{\star}$ and $[\,\frac{1}{c}\frac{\partial}{\partial t}\mathbb{1} + \boldsymbol{\nabla}\cdot\boldsymbol{\sigma}\,]\,\Psi^{\star} = -\imath\frac{m_0 c}{\hbar}\Psi$. To decouple these two equations it is necessary to operate with $\frac{1}{c}\frac{\partial}{\partial t}\mathbb{1} + \boldsymbol{\nabla}\cdot\boldsymbol{\sigma}$ on $[\,\frac{1}{c}\frac{\partial}{\partial t}\mathbb{1} - \boldsymbol{\nabla}\cdot\boldsymbol{\sigma}\,]\,\Psi = -\imath\frac{m_0 c}{\hbar}\Psi^{\star}$, which leads to a Klein-Gordon equation $\Box\Psi = -\frac{m_0^2 c^2}{\hbar^2}\,\Psi$ in Ψ. Similarly it is possible to obtain a Klein-Gordon equation for $\Psi^{\star}$. It is in this sense that the Dirac equation "squares" to the Klein-Gordon equation, and this is needed to decouple the two equations.

As will be seen below, it is due to the structure "vector × spinor" on both sides that this equation will be covariant and (5.36) is only a special case of (5.38).

5.5.2.3 *Covariance*

Let (5.38) be noted in short-hand as follows: $\underset{\sim}{\mathbf{D}}\,\mathbf{\Psi} = -\imath\frac{m_0 c}{\hbar}\underset{\sim}{\mathbf{S}}\,\mathbf{\Psi}$. To prove that it is covariant it is necessary to show that $\underset{\sim}{\mathbf{L}}\,\underset{\sim}{\mathbf{D}}\,\underset{\sim}{\mathbf{L}}^{-1}\underset{\sim}{\mathbf{L}}\mathbf{\Psi} = -\imath\frac{m_0 c}{\hbar}\underset{\sim}{\mathbf{L}}\,\underset{\sim}{\mathbf{S}}\,\underset{\sim}{\mathbf{L}}^{-1}\underset{\sim}{\mathbf{L}}\mathbf{\Psi}$ obtained from $\underset{\sim}{\mathbf{D}}\,\mathbf{\Psi} = -\imath\frac{m_0 c}{\hbar}\underset{\sim}{\mathbf{S}}\,\mathbf{\Psi}$ by left-multiplication with $\underset{\sim}{\mathbf{L}}$ and inserting $\underset{\sim}{\mathbf{L}}^{-1}\underset{\sim}{\mathbf{L}} = \underset{\sim}{\mathbb{1}}$, reduces to $\underset{\sim}{\mathbf{D}}'\,\mathbf{\Psi}' = -\imath\frac{m_0 c}{\hbar}\underset{\sim}{\mathbf{S}}'\,\mathbf{\Psi}'$. This is indeed the case due to the fact that four-vectors in the Dirac equation transform according to $\underset{\sim}{\mathbf{V}} \to \underset{\sim}{\mathbf{L}}\,\underset{\sim}{\mathbf{V}}\,\underset{\sim}{\mathbf{L}}^{-1}$. The result can be noted as:

$$\begin{aligned}&\begin{pmatrix} & \frac{\partial}{\partial ct'}\mathbb{1} + \mathbf{\nabla}'\boldsymbol{\cdot}\boldsymbol{\sigma} \\ \frac{\partial}{\partial ct'}\mathbb{1} - \mathbf{\nabla}'\boldsymbol{\cdot}\boldsymbol{\sigma} & \end{pmatrix}\begin{pmatrix}\mathbf{L}\Psi & \\ & \mathbf{L}^{\dagger-1}\Psi^{-1\dagger}\end{pmatrix} = \\ -\imath\frac{m_0 c}{\hbar}&\begin{pmatrix} & \mathbf{L}\,[\,\mathbf{s}\boldsymbol{\cdot}\boldsymbol{\sigma}\,]\,\mathbf{L}^{\dagger} \\ -\mathbf{L}^{\dagger-1}\,[\,-\mathbf{s}\boldsymbol{\cdot}\boldsymbol{\sigma}\,]\,\mathbf{L}^{-1} & \end{pmatrix}\begin{pmatrix}\mathbf{L}\Psi & \\ & \mathbf{L}^{\dagger-1}\Psi^{-1\dagger}\end{pmatrix},\end{aligned} \tag{5.40}$$

where accents have been used to indicate that the Dirac operator has also been Lorentz-transformed by transforming it as a four-vector. Ψ can be considered as the original rotation around the z-axis, for which a Lorentz-covariant formulation has been obtained.

5.5.2.4 *Simplifications in SL(2,ℂ)*

This equation is actually fairly useless because $\underset{\sim}{\mathbf{S}}$ and $\underset{\sim}{\mathbf{D}}$ must be transformed by the same Lorentz transformation, and it is not known which value of $\underset{\sim}{\mathbf{S}}$ will correspond to a given value of $\underset{\sim}{\mathbf{D}}$. There is in this respect no liberty of choice for the explicit arbitrary value of $\underset{\sim}{\mathbf{S}}(\mathbf{r}, t)$ that must be associated with $\underset{\sim}{\mathbf{D}}$. Now it is known that the equation is covariant, the quantity $\underset{\sim}{\mathbf{S}}(\mathbf{r}, t)$ must be rendered implicit again by expressing it in terms of $\mathbf{\Psi}$. This can be done in a covariant way and it leads to a difficult non-linear equation, in the same way as (5.28) is not linear in ψ. Fortunately, as already promised, the result allows for a simplification that is different from the one that would lead to the Dirac equation, and which is now rigorously covariant.

Let Ψ be the original rotation around the z-axis in the rest frame. Note that $\mathbf{s}\boldsymbol{\cdot}\boldsymbol{\sigma} = \mathbf{S}$. The various transformations of the equations that will be

Table 5.1 Outline of the operations intervening in the simplification of (5.38)

Equation			Comments
$\mathbb{1}\frac{d}{d\tau}\Psi$	$=$	$-\imath\frac{m_0 c}{\hbar}\mathbf{S}\Psi$	$\Psi = \Psi^{\dagger-1}$ is a rotation
$\downarrow$		$\downarrow$	Left multiplication by $\mathbf{L}$
$\mathbf{L}\mathbb{1}\frac{d}{d\tau}\Psi$	$=$	$-\imath\frac{m_0 c}{\hbar}\mathbf{L}\mathbf{S}\Psi$	
$\downarrow$		$\downarrow$	$\mathbf{L}^{\dagger}\mathbf{L}^{\dagger-1} = \mathbb{1}$, $\Psi = \Psi^{\dagger-1}$
$[\mathbf{L}\mathbb{1}\frac{d}{d\tau}\mathbf{L}^{\dagger}]\,[\mathbf{L}^{\dagger-1}\Psi^{\dagger-1}]$	$=$	$-\imath\frac{m_0 c}{\hbar}\mathbf{L}\mathbf{S}\Psi$	
$\downarrow$		$\downarrow$	Definitions $\mathbf{D}'_R$ and Ψ'_L
$\mathbf{D}'_R\Psi'_L$	$=$	$-\imath\frac{m_0 c}{\hbar}\mathbf{L}\mathbf{S}\Psi_R$	Ψ'_L may not be a rotation
$\downarrow$		$\downarrow$	taking the first columns
$\mathbf{D}'_R\psi'_L$	$=$	$-\imath\frac{m_0 c}{\hbar}\mathbf{L}\mathbf{S}\psi_R$	
$\downarrow$		$\downarrow$	$\mathbf{S}\psi_R = \psi_R$, definition ψ'_R
$\mathbf{D}'_R\psi'_L$	$=$	$-\imath\frac{m_0 c}{\hbar}\psi'_R$	Ψ'_R may not be a rotation

described are summarized in Table 5.1. Start with $[\,\mathbb{1}\frac{d}{dc\tau}\,]\,\Psi = -\imath\frac{m_0 c}{\hbar}\mathbf{S}\Psi$ (from (5.26)) and multiply it by $\mathbf{L}$ on both sides.

- *Left-hand side.* The four-gradient $\mathbf{D} = \frac{\partial}{\partial ct}\mathbb{1} + \boldsymbol{\nabla}\cdot\boldsymbol{\sigma}$ is a four-vector. It will transform as $\mathbf{D} \to \mathbf{L}\mathbf{D}\mathbf{L}^{\dagger}$, therefore insert $\mathbf{L}^{\dagger}\mathbf{L}^{\dagger-1} = \mathbb{1}$. For the rotation Ψ, we have $\Psi^{\dagger-1} = \Psi$. Replace thus Ψ with $\Psi^{\dagger-1}$ because it is $\Psi^{\dagger-1}$ which transforms with the operator $\mathbf{L}^{\dagger-1}$ that has been inserted.
- *Right-hand side 1.* On the right-hand side, the same operation can be performed. The quantity $\mathbf{S}$ is a four-vector, and it will thus transform as $\mathbf{S} \to \mathbf{L}\mathbf{S}\mathbf{L}^{\dagger}$. One can therefore insert here also $\mathbf{L}^{\dagger}\mathbf{L}^{\dagger-1} = \mathbb{1}$, and use $\Psi^{\dagger-1} = \Psi$ because it is $\Psi^{\dagger-1}$ which transforms with $\mathbf{L}^{\dagger-1}$.
- With these transformations the equation will become: $[\,\mathbf{L}\mathbb{1}\frac{d}{d\tau}\mathbf{L}^{\dagger}\,]$ $[\,\mathbf{L}^{\dagger-1}\Psi^{\dagger-1}\,] = [\,\mathbf{L}\mathbf{S}\mathbf{L}^{\dagger}\,]$ $[\,\mathbf{L}^{\dagger-1}\Psi^{\dagger-1}\,]$, which corresponds then exactly to the result of the operations carried out on the block $\mathbf{L}^{\dagger-1}\Psi^{-1\dagger}$ within (5.40).

- *Right-hand side 2.* The right-hand side will be transformed differently. It is pointless to insert $\mathbf{L}^{\dagger}\mathbf{L}^{\dagger -1} = \mathbb{1}$, because the simplification aimed at now is $\mathbf{S}\psi_R = \psi_R$. After inserting $\mathbf{L}^{\dagger}\mathbf{L}^{\dagger -1} = \mathbb{1}$ one would obtain two terms $\mathbf{S}'_R = \mathbf{LSL}^{\dagger}$ and $\mathbf{L}^{\dagger -1}\Psi^{\dagger -1}$, but the first column of $\mathbf{L}^{\dagger -1}\Psi^{\dagger -1}$ is ψ'_L. This would lead to two quantities $\mathbf{S}'_R\psi'_L$ and it would no longer be possible to simplify this product as can be done with $\mathbf{S}\psi_R = \psi_R$.
- In the last line of Table 5.1, indices R and L have been added in order to indicate to which representations the quantities belong.
- *Equation for the spinors.* The first columns of Ψ_R and Ψ_L are called respectively ψ_R and ψ_L. We have thus proved the covariance of $\mathbf{D}_R\psi_L = -\imath\frac{m_0c}{\hbar}\psi_R \Rightarrow \mathbf{D}'_R\psi'_L = -\imath\frac{m_0c}{\hbar}\psi'_R$, where $\mathbf{D}'_R = \mathbf{L}\frac{d}{dc\tau}\mathbb{1}\,\mathbf{L}^{\dagger} = \frac{\partial}{\partial ct'}\mathbb{1} + \boldsymbol{\nabla}'{\cdot}\boldsymbol{\sigma}$.
- *Companion equation.* The companion equation is $\mathbf{D}_L\psi_R = -\imath\frac{m_0c}{\hbar}\psi_L$. The development is, *mutatis mutandis*, completely analogous. To see this we start from $[\,\mathbb{1}\frac{d}{dc\tau}\,] = +\imath\frac{m_0c}{\hbar}\mathbf{S}^{\star}\Psi$, which is equivalent to (5.26), as $\mathbf{S}^{\star} = -\mathbf{S}$. We multiply both sides to the left with $\mathbf{L}^{\dagger -1}$. The left-hand-side then becomes $\mathbf{L}^{\dagger -1}[\,\mathbb{1}\frac{d}{dc\tau}\,]\mathbf{L}^{-1}\mathbf{L}\Psi = \mathbf{D}'_L\Psi'$, where $\mathbf{L}^{-1}\mathbf{L} = \mathbb{1}$ has been inserted. The resulting equation corresponds then exactly to what happens to the block $\mathbf{L}\Psi$ in (5.40). Again $\mathbf{L}^{-1}\mathbf{L}$ is not inserted on the right-hand side in order to be able to simplify. The right-hand side then stays $\mathbf{L}^{\dagger -1}\,\mathbf{S}^{\star}\,\Psi$. This is then changed to $\mathbf{L}^{\dagger -1}\,\mathbf{S}^{\star}\,\Psi^{\dagger -1}$ using $\Psi^{\dagger -1} = \Psi$, which allows $\mathbf{S}^{\star}\,\psi_R = -\psi_R$ to be simplified, such that $\mathbf{D}'_L\psi'_R = -\imath\frac{m_0c}{\hbar}\mathbf{L}^{\dagger -1}\psi_L = -\imath\frac{m_0c}{\hbar}\psi'_L$ is obtained. Here, $\mathbf{D}'_L = \mathbf{L}^{\dagger -1}\frac{d}{dc\tau}\mathbb{1}\,\mathbf{L}^{-1} = \frac{\partial}{\partial ct'}\mathbb{1} - \boldsymbol{\nabla}'{\cdot}\boldsymbol{\sigma}$.

This shows that the equations $[\,\frac{\partial}{\partial ct}\mathbb{1} + \boldsymbol{\nabla}{\cdot}\boldsymbol{\sigma}\,]\,\psi_L = -\imath\frac{m_0c}{\hbar}\psi_R$ and $[\,\frac{\partial}{\partial ct}\mathbb{1} - \boldsymbol{\nabla}{\cdot}\boldsymbol{\sigma}\,]\,\psi_R = -\imath\frac{m_0c}{\hbar}\psi_L$ are covariant. It can be seen from this that the simplified covariant equation is not $[\,\frac{\partial}{\partial ct}\mathbb{1} - \boldsymbol{\nabla}{\cdot}\boldsymbol{\sigma}\,]\,\psi = -\imath\frac{m_0c}{\hbar}\,\psi$, which would lead to the Dirac equation, but the equation:

$$\boxed{\left[\frac{\partial}{\partial ct}\mathbb{1} - \boldsymbol{\nabla}{\cdot}\boldsymbol{\sigma}\right]\,\psi_R = -\imath\frac{m_0c}{\hbar}\psi_L,} \tag{5.41}$$

and its companion equation:

$$\boxed{\left[\frac{\partial}{\partial ct}\mathbb{1} + \boldsymbol{\nabla}{\cdot}\boldsymbol{\sigma}\right]\,\psi_L = -\imath\frac{m_0c}{\hbar}\psi_R,} \tag{5.42}$$

which shows that (5.38) squares to the Klein-Gordon equation. These two equations share a feature with Majorana's equation in that they contain a "twist", linking a spinor on one side of the equation to another "conjugated" spinor on the other side of the equation. This necessitates moving to and fro between the two SL(2,$\mathbb{C}$) representations. But the conjugation at stake here is not charge conjugation.

This equation has not been guessed, but derived from a well-defined set of assumptions. In (5.41), the first column ψ_L of $\Psi_L = \Psi^{\dagger -1}$ is a kind of "conjugated" spinor. The operation $\Psi_R \rightarrow \Psi_L$ leaves rotation matrices invariant, but does not leave general homogeneous Lorentz transformations invariant. Conjugation is an expedient that allows us to work on $\Psi^\dagger$ (which transforms by right-hand multiplication: $\Psi^\dagger \rightarrow \Psi^\dagger \mathbf{L}^\dagger$) by left-hand multiplication. Right-hand multiplication is transformed into left-hand multiplication by taking the inverses. The conjugation used here is not charge conjugation. Let the rotation matrix in (3.4) be written as a lexicographic juxtaposition of two 2×1 spinors: $[\,\psi_1\ \psi_2\,]$. For the restriction of the charge conjugation $C : \psi \rightarrow -\imath\sigma_y\psi^*$ to SU(2) we would have: $C\psi_2 = -\psi_1$ and $C\psi_1 = \psi_2$. For the different type of conjugation P encountered here we have $P\psi_1 = \psi_1$ and $P\psi_2 = \psi_2$ in the restriction to SU(2). The conjugation here is a parity transformation that transforms left- into right-handed representations and *vice versa*.

5.5.2.5 *Most general non-linear form of the equation*

(5.38) must now be generalized to its most general form whereby we also express how $\mathbf{e}'_z$ is related to Ψ according to (5.27). Again, as in Chapter 4:

$$\Psi = \begin{pmatrix} a & b \\ c & d \end{pmatrix}. \tag{5.43}$$

The derivation of (5.27) is based on the introduction of $\mathbb{1} + \mathbf{e}'_z{\cdot}\boldsymbol{\sigma}$. The fuss of finding an expression for $\mathbb{1} + \mathbf{e}'_z{\cdot}\boldsymbol{\sigma}$ can be avoided by the following considerations. (5.33) shows that a general spinor will be of the type:

$$\boldsymbol{\Psi} = \begin{pmatrix} a & b & & \\ c & d & & \\ & & d^* & -c^* \\ & & -b^* & a^* \end{pmatrix}. \tag{5.44}$$

As $\mathbf{e}'_z$ is a vector, its generalization in the Dirac-like equation will thus automatically enter into a covariant formulation of the Rodrigues formula.

The generalization of $\mathbf{e}'_z$ can be calculated from the rules $\mathbf{V} \to \mathbf{LVL}^\dagger$ and $\mathbf{V}^{-1} \to \mathbf{L}^{-1\dagger}\mathbf{V}^{-1}\mathbf{L}^{-1}$. This leads to the following generalization:

$$\mathbf{e}'_z \cdot \boldsymbol{\sigma} \to \begin{pmatrix} & & aa^* - bb^* & ac^* - bd^* \\ & & ca^* - db^* & cc^* - dd^* \\ cc^* - dd^* & bd^* - ac^* & & \\ db^* - ca^* & aa^* - bb^* & & \end{pmatrix}. \tag{5.45}$$

It is also known how $\frac{d}{d\tau}\mathbf{e}_{ct}$ should be written in terms of the four-gradient using the Dirac matrices. We can thus write the non-linear generalization of the Rodrigues equation that corrects the Dirac equation. Using the spinors:

$$\psi_1 = \begin{pmatrix} a \\ c \end{pmatrix}, \quad \psi_2 = \begin{pmatrix} b \\ d \end{pmatrix}, \quad \psi_1^c = \begin{pmatrix} -c^* \\ a^* \end{pmatrix}, \quad \psi_2^c = \begin{pmatrix} -d^* \\ b^* \end{pmatrix}, \tag{5.46}$$

(5.45) can be rewritten elegantly as:

$$\mathbf{e}'_z \cdot \boldsymbol{\sigma} \to \begin{bmatrix} & \psi_1 \otimes \psi_1^\dagger - \psi_2 \otimes \psi_2^\dagger \\ \psi_1^c \otimes \psi_1^{c\dagger} - \psi_2^c \otimes \psi_2^{c\dagger} & \end{bmatrix}, \tag{5.47}$$

while (5.44) becomes:

$$\boldsymbol{\Psi} = \begin{bmatrix} \psi_1\,,\,\psi_2 & \\ & -\psi_2^c\,,\,\psi_1^c \end{bmatrix}, \tag{5.48}$$

where $\psi_1\,,\,\psi_2$ just notes the 2×2 matrix obtained by the juxtaposition of ψ_1 and ψ_2. We obtain then:

$$\left[\imath\hbar \sum_{\mu=0}^{3} \gamma_\mu \partial_\mu\right] \boldsymbol{\Psi} = -m_0 c \begin{bmatrix} & \psi_1 \otimes \psi_1^\dagger - \psi_2 \otimes \psi_2^\dagger \\ \psi_2^c \otimes \psi_2^{c\dagger} - \psi_1^c \otimes \psi_1^{c\dagger} & \end{bmatrix} \boldsymbol{\Psi}, \tag{5.49}$$

where it has been taken into account that the true vector $\mathbf{e}'_z$ must be replaced by the corresponding axial vector. This gives a modified Dirac equation that keeps track of the spin.[26] (5.49) can be considered as the relativistic counterpart of (5.27) in SU(2).

---

[26] The meaning of the one-column spinor quantities used as solutions for the conventional free-space Dirac equation can be addressed by inspecting their meaning in the Cartan representation, which directly works on decoupled SL(2,$\mathbb{C}$) spinors. In the decoupled equations, one must ultimately make a special linear combination of two one-column solutions in order to be able to reconstruct a meaningful spinor. But by using $\mathbf{e}'_z$ instead of $\mathbf{n}$, only one frequency will occur in the solutions, such that a single one-column solution then obtains a meaning (as long as we restrict ourselves to SU(2), as in SL(2,$\mathbb{C}$) single-column quantities do no longer contain the complete information). This must then also be correct in the Dirac representation which is equivalent to the Cartan representation by a

5.5.3 The Dirac equation: At last!

An inspection of (5.38), with its shorthand notation $\underset{\sim}{\mathbf{D}}\,\mathbf{\Psi} = -\imath \frac{m_0 c}{\hbar} \underset{\sim}{\mathbf{S}}\,\mathbf{\Psi}$, reveals quickly that it would be absurd to change it into the Dirac equation. The substitution:

$$-\imath\frac{m_0 c}{\hbar}\begin{pmatrix} \mathbf{0} & s_{ct}\mathbb{1} + \mathbf{s}{\cdot}\boldsymbol{\sigma} \\ -(s_{ct}\mathbb{1} - \mathbf{s}{\cdot}\boldsymbol{\sigma}) & \mathbf{0} \end{pmatrix}\begin{pmatrix} \Psi & \mathbf{0} \\ \mathbf{0} & \Psi^{-1\dagger} \end{pmatrix} \rightarrow -\imath\frac{m_0 c}{\hbar}\begin{pmatrix} \Psi & \mathbf{0} \\ \mathbf{0} & \Psi^{-1\dagger} \end{pmatrix}, \tag{5.50}$$

that would permit us to recover the genuine Dirac equation, can never be made because on the left-hand side the non-zero blocks are off-diagonal, while on the right-hand side they are on-diagonal. The reader may recognize here an analogy with the remark made about the definition of the spin in Subsection 5.4.1: a reflection operator can never have a spinor as an eigenvector. Nevertheless, it was possible to give a sense to the eigenvalue equation for the spin by introducing sets of spinors. A completely analogous development can be used here, which will finally reveal the true meaning of the Dirac equation. Let us note $\mathbf{L} = \Psi$ and introduce the two-column quantities:

$$\mathbf{E}_{\mathbb{1}} = \begin{pmatrix} \mathbb{1} \\ \mathbf{0} \end{pmatrix}, \quad \mathbf{E}_{\Psi} = \underset{\sim}{\mathbf{L}}\mathbf{E}_{\mathbb{1}} = \begin{pmatrix} \Psi \\ \mathbf{0} \end{pmatrix}, \tag{5.51}$$

and construct the sets $\mathscr{S} = \{\,\mathbf{E}_{\Psi}, \underset{\sim}{\mathbf{S}}\,\mathbf{E}_{\Psi}\,\}$ and $\mathscr{S}_0 = \{\,\mathbb{1}, \underset{\sim}{\mathbf{S}}\,\}$. As $\underset{\sim}{\mathbf{S}}^2 = \underset{\sim}{\mathbb{1}}$, then $\underset{\sim}{\mathbf{S}}\mathscr{S} = \mathscr{S}$, $\underset{\sim}{\mathbf{S}}\mathscr{S}_0 = \mathscr{S}_0$. Simultaneously $\underset{\sim}{\mathbf{S}}(\underset{\sim}{\mathbb{1}} + \underset{\sim}{\mathbf{S}}) = (\underset{\sim}{\mathbb{1}} + \underset{\sim}{\mathbf{S}})$ and $\underset{\sim}{\mathbf{S}}(\underset{\sim}{\mathbb{1}} + \underset{\sim}{\mathbf{S}})\mathbf{E}_{\Psi} = (\underset{\sim}{\mathbb{1}} + \underset{\sim}{\mathbf{S}})\mathbf{E}_{\Psi}$, such that we can consider representing the sets

similarity transformation, as discussed in Footnote 3 of Chapter 4. One could also build one-column quantities by using a tensor product containing dotted spinors, as introduced in Section 4.7. To make the transition to a rank-2 representation, in principle also the 4×4 reflection matrices would have to be formulated as tensor products $\mathbf{L}\otimes\dot{\mathbf{L}}$. But as discussed in Subsection 6.2.6 we can give a meaning to such one-column matrices of rank 2 without reformulating the 4×4 reflection matrices as tensor products. This gives then a new meaning to the equation. However, in its original meaning the Dirac equation definitely corresponds to a rank-1 equation. This will even remain true for the bi-vector transformations that will be discussed in deriving (5.67). In the Dirac equation we are thus not dealing with a tensor product of two SL(2,$\mathbb{C}$) representations but with a direct sum of two SL(2,$\mathbb{C}$) representations. Such a direct sum permits then to accommodate for reversals. The same result will be reached by a completely different reasoning in Subsection 5.5.3.

by the sums of their elements. We will have:

$$\overline{\Psi} = (\underset{\sim}{\mathbb{1}} + \underset{\sim}{\mathbf{S}})\mathbf{E}_{\Psi} = \begin{pmatrix} \Psi \\ -\mathbf{S}^{\star}\Psi \end{pmatrix}. \tag{5.52}$$

Under a Lorentz transformation $\underset{\sim}{\mathbf{K}}$ we will have $\Psi \to \mathbf{K}\Psi = \Psi'$, and $-\mathbf{S}^{\star}\Psi \to \mathbf{K}^{\dagger-1}[-\mathbf{S}^{\star}]\mathbf{K}^{-1}\mathbf{K}\Psi = -\mathbf{S}^{\star\prime}\Psi'$. The quantity $-\mathbf{S}^{\star}\Psi$ transforms thus like $\Psi^{\star} = \Psi^{\dagger-1}$ with $\mathbf{K}^{\dagger-1}$, and is therefore noted as $\Upsilon^{\star}$. This leads to the correspondences:

$$\Psi \leftrightarrow \overline{\Psi} = (\underset{\sim}{\mathbb{1}} + \underset{\sim}{\mathbf{S}})\,\mathbf{E}_{\Psi} = \begin{pmatrix} \Psi \\ \Upsilon^{\star} \end{pmatrix}, \qquad \mathbb{1} \leftrightarrow \overline{\mathbb{1}} = (\underset{\sim}{\mathbb{1}} + \underset{\sim}{\mathbf{S}})\,\mathbf{E}_{\mathbb{1}} = \begin{pmatrix} \mathbb{1} \\ -\mathbf{S}^{\star} \end{pmatrix}. \tag{5.53}$$

Under a Lorentz transformation $\mathbf{K}$, and noting $\mathbf{K}^{\dagger-1} = \mathbf{K}^{\star}$, then:

$$\underset{\sim}{\mathbf{K}}\overline{\Psi} = \overline{\Psi}' = \begin{pmatrix} \mathbf{K}\,\Psi \\ \mathbf{K}^{\star}\Upsilon^{\star} \end{pmatrix} = \begin{pmatrix} \Psi' \\ \Upsilon^{\star\prime} \end{pmatrix}. \tag{5.54}$$

Here $\overline{\Psi}'$ can be defined through the relation $\underset{\sim}{\mathbf{K}}\overline{\Psi} = \overline{\Psi}'$, but also by a construction of the type given in (5.52), as the Lorentz covariance of this procedure has been proved. We have now: $\underset{\sim}{\mathbf{S}}\overline{\Psi} = \overline{\Psi}$ by construction. Let us check the covariance of this result. From $\underset{\sim}{\mathbf{K}}\underset{\sim}{\mathbf{S}}\overline{\Psi} = \underset{\sim}{\mathbf{K}}\overline{\Psi}$ we derive:

$$\overline{\Psi}' = \underset{\sim}{\mathbf{K}}\overline{\Psi} = \underset{\sim}{\mathbf{K}}\underset{\sim}{\mathbf{S}}\overline{\Psi} = \underset{\sim}{\mathbf{K}}\underset{\sim}{\mathbf{S}}\underset{\sim}{\mathbf{K}}^{-1}\underset{\sim}{\mathbf{K}}\overline{\Psi} = \underset{\sim}{\mathbf{S}}'\overline{\Psi}'. \tag{5.55}$$

Finally, one can show that $\Upsilon^{\star} = \Psi^{\star}\sigma_z$.[27] This will show that (5.38) can always be simplified to the Dirac equation. In fact, both sides of (5.38) can be multiplied to the right with:

$$\begin{pmatrix} \mathbb{1} \\ \sigma_z \end{pmatrix}, \tag{5.56}$$

---

[27]One considers a spinning motion around the z-axis given by (5.7). For this rotation the spin vector is $-\mathbf{S}_0^{\star} = \sigma_z$. A general Lorentz spinor will now be given by $\Psi = \mathbf{LR}$. All calculations must now be made for this quantity. Both the new spin vector $-\mathbf{S}^{\star} = \mathbf{L}^{\dagger-1}[-\mathbf{S}_0^{\star}]\,\mathbf{L}^{-1}$, and the new left-handed representation matrix $\Psi^{\star} = \Psi^{\dagger-1}$ must be calculated. Then $\Upsilon^{\star} = -\mathbf{S}^{\star}\,\Psi$ must be calculated as well. But this is: $\Upsilon^{\star} = \mathbf{L}^{\dagger-1}[-\mathbf{S}_0^{\star}]\,\mathbf{L}^{-1}\mathbf{LR} = \mathbf{L}^{\dagger-1}\sigma_z\mathbf{R}$. We must now check that $\Upsilon^{\star} = \Psi^{\star}\sigma_z$. But $\sigma_z\mathbf{R} = \mathbf{R}\sigma_z$ such that $\Upsilon^{\star} = \mathbf{L}^{\dagger-1}\sigma_z\mathbf{R} = \mathbf{L}^{\dagger-1}\mathbf{R}\sigma_z$. As $\mathbf{L}^{\dagger-1}\mathbf{R} = \mathbf{L}^{\dagger-1}\mathbf{R}^{\dagger-1} = \Psi^{\star}$ this completes the proof. With the notation of (4.9) for $\mathbf{L}$, the final expression for $\Upsilon^{\star}$ is: $\Upsilon^{\star} = \begin{pmatrix} d^{*}e^{-\imath\omega_0\tau/2} & c^{*}e^{+\imath\omega_0\tau/2} \\ -b^{*}e^{-\imath\omega_0\tau/2} & -a^{*}e^{+\imath\omega_0\tau/2} \end{pmatrix}$. Note that the gimmick of not transforming the factor σ_z that occurs in $\Upsilon^{\star}$ is also used in the derivation in Subsection 5.5.2.4 as can be seen from the last transition in the overview of this derivation given in Table 5.1.

such that $\overline{\Psi}$ is always obtained by the procedure:

$$\overline{\Psi} = \begin{pmatrix} \Psi & \\ & \Psi^\star \end{pmatrix} \begin{pmatrix} \mathbb{1} \\ \sigma_z \end{pmatrix}. \tag{5.57}$$

Note that the term σ_z occurs here only as the proof was based on covariance and started from a rotation around the z-axis. If we had started with another spin vector $\mathbf{S}$ in the rest frame of the electron we should have replaced σ_z by $\mathbf{S}$. This shows that the 4×2 matrix in (5.57) is a kind of initial condition, very much in the same way as $\psi(\tau) = \mathbf{R}(\tau)\psi(0)$ could be used for the time evolution of a 2×1 spinor in SU(2). (It is different in the fact that it does not relate the situation at t to $t = 0$ in the lab frame but to $\tau = 0$ in the rest frame.) This way, the quantity $\overline{\Psi}$ can be introduced also on the left-hand side of (5.38). We have thus finally derived the Dirac equation (the necessary steps are summarized in Table 5.2).

For the spin in SU(2), it was shown that it was possible to multiply the eigenvector by $e^{\imath\chi}$ and that this led to another meaningful two-element set. However, in SL(2,$\mathbb{C}$), multiplying Ψ by a phase factor $e^{\imath\chi}$ no longer leads to a meaningful result. It is also meaningless to define spinors as one-column quantities in SL(2,$\mathbb{C}$). The solutions of the Dirac equation thus describe sets that contain two elements. A set contains a left-handed and a right-handed spinor, that can be imagined to represent states with a same rest mass. It is this move that finally completes the validation of the Dirac equation. But it can be noted that the sets $\mathscr{S}$ and $\mathscr{S}_0$ are different solutions of the same Dirac equation. Therefore, the sets defined by the Dirac equation are not defined up to a phase factor, but up to a spinor Ψ. This corresponds to the fact that the eigenvalues are degenerate and define two-dimensional vector spaces of eigenvectors. The degeneracy explains why it is preferable to write the solutions using a two-column matrix formalism, in conformity with Section 2.13. As already expressed, the Dirac equation does not specify the direction of the spin axis. In this sense it is thus not complete. But there is no obligation to make the simplification to the Dirac equation.

The two-column quantity $\overline{\mathbb{1}}$ is an eigenvector of $\underset{\sim}{\mathbf{S}}$. However, it is not the eigenvector $\overline{\Psi}$ that would be needed to reduce the Dirac-like equation (5.38) to the traditional Dirac form. (Note for comparison that in the solution of (5.38) there would be no blocks $\mathbb{1}$, and there would be null blocks $\mathbf{0}$.)

In the standard form of the Dirac equation one uses a different set of gamma matrices as in the Weyl representation used earlier. Actually, the

Table 5.2 Overview of the various steps required in the proof of the Dirac equation

Rodrigues formula, (5.2)	$\boxed{\psi = \cos(\varphi/2)\mathbb{1} - \imath\sin(\varphi/2)[\,\mathbf{n}\cdot\boldsymbol{\sigma}\,]}$
Substitution $\varphi = \omega_0\tau$	$\downarrow$
Rotating frame, (5.8)	$\psi = \cos(\omega_0\tau/2)\mathbb{1} - \imath\sin(\omega_0\tau/2)[\,\mathbf{n}\cdot\boldsymbol{\sigma}\,]$
Derivation	$\downarrow$
Differential form, (5.10)	$\frac{d\psi}{d\tau} = -\imath\omega_0[\,\mathbf{n}\cdot\boldsymbol{\sigma}\,]\,\psi$
Einstein-Planck relations	$\downarrow$
Spinning-frame model	$\frac{d}{dc\tau}\psi = -\imath\frac{m_0c}{\hbar}[\,\mathbf{n}\cdot\boldsymbol{\sigma}\,]\,\psi$
$\mathbf{n}$ not covariant in SU(2)	$\downarrow$
Spinning-top model, (5.26)	$\boxed{\frac{d}{dc\tau}\psi = -\imath\frac{m_0c}{\hbar}[\,\mathbf{s}\cdot\boldsymbol{\sigma}\,]\,\psi}$
$[\mathbf{s}\cdot\boldsymbol{\sigma}\,]\,\psi = \psi$ not covariant in SL(2,$\mathbb{C}$)	$\downarrow$
Dirac-like equation, (5.39)	$\boxed{[\,\gamma^{\mu}c\hat{\mathrm{p}}_{\mu}\,]\,\boldsymbol{\Psi} = m_0c^2\,[\,\gamma_5\gamma_{\mu}s_{\mu}\,]\,\boldsymbol{\Psi}}$
Introducing sets through $\overline{\boldsymbol{\Psi}}$	$\downarrow$
Dirac equation	$\boxed{[\,\gamma^{\mu}c\hat{\mathrm{p}}_{\mu}\,]\,\overline{\boldsymbol{\Psi}} = m_0c^2\,\overline{\boldsymbol{\Psi}}}$

Dirac equation uses γ_5 as its γ_0 and *vice versa.* It is well known that all choices for the gamma matrices lead to equivalent solutions. The paradox that it was not possible to make the substitution of (5.50) occurs thus also in the Dirac representation. But there it may go unnoticed due to the fact that the standard representation also contains block matrices on the diagonal.

Note that there is absolutely no obstacle to adding the spinors of the two elements that belong to $\mathscr{S}$. One of them corresponds to a block matrix along the main diagonal, the other to a block matrix on the secondary diagonal. Therefore, after any odd or even combination $\mathbf{L}$ of Lorentz reflections, the blocks $\mathbf{L}\Psi$ and $\mathbf{L}^{\star}\Psi^{\star}$ will always occupy non overlapping positions in the formalism, such that they can always be seperated out from the sum $\overline{\Psi}$. The blocks Ψ and $\Psi^{\star}$ will also always be operated on by the right representation matrices. It is like adding up $\mathbf{e}_x + \imath\mathbf{e}_y$ in SU(2): both contributions can always be recovered from their sum.

For the moment, the solutions of the Dirac equation are not spinors, but two-element sets of spinors. By proving the covariance in (5.55), it is shown that these solutions are always of the form:

$$\overline{\Psi} = \begin{pmatrix} \Psi \\ \Psi^{\star}\sigma_z \end{pmatrix} = \begin{pmatrix} a & b \\ c & d \\ d^* & c^* \\ -b^* & -a^* \end{pmatrix}. \tag{5.58}$$

But this shows now that a one-column quantity $\tilde{\psi}$ taken from $\overline{\Psi}$ contains the complete information of a spinor. This is somewhat unexpected and it is satisfying to see how the pieces of the jigsaw puzzle finally come together. It is in line with the result of the discussion in Footnotes 2 of Chapter 4 and 26 of the present chapter. Using the analogy of the Meccano game introduced in Section 2.13, it can be said that Ψ has been taken apart and reassembled into an equivalent Meccano construction, *viz.* a single column vector taken out of the construction $\overline{\Psi}$. We see that this corresponds very much to the philosophy promoted in Section 2.13. We take matrices apart and then put the individual column matrices together in other ways, ending up with different matrices containing the same or equivalent *geometrical* information. The "superposition" used to define $\overline{\Psi}$ eventually serves only to extend the possibilities of the Meccano game. It makes sense after all to calculate with one-column quantities rather than with SL(2,$\mathbb{C}$) matrices, and the traditional Dirac equation thus perfectly keeps track of the spin of the electron without any loss of information. The freedom to choose values for the spin that caused so much concern is thus just a freedom to choose initial conditions for the spin.[28] However, by "squaring" the Dirac equation to decouple the parts coming from Ψ and $\Psi^{\star}$, in the form of two decoupled Klein-Gordon equations in 2×1 matrices, information will be lost. The two

[28]The quantity $\tilde{\psi} = [a, c, d^*, -b^*]^{\top}$ is often called a *bi-spinor* in the literature. This is actually a misnomer, because it is the combination of two semi-spinors containing each only half of the information into a true spinor quantity that contains the full information. It is also often encrypted in the abstract statement that the bi-spinor is a representation that belongs to $\mathbf{D}^{(\frac{1}{2},0)} \oplus \mathbf{D}^{(0,\frac{1}{2})}$, which is perhaps easier to understand with hindsight than when first confronted with it. Note that in the Dirac representation, it takes more effort to grasp the meaning of the bi-spinor $\mathbf{D}^{(\frac{1}{2},0)} \oplus \mathbf{D}^{(0,\frac{1}{2})}$ because the essence of the structure becomes blurred by the fact that the Lorentz transformation matrices are not block-diagonal as in the Weyl representation.

coupled equations are:

$$\boxed{\begin{aligned} \left[\frac{\partial}{\partial ct}\mathbb{1} - \nabla\cdot\boldsymbol{\sigma}\right]\begin{pmatrix} a \\ c \end{pmatrix} &= -\imath\frac{m_0 c}{\hbar}\begin{pmatrix} d^* \\ -b^* \end{pmatrix}, \\ \left[\frac{\partial}{\partial ct}\mathbb{1} + \nabla\cdot\boldsymbol{\sigma}\right]\begin{pmatrix} d^* \\ -b^* \end{pmatrix} &= -\imath\frac{m_0 c}{\hbar}\begin{pmatrix} a \\ c \end{pmatrix}. \end{aligned}} \tag{5.59}$$

But after squaring they will become:

$$\begin{aligned} \begin{pmatrix} \Box & \\ & \Box \end{pmatrix}\begin{pmatrix} a \\ c \end{pmatrix} &= -\frac{m_0^2c^2}{\hbar^2}\begin{pmatrix} a \\ c \end{pmatrix}, \\ \begin{pmatrix} \Box & \\ & \Box \end{pmatrix}\begin{pmatrix} d^* \\ -b^* \end{pmatrix} &= -\frac{m_0^2c^2}{\hbar^2}\begin{pmatrix} d^* \\ -b^* \end{pmatrix}, \end{aligned} \tag{5.60}$$

where $\Box = \frac{1}{c^2}\frac{\partial^2}{\partial t^2} - \frac{\partial^2}{\partial x^2} - \frac{\partial^2}{\partial y^2} - \frac{\partial^2}{\partial z^2}$. This way a single equation will only determine, for example, a and c, or only b and d while this cannot be the complete solution. The loss of information can be avoided by using the two-column spinors rather than the single-column spinors. For a rotation, the two 2×1 spinors become identical, because $b = -c^*$ and $d = a^*$. The spin-up solutions will both become:

$$\psi_+ = \begin{pmatrix} a \\ c \end{pmatrix} = \begin{pmatrix} d^* \\ -b^* \end{pmatrix} = e^{-\imath\omega_0\tau/2}\begin{pmatrix} 1 \\ 0 \end{pmatrix}. \tag{5.61}$$

It is only after also applying boosts to them that the spinors become different.[29] It follows thus that the exact equation (5.38) does not stand in exact

[29] Note that in the physics literature, the one-column quantity $\tilde{\psi}$ is usually normalized according to $\frac{1}{2}\tilde{\psi}^\dagger\tilde{\psi} = \frac{1}{2}(aa^* + bb^* + cc^* + dd^*) = \gamma$ (see (C.12)). From the viewpoint of pure group theory, the normalization condition should be always $ad - bc = 1$, i.e. $\frac{1}{2}\tilde{\psi}^\dagger\gamma_0\tilde{\psi} = 1$. The alternative normalization procedure used in physics is suggested by the derivation of a continuity equation for the probability charge-current density from the Dirac equation $-\frac{\hbar}{\imath}\gamma_0\frac{\partial}{\partial t}\tilde{\psi} - \frac{\hbar c}{\imath}\boldsymbol{\gamma}\cdot\nabla\tilde{\psi} = m_0c^2\tilde{\psi}$ (a), where $\boldsymbol{\gamma} = (\gamma_1, \gamma_2, \gamma_3)$. By Hermitian conjugation and using the identities $\gamma_0^\dagger = \gamma_0$ and $\boldsymbol{\gamma}^\dagger = -\boldsymbol{\gamma}$ this yields: $\frac{\hbar}{\imath}\frac{\partial}{\partial t}\tilde{\psi}^\dagger\gamma_0 - \frac{\hbar c}{\imath}(\nabla\tilde{\psi}^\dagger)\cdot\boldsymbol{\gamma} = m_0c^2\tilde{\psi}^\dagger$ (b). Combining the equations according to $[\tilde{\psi}^\dagger\gamma_0]$ $[(a)] - [(b)]$ $[\gamma_0\tilde{\psi}]$ and using $\boldsymbol{\gamma}\gamma_0 = -\gamma_0\boldsymbol{\gamma}$ yields then an equation that we can interpret as a continuity equation, whereby $\tilde{\psi}^\dagger\tilde{\psi}$ is identified with a probability density. For the complete probability charge-current density one finds $j_\mu = \overline{\tilde{\psi}}\gamma_\mu\tilde{\psi}$, with the definition $\overline{\tilde{\psi}} = \tilde{\psi}^\dagger\gamma_0$. In SU(2), a spinor $\psi = [\,\xi_0\ \ \xi_1\,]^\top$ is normalized according to $\psi^\dagger\psi = \xi_0\xi_0^* + \xi_1\xi_1^* = 1$. This normalization condition is the counterpart of $\frac{1}{2}\tilde{\psi}^\dagger\gamma_0\tilde{\psi} = 1$ for the spinor $\tilde{\psi}$. From

correspondence with the 2×1 single-column solutions of the Dirac equation. We must solve two equations, and we must combine the two solutions in such a way that it does not result in a spurious solution, such as:

$$\begin{pmatrix} \psi_+ \\ \mathbf{0} \end{pmatrix}. \tag{5.62}$$

This is not too much of a concern, because it is an easily avoidable error. It is necessary to check that the quantities a, c, d^* and $-b^*$ satisfy $ad - bc = 1$. As seen earlier, a three-dimensional approximation is also apt to produce the Dirac equation, while the three-dimensional approximation is not exact. One must thus always keep an eye open to avoid picking inadvertently a spurious solution.

It may finally be noted that the derivation shows that the Dirac equation and the Dirac-like (5.38) are both correct. The Dirac-like equation applies to single particles, while the Dirac equation applies to sets of particles, coded as sums of spinors. The probabilistic character of the Dirac equation will become even more clear in the next subsection.

5.5.4 *From orbits to orbitals*

We may jubilate now at the idea that we have proved the Dirac equation by deriving it deductively from the image of a spinning top, thus giving the equation an exact intuitive interpretation. But the ultimate piece of the proof is still lacking! We have considered an electron at rest and derived an equation for it that defines the spin. In the overview in Table 5.2, this is (5.26) for the spinor $\psi(\tau)$. We have then generalized this equation by

---

the Rodrigues-like equation $\frac{d}{d\tau}\psi = -\imath\frac{\omega_0}{2}\,[\mathbf{s}{\cdot}\boldsymbol{\sigma}]\psi$ (see (5.9)) one can derive $\frac{d}{d\tau}\psi^\dagger\psi = 0$ in the same way as one derives the continuity equation from the Dirac equation. This equation is thus the counterpart of the continuity equation obtained from the Dirac equation. According to (5.61), it corresponds to $\psi^\dagger\psi = 1$ in a rest frame. The Rodrigues equation describes a spinning electron at rest, whose position is not known as it has not been specified. This complete lack of information corresponds to a probability density that is constant over the whole of $\mathbb{R}^3$. Because the probability density and $\psi^\dagger\psi$ are both constants in the rest frame and are both parts of a four-vector that satisfies a continuity equation, they will always transform the same way under Lorentz transformations and always remain proportional. It is thus logical to identify the probability density with $\psi^\dagger\psi$, where ψ contains a normalization constant that can be fixed by introducing a condition of the form $\int \psi^\dagger(\mathbf{r},t)\psi(\mathbf{r},t)\, d\mathbf{r} = 1$ for it. This analysis leads to an extension of the definition of a spinor to that of a probability charge-current density, whereby the generalization starts from the identity $\frac{1}{2}\tilde{\psi}^\dagger\tilde{\psi} = \gamma$ rather than from the identity $\frac{1}{2}\,\tilde{\psi}^\dagger\,\gamma_0\,\tilde{\psi} = 1$.

covariance to a moving frame. These are the further steps in the overview in Table 5.2. Transforming $\psi(\mathbf{r}_0, \tau)$ to a moving frame by using the Lorentz transformation changes it into $\psi(\mathbf{r}, t)$. Here, the position vectors $\mathbf{r}_0$ in the frame at rest and $\mathbf{r}$ in the moving frame enter the scene because it is necessary to know the position of the electron to make the Lorentz transformation for the time correctly. In principle, this construction only defines the wave functions $\psi(\mathbf{r}, t)$, $\Psi(\mathbf{r}, t)$, $\Psi^\star(\mathbf{r}, t)$ and $\overline{\Psi}(\mathbf{r}, t)$ on the path $\mathbf{r}(t)$ of the electron. But this is not the way in which wave functions are used in quantum mechanics. They are considered as defined over the whole of space-time. We must thus still introduce a change of the definition domain for the wave function and prove that it is self-consistent. It is not always common practice in physics to worry about definition domains of functions, even if it is important in mathematics, as observed in Footnote 1 of Chapter 1. We have therefore always been rigorous about defining the definition domains of the functions we used, even if this might have come over as pedantic. But in the specific case of the wave function it will turn out to be extremely important to be rigorous about what the definition domains are. Fortunately, this extension of the definition domain of the wave function from the path of the electron to the whole of space-time comes about naturally. It intervenes on two occasions in different guises: in trying to define the wave function in a moving frame by covariance, and when trying to solve the differential equation (which can be considered as an alternative definition of the wave function):

• When we solve the differential equation, it is done over the whole of space-time, without any forethought about the question what the definition domain is supposed to be. This is really generous as the equation needs only to be solved on the path of the electron. Even if it might be done unwittingly, this transgression of the conceptual boundaries of the initially intended definition domain of the wave function defines a natural extension of it. This extension is self-consistent and can be considered as natural because it satisfies the differential equation over the whole of space-time. That the definition domain can this way become extended beyond the classically expected definition domain is well illustrated by the tunnel effect.

• When the Lorentz transformation is introduced for the actual position $(\mathbf{r}_0, \tau_0)$ of the electron on its path, the same Lorentz transformation applies to any other possible position $(\mathbf{r}_1, \tau_1)$ that is not on that path. When we write $(\mathbf{r}, t)$ in the moving frame we may then think in complete ignorance that it applies already for any value of $(\mathbf{r}, t)$ in space-time. Again this defines inadvertently a natural extension of $\psi(\mathbf{r}, t)$, $\Psi(\mathbf{r}, t)$, $\Psi^\star(\mathbf{r}, t)$ and $\overline{\Psi}(\mathbf{r}, t)$ to the whole of space-time. This way other possible positions for the electron are added to the definition domain. It is the inclusion of these additional possible positions in the definition domain that will render the wave function probabilistic. In the extension we lose track of the actual path of the electron that gave rise to it. We cease to describe the actual positions and start to describe possible positions. This is why the issue of the definition domain is crucial to the meaning of the wave function in quantum mechanics.

In fact, after the extension of the definition domain of the wave function, one can ask if the single-valued extension constructed this way has a meaning, and if so, what this meaning would be. One can then argue that the extension describes a probability distribution. This intuition can be further justified by deriving a continuity equation for the probability charge-current density from the wave equation, as is done in textbooks. The continuity equation ensures that the probability calculus will be self-consistent over the whole of space-time.

The consequences of the extension in the context of the free-space Dirac equation will be discussed in more depth in Chapter 6. When potentials are introduced, defining the extension in a self-consistent way will become truly non-trivial in the second procedure. Solving this problem leads to the Bohr-Sommerfeld quantization conditions and becomes tied up with applying the procedure of separating the variables in partial differential equations to the Dirac equation. This will be discussed in Chapter 8. The complete rigorous construction of the wave function and its corresponding wave equation is thus an elaborate and dazzling piece of mathematics. Quantum mechanics is a poem of mathematics. We can only marvel with awe at the thought that physicists have been able to guess this mathematical construct with all its inherent subtleties by mere intuition. To seize the beauty of a poem we must have an excellent command of the language it has been written in. In the case of quantum mechanics, we only poorly understood this language. We are not all gifted with Dirac's phenomenal intuition, such that to most of us the equation comes out of the blue and looks beyond our grasp. Filling this gap in our understanding by bringing down the language barrier of group theory has been the leitmotiv behind the present derivation, that, for the reasons explained in Chapter 2, wanted to be at once rigorous and intuitive.

(On a first reading the reader may jump from here directly to Section 5.6.)

5.5.5 *Further justification of the criterion for Lorentz covariance*

The similarity transformations used to prove the covariance of the Dirac-like equation (5.38) correspond to Lorentz transformations for four-vectors. Let us check the consequences of the Lorentz covariance for a similarity transformation based on a boost. The form of the boost with velocity $\mathbf{v} = v\mathbf{u}$ in a right-handed SL(2,$\mathbb{C}$) representation is $\mathbf{B}(\mathbf{v}) = \sqrt{\frac{\gamma+1}{2}}\,\mathbb{1} - \sqrt{\frac{\gamma-1}{2}}\,\mathbf{u}{\cdot}\boldsymbol{\sigma}$. Such matrices correspond to unit vectors since $\mathbf{B}(\mathbf{v})\mathbf{B}^\star(\mathbf{v}) = \mathbb{1}$. Moreover, $\mathbf{B}^\dagger(\mathbf{v}) = \mathbf{B}(\mathbf{v})$. Here $\mathbf{u}$ is of course defined as $\mathbf{u} = \mathbf{v}/v$. The matrices:

$$\begin{pmatrix} \mathbf{B} & \mathbf{0} \\ \mathbf{0} & \mathbf{B}^\star \end{pmatrix} \quad \text{and} \quad \begin{pmatrix} \mathbf{B}^\star & \mathbf{0} \\ \mathbf{0} & \mathbf{B} \end{pmatrix} \tag{5.63}$$

are respectively a boost and its inverse in the Dirac representation. Operating with a boost on a four-vector we obtain $\mathbf{V} \to \mathbf{BVB}$, $\mathbf{V}^\star \to \mathbf{B}^\star\mathbf{V}^\star\mathbf{B}^\star$. For an axial vector we have $-\mathbf{V}^\star \to -\mathbf{B}^\star\mathbf{V}^\star\mathbf{B}^\star$, such that the difference between true and axial vectors does not affect their transformation properties. As for a boost $\mathbf{B}^\dagger = \mathbf{B}$, the transformation of the vector $\mathbf{V} = s_{ct}\mathbb{1} + \mathbf{s}{\cdot}\boldsymbol{\sigma} \to \mathbf{BVB} = \mathbf{BVB}^\dagger$ corresponds to a general boost. Explicit calculation shows that it is:

$$\mathbf{s} \to \gamma\left[s_{ct} - \frac{\mathbf{v}\cdot\mathbf{s}}{c}\right]\mathbb{1} + \gamma\left[\mathbf{s}_{\parallel} - \frac{\mathbf{v}s_{ct}}{c}\right]{\cdot}\boldsymbol{\sigma} + \mathbf{s}_\perp{\cdot}\boldsymbol{\sigma}, \tag{5.64}$$

where $\mathbf{s}_\parallel$ is the part of $\mathbf{s}$ that is parallel with $\mathbf{u}$ and $\mathbf{s}_\perp$ is the part of $\mathbf{s}$ that is perpendicular to $\mathbf{u}$. This expression corresponds to the Lorentz transformation of the vector $s_{ct}\mathbb{1} + \mathbf{s}{\cdot}\boldsymbol{\sigma}$ under the boost with boost vector $\mathbf{v}$. This is a traditional Lorentz transformation where s_{ct} plays the role of ct and $\mathbf{s}$ the role of $\mathbf{r}$. The calculation on $\mathbf{B}^\star\mathbf{V}^\star\mathbf{B}^\star$ shows that it corresponds to the Lorentz transformation of the vector $s_{ct}\mathbb{1} - \mathbf{s}{\cdot}\boldsymbol{\sigma}$ under the same boost. The point of interest here is the special case where $s_{ct} = 0$, such that the Lorentz transformations of $\mathbf{S} = \mathbf{s}{\cdot}\boldsymbol{\sigma}$ and $\mathbf{S}^\star = -\mathbf{s}{\cdot}\boldsymbol{\sigma}$ are obtained. The very same derivation applies also to the operator $(\frac{\partial}{\partial ct}, \boldsymbol{\nabla})$, since it applies to a general four-vector (e.g. $(E, c\mathbf{p})$). A similar operation can be performed for the transformations under rotations, working again on a general four-vector. For a rotation matrix the simplifying relation is different from that for a boost, as here $\mathbf{R}^\dagger = \mathbf{R}^{-1}$. The rotation matrix and its inverse are here:

$$\begin{pmatrix} \mathbf{R} & \mathbf{0} \\ \mathbf{0} & \mathbf{R} \end{pmatrix} \quad \text{and} \quad \begin{pmatrix} \mathbf{R}^\dagger & \mathbf{0} \\ \mathbf{0} & \mathbf{R}^\dagger \end{pmatrix} \tag{5.65}$$

respectively, where the rotation with axis $\mathbf{a}$ and angle φ is given by: $\mathbf{R} = \cos(\varphi/2)\,\mathbb{1} - \imath\sin(\varphi/2)\,[\,\mathbf{a}\cdot\boldsymbol{\sigma}\,]$. The calculation on $s_{ct}\mathbb{1} + \mathbf{s}\cdot\boldsymbol{\sigma}$ yields:

$$\mathbf{s} \rightarrow \mathbf{s}' = s_{ct}\mathbb{1} + (\mathbf{s}_{\|} + \cos\varphi\,\mathbf{s}_{\perp} - \sin\varphi\,(\mathbf{a}\wedge\mathbf{s}_{\perp}))\cdot\boldsymbol{\sigma}. \tag{5.66}$$

This presents the image under the rotation with axis $\mathbf{a}$ and angle φ of the vector $\mathbf{s}$. The idea is thus to take the initial value $\mathbf{s}$, search for its component $\mathbf{s}_{\perp} = \mathbf{s} - (\mathbf{a}\cdot\mathbf{s})\,\mathbf{a}$ that is perpendicular to $\mathbf{a}$, calculate $\mathbf{a}\wedge\mathbf{s}_{\perp}$, and then use these calculated values to calculate the value of the image $\mathbf{s}'$ using the value of φ. If $\mathbf{a}\cdot\mathbf{s} = \cos\chi$, then both $\mathbf{s}_{\perp}$ and $\mathbf{a}\wedge\mathbf{s}_{\perp}$ will have length $\sin\chi$, such that $\mathbf{s}'$ still has indeed unit length. In summary, this means that for vectors the similarity transformation corresponds to a Lorentz transformation. In the same way, it is possible to check that $\underset{\sim}{\mathbf{L}}(\gamma_{ct}\frac{d}{dc\tau})\underset{\sim}{\mathbf{L}}^{-1}$ indeed corresponds to $\sum_{\mu}\gamma_{\mu}\frac{\partial}{\partial x_{\mu}}$.

As already mentioned, in the new Dirac-like equation (5.38) it is tacitly implied that the Lorentz transformation $\underset{\sim}{\mathbf{L}}$ that permits $\sum_{\mu}\gamma_{\mu}\frac{\partial}{\partial x_{\mu}}$ to be obtained as $\underset{\sim}{\mathbf{L}}(\gamma_{ct}\frac{d}{d\tau})\underset{\sim}{\mathbf{L}}^{-1}$, is the same one that permits $\underset{\sim}{\mathbf{L}}\underset{\sim}{\mathbf{S}}\underset{\sim}{\mathbf{L}}^{-1}$ to be obtained from $\underset{\sim}{\mathbf{S}}$, and transforms $\underset{\sim}{\mathbf{S}}$ into $\underset{\sim}{\mathbf{S}}'$. Here, $\underset{\sim}{\mathbf{S}} = \sum_{\mu} s_{\mu}\gamma_{\mu}$. Only when this condition is satisfied will the equation "square" to a Klein-Gordon equation for $\underset{\sim}{\mathbf{L}}\boldsymbol{\Psi}$. (In other words, the Klein-Gordon equation can only be derived from from (5.49) if $\sum_{\mu}\gamma_{\mu}\frac{\partial}{\partial x_{\mu}}$ is expressed as $\underset{\sim}{\mathbf{L}}_1(\gamma_{ct}\frac{d}{d\tau})\underset{\sim}{\mathbf{L}}_1^{-1}$, if the the matrix $\underset{\sim}{\mathbf{S}}'$ that contains the entries $\mathbf{S}'$ is expressed as $\underset{\sim}{\mathbf{L}}_2\underset{\sim}{\mathbf{S}}\underset{\sim}{\mathbf{L}}_2^{-1}$ and if it is postulated that $\underset{\sim}{\mathbf{L}}_1 = \underset{\sim}{\mathbf{L}}_2$.)

5.5.6 *Lorentz covariance of the equation: Unitary matrices*

In textbooks, the problem of covariance is discussed in terms of unitary matrices, using arguments that the measured quantities must be real in all frames. This book, however, has pursued an approach to the same covariance by using similarity transformations. It must now be explained why this difference in approaches is not a contradiction.

The true covariance argument is one based on similarity transformations. To make the argument in terms of unitary matrices, one first multiplies both sides of the Dirac equation by γ_{ct}. This is performed on the final result of derivation, not on one of the embryonic equations presented along the way. Once the Dirac equation has been multiplied on both sides with γ_{ct}, the operator $\frac{d}{dct}\mathbb{1}$ (which is part of the operator that projects out the total energy E from the equation) is diagonal. Whatever similarity transformation is applied now on the equation, it will not be possible to change the operator $\frac{d}{dct}\mathbb{1}$, because it is a scalar. This implies that the total energy is kept constant. This is puzzling because it is difficult to see how one would not change the energy by making Lorentz transformations. The answer is given by the operations on the right-hand side of the equation on which the Lorentz transformations are made. Working at constant energy is of course useful in calculating stable radiation-less orbits for the hydrogen atom, for example, because by keeping the energy constant, it is certain that there can be no loss of energy due to radiation. But proving Lorentz covariance for transformations at fixed total energy is not general: in the textbook case without electromagnetic fields, the only elements of the Lorentz group that keep the energy fixed are of a rotational type. Let us now show that by multiplying both sides of the equation by γ_{ct} the operations on the right-hand side have been limited to rotations.

The operation that consists in multiplying both sides of the Dirac equation will be called the γ_{ct} swap. In the Dirac equation, the matrices associated with $(m_0c^2, c\mathbf{p})$ used after the γ_{ct} swap are $(\alpha_{ct}, \alpha_x, \alpha_y, \alpha_z) = (\gamma_{ct}, \gamma_{ct}\gamma_x, \gamma_{ct}\gamma_y, \gamma_{ct}\gamma_z)$, which satisfy the commutation relations $\alpha_\mu\alpha_\nu + \alpha_\nu\alpha_\mu = 2\delta_{\mu\nu}\mathbb{1}$. In other words, in these cases it is possible to identify the group of operations used with the group SO(4) of rotations of Euclidean space $\mathbb{R}^4$ with the ordinary metric signature $+++\,+$. The intervening matrices α_μ are therefore unitary. This shows that only rotations have been treated in this kind of proof of Lorentz covariance. Before the γ_{ct} swap, there is a situation with a constant rest mass m_0, such that $E^2 - c^2\mathbf{p}^2 = (m_0c^2)^2$, leading to the metric signature $+---$ for the left-hand side. After the γ_{ct} swap, the total energy E is a constant, such that $E^2 = c^2\mathbf{p}^2 + (m_0c^2)^2$, leading to the metric signature $++++$ for the right-hand side. In the free-space Dirac equation the rest mass remains a constant anyway, such that the treatment is restricted to three-dimensional rotations. This is then not truly a four-dimensional rotation group with a signature $++++$. It will be explained later that within a potential the velocity can be changed even when the total energy is fixed. The paths then no longer need to be circular. It will be discussed in Subsection (6.2.10.2) how for a central potential with rotational symmetry in the context of the Schrödinger equation only three-dimensional rotations are needed. Note that in the situation with the signature $++++$ the potential V is coupled to the unit matrix, rather then to γ_{ct}. It can thus be stated that we have two different metrics, with signature $++++$ for radiationless motion and with signature $+---$, when radiation can be present.

Yet another way of seeing this is as follows. In the Dirac-like equation (5.38), m_0c^2 is no longer associated with a unit matrix before the γ_{ct} swap. The quantity m_0c^2 does not appear in an association with a unit matrix, but with the three matrices $\gamma_x, \gamma_y, \gamma_z$ in a rest frame, and with the four matrices $\gamma_{ct}, \gamma_x, \gamma_y, \gamma_z$ in a moving frame. After the γ_{ct}-swap, the term $s_{ct}\gamma_{ct}$ will have been transformed to $s_{ct}\mathbb{1}$. In the absence of a potential, the motion will be uniform such that $s_{ct}\gamma_{ct}$ will be a constant. Hence, in the absence of a potential there will be, after the γ_{ct} swap, only three anti-commuting matrices present in the equation, namely $(\alpha_x, \alpha_y, \alpha_z)$. In other words, the system could be described within SO(3).

An inspection of (5.49) prior to the γ_{ct} swap reveals that the matrix $\sum_\mu cp_\mu\gamma^\mu$ is not unitary. However, multiplication by γ_{ct} changes the structure of the matrices. The proof of Lorentz covariance that was made on the seeding equation (5.26) (with $(\frac{\partial}{\partial c\tau}, \mathbf{0})$ rather than $(\frac{d}{dct}, \boldsymbol{\nabla})$ on the left-hand side) would simply break down if it was first multiplied by γ_{ct}.

The solution to this apparent contradiction with textbook proofs is that they treat Lorentz covariance for bi-vectors rather than for vectors. The structure of a matrix corresponding to a vector becomes block-diagonal after multiplication with γ_{ct}. A block-diagonal structure corresponds to the structure for the transformation of bi-vectors. Hence, the argument about unitary matrices after performing the γ_{ct} swap on the Dirac-like equation (5.39) (with $(\frac{d}{dct}, \boldsymbol{\nabla})$ on the left-hand side) applies to representations of bi-vectors, rather than of vectors. As already mentioned, the approach to the proof of Lorentz covariance made above breaks down after the γ_{ct} swap. But the transformation laws that can be derived for the modified equation do make sense for bi-vectors. If $\mathbf{s}$ were treated as part of a bi-vector, we would obtain:

$$\begin{aligned}\mathbf{s} \to \mathbf{s}' &= \frac{\gamma+1}{2}\,\mathbf{s} - \frac{\gamma-1}{2}\mathbf{u} - \imath\beta\gamma(\mathbf{u}\wedge\mathbf{s}) - \frac{\gamma-1}{2}(\mathbf{u}\wedge(\mathbf{u}\wedge\mathbf{s}))\\ &= \mathbf{s}_{\parallel} + \gamma\mathbf{s}_{\perp} - \imath\beta\gamma(\mathbf{u}\wedge\mathbf{s}_{\perp}).\end{aligned} \tag{5.67}$$

This can be completed by the equation for the transformation of $\mathbf{u}\wedge\mathbf{s}$, which is the second vector component of the bi-vector, and the whole shows then emphatically that bi-vectors do not transform like vectors. An example of how a Lorentz transformation of a bi-vector looks is given by the electromagnetic-field tensor (see Section C.5 of Appendix C). The components of a bi-vector that are parallel to the spatial part $\mathbf{v}$ of the velocity four-vector $(v_{ct}, \mathbf{v})$ of a boost are not affected by Lorentz transformations, while the perpendicular components are. Bi-vectors appear like a complex vector of the type $\mathbf{E} + \imath\mathbf{B}$ in this formalism, but a bi-vector is a tensor, not a complex vector.

For a single vector component $\mathbf{V}$ of a bi-vector, one finds that it is no longer transformed to $\mathbf{MVM}^\dagger$, but to $\mathbf{MVM}^{-1}$. For rotations, where $\mathbf{M}^\dagger = \mathbf{M}^{-1}$, this does not introduce changes, but for boosts, where $\mathbf{M}^\dagger \neq \mathbf{M}^{-1}$, there is a real change between the situations before and after the γ_{ct} swap. Therefore, this swap installs a difference between rotations and boosts. For a rotation with axis $\mathbf{a}$ after the swap, one obtains the result of (5.66) again. This equation presents analogies with (5.67). We can add to (5.66) the equation for the transformation of $\mathbf{a} \wedge \mathbf{s}$, and the set of two equations can then be interpreted as describing the rotation of two vectors $\mathbf{s}$ and $\mathbf{a} \wedge \mathbf{s}$ around $\mathbf{a}$, whereby $\mathbf{s}_{\parallel}$ remains constant. In summary, without the γ_{ct} swap, the covariance is proved for vector solutions, while with the γ_{ct} swap the covariance is proved for bi-vector solutions at constant energy.

In the SL(2,$\mathbb{C}$) representations, the transformation of a vector $\mathbf{V}$ obeys the rule $\mathbf{V}' = \mathbf{LVL}^\dagger$, with $\mathbf{L}^\dagger \neq \mathbf{L}^{-1}$. But in the Dirac representation, the transformation rule for vectors becomes $\underset{\sim}{\mathbf{V}}' = \underset{\sim}{\mathbf{L}}\underset{\sim}{\mathbf{V}}\underset{\sim}{\mathbf{L}}^{-1}$ for the 4×4 Dirac matrices $\underset{\sim}{\mathbf{L}}$ constructed from $\mathbf{L}$, and $\underset{\sim}{\mathbf{V}}$ constructed from $\mathbf{V}$ such that the vectors transform conveniently by similarity transformation, according to unitary transformations in SL(2,$\mathbb{C}$). Unitary transformations in SL(2,$\mathbb{C}$) are at a fixed value m_0c^2 with a variable value for the total energy E (which is just Lorentz invariance), while the unitary transformations discussed in the Dirac representation keep both the total energy E and the rest energy m_0c^2 simultaneously constant. True Lorentz covariance where the total energy E can vary and only the rest energy m_0c^2 is kept fixed is taken care of by similarity transformations. The differences between the rules in SL(2,$\mathbb{C}$) and in the Dirac equation, and the subtlety that the four matrices from (4.10) can in principle all be different, is something that requires great care in using the formalism. Both for rotations and boosts, the set of the four matrices defined by (4.10) contains eventually only two different matrices. But the way the four matrices combine two by two to pairs of equal values is different for rotations and boosts.

Note that in the frame at rest, $\mathbf{s}$ is fixed, such that $\frac{d\mathbf{s}}{d\tau} = 0$. When this is expressed in a moving frame by replacing $\frac{d}{dc\tau}$ by $\frac{\partial}{\partial ct}\mathbb{1} - \boldsymbol{\sigma}\cdot\boldsymbol{\nabla}$, and $\mathbf{s}\cdot\boldsymbol{\sigma}$ by $s_{ct}\mathbb{1} + \mathbf{s}\cdot\boldsymbol{\sigma}$ this leads to a whole set of equations. Operating, once more $\frac{\partial}{\partial ct}\mathbb{1} + \boldsymbol{\sigma}\cdot\boldsymbol{\nabla}$ on these equations we obtain a full set of Maxwell equations, including an equation that is analogous to Faraday's law of induction $\frac{1}{c}\frac{\partial}{\partial t}\mathbf{B} - \boldsymbol{\nabla} \wedge \mathbf{E} = 0$ (see Appendix C). From this approach it can be appreciated that the rotation axis behaves like an electromagnetic potential. A magnetic dipole associated with **s** induces an electric dipole in a moving frame.

5.6 The Dirac equation in the presence of an electromagnetic potential

To justify the minimal substitution requires two ideas:

• *Idea 1: Coding Lorentz transformations by using E and $c\mathbf{p}$ as parameters.* In free space, it is possible to calculate quantities in other frames by using Lorentz transformations. One can ask if the converse is true. In fact, when E and $c\mathbf{p}$ are known in some frame for a particle with rest energy m_0c^2 it is possible to calculate the time part of a pure Lorentz boost that will yield the values E and $c\mathbf{p}$. This can be illustrated by multiplying the textbook equations for the transformation of the time under a simple boost with velocity $\mathbf{v}$ by m_0c^2:

$$\tau = \gamma(t - \mathbf{v}\cdot\mathbf{r}/c^2) \quad \rightarrow \quad m_0c^2\tau = Et - \mathbf{p}\cdot\mathbf{r}. \tag{5.68}$$

Here, the proper time τ is calculated in the frame of the travelling electron in terms of the coordinates $(t, \mathbf{r})$ in the lab frame in which the electron has a velocity $\mathbf{v}$. The velocity parameters are γ and $\boldsymbol{\beta} = \mathbf{v}/c$. Here γm_0c^2 is substituted by E, and $\boldsymbol{\beta}\gamma m_0c^2$ by $c\mathbf{p}$. Therefore, in general E and $c\mathbf{p}$ can be considered as boost parameters.

Of course, the quantities E and $c\mathbf{p}$ contain only three independent parameters, such that the information about the Lorentz transformation will not be complete. A general Lorentz transformation is the composition of a boost and a rotation. Supplementary information (about angular momentum, for instance) may be needed to reconstruct the full transformation. However, this is not the case when only the time coordinate needs to be Lorentz transformed, as then the knowledge of E and $c\mathbf{p}$ is sufficient. In conclusion, the quantities E and $c\mathbf{p}$ can be seen as parameters that contain all the information needed to Lorentz transform the time coordinate.[30]

As all four-vectors transform the same way, the values of the quantities E and $c\mathbf{p}$ contain all the information we need to make the Lorentz transformation of m_0c^2.

Similarly, the value of the four-gradient $\partial_\mu = (\frac{\partial}{\partial ct}, \frac{\partial}{\partial x}, \frac{\partial}{\partial y}, \frac{\partial}{\partial z})$, which is coded as $\sum_\mu \gamma_\mu \partial_\mu$ and transforms like $\gamma^\mu c\hat{p}_\mu$ also defines everything

[30]This reasoning will be questioned in Chapter 8.

needed to make the Lorentz transformation of $\frac{d}{d\tau}$. This is seen when transforming $-\frac{\hbar}{\imath}\frac{d}{dc\tau}\gamma_{ct}$ to a moving frame by $\underset{\sim}{\mathbf{L}}\,(-\frac{\hbar}{\imath}\frac{d}{dc\tau}\gamma_{ct}\,)\,\underset{\sim}{\mathbf{L}}^{-1}$ and writing the result as $-\frac{\hbar}{\imath}\sum_{\mu}\gamma_{\mu}\partial_{\mu}$. Here, the coordinates (ct,x,y,z) used in the differentiation are the coordinates in the moving frame. The operators $(\frac{\partial}{\partial ct},\frac{\partial}{\partial x},\frac{\partial}{\partial y},\frac{\partial}{\partial z})$ are thus frame-dependent values, because in another frame they would be $(\frac{\partial}{\partial ct'},\frac{\partial}{\partial x'},\frac{\partial}{\partial y'},\frac{\partial}{\partial z'})$. The four-gradient $\partial_{\mu}=(\frac{\partial}{\partial ct},\boldsymbol{\nabla})$ and the energy-momentum $cp_{\mu}=(E,c\mathbf{p})$ are four-vectors. We can obtain the transformation for $(\frac{\partial}{\partial ct},\boldsymbol{\nabla})$ from the transformation for $(E,c\mathbf{p})$ by making the substitution $\gamma^{\mu}\,\hat{\mathrm{p}}_{\mu}=-\frac{\hbar}{\imath}\gamma_{\mu}\partial_{\mu}$, and *vice versa*.

• *Idea 2: The rest mass changes in a potential.* Imagine that the electron is at rest within a Coulomb potential V of a point charge. There is no magnetic field, such that we can put $\mathbf{A}=\mathbf{0}$. Suppose that the electron is at a finite distance $r=r_0$ from the point charge such that $V(r)\neq 0$. Of course, at infinite distance $r\to\infty$ the electron would feel a potential $\lim_{r\to\infty}V(r)=0$. At infinite distance, the rest mass of the electron is m_0 such that its rest energy is m_0c^2. At finite distance $r=r_0$, the rest energy will now no longer be m_0c^2 but $m_*c^2=m_0c^2+qV$, because the potential energy contributes to the total energy. Here, q is the charge of the electron. For example, in the Coulomb potential of a proton, $qV<0$ and $m_*c^2<m_0c^2$. The difference in energy $m_0c^2-m_*c^2$ has been radiated away. When an electron comes from a situation at rest at $r\to\infty$ to settle on a stable orbit, it must emit radiation. On this stable orbit the electron will not be at rest but have some finite velocity. To come to absolute rest, it would have to emit further radiation. It would be elegant physically to explain the whole change in rest energy in terms of radiation, but this is complicated by the fact that it is not possible to go from a situation at rest at $r\to\infty$ to another situation at rest in $r=r_0$ by a single radiative process. The conservation laws of energy and momentum must be applied simultaneously, and this leads in general to the result that in a single radiation process the particle must recoil to satisfy the conservation of both energy and momentum simultaneously. The description in terms of radiative processes can be made correct by assuming a *globally recoilless series of photon emissions* from the initial state where the electron is at rest at infinite distance $r\to\infty$ to a final state where the electron is again at rest at a finite distance $r=r_0$ from the nucleus. The treatment in terms of radiation could also become more complicated for a Dirac equation describing a charged particle that would be different from the electron and have internal rotational degrees of freedom. The laws of conservation of energy and

momentum would then also have to allow for the possibility of "inelastic processes" with internal transitions between the rotational energy levels of the charged particle.

It may be asked in what respect this lower-energy electron at rest would look different from the electron at rest at $r \to \infty$. The answer is probably that it has a polarized charge distribution, analogous to the way a dielectric solid lowers its energy within an external electric field. The electron may contain only charges of one sign, such that it would have to be the shape of the charge distribution that has to be changed, but this level of speculations is unnecessary; it suffices to know that the rest energy is now $m_* c^2 = m_0 c^2 + qV < m_0 c^2$.

• *Combining the two ideas.* The problem in a potential is now that the relation between E, $c\mathbf{p}$, and $m_0 c^2$ used to define the Lorentz transformation will no longer correctly determine the parameters γ and $\boldsymbol{\beta}$, because E and $c\mathbf{p}$ are no longer purely kinetic. However, it is possible to maintain the correct relation between the parameters by Lorentz transforming $m_* c^2 = m_0 c^2 + qV$. Under a Lorentz transformation, both $m_* c^2$ and qV will be transformed, while $m_0 c^2$ is a constant of nature. The total energy $m_* c^2$ will transform to the four-vector $(E, c\mathbf{p})$, and the static Coulomb potential will transform to the four-potential $(qV, q\mathbf{A})$. Putting the quantities that transform on one side of the equation and the quantities that are constants on the other, it becomes thus:

$$\sum_\mu (cp_\mu - qA_\mu)\gamma^\mu, \tag{5.69}$$

which will be the four-vector that is the correct Lorentz transformation of $m_0 c^2 \gamma^{ct}$. In order to correctly describe the Lorentz transformation it is thus necessary to make the substitutions $E \to E - qV$, $c\mathbf{p} \to c\mathbf{p} - q\mathbf{A}$. Furthermore, from classical mechanics it is known that this scheme works perfectly to calculate the dynamics of a charged particle within an electromagnetic field, such that the same conclusion is reached in two different ways.

In free space, $(E, c\mathbf{p})$ transform the same way as $(\frac{\partial}{\partial ct}, \boldsymbol{\nabla})$. In the free-space Dirac equation, $\sum_\mu \gamma_\mu \partial_\mu = \gamma_{ct}\frac{\partial}{\partial ct} + \boldsymbol{\nabla}\cdot\boldsymbol{\gamma}$ is the correct Lorentz transformation of $\gamma_{ct}\frac{d}{dc\tau}$, and from this it can be inferred that $\sum_\mu \gamma^\mu c p_\mu$ is the correct Lorentz transformation of $m_0 c^2 \gamma^{ct}$, or *vice versa.* This idea can now also be applied the other way around to derive from the rules $E \to E - qV$, $c\mathbf{p} \to c\mathbf{p} - q\mathbf{A}$ the substitution needed for the four-gradient such as to obtain its correct Lorentz-transformed expression in the presence of a potential. In fact, the whole derivation of the Dirac equation in this book has been made within a classical framework (see for example the conclusion in Section 5.2). There is no reason to doubt the validity of the classical rules $E \to E - qV$, $c\mathbf{p} \to c\mathbf{p} - q\mathbf{A}$ within this framework. This justifies the minimal substitution, at least partly.

• *The magnetic field.* We must insist on the qualifier "partly". It has not been possible, for example to justify it for the case $V = 0$, $\mathbf{A} \neq \mathbf{0}$, where in the rest frame there is no electric field and only a non-zero magnetic field. Such a frame cannot be obtained by covariance from a frame where $V \neq 0$ & $\mathbf{A} = \mathbf{0}$. Our demonstration is thus not complete. However, as the context wherein the Dirac equation has been derived is entirely classical, one can invoke the argument that the rule has already been established to be generally valid within classical mechanics. But the arguments can also be improved with the approach described here. In fact, the objection about the failure of covariance is misleading as every magnetic field is produced by a moving charge. In the case $V = 0$ & $\mathbf{A} \neq \mathbf{0}$, in reality $(V, \mathbf{A}) = (V_1, \mathbf{A}_1) + (V_2, \mathbf{A}_2)$, where $(V_1, \mathbf{A}_1)$ is the four-potential of the moving electrons within the atoms and $(V_2, \mathbf{A}_2)$ is the four-potential of the nuclei of the atoms which are at rest, such that $\mathbf{A}_2 = \mathbf{0}$. The electric field V_1 of the moving electrons is thus compensated by the electric field of the nuclei at rest. This partly resolves the problem, however it must be noted that certain magnetic fields are not attributed to the orbital motion of the electrons but to their spin. With respect to this point, it could be argued that the rule is expected only to depend on $\mathbf{A}$, not on the mechanism responsible for $\mathbf{A}$.

• *Minimal coupling.* The substitution is called minimal because the coupling introduced is minimal, considering that the electron only interacts with the electromagnetic field as a point charge, and neglecting its spin and current. It has not been taken into account that it could have a magnetic dipole moment related to its spin that could couple to a magnetic field. Nor has it been discussed that such a moving magnetic dipole could give rise to an electric dipole, or that the electron could possess higher-order multipole moments.

• *Change in the meaning of the partial derivatives.* The aim of the minimal substitution is to obtain the kinetic part $(E_{kin}, c\mathbf{p}_{kin})$ of the energy-momentum four-vector $(E, c\mathbf{p})$, because it is the kinetic part that can be used to write the instantaneous Lorentz transformation. What is needed are the instantaneous boost parameters. These are instantaneous, local numerical values and it is irrelevant how they are related to the values $(E(\mathbf{r}, t), c\mathbf{p}(\mathbf{r}, t))$ of these parameters in other points $(\mathbf{r}, t)$ of space-time.

This is the reason why the partial derivatives should be defined in a way that does not consider the functional dependence $(E(\mathbf{r},t), c\mathbf{p}(\mathbf{r},t))$. This is different from the true definition, which was just based on the fact that ψ is a function of four arguments, $\psi \in F(\mathbb{R}^4, \mathbb{C})$. With the pristine definition it was thus conceivable that the partial derivatives of $(E(\mathbf{r},t), c\mathbf{p}(\mathbf{r},t))$ had to be drawn into the calculations, but this is here definitely no longer the case. This is the change of definition of the partial derivatives anticipated in Subsection 3.10.5.5.

• *The laboratory frame.* In reality, the term *"lab frame"* that has been used here regularly, is not exactly the lab frame, but the frame of the nucleus. In fact, the minimal substitution for the case $\mathbf{A} \neq \mathbf{0}$ can only be correct in the frame of the nucleus, as this is the only frame wherein the electromagnetic potential can be constant and attributed the values given in these calculations. The frame of the nucleus is not the lab frame. The frame of the nucleus is the only one wherein the potential is the same before and after the radiative process, provided that the transitions do not affect the internal state of the nucleus, which is fortunately the case. The frame of the nucleus may itself recoil a little, it may also jiggle due to the electron motion. But this will be ignored in the calculations undertaken in this book by considering the nucleus so heavy that one can overlook the fact that the rest frame of the nucleus, wherein the potential of the nucleus remains fixed, is not just a single fixed frame but a whole set of them with different velocities at different times. This is, however, an approximation.

We can Lorentz-transform the Dirac rule, and in doing so see that it automatically corrects for previous radiation effects. Following the classical knowledge, the particle emits radiation all of the time. However, this may be due to the fact that only three degrees of freedom are attributed to the electron, *viz.* the translational ones. By adding the rotational degrees, the necessity of radiation loss can be avoided. It is by taking care of the parameter $\mathbf{s}$ that this can be achieved.

5.7 The g-factor of the electron

As many know, the Chinese expression for "crisis" consists of two characters side by side. The first is the symbol for "danger", the second the symbol for "opportunity".

— Al Gore

The fact that the Dirac equation has now been derived from a set of well-defined assumptions introduces a completely new situation in physics. The better understanding of the Dirac equation can now be used as a bistoury with which one can analyze the meaning of the calculations. The historical situation was different. In a sense, Dirac's equation had to be accepted as God-given. Dirac just guessed his equation. The equation was then validated by comparing its predictions with the experimental results and it passed the test with flying colours. It was therefore a phenomenal breakthrough. But such an approach has also its limitations. With such an equation we are entitled to conjecture that the set of axioms that would be necessary to derive it might contain a number of mystery axioms that manage to capture some physical truth that is beyond intuition in a miraculous way. One could marvel at the magic. But in the new situation there is not too much space for such beliefs as the equation has been derived classically. The mystery resides in the non-classical solutions that are adopted for the equation.

It may be pertinent to add here perhaps a remark on the magnetic moment of the electron within the context of the textbook Dirac equation. In the presence of an external magnetic field, one can consider the non-relativistic limit of this Dirac equation, which results in a term that contains $\imath \frac{\hbar q}{2m_0 c} [\mathbf{B} \cdot \boldsymbol{\sigma}] \Psi$.

As explained in Subsection 3.10.2, the term $\mathbf{B} \cdot \boldsymbol{\sigma}$ is merely the coding of a single vector, *viz.* the magnetic field $\mathbf{B}$. In fact, within SU(2) $\subset$ SL(2,$\mathbb{C}$) any vector $\mathbf{a} = (a_x, a_y, a_z)$ is just coded as $\mathbf{a} \cdot \boldsymbol{\sigma}$. The "vector" $\boldsymbol{\sigma}$ that occurs in this notation is only a useful convention introduced to write the three Pauli matrices $\sigma_x, \sigma_y, \sigma_z$ simultaneously in a more compact form. These matrices do not correspond to vector components, but rather to vectors in their own right. One can consider them as the codings for the three basis vectors $\mathbf{e}_x$, $\mathbf{e}_y$, and $\mathbf{e}_z$, such that $\boldsymbol{\sigma}$ corresponds to a notation for the full basis. The term $\mathbf{B} \cdot \boldsymbol{\sigma}$ is thus not a genuine scalar product between the vector $\mathbf{B}$ and some hypothetical vector $\boldsymbol{\sigma}$. The quantity $\frac{\hbar q}{2m_0 c} \mathbf{B} \cdot \boldsymbol{\sigma}$ can be rewritten as $-\mu_B \boldsymbol{\sigma} \cdot \mathbf{B}$, where μ_B is the Bohr magneton. It is then possible to write $\mu_B \boldsymbol{\sigma} = \mu \hbar \boldsymbol{\sigma} = (2\mu) \frac{\hbar}{2} \boldsymbol{\sigma}$, where $\mu = \frac{q}{2m_0 c}$ is the classical gyromagnetic ratio.[31]

A very inconvenient truth must now be addressed, *viz.* that Dirac very unfortunately got lost in Galileo's dark labyrinth when he misinterpreted

[31]Sometimes one encounters $\mu = \frac{q}{2m_0}$ rather than $\mu = \frac{q}{2m_0 c}$. The factor $\frac{1}{c}$ just represents the transition from electrostatic to electromagnetic units.

the notation as a genuine scalar product, whereby the term $\mathbf{S} = \frac{\hbar}{2}\boldsymbol{\sigma}$ would correspond to the electron spin. This is incorrect; being a vector in its own right, the spin $\mathbf{S} = \frac{\hbar}{2}\,\mathbf{s}$ is coded as $\frac{\hbar}{2}\,\mathbf{s}{\cdot}\boldsymbol{\sigma}$. The spin is thus proportional to $\mathbf{s}{\cdot}\boldsymbol{\sigma}$, where $\mathbf{s}$ is the unit vector along the spin axis. As already stated, every vector quantity $\mathbf{a}$ is coded as $\mathbf{a}{\cdot}\boldsymbol{\sigma}$ in SU(2), and $\boldsymbol{\sigma}$ here has nothing to do with spin; there is no direction (of spin) inside $\boldsymbol{\sigma}$. In the term $\mathbf{s}{\cdot}\boldsymbol{\sigma}$, it is $\mathbf{s}$ which codes the spin, not $\boldsymbol{\sigma}$. The coding of the spin vector must correspond to one matrix, not to three matrices σ_x, σ_y, σ_z. The triplet of matrices $\boldsymbol{\sigma}$ should be considered as the representation of a basis of three vectors $\mathbf{e}_x$, $\mathbf{e}_y$, $\mathbf{e}_z$ for $\mathbb{R}^3$, rather than a single vector. As such it only serves to denote a choice of reference frame. The term $-\mu_B\boldsymbol{\sigma}{\cdot}\mathbf{B}$ represents a vector, while the energy term $-\boldsymbol{\mu}{\cdot}\mathbf{B}$ is a scalar. Hence, $-\mu_B\boldsymbol{\sigma}{\cdot}\mathbf{B}$ can never be identified with $-2\mu\mathbf{S}{\cdot}\mathbf{B}$.

It is the notation $\mathbf{a}{\cdot}\boldsymbol{\sigma}$ for the coding of the vector quantity $\mathbf{a}$ within SU(2) that misleadingly looks like a scalar product between the vector $\mathbf{a}$ and some vector $\boldsymbol{\sigma}$, which fooled Dirac into believing that $\mathbf{B}\cdot\frac{\hbar q}{2m_0}\boldsymbol{\sigma}$ would correspond to the coding of a scalar product that would be the quantum mechanical counterpart of the classical quantity $\frac{\hbar q}{2m_0 c}\mathbf{s}\cdot\mathbf{B}$. Through the over-interpretation, $\frac{\hbar q}{2m_0 c}\mathbf{B}\cdot\boldsymbol{\sigma}$ can then be replaced by $-2\mu\mathbf{S}{\cdot}\mathbf{B}$, such that one has to postulate the existence of a g-factor $g = 2$ inside the "magnetic moment" $-2\mu\mathbf{S}$. This fiddle factor $g = 2$, is thus just a gimmick that must be introduced to make sure that the calculations keep producing the correct results within the context of the wrong interpretation. It is needed to compensate for the term $\frac{\hbar}{2}$ introduced by error to get $\frac{\hbar}{2}\boldsymbol{\sigma}$ into the equation. There is thus absolutely no real physics related to it. This can be seen as a mystifying, *ad hoc* physical argument intended to explain away the difficulties produced by the over-interpretation $\mathbf{S} = \frac{\hbar}{2}\boldsymbol{\sigma}$. The *ad hoc* argument leaves the uneasy impression that there is something that can never be understood, while there is in reality nothing to understand. Despite the incorrect interpretation that is given here for the rules that must be followed to carry out the calculations, the rules themselves are correct.[32]

[32]A very scary problem must now be evoked. The whole traditional theory for magnetism in the solid state capitalizes on Dirac's picture of a magnetic dipole moment associated with the spin $\frac{\hbar}{2}\,\mathbf{s}$ whereby $\mathbf{B}{\cdot}\boldsymbol{\sigma}$ is loosely identified with $\mathbf{B}\cdot\mathbf{s}$, and it does this with satisfactory results. An attempt will be made to find a solution that restores that picture in Chapter 9.

There are two problems with the traditional treatment of the spin in a magnetic field[33]:

- Within the context of the Dirac equation, the spin is described by an equation wherein the spin is not allowed to move in the rest frame of the electron, because it is assumed that $\mathbf{s}$ is constant in the Rodrigues equation it all started from.
- In the derivation of the minimal substitution no allowance has been made for some possible interaction between the magnetic field and the spin of the electron by introducing a corresponding coupling term. The minimal substitution only describes the coupling of the electromagnetic field to the charge of the electron.

One may think then that it is possible to restore the picture and get a coupling between the spin and the magnetic field by getting the spin into the equations using the Dirac-like equation (5.38) that contains the spin explicitly.[34]

---

[33]Within the context of the traditional Dirac equation, the solution to the paradox created by the fact that $\mathbf{B}\cdot\boldsymbol{\sigma} \neq \mathbf{B}\cdot\mathbf{s}$ will (within a non-relativistic context of SU(2)) come in Section 9.6, where it will be shown that the term $\frac{\hbar q}{2m_0}\mathbf{B}\cdot\boldsymbol{\sigma}$ contains two ingredients:

(1) The fact that it behaves as a vector forces us to align the spin $\frac{\hbar}{2}\mathbf{s}$ with the magnetic field $\mathbf{B}$ to obtain a solution that is stationary with respect to the energy. The equations show, in fact, that the wave function must be an eigenfunction of $\mathbf{B}\cdot\boldsymbol{\sigma}$ in order to be stationary with respect to the energy.

(2) The alignment condition is certainly a consequence of the way the equation is set up. It is assumed that $\mathbf{s}$ does not vary with time in the rest frame of the electron. If a stick is placed in a gravitational field, the only way to prevent its orientation from being varied with time is to put it in perfectly vertical position. To find a more general situation for the electron spin, both the possibility that $\mathbf{s}(\tau)$ varies with time and a coupling term that will define how it will vary must be allowed for.

It is possible to reconstruct the unit vector $\mathbf{s}(\tau)$ along the spin axis from the wave function. In a rest frame this unit vector calculated from the wave function and the wave function itself are related through $[\,\mathbf{s}(\tau)\cdot\boldsymbol{\sigma}\,]\,\psi = \psi$, such that we must have $\mathbf{s}(\tau) \parallel \mathbf{B}$. As it is assumed that $\mathbf{B} = B\mathbf{e}_z$, this proves at once that we must have $\mathbf{s}(\tau) = \mathbf{e}_z$ to obtain a state that is stationary with respect to the energy.

[34]It would be possible to modify the Dirac-like equation (5.38), knowing now exactly what we are doing. Both the Dirac equation and the Dirac-like equation are set up to yield states that are stationary with respect to the energy and they only describe such states. If $\mathbf{s} \parallel \mathbf{B}$ must be true to obtain a stationary state, then a state with $\mathbf{s} \nparallel \mathbf{B}$ cannot be entered meaningfully into the equations, as the equations are only correct for the stationary case $\mathbf{s} \parallel \mathbf{B}$. It is thus necessary to set up a new equation that allows for $\mathbf{s}$ to change with time in the frame of the electron and for a coupling between the magnetic field $\mathbf{B}$ and $\mathbf{s}(t)$. But it would be impossible to guess the magnitude of this coupling.

But as will be discovered in Subsection 9.2.1, the main term in the new equation will still be the term $\frac{\hbar q}{2m_0}\mathbf{B}{\cdot}\boldsymbol{\sigma}$.[35]

5.8 More over-interpretations in traditional quantum mechanics

The same mistake of over-interpreting $\boldsymbol{\sigma}$ as the unit vector along the spin axis arises when one interprets $\hat{\mathbf{L}}{\cdot}\boldsymbol{\sigma}$ as a "spin-orbit coupling", rather than just the coding of the vector quantity $\hat{\mathbf{L}}$. This "spin-orbit" coupling has nothing to do with coupling between orbital motion and spin, even if such a coupling is a physically plausible effect. The misinterpretation of $\frac{\hbar}{2}\boldsymbol{\sigma}$ as a spin operator will be discussed in Subsection 9.2.2.

Another misinterpretation of the same type occurs in particle physics with the definition of the "helicity" $\mathbf{u}{\cdot}\boldsymbol{\sigma}$, where $\mathbf{u} = \mathbf{p}/|\mathbf{p}|$. Again, this is just the transcription of the unit vector $\mathbf{u}$ in the language of SU(2), not the projection of the spin on the direction of motion. This matter of pure notation cannot be used to claim that the spin should be parallel with the direction of motion, nor that the difference between chirality and helicity would be a subtle issue. The notation says nothing more than that $\mathbf{p}$ is a vector. Moreover, introducing $\mathbf{u}$ is not Lorentz covariant as $\mathbf{p}$ is part of a four-vector rather than a Euclidean vector.

Finally, the incorrect interpretation of $\boldsymbol{\sigma}$ also leads to the erroneous idea that the various components of the spin cannot exist or not be measured

---

[35] As the term $\frac{\hbar q}{2m_0}\mathbf{B}{\cdot}\boldsymbol{\sigma}$ forces the spin to be aligned with the magnetic field, it is impossible to check the veracity of the "potential-energy" term $-g\boldsymbol{\mu}{\cdot}\mathbf{B}$. The interpretation of $\frac{\hbar q}{2m_0}B = \frac{\hbar q}{2m_0 c}\mathbf{s}\cdot\mathbf{B}$ is thus not a strict logical necessity. It could be argued that everything just works *as though* some magnetic dipole moment is associated with the spin, and that the picture of a magnetic dipole moment can therefore be preserved as it is *compatible* with the experimental results. But the pictorial interpretation raises the very difficult question how it is possible that the calculations treat the corresponding physics correctly, while it is impossible to spot any underlying assumptions in the derivation of the equations that could be responsible for the success of the calculations. First of all, the minimal substitution does not introduce a coupling term between a hypothetical magnetic dipole moment and the magnetic field. Secondly, it could be imagined that the hypothetical magnetic dipole moment of the electron is due to internal current loops, but no information about such currents has been introduced into the equations. According to Lorentz, such a mechanism can never explain the magnitude of the anomalous Zeeman effect observed. One is forced to conclude that the term $\mathbf{B}{\cdot}\boldsymbol{\sigma}$ only describes a coupling between the charge of the electron and the magnetic field, as this is all the minimal coupling can introduce. This is also confirmed by the derivation of the anomalous Zeeman term $q\mathbf{B}{\cdot}\boldsymbol{\sigma}$ and the spin-orbit coupling term $(q/c)(\mathbf{v}\wedge\mathbf{E}){\cdot}\boldsymbol{\sigma}$ in (C.24) which does not rely on any notion of spin or quantum mechanics and only couples the charge-current four-vector to the electromagnetic-field tensor. This remark may even apply to the result from quantum electrodynamics.

simultaneously, because the operators do not commute. The operator $\hat{\mathrm{S}}_z = \frac{\hbar}{2}\sigma_z$ does not express the z-component of the spin, but the spin when it is aligned along the z-axis, as the general expression $\frac{\hbar}{2}\,[\,\mathbf{s}{\cdot}\boldsymbol{\sigma}\,]$ clearly shows, by putting $\mathbf{s} = \mathbf{e}_z$. The x- and y-components of $\mathbf{s}{\cdot}\boldsymbol{\sigma}$ are then just zero. All that $[\,\hat{\mathrm{S}}_z, \hat{\mathrm{S}}_x\,] \neq 0$ expresses is a tautology: when the spin is aligned with $\mathbf{e}_z$ then it cannot be aligned with $\mathbf{e}_x$, and *vice versa*. From what has been explained in Chapter 3, it is easy to see that $\sigma_x^2+\sigma_y^2+\sigma_z^2 = (\sigma_x+\sigma_y+\sigma_z)^2 = [\,(1,1,1){\cdot}\boldsymbol{\sigma}\,]^2 = 3\,\mathbb{1}$ just expresses $(\mathbf{e}_x + \mathbf{e}_y + \mathbf{e}_z)^2 = 3$. The square of the unit vector $\mathbf{s}$ is obtained from the algebra: $[\,\mathbf{s}{\cdot}\boldsymbol{\sigma}\,]^2 = s^2\,\mathbb{1} = \mathbb{1}$. Both sides in these two equations can now be multiplied by $\frac{\hbar^2}{4}$ to show that the square of the length of the spin vector $\frac{\hbar}{2}\mathbf{s}$ is $\hbar^2/4$ rather than $3\hbar^2/4$, as claimed in textbooks, based on the incorrect interpretation of the algebra.

5.9 Conclusion

The results obtained up to now form a solid basis for further exploring quantum mechanics. The master equations have been derived, it is known exactly what they mean and there is nothing counter-intuitive about them. There is evidence enough that the approach is reliable and innovating. Very obviously it could never have been successful without the constant concern of being meticulous about the mathematics. To dilineate the frontier between classical mechanics and quantum mechanics it is vital to continue working classically in the further development and act as though the task were to prove that quantum mechanics is just a part of the theory of relativity. It must thus be avoided at all price to introduce special new assumptions just to make it easier to obtain the proofs. The aim of this working philosophy is to spot the counter-intuitive physical effects that cannot possibly be explained without introducing special non-classical assumptions. These special assumptions will then really be axioms of quantum mechanics. Moreover, working with classical ideas has the advantage that it does not require the genius intuition of a "quantum mechanic".

5.10 Complementary remarks on mathematics

5.10.1 *Relation with the Lie algebra*

5.10.1.1 *Operators*

It has been possible to introduce $m_0c^2\tau = Et - \mathbf{p}\cdot\mathbf{r}$ as a phase of a spinor quantity ψ that looks like a wave. It is based on this fact that one traditionally introduces the operators $\hat{\mathrm{E}} = -\frac{\hbar}{\imath}\frac{\partial}{\partial t}$ and $\hat{\mathbf{p}} = \frac{\hbar}{\imath}\boldsymbol{\nabla}$ for energy and momentum.[36] The Schrödinger equation is often "derived" by introducing such operators and requiring that an equation for the conservation of energy must result. Analogously, the Dirac equation is "traditionally" derived starting from $\sum \gamma^{\mu} c\hat{\mathrm{p}}_{\mu}$ with the condition that squaring it must lead to $E^2 - c^2\mathbf{p}^2 = (m_0c^2)^2$. But in this kind of derivation, it is not clear why operators should be introduced in the first place, and also why the operator corresponding to the potential energy in the Schrödinger equation is just a multiplication with V. The development requires the introduction of the notion that there is a wave associated with a particle, and one must explain why this wave is travelling faster than light.

The derivation of the Dirac equation based on the Rodrigues formula and $\gamma_{ct}\frac{d}{dc\tau} = \sum \gamma_{\mu}\frac{\partial}{\partial x_{\mu}}$ comes about naturally while trying to express that the electron is spinning. It is less puzzling than the traditional derivations and gives meaning to them. The problem of the superluminal velocities will receive a much simpler solution. We can appreciate that our derivation improves our understanding of the traditional "derivations". It provides so to say a missing link. Taking the step from the derivation proposed in this book with its introduction of the gamma matrices to Dirac's postulate that also introduces the gamma matrices just hinges on the fact that the energy-momentum four-vector $(E, c\mathbf{p})$ and the four-gradient $(\frac{\partial}{\partial ct}, \boldsymbol{\nabla})$ transform the same way under Lorentz transformations, because they are both four-vectors. The traditional substitutions $\hat{\mathrm{E}} = -\frac{\hbar}{\imath}\frac{\partial}{\partial t}$ and $\hat{\mathbf{p}} = \frac{\hbar}{\imath}\boldsymbol{\nabla}$ allow then a return to the original result.[37]

As by this an operator formalism is introduced, it is now possible to continue the discussion of angular momentum started in Subsection 3.10.5. The

---

[36]The fact that the expression of $\hat{\mathrm{E}}$ contains a minus sign while the one for $\hat{\mathbf{p}}$ does not means that it is possible to write $-\frac{\hbar}{\imath}\gamma_{ct}\frac{d}{dc\tau} = -\frac{\hbar}{\imath}\gamma_{ct}\frac{\partial}{\partial c\tau} - \frac{\hbar}{\imath}\gamma_{x}\frac{\partial}{\partial x} - \frac{\hbar}{\imath}\gamma_{y}\frac{\partial}{\partial y} - \frac{\hbar}{\imath}\gamma_{z}\frac{\partial}{\partial z}$ as $-\frac{\hbar}{\imath}\gamma_{ct}\frac{d}{dc\tau} = \sum \gamma^{\mu}\hat{\mathrm{p}}_{\mu}$. Note that both γ^{μ} and γ_{μ} satisfy all the commutation relations that define the gamma matrices. Therefore, a confusion between them does not lead to equations that are *a priori* incorrect when one considers them out of context, because it simply corresponds to a different choice for the gamma matrices. But within a context it jeopardizes the overall consistency. Finally, it may be noted that the Dirac equation is historically derived from $(\sum \gamma_{\mu} c\hat{\mathrm{p}}_{\mu})^2 = (m_0c^2)^2\mathbb{1}$, (i.e. with opposite conventions for γ^{μ} and γ_{μ} to those described here, as noted in Footnote 7), which is why we have put "traditionally" between quotes in the main text.

[37]The different signs in the operators $-\frac{\hbar}{\imath}\frac{\partial}{\partial t}$ and $\frac{\hbar}{\imath}\boldsymbol{\nabla}$ come from the fact that these four-gradient operators belong to a left-handed representation in the equations. We must work on the right-handed spinor Ψ with the left-handed four-gradient $\frac{\partial}{\partial ct}\mathbb{1} - \boldsymbol{\nabla}\cdot\boldsymbol{\sigma}$. The best way to see this is perhaps by inspection of (5.38).

operator formalism makes it possible to redefine the general mathematical operators $\hat{\mathrm{L}}_\mu$ in terms of the more specific angular-momentum operators. The relationship between spin and angular momentum can now also be discussed.

In SU(2), there is the particularity that the role of the operator $\frac{1}{2}\sigma_z$ can also be fulfilled by the operator $\hat{\mathrm{L}}_z\sigma_z = \imath(x\frac{\partial}{\partial y} - y\frac{\partial}{\partial x})\sigma_z$ working on the spinor. This is generally true for the spinor $[\xi_0, \xi_1]^\top$ as a function of $(x, y, z) \in \mathbb{C}^3$. We must then express (ξ_0, ξ_1) in terms of (x, y, z) according to (3.11)–(3.13). Operating with $\hat{\mathrm{L}}_z\sigma_z$ on $[\xi_0, \xi_1]^\top$ we obtain the same result as with $\frac{1}{2}\,\sigma_z\,[\xi_0, \xi_1]^\top$. The differentiation calculus remains the same when (x, y, z) are restricted to $\mathbb{R}^3$.

But in considering $(x, y, z) \in \mathbb{R}^3$, an alternative derivation can be used, based on the fact that $\hat{\mathrm{L}}_z$ reduces to $\imath\frac{\partial}{\partial \phi}$ when spherical coordinates (r, θ, ϕ) are introduced for (x, y, z). Even more derivations become possible when the spinor corresponds to a rotation around the z-as, such that its ξ_1-component is zero. Introducing spherical coordinates (r, θ, ϕ), the spinor will then be $[e^{-\imath\phi/2}, 0]^\top$. Using the Rodrigues formula with rotation angle φ and axis $\mathbf{n} = \mathbf{e}_z$, this will become $[e^{-\imath\varphi/2}, 0]^\top$, such that $\varphi = \phi$. Then $\hat{\mathrm{L}}_z\sigma_z$, $\imath\frac{\partial}{\partial\phi}\sigma_z$, $\imath\frac{\partial}{\partial\varphi}\sigma_z$, and $\frac{1}{2}\sigma_z$ can be used on the spinor $[e^{-\imath\phi/2}, 0]^\top$, and with all the different forms the same result will be obtained, whereby the spinor turns out to be an eigenvector. The other eigenvector is the conjugate spinor $[\xi_1, -\xi_0]^\top$.

There is thus no real need to introduce a different operator for spin and angular momentum.[38] It can be seen from the form $\frac{1}{2}\sigma_z = \frac{1}{2}\mathbf{e}_z{\cdot}\boldsymbol{\sigma}$ that $(\sigma_x, \sigma_y, \sigma_z)$ are a basis for the spin operators $\frac{1}{2}\mathbf{s}{\cdot}\boldsymbol{\sigma}$. The physically meaningful choice is $\mathbf{s} = \mathbf{e}'_z$. The Pauli matrices fulfill thus exactly the same role in SU(2) as $\mathbf{e}_x, \mathbf{e}_y, \mathbf{e}_z$ in $\mathbb{R}^3$.

The Pauli matrices are reflection operators. The three fundamental spin operators $\sigma_x, \sigma_y, \sigma_z$ are thus reflection operators. The Pauli matrices constitute a canonical basis for the general reflection operators $\mathbf{e}'_z{\cdot}\boldsymbol{\sigma}$. The reflections are the generators of the rotation group, which is true in general.

The infinitesimal generators $\hat{\mathrm{L}}_\mu$ of the Lie algebra satisfy commutation relations that are similar to those for the matrices $\frac{1}{2}\sigma_\mu$; they are thus very similar to the reflection operators in the representation theory of SU(2).

[38]This will be the key to Dirac's statement that it is necessary to use an operator $\hat{\mathbf{J}} = \hat{\mathbf{L}} + \frac{1}{2}\hat{\boldsymbol{\sigma}}$ in the hydrogen atom. (Here, the pre-factor $\hbar$ is ignored in the operators.) In reality, apart from the fact that they are both wrongly coded, $\hat{\mathbf{L}}$ and $\frac{1}{2}\hat{\boldsymbol{\sigma}}$ are the same operators.

They are, however, not true reflection operators as in general they do not square to the identity operator (due to the presence of a factor $\imath$). It will now be discussed how the differential form of the Rodrigues formula looks like an operational definition of an infinitesimal generator of the Lie algebra.

5.10.1.2 *Lie algebras*

The term "infinitesimal generator" within the Lie algebra is somewhat unfortunate, as it is not infinitesimal. It is also in general not a generator of the group in the way the Pauli matrices can be considered as generators for the rotation group. In the rotation group, the Pauli matrices can be used to define reflection operators that generate the group of rotations and reversals. The best way to understand conceptually what the "infinitesimal generators" are about is to illustrate it on the homogeneous Lorentz group. The Lie group operates on the four-dimensional four-vectors of Minkowski space-time $\mathbb{R}^4$ which is a vector space (with a metric). The Lie group of the homogeneous Lorentz transformations itself is a six-dimensional manifold. However, the playground of the Lie algebra is yet another mathematical space: the six-dimensional tangent space $\mathbb{R}^6$ to the six-dimensional manifold at the identity element. The choice of the identity element is arbitrary, but it has the advantage that it makes the expressions simpler.

Finding the "infinitesimal generators" consists only in choosing an appropriate basis for this six-dimensional tangent space, which (in marked contrast with the group manifold) is a true vector space. To find such a basis it is not really necessary to introduce a number of well-chosen one-parameter sets of infinitesimal group elements and deriving the tangent vectors that correspond to them by a procedure of taking a Lie derivative. The one-parameter sets are well chosen in the sense that they are "mutually" orthogonal. One can choose any normalized orthogonal basis of the tangent space right ahead. Any choice will do. It is for instance this approach that is used in introducing the "infinitesimal operators" for SU(3). It is then not immediately obvious how this can be related to the procedure based on Lie derivatives. The number of "infinitesimal generators" will be the number of independent parameters that are needed to specify the full *Vielbein*. In SU(n), with its complex metric, this is $n^2 - 1$.

Textbook introductions to Lie algebras *do not define explicitly* what an infinitesimal generator is. They proceed by giving examples for the simple cases of the rotation group and the Lorentz group, from which the reader is supposed to be able to pick up the idea. The approach consists in taking the

Lie derivatives of well-selected one-parameter sets that lead to the correct result. But it is not discussed how this selection of one-parameter sets has to be performed. One may at that stage not feel the need for more clarification as it all looks ever so easy. But in trying to work it out for a less obvious case like SU(3) after finding a parameterization for it, it may not be obvious how to express the one-parameter sets in terms of the parameters introduced.[39] The basic idea that one wishes to define a normalized orthogonal basis for tangent space is needed to understand how one can make a proper selection of one-parameter sets. One also needs a definition of this orthogonality. The only valid definition is the one given above: the infinitesimal generators are a normalized orthogonal basis for the tangent space at the identity element.

In other words, the one-parameter sets have to be chosen in such a way that they define the proper complete normalized orthogonal basis for tangent space evoked in the definition. Moreover, the orthogonality of the six basis vectors is not an orthogonality in space-time but one in a six-dimensional space of matrices that belong to the linear group L(4,$\mathbb{C}$). The reader is invited to consider a trivial basis of matrices $\mathbf{e}_{jk}$, with $(j,k) \in ([1,n] \cap \mathbb{N})^2$ for the n^2-dimensional vector space L(n,$\mathbb{C}$), defined by $(\mathbf{e}_{jk})_{\ell m} = \delta_{j\ell}\delta_{km}$ and to verify that:

$$< \mathbf{A}, \mathbf{B} > = \sum_{j=1}^{n}\sum_{k=1}^{n} a_{jk}^{*} b_{jk} = \mathrm{Tr}(\mathbf{A}^{\dagger}\mathbf{B}) \tag{5.70}$$

defines a scalar product for two vectors $\mathbf{A} \in$ L(n,$\mathbb{C}$) and $\mathbf{B} \in$ L(n,$\mathbb{C}$). Here the "vectors" $\mathbf{A}$ and $\mathbf{B}$ are actually complex $n \times n$ matrices. When the matrices are real, one can replace $\mathbf{A}^{\dagger}$ by $\mathbf{A}^{\top}$. It is with respect to this scalar product that the orthogonality of the basis of tangent space is defined.

Here again the exception that the dimensions of the vector space $\mathbb{R}^3$ and of the manifold SU(2) are the same for the group of the rotations in $\mathbb{R}^3$ can be a source of confusion, leading to the misconception that generators and infinitesimal generators would always coincide. The Pauli matrices can be considered as a basis for the reflection operators that generate the rotation

[39] With certain parameterizations of the manifold it can become very difficult to express the choices to be made in terms of special combinations of the given parameters. The parameter sets correspond to curvilinear coordinates that may contain singularities. It is tempting to take the coordinate lines as the one-parameter sets for which one could calculate the tangent vectors by taking the Lie derivatives. But the coordinate lines are not necessarily all mutually orthogonal, and the system of curvilinear coordinates may even have a singularity at the identity element, just as spherical coordinates for a sphere present singularities at the poles of the sphere. This results in confusion, because we do not find a complete basis.

group. In this sense, the Pauli matrices are a vector basis for the true generators of the rotation group. They can also be considered as a basis for the infinitesimal generators of the rotation group. In the rotation group the two concepts coincide, but not in the Lorentz group. The fact that both concepts go by the name "generator" can then only amplify the confusion.

When we generalize towards the Lorentz group, the resulting vector space $\mathbb{R}^4$ is four-dimensional while the resulting manifold SL(2,$\mathbb{C}$) is six-dimensional. The reflection operators are defined by unit vectors that are normal to the reflection planes. When the Pauli matrices are considered as a vector basis for the reflection operators that generate the rotation group, it can thus be expected that their generalization will be a set of four Dirac gamma matrices that will play the role of a vector basis for the reflection operators of space-time that generate the Lorentz group. This is of course because space-time $\mathbb{R}^4$ is four-dimensional. But when the Pauli matrices are considered as a vector basis for the infinitesimal generators of the rotation group, then in the generalization we will need a set of six matrices, as the manifold is six-dimensional.

Based on the way the Pauli matrices can be obtained as infinitesimal generators for the rotation group by expressing that they are the true generators defined by a rule of the type $\sigma_\mu\sigma_\nu + \sigma_\nu\sigma_\mu = 2\delta_{\mu\nu}\mathbb{1}$, one could expect that the infinitesimal generators of the Lorentz group could be orthogonal with respect to a rule of the type $\gamma_\mu\gamma_\nu + \gamma_\nu\gamma_\mu = 2k_{\mu\nu}\mathbb{1}$, with $|k_{\mu\nu}| = 1$. In reality, this expresses orthogonality of vectors of $\mathbb{R}^4$ when we take $k_{\mu\nu} = g_{\mu\nu}$. One can perhaps discover rather quickly that it is impossible to find the infinitesimal generators by the Dirac trick, as the Dirac representation contains only five gamma-matrices γ_μ that satisfy $\gamma_\mu\gamma_\nu + \gamma_\nu\gamma_\mu = 2k_{\mu\nu}\mathbb{1}$, while a six-dimensional basis is needed here.

What is needed is not to have an orthogonality of vectors in the vector space $\mathbb{R}^4$ on which the transformation matrices are working, but an orthogonality of the transformation matrices themselves in their own vector space. This vector space is a subspace of SL(4,$\mathbb{C}$), and has a different scalar product than the vector space $\mathbb{R}^4$. There are thus two types of scalar products for the 4×4 matrices that may occur in the Dirac representation: one for four-vectors of space-time that can be coded by 4×4 matrices, and one for Lorentz transformations of the group manifold that can also be coded by 4×4 matrices. The two scalar products are different. The scalar product that can be used for the infinitesimal generators is defined in (5.70).

The infinitesimal generators of the Lorentz group are all mutually orthogonal with respect to that scalar product in their six-dimensional vector space of matrices. The orthogonality in $\mathbb{R}^6$ of the infinitesimal generators of the Lorentz group is based on the definition of (5.70) for the 4×4 matrices. Similarly, the infinitesimal generators of SU(3) are all mutually orthogonal, not in $\mathbb{C}^3$, but in SL(3,$\mathbb{C}$) according to (5.70). It is for this reason that they do not look orthogonal with respect to a rule of the type $\gamma_\mu \gamma_\nu + \gamma_\nu \gamma_\mu = 2\delta_{\mu\nu}\mathbb{1}$.

The discussion which shows that we must use different scalar products for the vector space and for the manifold clearly illustrates the difference between "generators" and "infinitesimal generators". The six-dimensional "infinitesimal generators" that build the required normalized orthogonal basis for the Lie algebra are not the "generators" of the group in the sense that they could serve as a basis for reflection operators that generate the group. The true generators of the Lorentz group that generalize the idea of the Pauli matrices in the form of reflection operators are defined by a normalized four-dimensional four-vector of Minkowski space-time.

5.10.1.3 *The differential form of the Rodrigues formula as an operational definition of the infinitesimal generators*

This chapter began with a derivation of the Rodrigues formula:

$$\frac{d}{d\varphi}\psi = -\frac{\imath}{2}\,[\,\mathbf{n}{\cdot}\boldsymbol{\sigma}\,]\,\psi. \tag{5.71}$$

From this it can be appreciated that up to some factors, this corresponds just to defining an infinitesimal generator. By choosing $\mathbf{n} = \mathbf{e}_x$, $\mathbf{n} = \mathbf{e}_y$, $\mathbf{n} = \mathbf{e}_z$, this leads (after normalization) to a proper basis of infinitesimal generators $\sigma_x, \sigma_y, \sigma_z$. As for $\mathbf{e}_z$, the operator $\frac{d}{d\varphi}$ can be linked to $x\frac{\partial}{\partial y} - y\frac{\partial}{\partial x}$; the angular momentum operators can be shown to correspond to generators of the Lie algebra, according to the description given to derive the expressions for the generators. This prescription starts, however, from infinitesimal rotations, such that at first sight it may look confusing that they seem to correspond to reflection operators (when one restricts them to SU(2)).

The infinitesimal rotation is not a reflection operator because there is a factor $\imath$ that intervenes in the definition of an infinitesimal generator. As already stated, the identity of the operators $\hat{\mathrm{L}}_z\sigma_z$ and $\frac{1}{2}\sigma_z$ only holds in SU(2).

The angular-momentum operators can also be used in higher-degree representations based on spherical harmonics obtained from the spinors by tensor products. Eventually, in such representations, the angular momentum states will also be described in terms of two variables (the spherical coordinates (θ, ϕ) rather than three.

Just as in the case of SU(2), the angular-momentum operators $\hat{\mathrm{L}}_x\sigma_x$, $\hat{\mathrm{L}}_y\sigma_y$, $\hat{\mathrm{L}}_z\sigma_z$, fulfill the role of a basis, for general angular-momentum operators with an axis $\mathbf{s} = \mathbf{e}'_z$. The corresponding operator will be $s_x\hat{\mathrm{L}}_x\sigma_x + s_y\hat{\mathrm{L}}_y\sigma_y + s_z\hat{\mathrm{L}}_z\sigma_z$. In spherical coordinates, these operators project out the degree 1/2 in $e^{\imath\varphi}$ from the expression $[\xi_0, \xi_1]^\top$. In higher-order representations they therefore also project the degrees of the polynomials, because these polynomials are obtained from tensor products of the spinors. As already suggested in Subsection 3.10.5.5, the concept of degree operator is more fundamental than the concept of angular momentum. This will be discussed in more detail in Chapter 6 and in Chapter 12.

It is easy to check from a Taylor expansion of the Rodrigues formula $\cos(\varphi/2)\mathbb{1} - \imath\sin(\varphi/2)\mathbf{n}\cdot\boldsymbol{\sigma}$ for small values of φ, that an infinitesimal rotation has the form $\mathbb{1} - \frac{\imath}{2}\,\varphi\,\mathbf{n}\cdot\boldsymbol{\sigma}$. From this it is obvious that in order to recover the generator $\frac{1}{2}\mathbf{n}\cdot\boldsymbol{\sigma}$ from this infinitesimal rotation, it is necessary to drop the unit matrix, multiply by $\imath$ and then take the factor that goes with φ. But this prescription is exactly the definition of how to obtain a generator in general for the Lie algebra. The operations involved in the whole procedure in the case $\mathbf{n} = \mathbf{e}_z$ can be expressed as $\imath\frac{\partial}{\partial\varphi}$, which is nothing else than $\hat{\mathrm{L}}_z$.[40]

5.10.1.4 *Transporting vectors*

Introducing a differential form of the Rodrigues formula is thus similar to introducing the Lie algebra. But it was important to render this differential equation Lorentz covariant. In fact, taking the Lie derivatives at the identity element may simplify the expressions so much that it conceals the true symmetry. To clearly display the full symmetry it is therefore useful to write the equations in their most general covariant form. In a sense, this permits co-transportation of vectors of space-time when moving on the manifold. This procedure could perhaps be considered as mimicking

[40]Note that while an isotropic vector is used to code a rotation within SU(2), the matrix representation of the isotropic vector is not a rotation matrix. This is obvious from the fact that their determinants are different and it is due to the re-normalization procedure that is needed to make the transition from an isotropic vector to a spinor. The procedure of subtracting the unit matrix in the derivation of the expressions for the generators of the Lie group has therefore no relationship with what occurs in the isomorphism $\mathbb{1} + \mathbf{e}_z\cdot\boldsymbol{\sigma} \leftrightarrow \mathbf{e}_z\cdot\boldsymbol{\sigma}$.

the concept of parallel transport on a manifold. But as already identified, we lift a degeneracy in the dimensions that exists in the rotation group when we go to the Lorentz group. In the Lorentz group the dimension of tangent space is six while the dimension of space-time is four. The real case of parallel transport on the Lorentz group manifold would thus be one from six-dimensional tangent space to six-dimensional tangent space. But when covariance of spin is required, we are transporting four-dimensional vectors. (To also incorporate the vectors of $\mathbb{R}^4$ into the tangent space of the Lie group, it may be necessary to use the Poincaré group rather than the homogeneous Lorentz group. This may then serve to explain why the vectors of $\mathbb{R}^4$ also transform by similarity transformations. These are points not yet studied and must therefore be considered as guesses.)

It may be noted for fun that it is possible to render the differential form of the Rodrigues formula $\frac{d}{d\tau}\psi = -\imath\frac{\omega_0}{2}\,[\,\mathbf{e}_z\cdot\boldsymbol{\sigma}\,]\,\psi$ covariant within SL(2,$\mathbb{C}$) by considering $\mathbf{e}_z$ as a locally simplified form of a complex quantity $\mathbf{s} \in \mathbb{C}^3$ that transforms under Lorentz transformations like an electromagnetic tensor $\mathbf{E} + \imath c\mathbf{B}$.[41] The special local value of $\mathbf{e}_z$ for $\mathbf{s}$ is then a nice example of a case where the simplification conceals the true symmetry of $\mathbf{s} \in \mathbb{C}^3$. One can see this by just stipulating that $\mathbf{s}$ should transform according to: $\mathbf{s}\cdot\boldsymbol{\sigma} \rightarrow \mathbf{L}\,[\,\mathbf{s}\cdot\boldsymbol{\sigma}\,]\,\mathbf{L}^{-1}$ rather than $\mathbf{s}\cdot\boldsymbol{\sigma} \rightarrow \mathbf{L}\,[\,\mathbf{s}\cdot\boldsymbol{\sigma}\,]\,\mathbf{L}^{\dagger}$.[42] This demonstrates that there is a lot of ambiguity in a Rodrigues formula with $\mathbf{n} = \mathbf{e}_z$. In fact, $\mathbf{e}_z$ can be interpreted as $\mathbf{n}$, as a special value of $\mathbf{e}'_z$, and even as a special

---

[41]The reason why the six-component quantity appears like a three-dimensional complex vector from $\mathbb{C}^3$ can be traced back to the fact that the basis elements $\mathbf{e}_\mu$ and $\mathbf{e}_{\mu+3}$ of the Lie algebra sl(2,$\mathbb{C}$) are related by $\mathbf{e}_{\mu+3} = \imath\mathbf{e}_\mu$. (See equation 17.16 of [Cornwell (1984)].)

[42]Angular momentum can also be seen as a six-component real quantity (that can be coded as a three-component complex quantity). In fact, by generalizing the vectors $\mathbf{r}$ and $\mathbf{p}$ that occur within the matrix product $[\,\mathbf{r}\cdot\boldsymbol{\sigma}\,]\,[\,\mathbf{p}\cdot\boldsymbol{\sigma}\,]$ to four-vectors we obtain a generalization of the concept of angular momentum. The algebra shows that the product contains a scalar invariant $(Et - \mathbf{p}\cdot\mathbf{r})\mathbb{1}$ and a complex "vector" of $\mathbb{C}^3$. The real part of this quantity corresponds to the angular momentum, such that the six real components of the complex "vector" could be interpreted as a generalization of angular momentum. Interpreting $\mathbf{e}_z$ as the orientation of some angular momentum in $\mathbb{R}^3$, we obtain then a six-component generalization of angular momentum, and the corresponding real and imaginary components could give rise to the magnetic and induced electric dipole moment of the electron. Note however, that we have not at all proved yet that the six real components of angular momentum obtained this way transform according to a rule $\mathbf{s}\cdot\boldsymbol{\sigma} \rightarrow \mathbf{L}\,[\,\mathbf{s}\cdot\boldsymbol{\sigma}\,]\,\mathbf{L}^{-1}$, where $\mathbf{s}$ is complex. This raises the question of whether we have $[\,\mathbf{r}\cdot\boldsymbol{\sigma}\,]\,[\,\mathbf{p}\cdot\boldsymbol{\sigma}\,] \rightarrow \mathbf{L}\,[\,\mathbf{r}\cdot\boldsymbol{\sigma}\,]\,[\,\mathbf{p}\cdot\boldsymbol{\sigma}\,]\,\mathbf{L}^{-1}$, which is not obvious. It can be proved by: $[\,\mathbf{r}\cdot\boldsymbol{\sigma}\,]\,[\,\mathbf{p}\cdot\boldsymbol{\sigma}\,] \rightarrow \mathbf{L}\,[\,\mathbf{r}\cdot\boldsymbol{\sigma}\,]\mathbf{L}^{\dagger}\mathbf{L}^{\dagger-1}\,[\,\mathbf{p}\cdot\boldsymbol{\sigma}\,]\,\mathbf{L}^{-1}$. From this it transpires that the alternative ways to render the differential form of the Rodrigues formula covariant are full of non-trivial technicalities.

value of $\mathbf{r} \wedge \mathbf{p}/|\mathbf{r} \wedge \mathbf{p}|$. The interpretation in terms of $\mathbf{n}$ is not covariant, but the other two interpretations can be rendered covariant. Here, the fact that $\mathbf{e}_z$ is a special value of $\mathbf{e}'_z$ has served as a guiding principle. One of the reasons for this was that it was not clear how the four-gradient ∂_μ and the four-potential A_μ could be generalized to some hypothetical six-dimensional counterparts. With hindsight there is another reason: a six-parameter spin would have to be block-diagonal in (5.38), while the four-gradient has a block structure that is off-diagonal. The block-structures simply do not fit into the same equation.

5.10.1.5 *Bilinear covariants*

The relationship between angular momentum and spin can be understood by considering what the possible bilinear covariants of the Lorentz group are. In $\mathbb{R}^3$ the vectors constitute by definition a three-dimensional vector space. These vectors are associated with reflection operators that generate the rotation group. It can be asked if there exist representations that are rank-0 in the vectors (these are the scalars), rank-1 (the vectors), rank-2 (the bi-vectors), and rank-3 (the tri-vectors). The number of basis vectors in these representations are given by the binomial formula 1, 3, 3, 1, because these numbers correspond to the ways we can choose zero, one, two, and three different vectors in the set of three basis vectors. The rotations are obtained as a product of two reflections. The reflections are thus rank-1, while the rotations are rank-2. But the vectors and the bi-vectors can at first sight be confused as their representations have the same dimensions. In reality, they are different as the vector changes sign under a parity transformation, while the bi-vector does not. For the same reason, a tri-vector is not identical to a scalar as it changes sign under a parity transformation.

This can be seen at work algebraically in (5.1). In $[\mathbf{a}\cdot\boldsymbol{\sigma}][\mathbf{b}\cdot\boldsymbol{\sigma}] = (\sum_j a_j\sigma_j)(\sum_j b_k\sigma_k) = \sum_{jk} a_j b_k \sigma_j\sigma_k$, there are contributions where $j = k$. We will have then $\sigma_j\sigma_k = \mathbb{1}$, and this leads to the scalar contribution $\mathbf{a}\cdot\mathbf{b}\mathbb{1}$. In this expression the unit vectors $\mathbf{e}_j$ coded by σ_j have disappeared because σ_j has cancelled with another σ_j. It is therefore an expression with zero vectors. In the other terms, $\sigma_j\sigma_k$ can be rewritten in terms of σ_l: $\sigma_x\sigma_y = \imath\sigma_z(cycl.)$. Doing this yields the bi-vector $\mathbf{a}\wedge\mathbf{b}$. Thus, the bi-vector is of the type $\sum_{j\neq k} c_{jk}\sigma_j\sigma_k$ (with an additional factor $\imath$). It is an expression based on all combinations $\mathbf{e}_j\mathbf{e}_k$ with two vectors, but due to the relations $\sigma_x\sigma_y = \imath\sigma_z(cycl.)$ it can be written as a linear combination of the vectors. The tri-vectors will be expressions of the type $\sum_{j\neq k\neq l\neq j} c_{jkl}\sigma_j\sigma_k\sigma_l$. There will be only one such term. It is a pseudo-scalar because it changes sign under a parity transformation. There are no other quantities, as adding one more Pauli matrix to a product of three will lead to a simplification.

In $\mathbb{R}^4$, the numbers in the binomial formula are 1, 4, 6, 4, and 1, such that rank-1 and rank-2 representations no longer have the same components. They are four- and six-dimensional respectively. We have scalars, vectors $\sum_j c_j \gamma_j$, tensors $\sum_{j \neq k} c_{jk} \gamma_j \gamma_k$, tri-vectors $\sum_{j \neq k \neq l \neq j} c_{jkl} \gamma_j \gamma_k \gamma_l$, and pseudo-scalars $c_{0123} \gamma_0 \gamma_1 \gamma_2 \gamma_3$. As in the Weyl representation, the gamma matrices have their non-zero blocks all off-diagonal, products of two gamma matrices will be all block-diagonal, such that they cannot be expressed as linear combinations of γ_μ. The combinations $\mathbf{e}_j \mathbf{e}_k$, or $\gamma_j \gamma_k$ with $j \neq k$, can thus be considered as an additional set of basis vectors, with matrix representation $\sigma_{jk} = \gamma_j \gamma_k$. It is possible to express the products of three different gamma matrices as products of two gamma matrices by using the identity $\gamma_5 = \gamma_0 \gamma_1 \gamma_2 \gamma_3$ to simplify the equations. This way the possible expressions are all of the form $\Gamma = \sum_{\alpha\beta} c_{\alpha\beta} \gamma_\alpha \gamma_\beta$, where both α and β can run from 0 to 3, keep a constant value (e.g. $\alpha = 5$), or can be totally absent.

All possible bilinear covariants of the Lorentz group can then be created by building expressions of the type $\overline{\psi}_1 \Gamma \psi_2$, where $\overline{\psi}_1 = \psi^\dagger \gamma_0$ and ψ_1, ψ_2 are one-column 4×1 matrices. The transformation properties of the expressions can be checked in Table 5.3.

In some of the expressions a factor $\imath$ appears as for $\mathbf{a} \wedge \mathbf{b}$ in SU(2). They can be multiplied with $\imath$ to obtain expressions that are all real. To be sure of obtaining a real expression for rank 0, we take $\psi_1 = \psi_2 = \psi$. To understand the choice of the expression $\overline{\psi}_1 \Gamma \psi_2$ as the starting point, it is necessary to check the result of the operation $\overline{\Psi} \rightarrow (\gamma_0 \overline{\Psi})^\dagger = \overline{\Psi}^\dagger \gamma_0$ that consists in multiplying by γ_0 and taking the Hermitian conjugate on the matrix $\overline{\Psi}$ in (5.52). It can be seen then that $(\gamma_0 \overline{\Psi})^\dagger = (\, \mathbf{L}^{-1} \,;\, \mathbf{L}^\dagger \,)$ transforms with $\underset{\sim}{\mathbf{L}}^{-1}$, which is a feature needed for the covariance. From the discussion of (5.58) it is known that it is physically meaningful to take one-column quantities in the expressions.

Subsection 5.5.2.2 showed that in space-time the spin corresponds to a four-component axial vector. The angular-momentum bi-vector becomes a six-component anti-symmetric tensor. Axial vectors (i.e. spin) and bi-vectors (corresponding to generalized angular momentum) have then no longer the same numbers of components as in $\mathbb{R}^3$. Relativistic spin and relativistic angular momentum can thus not be added as claimed in the solution of the Dirac equation for the hydrogen problem by postulating $\hat{\mathbf{J}} = \hat{\mathbf{S}} + \hat{\mathbf{L}}$.

Table 5.3 Bilinear covariant

rank 0:	$\overline{\psi}\psi$	$=$	$\overline{\psi}\, \underset{\sim}{\mathbf{L}}^{-1}\underset{\sim}{\mathbf{L}}\psi$	$=$	$\overline{\psi}'\psi'$
rank 1:	$\overline{\psi}_1\,(\sum_\mu c_\mu\gamma_\mu)\,\psi_2$	$=$	$\overline{\psi}_1\, \underset{\sim}{\mathbf{L}}^{-1}[\,\underset{\sim}{\mathbf{L}}\,(\sum_\mu c_\mu\gamma_\mu)\,\underset{\sim}{\mathbf{L}}^{-1}]\,\underset{\sim}{\mathbf{L}}\psi_2$	$=$	$\overline{\psi}_1'\,(\sum_\mu c_\mu\gamma_\mu)'\,\psi_2'$
rank 2:	$\overline{\psi}_1\,(\sum_\mu c_{\mu\nu}\gamma_\mu\gamma_\nu)\,\psi_2$	$=$	$\overline{\psi}_1\, \underset{\sim}{\mathbf{L}}^{-1}[\,\underset{\sim}{\mathbf{L}}\,(\sum_\mu c_{\mu\nu}\gamma_\mu\gamma_\nu)\,\underset{\sim}{\mathbf{L}}^{-1}]\,\underset{\sim}{\mathbf{L}}\psi_2$	$=$	$\overline{\psi}_1'\,(\sum_{\mu\nu} c_{\mu\nu}\gamma_\mu\gamma_\nu)'\,\psi_2'$
rank 3:	$\overline{\psi}_1\,(\sum_\mu c_\mu\gamma_\mu\gamma_5)\,\psi_2$	$=$	$\overline{\psi}_1\, \underset{\sim}{\mathbf{L}}^{-1}[\,\underset{\sim}{\mathbf{L}}\,(\sum_\mu c_\mu\gamma_\mu\gamma_5)\,\underset{\sim}{\mathbf{L}}^{-1}]\,\underset{\sim}{\mathbf{L}}\psi_2$	$=$	$\overline{\psi}_1'\,(\sum_\mu c_\mu\gamma_\mu\gamma_5)'\,\psi_2'$
rank 4:	$\overline{\psi}_1\,\gamma_5\psi_2$	$=$	$\overline{\psi}_1\, \underset{\sim}{\mathbf{L}}^{-1}[\,\underset{\sim}{\mathbf{L}}\,\gamma_5\,\underset{\sim}{\mathbf{L}}^{-1}]\,\underset{\sim}{\mathbf{L}}\psi_2$	$=$	$-\overline{\psi}_1'\,\gamma_5\,\psi_2'$

Table 5.4 Construction of angular-momentum representations

$$\xi_0^{2\ell-k}\xi_1^k \quad = \quad \xi_0^{2\ell-2k}\xi_0^k\xi_1^k \quad \propto \quad \xi_0^{2\ell-2k}z^k$$

$$\underbrace{\xi_0\xi_0\cdots\xi_0}_{2\ell-k\text{ times}}\;\underbrace{\xi_1\xi_1\cdots\xi_1}_{k\text{ times}} \quad = \quad \underbrace{\xi_0\xi_0\cdots\xi_0}_{2\ell-2k\text{ times}}\;\underbrace{\xi_0\xi_0\cdots\xi_0}_{k\text{ times}}\;\underbrace{\xi_1\xi_1\cdots\xi_1}_{k\text{ times}} \quad \propto \quad \underbrace{\xi_0\xi_0\cdots\xi_0}_{2\ell-2k\text{ times}}\;\underbrace{zz\cdots z}_{k\text{ times}}$$

$$\Big\Downarrow \begin{array}{l}\xi_0 \propto e^{-\imath\phi/2} \equiv \uparrow \\ \xi_1 \propto e^{+\imath\phi/2} \equiv \downarrow\end{array} \qquad\qquad \Big\Downarrow \qquad\qquad \Big\Downarrow \uparrow\downarrow \text{ annihilate } (\boxtimes)$$

$$\underbrace{\uparrow\uparrow\cdots\uparrow}_{2\ell-k\text{ times}}\;\underbrace{\downarrow\downarrow\cdots\downarrow}_{k\text{ times}} \quad = \quad \underbrace{\uparrow\uparrow\cdots\uparrow}_{2\ell-2k\text{ times}}\;\underbrace{\uparrow\uparrow\cdots\uparrow}_{k\text{ times}}\;\underbrace{\downarrow\downarrow\cdots\downarrow}_{k\text{ times}} \quad \propto \quad \underbrace{\uparrow\uparrow\cdots\uparrow}_{2\ell-2k\text{ times}}\;\underbrace{\boxtimes\boxtimes\cdots\boxtimes}_{k\text{ times}}$$

total degree is ℓ		remaining degree in $e^{-\imath\phi}$ is $\ell-k$		degree in z is k
ℓ	$=$	$\ell-k$	$+$	k

Both spin and angular momentum cannot be three-component quantities, as is assumed in the solution of the Dirac equation, since there are no three-component bilinear covariants in the Lorentz group. This shows that the corresponding operators that occur in the algebra will have to be interpreted differently in terms of parameters that define features (such as the dimension) of the representations chosen.

Parity transformations $\mathbf{r} \to -\mathbf{r}$ have been used in $\mathbb{R}^3$ to distinguish between vectors and axial vectors or between scalars and pseudo-scalars. Transformations of the type $\mathbf{r} \to -\mathbf{r}$ in $\mathbb{R}^4$ are not able to distinguish between vectors and axial vectors or between scalars and pseudo-scalars. Another criterion is needed.

5.10.1.6 *General angular momentum*

Within higher-dimensional representations of the rotation group $\hat{\mathrm{J}}_z \propto \frac{d}{d\phi}$ can be considered as an operator for the number $\ell - k$, where k counts the number of single spin reflections $\uparrow \to \downarrow$ that are present in a polynomial $\xi_{2\ell-k}\xi_1^k$ according to the schema of Table 5.4.

Starting from $\xi_0^{2\ell}$, the polynomial $\xi_0^{2\ell-k}\xi_1^k$ is obtained by k flips. In the schema each symbol $\boxtimes$ stands for a combination $\uparrow\downarrow$ that annihilate each other (according to $-2\xi_0\xi_1 = z$ which reduces to $e^{\imath\phi/2}e^{-\imath\phi/2} = 1$ when $\theta = \pi/2$). Thus, $\ell - k$ corresponds then to the remaining spin within $\uparrow\uparrow\cdots\uparrow\boxtimes\boxtimes\cdots\boxtimes$. As in SU(2) (where $\ell = \frac{1}{2}$), the remaining spin $\ell - k$ can only be $\frac{1}{2}$ ($\uparrow$) or $-\frac{1}{2}$ ($\downarrow$), it can be understood why (apart from the constant $\hbar$) $\hat{\mathrm{L}}_z$ coincides with $\frac{1}{2}\sigma_z$ within SU(2).

Chapter 6

Towards a Better Understanding of Quantum Mechanics

6.1 The phase velocity of the de Broglie wave

In the very beginning of wave mechanics it was noted that the phase velocity of a de Broglie wave is larger than c. When the electron in its rest frame is pictured as a spinning top, then it can be considered both as a gyroscope and as a watch. The gyroscope comparison is obvious: the spin axis of the gyroscope is the spin axis of the electron. The watch comparison is also valid as the rotation angle $\varphi = \omega_0 \tau$ is a measure for the total time τ that has elapsed starting from the reference time $\tau = 0$. The time can then be measured by counting the number of turns the top has made around its rotation axis. This number of turns can be taken as a non-integer, real number $\omega_0 \tau / 2\pi$ and is, allowing for the conversion factor $\omega_0 / 2\pi$, a measure for the time. The electron clock works in this sense in a way that is perfectly analogous to the functioning of a regular clock. The rotation angle $\varphi = \omega_0 \tau$ around the spin axis corresponds to twice the phase $\varphi/2$ of the wave function as e.g. discussed in Section 3.8, and as transpires e.g. from (5.5). (See also (6.1) below.) The Dirac equation describes the time on this clock in the rest frame.

Note that when this equation was set up in the rest frame, for example in (5.26), the position of the particle was not specified. This is because in daily life we take it for granted that all clocks in different places are synchronized, such that the influence of the position of a clock on its reading is not an issue. The assumed synchronization of the clocks in different places $\mathbf{r}_1$ and $\mathbf{r}_2$ must follow the Einstein synchronization procedure, by sending light signals to and fro between observers in $\mathbf{r}_1$ and $\mathbf{r}_2$. If we wanted to synchronize clocks in $\mathbf{r}_1$ and $\mathbf{r}_2$ by a one-way signal, then this signal would have to travel at infinite speed. This is of course impossible and is the reason why it is necessary to use the Einstein synchronization procedure.

Due to the prior synchronization, the phases of the spinors $\psi(\mathbf{r}_j, \tau)$ in different places $\mathbf{r}_j$ are the same. The hypothetical one-way synchronization wave from $\mathbf{r}_1$ to $\mathbf{r}_2$ thus ensures that the phases in $\mathbf{r}_1$ and $\mathbf{r}_2$ are the same, such that there is no need to enter $\mathbf{r}$ in the expressions. This will change in a frame wherein the electron is moving. The time part of the Lorentz transformation $\tau = \gamma(t - \mathbf{v}\cdot\mathbf{r}/c^2)$ requires the information about the position $\mathbf{r}$, because the clock readings t will now depend on the position $\mathbf{r}$. In a frame wherein the electron appears to be moving, the clocks are no longer synchronized the same way as in the rest frame of the electron.

For an electron spinning around the z-axis we have in the rest frame:

$$\begin{aligned} \mathbf{R}(\tau) &= \begin{pmatrix} e^{-\imath\omega_0\tau/2} & 0 \\ 0 & e^{+\imath\omega_0\tau/2} \end{pmatrix}, \\ \psi_0(\tau) &= \mathbf{R}(\tau)\begin{pmatrix} 1 \\ 0 \end{pmatrix} = e^{-\imath\omega_0\tau/2}\begin{pmatrix} 1 \\ 0 \end{pmatrix}. \end{aligned} \tag{6.1}$$

After a general Lorentz transformation $\mathbf{L}$ of $\Psi(\mathbf{r}_0, \tau)$ for an arbitrary point $\mathbf{r}_0$ in the rest frame of the electron, we will obtain for a frame wherein the electron appears to be moving:

$$\begin{aligned} \Psi(\mathbf{r}, t) = \mathbf{L}\mathbf{R}(\mathbf{r}_0, \tau) &= \begin{pmatrix} a & b \\ c & d \end{pmatrix}\begin{pmatrix} e^{-\imath\omega_0\tau/2} & 0 \\ 0 & e^{+\imath\omega_0\tau/2} \end{pmatrix} \\ &= \begin{pmatrix} a\, e^{-\imath\omega_0\tau/2} & b\, e^{+\imath\omega_0\tau/2} \\ c\, e^{-\imath\omega_0\tau/2} & d\, e^{+\imath\omega_0\tau/2} \end{pmatrix}, \\ \psi(\mathbf{r}, t) &= e^{-\imath\omega_0\tau/2}\begin{pmatrix} a \\ c \end{pmatrix}. \end{aligned} \tag{6.2}$$

The same notations as in (4.9) have been adopted here. Here $(\mathbf{r}, t)$ is the Lorentz-transformed value of $(\mathbf{r}_0, \tau)$. This shows that the spinor continues to contain a single frequency. The spin vector corresponding to $\mathbf{R}$ in the rest frame is $\mathbf{S}_0 = \mathbf{e}_z{\cdot}\boldsymbol{\sigma} = \sigma_z$, such that $-\mathbf{S}_0^\star = \mathbf{S}_0 = \sigma_z$. The Lorentz transformation will transform $-\mathbf{S}_0^\star$ into:

$$-\,\mathbf{S}^\star = \mathbf{L}^{\dagger-1}\sigma_z\mathbf{L}^{-1} = \begin{pmatrix} dd^* - cc^* & -bd^* + ac^* \\ -bd^* + a^*c & bb^* - aa^* \end{pmatrix} \tag{6.3}$$

Rotations and boosts in the Oxy plane leave $\mathbf{e}_z$ invariant, such that $-\mathbf{S}^\star = \sigma_z$ and $\mathbf{S} = \sigma_z$ will remain true. As for these boosts $c \neq 0$, (6.2) shows that the spinor $\psi(\mathbf{r}, t)$ will now no longer be an eigenvector of the spin matrices $\mathbf{S}$ and $-\mathbf{S}^\star$. This may look paradoxical, but is due to an inherent weakness of the SL(2,$\mathbb{C}$) representation: it is not able to account properly for space-time reflections, because there are not enough 2×2 matrices to satisfy the conditions $\gamma_\mu\gamma_\nu + \gamma_\nu\gamma_\mu = 2g_{\mu\nu}\mathbb{1}$, and is therefore not able to express spin operators correctly. This is exactly what we discussed from Subsection 5.4.4.1 onwards. One needs the four-dimensional Dirac representation in order to be able to treat all space-time reflections correctly. The paradox that $\psi(\mathbf{r},t)$ seems no longer to be an eigenvector of its spin vector in the representation SL(2,$\mathbb{C}$) is therefore solved by lifting the calculations to the four-dimensional representation and constructing $\overline{\Psi}(\mathbf{r},t)$ from $\Psi(\mathbf{r},t)$ and $\Upsilon^\star(\mathbf{r},t) = -\mathbf{S}^\star\,\Psi(\mathbf{r},t)$ as explained in Subsection 5.5.3.

What these considerations show is that the components of the spinor ψ always have a common phase $\tilde{\psi}(\tau) = e^{-\imath\omega_0\tau/2}$. It can therefore be called *the phase of the wave function.* After a boost with velocity parameter $\mathbf{v} = v\mathbf{u}$, where $\mathbf{u} = \mathbf{v}/v$, the invariant scalar $e^{-\imath\omega_0\tau/2}$ can be rewritten as $e^{-\imath\frac{\omega}{2}(t-\frac{\mathbf{v}\cdot\mathbf{r}}{c^2})}$, by applying $\omega_0 = \omega/\gamma$ and $\tau = \gamma(t - \frac{\mathbf{v}\cdot\mathbf{r}}{c^2})$. As explained in the derivation of (5.68), the latter can be transformed into $e^{-\imath(Et-\mathbf{p}\cdot\mathbf{r})/\hbar}$ by using the substitution $\hbar\omega_0 = 2m_0c^2$ (or $\hbar\omega = 2mc^2$) introduced in Subsection 5.1.3. The phase of the wave function thus corresponds to the phase of the de Broglie wave. This phase can also be rewritten as $\tilde{\psi}(\mathbf{r},t) = e^{-\imath\omega_0\tau/2} = e^{-\imath\frac{\omega_0}{2}\gamma(t-\mathbf{v}\cdot\mathbf{r}/c^2)}$. This describes only the clock readings of the electron in a frame where the electron is moving.

Let us nevertheless try to interpret $\tilde{\psi}(\mathbf{r},t)$ as a wave travelling in space. For this purpose $\tilde{\psi}(\mathbf{r},t)$ can be rewritten as $\tilde{\psi}(\mathbf{r},t) = e^{+\imath\gamma\frac{\omega_0}{2}\,\mathbf{v}\cdot(\mathbf{r}-\mathbf{w}t)/c^2}$, where $\mathbf{w} = \frac{c^2}{v}\mathbf{u}$. By putting $\hbar\omega_0 = 2m_0c^2$ as in Subsection 5.1.3, we obtain then $\hbar\gamma\frac{\omega_0}{2}\,\mathbf{v}/c^2 = \mathbf{p}$. Introducing a second substitution $\mathbf{p} = \hbar\mathbf{k}$ we obtain $\tilde{\psi}(\mathbf{r},t) = e^{\imath\mathbf{k}\cdot(\mathbf{r}-\mathbf{w}t)}$. We have this way reproduced de Broglie's derivation that relates $\omega t - \mathbf{k}\cdot\mathbf{r}$ to $Et - \mathbf{p}\cdot\mathbf{r}$ by using his substitutions $E = \hbar\omega$ and $\mathbf{p} = \hbar\mathbf{k}$. It can then be said that $\tilde{\psi}(\mathbf{r},t)$ describes a matter wave travelling in space with a phase velocity $\mathbf{w}$ and a wave vector $\mathbf{k}$. The spinor field $\psi(\mathbf{r},t) = \psi(\mathbf{0},0)\,e^{-\imath(Et-\mathbf{p}\cdot\mathbf{r})/\hbar}$ is also interpreted this way. This interpretation is summarized in Figure 6.1.

The electron spin is like a clock. The spinor $\tilde{\psi}$ is thus describing the proper time as read on this clock using variables of the lab frame. But the formulation $\tilde{\psi}(\mathbf{r},t) = e^{\imath\mathbf{k}\cdot(\mathbf{r}-\mathbf{w}t)}$ over-interprets the true meaning of $\tilde{\psi}$ in a weird way as a "matter wave" propagating in space. In the most elementary, simplified form, this matter wave describes the probability density, which is subject to a continuity equation. This continuity equation expresses that

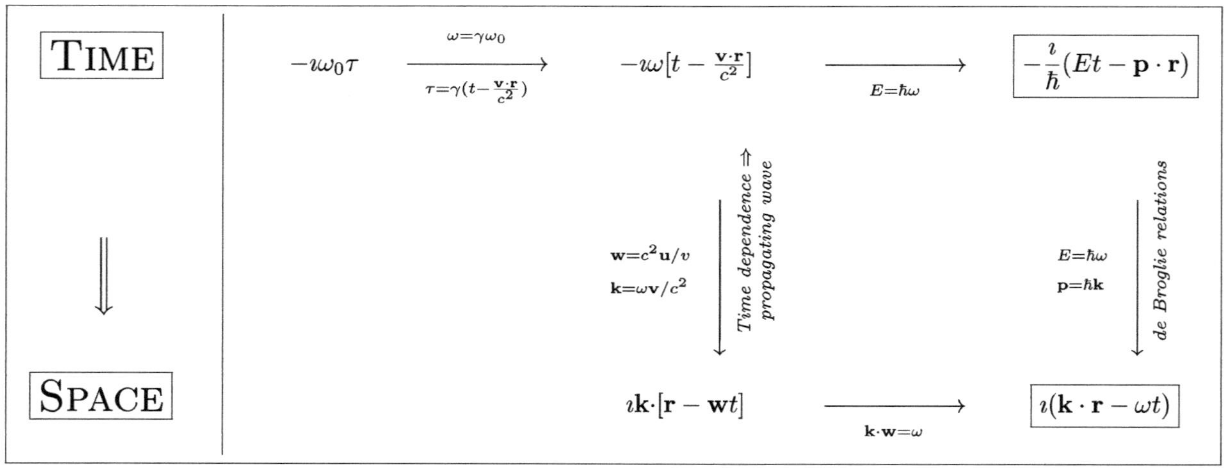

Fig. 6.1 Diagram for the phase of the wave function. The spinor $\psi(\mathbf{r},t)=\psi(\mathbf{0},0)\,e^{-\frac{\imath}{\hbar}(Et-\mathbf{p}\cdot\mathbf{r})}$, which describes the spinning motion of the moving electron in TIME, has historically been reinterpreted in terms of a motion in SPACE of a de Broglie wave $\psi(\mathbf{r},t)=\psi(\mathbf{0},0)\,e^{\imath(\mathbf{k}\cdot\mathbf{r}-\omega t)}$. This de Broglie wave is supposed to propagate through space at a superluminal phase velocity $w=c^2/v$ with respect to the lab frame.

the probability density of an electron (which behaves completely analogously to a charge density) is time-invariant in a reference frame wherein the electron is translationally at rest. The wave equation therefore expresses in spinor form translational invariance for the charge-current density vector with respect to time translations, just like Bloch waves in a crystal express translational invariance with respect to space translations. But this emphasizes in another way that the de Broglie wave is a time wave, rather than a wave in space.

The slope of a time axis in Minkowski space-time is $w = c^2/v$, which is exactly the speed of the de Broglie wave, which confirms the idea that the de Broglie wave should be understood as a time wave rather than a wave "propagating" in space. The de Broglie wave is nothing other than the hypothetical synchronization wave discussed earlier. In fact, $\mathbf{w}$ is the velocity of this one-way synchronization wave in the reference frame of the electron, transformed to a frame wherein the electron is moving with velocity $\mathbf{v}$. The synchronization wave serves to put the clock readings (i.e. the phases) equal in all positions within the rest frame of the electron. It is therefore a phase velocity. In the rest frame of the electron, this phase velocity is infinite.

There is, therefore, no conflict with the theory of relativity in this superluminal velocity $w = c^2/v$ of the de Broglie wave for an electron moving at a speed v. The apparent conflict is only due to over-interpreting the "time wave" as a physically meaningful synchronization wave that would propagate through space. It does not seem necessary to explain away these superluminal velocities by introducing considerations about group velocities and wave packets. These considerations have their own problems, as the different components of a wave packet should not move away from each other with time due to possible dispersion. It is also never checked in practice if the solutions of the Schrödinger equation for a specific problem really have all the necessary properties of a stable wave packet.

Note that in principle the spinor field only needs to be defined on the actual path of the electron, not on the whole of space-time as already identified in Subsection 5.5.4. It is the extrapolation to space-time that turns the spinor field into something that can be over-interpreted as a wave. In the approach of this book, this wave corresponds to the description of a set of possible electron histories. The electrons themselves are particles. This is not a postulate, merely a working assumption. The aim is to check the viability of this assumption.

6.2 Quantization as a pure consequence of Lorentz invariance

6.2.1 *A puzzling result*

By applying the Dirac equation to itself, we obtain the Klein-Gordon equation:

$$\left[\frac{1}{c^2}\frac{\partial^2}{\partial t^2} - \frac{\partial^2}{\partial x^2} - \frac{\partial^2}{\partial y^2} - \frac{\partial^2}{\partial z^2}\right]\Psi = -\frac{\omega_0^2}{4}\Psi. \tag{6.4}$$

It may be surprising that the Klein-Gordon equation and the Dirac equation are obtained starting from the same *ansatz* for a rotating frame, as the Klein-Gordon equation is known to apply to particles of spin 0, while the Dirac equation applies to particles of spin $\frac{1}{2}$. In obtaining the Klein-Gordon equation by applying the Dirac equation to itself, the Klein-Gordon equation applies then only to the individual components of the Dirac spinor.

Using the standard mathematical approach a solution for (6.4) can be proposed of the form $\Psi(\mathbf{r},t) = \psi(\mathbf{r})\, e^{\imath\omega t}$. This leads to:

$$\Delta\psi = [\,\omega^2 - \omega_0^2/4\,]\,\psi. \tag{6.5}$$

This equation is isomorphic to a time-independent Schrödinger equation for a particle. In this isomorphism the role of the counterpart of the total energy E is played by the quantity $\hbar^2\omega^2/2m$ while the role of the counterpart of the potential $V(r)$ is played by the constant $\hbar^2\omega_0^2/8$. Neglecting further the factors $\hbar^2/2m$, the quantity $\omega_0^2/4$ will be called in an abuse of langauge the "potential". This constant "potential" $\omega_0^2/4$ is spherically symmetric with respect to any point in $\mathbb{R}^3$. The solution of a time-independent Schrödinger equation for a spherically symmetric potential $V(r)$ is described in many textbooks about quantum mechanics. It starts from expressing the Laplacian in spherical coordinates, with respect to the centre of symmetry that is taken as origin for the reference frame. The rotational symmetry of the mathematical problem with respect to this point leads then to the final solution. As in the case presented here the potential is constant, the choice of the origin of the reference frame is arbitrary. The wave function solutions are spherical harmonics. As already noted in Subsection 3.9, spherical harmonics of degree n are obtained by taking tensor products of n identical spherical harmonics of degree 1. Mathematically, this corresponds to an image where one particle with coordinates (x, y, z) is replaced by n identical particles with the same coordinates. Actually, since (x, y, z) are quadratic in the spinor components, we have an image of $2n$ identically spinning particles. It is hard at this point to imagine a physical image that would correspond to this construction, such that the solutions remain rather abstract. As the solutions of (6.4) are quantized and have rotational symmetry, one may wonder what they mean. (This issue will be discussed in Section 6.4, by showing a relationship with Feynman's all-histories approach.) A full explanation for the fact that (6.5) allows for solutions with rotational symmetry will come in Subsection 6.2.10.4.

6.2.2 *The seeds for a probabilistic approach*

The quantities (x, y, z, t) occur in (6.4) as consequence of the fact that the clock readings τ and angular velocity ω_0 of the rotating triad are expressed via $\omega_0\tau$ as $\omega t - \mathbf{k}\cdot\mathbf{r}$ in a moving frame. The quantity $\mathbf{k}$ is then related to the velocity of the moving frame, and the selection of the origin is defined by the coinciding positions at $t = \tau = 0$ of the reference frames chosen to define the equation that links τ and t in the Lorentz transformation of the boost. This origin does not need to coincide with the position of the rotating triad. It may be noted that originally, the position of the triad is not specified, but as in a Lorentz transformation clock readings in a moving frame depend also on the position of the clock (as the transformation law of a boost is of the type $t' = \gamma(t - vx/c^2)$), it becomes necessary to introduce the position coordinates (x, y, z) of the triad in order to formulate $\omega_0\tau$ correctly. In general, a Lorentz transformation is used to calculate the new coordinates (x, y, z, t) in a moving frame of one specific particle at (x_0, y_0, z_0, t_0) in a rest frame. But the transformation law of a boost gives simultaneously the values of (x, y, z, t) for the whole of space-time, and thus of any other particle that would travel at the same speed on some different world line. This means that the coordinates (x, y, z) become hypothetical quantities. The quantity $\psi(x, y, z, t)$ will code the orientation of the triad at time t just in case the particle happens to be in (x, y, z). In all other points (x, y, z, t) than the actual position and time that correspond to (x_0, y_0, z_0, t_0), the quantity $\psi(x, y, z, t)$ will only be the hypothetical quantity that expresses the orientation the triad would have had if the particle had been in this point. This suggests that the situation carries the seeds for a probabilistic approach (see Section 6.3). The structure of the Lorentz transformation extends the definition of the spinor field $\psi(\mathbf{r}, t)$ from the world line of the particle to the whole of space-time. The true picture of this spinor field is thus different from that of a wave that would propagate in space. The "waves" are only the symmetry-adapted functions that one must use to express Lorentz symmetry.

6.2.3 *Why is the wave function a function? Part 1*

It has been shown that there are (incomplete) representations of the rotation group in terms of sets of harmonic polynomials $\psi \in F(\mathbb{R}^3, \mathbb{C})$ of real variables $(x, y, z) \in \mathbb{R}^3$, that take complex values $\psi(x, y, z)$. The complex values of these polynomials no longer allow recovery of the full contents of the information contained in a spinor (ξ_0, ξ_1) that represents a rotation or a triad. This is a consequence of a special isomorphism discussed in Section 3.10.

The use of these representations based on spinors and harmonic polynomials in quantum mechanics (e.g. in the calculation of the energy levels of the hydrogen atom) raises some issues:

(1) A spherical harmonic of degree N is a component of the tensor constructed starting from the tensor product of $2N$ identical copies of the basic spinor. This creates the impression that a coherent state of $2N$ identical electrons is being described rather than a single electron. This is certainly not the case, because it does not tally with the observed mass of an atom (e.g. the hydrogen atom).
(2) Is it not necessary to use covariant derivatives for the spinors when transforming to the *curvilinear* set of spherical coordinates? The same question arises also for the spherical harmonics, as they are components of a tensor.
(3) Moreover, the covariant derivatives would be different for the various tensor products.

This problem of the covariant derivatives raises an even more serious issue that can be related to a remark due to Cartan, who pointed out that spinor fields cannot be defined in curved space. The idea is that a rotation of an angle φ around an axis $\mathbf{s}$ is given by the matrix $\mathbf{R}(\varphi) = \cos(\varphi/2)\mathbb{1} - \imath \sin(\varphi/2)\,[\mathbf{s}{\cdot}\boldsymbol{\sigma}]$. Now, after a rotation over 2π, we obtain for the rotation matrix: $\mathbf{R}(2\pi) = -\mathbb{1}$. Hence, in trying to describe the orientation of a triad along an orbit as a single-valued spinor field $\psi(x, y, z)$ in terms of the coordinates (x, y, z) a problem is encountered. The quantity $\mathbf{R}$ will not be a function because it is not single-valued. This quantity $\mathbf{R}$ is equivalent to ψ, and the argument equally applies to ψ. Hence, the spinor field is *a priori* not a function. This is a mathematical problem that goes beyond the fact that the wave functions used in quantum mechanics are fields of unit rays. The problem exists independently of the way spinors are used in quantum mechanics.

The whole problem can be traced back to the difference that exists between vectors in $\mathbb{R}^3$ and reflections (and normals defining reflection planes) within the rotation-reflection group. The rotation group has a different topology from that of $\mathbb{R}^3$. The two topologies are incompatible, just as the topology of the Moebius ring is incompatible with that of a normal ring. In the topology of the rotation group, $\mathbf{s}$ and $-\mathbf{s}$ cannot be distinguished, because they define the same reflection. In the representation SO(3), with its different, Moebius-like topology, there is therefore only one copy of the rotation group. But by defining the spinor quantities (ξ_0, ξ_1) in terms of (X, Y, Z) through a square root, it is possible to distinguish again between the two possible values $\mathbf{s}$ and $-\mathbf{s}$. Making a turn of 2π, $-(\xi_0, \xi_1)$ is obtained instead of (ξ_0, ξ_1). The situation with $-(\xi_0, \xi_1)$ corresponds exactly to the situation with $-\mathbf{s}$. But meanwhile, the rotation angle is no longer π but 2π. The image in SU(2) is thus closer to the image in $\mathbb{R}^3$, but there is a factor 2 between the rotation angles that are being used. A rotation angle of 4π in SU(2) corresponds to an angle 2π in $\mathbb{R}^3$.

The incomplete spherical harmonics of $F(\mathbb{R}^3, \mathbb{C})$ can be used to describe the position of a particle on an orbit. This corresponds then to the angle of rotation. In the same way, one can use such spherical harmonics of $F(\mathbb{R}^3, \mathbb{C})$ to describe the rotation angle of a triad. Note that wave functions and orbitals may look superficially very different from classical orbits, but can nevertheless be discussed in terms of orbits based on the relationship described in Subsection 5.5.4. This will be developed in Chapter 8.

6.2.4 *Why is the wave function a function? Part 2*

In the solution of (6.5), one has at first sight the impression that the quantization conditions emerge as an unavoidable part of the mathematics. This may appear startling, as it relies only on the assumption of Lorentz invariance and rotational symmetry with respect to some special point (which serves as the origin of a reference frame in which the position of the triad that describes the rotating frame is (x, y, z)). It is easy to pinpoint where the quantization is introduced. At a certain stage in the development of the solutions for (6.5), one finds $\Phi(\phi) = e^{\imath m\phi}$ for the part of the wave function that contains its dependence on the angle ϕ. It is then postulated that Φ must be single-valued. Of course, in stating this one makes abstraction of the fact that the wave function is in reality a unit ray. The wave function should be single-valued for any phase factor that one might have selected within the unit ray. The wave function is certainly single-valued for the free-space solution, as it was defined from a Lorentz transformation applied to the single-valued function $e^{\imath\omega_0\tau/2}$.

There is, however, no reason why a particle in uniform rectilinear motion should be described by a spinor field with rotational invariance. Hence, the solutions with rotational invariance are only potentialities that are of no use in the original problem. The subset of the solutions of (6.5) that are relevant for the initial differential equation must be selected. Eventually, only the meaningful solutions $\psi_+ \, e^{-\imath\frac{\omega t}{2}}$ and $\psi_- \, e^{+\imath\frac{\omega t}{2}}$ (with "spin $\frac{1}{2}$") will be retained, as already discussed. All the rest must be thrown out again.

The additional solutions with rotational symmetry become essential and can no longer be ignored when considering the presence of a $1/r$-potential, which imposes rotational symmetry (which is not necessarily the same thing as 2π-turn symmetry). It is in this context that postulating that the wave function should be a function raises many questions. Postulating that ψ is a single-valued function of $\mathbf{r}$, implies that the set of couples $(\mathbf{r}, \psi(\mathbf{r}))$ corresponding to all the points $\mathbf{r}$ that constitute a closed loop becomes a fibre. Here, $\psi(\mathbf{r})$ can be of the complete (triad) type ($\psi \in F(\mathbb{C}^3, \mathbb{C})$) or of the incomplete (vector) type ($\psi \in F(\mathbb{R}^3, \mathbb{C})$). As shown by Chern [Frankel (1997)] this leads to "topological quantization". It should be realized that there is *a priori* no reason to postulate the structure of a fibre by suggesting that $\psi(\mathbf{r})$ is a (wave) *function*. Hence, the postulate that ψ is single-valued is a crucial ingredient for obtaining quantization.

6.2.5 *Why is the wave function a function? Part 3: A plethora of physical effects*

There are several arguments to show that the 2π-turn symmetry introduced by the postulate that the wave function should be single-valued is arcane. There are both classical and relativistic arguments.

(1) Classical objection. Imagine classically that we put a gyroscope on board a space station in orbit around the Earth. The fact that ψ is a function implies then that the gyroscope in the analogy will have made exactly an integer number of revolutions around its axis when the space station has made a full orbit around the Earth. This is exactly what Chern's theorem means and implies a phase lock that is reminiscent of the situation of the Moon which in its orbit around the Earth always shows us the same face. This phase lock condition turns out to be exactly equivalent to the Bohr quantization condition. This has been expressed in a different way by

de Broglie who observed that the length of an orbit in the hydrogen atom is always a multiple of the wave length of the de Broglie wave [Bohm (1951)].

Classically, there is no reason why the gyroscope should have made an integer number of revolutions after completing a full orbit. Of course, it could be postulated that the electron has some internal distribution of charges that renders it asymmetrical, and this could explain the phase lock in a way completely analogous to the mechanism that is responsible for the phase lock of the Moon. But this would require the introduction of additional assumptions about the asymmetry of the charge distribution within the electron, about which nothing is known. The validity of such assumptions would be completely uncertain. This kind of uncertainty could very quickly become a profound obstacle to further progress. The essential point here is that by introducing the technique of separating the variables within the Schrödinger or Dirac partial differential equations, we stipulate that the motion is truly periodic. This contains the *ansatz* that the history can be described by a truly periodic spinor field $\tilde{\psi}(x, y, z, t) = \psi(x, y, z)e^{\imath\omega t}$, whereby $\psi(x, y, z)$ is a single-valued function. In the language of quantum mechanics, the existence of the frequency ω that is needed for this corresponds to the existence of a fixed total energy. But the difficulty is that there is no reason why the periods of the gyroscope and of the orbit should match. Relativistically, there are a host of other objections. The relativistic objections can also be visualized with such a gyroscope analogy.

(2) Local Lorentz contraction and time dilatation. Relativistically, a distinction must be made between a purely geometrical rotation and a physical rotation of the type that occurs in the kinematics on a circular orbit. For uniform motion along a circular orbit, the instantaneous distances $d\varsigma$ covered and the times dt elapsed could be written: $d\varsigma' = \gamma_v(d\varsigma - vdt)$, $dt' = \gamma_v(dt - vd\varsigma/c^2)$. Here, the distances are not integrable, only the time is. This is due to the fact that time is a local quantity, while distance is not. The integrated covered distance is always subject to new instantaneous Lorentz transformations, therefore only integrated time can be used.

The starting point for the derivation of the Dirac equation was expressing a rotation with angular frequency ω_0 in the electron's own rest frame, using the quantity $\omega_0\, d\tau$ to do so. Fortunately, the quantity $\omega_0\, d\tau$ corresponds to a relativistic invariant $\omega dt - \mathbf{k}{\cdot}d\mathbf{r}$. From this it can be appreciated that if $\omega_0\tau = 2\pi$, then at least in principle $\omega t \neq 2\pi$. It is then surprising that the solutions can be factorized as $\psi(\mathbf{r})e^{\imath\omega t}$, where the spatial part ψ is a *function* that uniquely depends on $\mathbf{r}$, because the total angle during one turn along the orbit will *a priori* no longer be 2π; what is 2π for one observer will no longer be 2π for another observer. This argument of Lorentz contraction and time dilatation along a circular orbit has been used by Einstein to motivate the notion that space-time geometry is not Euclidean in general relativity and must be curved.

(3) Thomas precession. There is a corollary to the relativistic argument that it is not certain that the gyroscope should make an integer number of revolutions in one orbit. The composition of two boosts that are not collinear gives rise to a Lorentz transformation that is composed of a boost followed by a rotation, called *Thomas precession.* Due to this Thomas precession, the rotation axis of a gyroscope in an orbit will in general no longer be fixed. The only way to escape from this verdict is to postulate that the orbit is planar and the rotation axis strictly perpendicular to the orbital plane. (One can then prove using (5.64) that it will remain perpendicular to this plane during the whole motion.) Hence, in general, not only the phase ωt in $\omega t - \mathbf{k} \cdot \mathbf{r}$, but also the orientation of the rotation axis $\mathbf{s}$ must return to the initial position to support the postulation that ψ is single-valued.

(4) Perihelion precession. Finally, relativistic orbits are no longer true ellipses, but ellipses that undergo perihelion precession as observed for Mercury. This type of effect exists already within the framework of special relativity, and does not only change the periodicity of the orientation of the triad. When the orbit exhibits precession it will intersect itself. At the points of intersection, again the problem that the function $\psi(x, y, z)$ is *a priori* not single-valued will arise.

In summary, there are many arguments to show that the postulate that the wave function should be single-valued is paradoxical: classically (as in the example of the gyroscope on a space ship) and relativistically (time dilatation, Thomas precession, perihelion precession). In establishing this list, these effects have only been described geometrically. But it is because these geometrical effects produce measurable physical interactions which change the total energy of the system, that they must be described. Why else would this be of any concern! In a general situation, these effects will become physically coupled, resulting in a very complicated interplay. For instance, it is generally admitted that the spin is equivalent to a magnetic dipole. This magnetic dipole will not interact with the electric field of the nucleus in the hydrogen atom, but it will interact with an external magnetic field (Zeeman effect) and with the magnetic field of the nucleus produced by its spin (hyperfine interaction).

In a first approach, the magnetic effects could be ignored. The external magnetic field could be switched off, arguing that the hyperfine splitting is a minute effect. But some other physical effects can spoil the party. The travelling magnetic dipole will give rise to an induced electric dipole that will interact with the electric field of the nucleus, such that it will be necessary to change the equations.

6.2.6 *Solution to the paradox – part 1: The need for a manifold* $\mathbb{M}_N$

The solution to this problem of why the wave function is a single-valued function is again not physical but purely mathematical. In this solution the isomorphism defined in Section 3.10 does not map (X, Y, Z) onto points (x, y, z) of the usual set $\mathbb{R}^3$, but onto points of a manifold, that is a multiple copy of $\mathbb{R}^3$. Let us first address the problem raised by the fact that it takes a rotation angle 4π to get a spinor back to its initial value. Consider the restriction of the spinor field to $\mathbb{R}^2$ when the orbit is planar. Let us introduce a manifold $\mathbb{M}_2 = \mathbb{R}^2 \times \{-1, 1\}$, whereby the connectivity between the elements $(x, y, j) \in \mathbb{M}_2$ is defined through the use of polar coordinates (r, ϕ) for (x, y). In fact, $(r, 2\pi, -1) = (r, 0, +1)$ and $(r, 2\pi, +1) = (r, 0, -1)$. This is a Riemann surface (see Figures 6.2–6.4) and may be visualized by a helicoidal ramp in a parking building with two levels, whereby the upper level is back-connected to the lower level as in the visual paradox within Escher's drawing *Waterfall*. There is a much simpler way to visualize this in the plane on a sheet of paper. It can be postulated that the physical angle on the sheet of paper corresponds to twice the mathematical angle, such that a full turn on the sheet of paper visualizes a mathematical angle of 4π.

On $\mathbb{M}_2$, one can now define a spinor field $\psi(x, y, j)$ unambiguously. The spinor field is then periodic over $\mathbb{M}_2$, rather than over $\mathbb{R}^2$, and it can serve as a wave function. This means using spinor fields that are functions which belong to the set $F(\mathbb{M}_2, \mathbb{C})$ rather than $F(\mathbb{R}^3, \mathbb{C})$.

The same approach could be used in the case of a periodic orbit that crosses itself, when its true orbital period is $2\pi N$ rather than 2π. This could happen for example when the perihelion shift of an "elliptical" orbit is of magnitude $2\pi\ell/n$, with $(\ell, n) \in \mathbb{N}^2$, $\ell < n$, $\ell \neq 0$, such that it is only after going n times around the ellipse, that the starting position

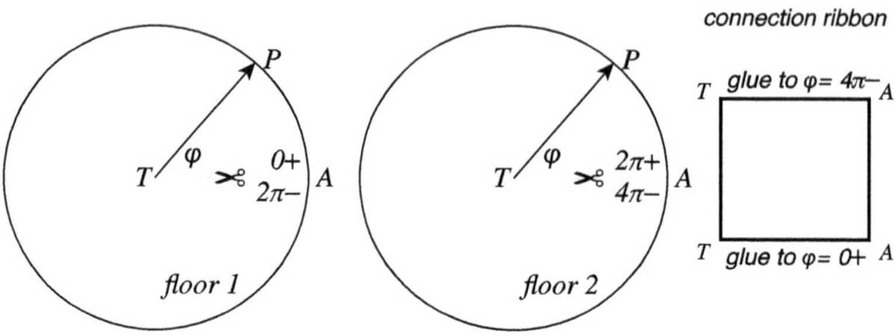

Fig. 6.2 Instructions for constructing a paper model of the Riemann surface for the two-valued square root "function" $f \in F(\mathbb{C}, \mathbb{C}) : z \rightarrow f(z) = z^{1/2}$. Start with two sheets of paper labelled "floor 1" and "floor 2". Both sheets of paper represent the complex plane with origin T. Make a cut along the positive x-axis TA in both sheets and lift upwards the part on the side of the cut that corresponds to $z = re^{\imath\varphi}$ with $\varphi \xrightarrow{<} 2\pi$ (noted as $2\pi-$), such that it is higher then the part on the other side of the cut where $\varphi \xrightarrow{>} 0$ (noted as $0+$). In consequence of this lift, somebody travelling along the unit circle in the sense of increasing φ-angles would then travel upwards in the same way as a car moves up to a higher floor in a car park with several floors. Such a car park is in a sense a giant helicoidal staircase. Next, repeat the action with the part where $\varphi \xrightarrow{<} 4\pi$ on the second floor. The point $\varphi \xrightarrow{<} 2\pi$ marks the point where one reaches the second floor and one can start the ascent from $\varphi \xrightarrow{>} 2\pi$ to the third floor that could start at $\varphi \xrightarrow{<} 4\pi$. This motion from $\varphi \xrightarrow{>} 2\pi$ to $\varphi \xrightarrow{<} 4\pi$ is represented by the second sheet of paper. The interval of angles is noted as $[0, 2\pi[$ on the first sheet and $[2\pi, 4\pi[$ on the second sheet. The point is now that there is no further ascent from $\varphi = 4\pi$ onwards because the second floor corresponds to the ground floor again, such that the region $\varphi \xrightarrow{<} 4\pi$ must connect back to the region $\varphi \xrightarrow{>} 0$. A third sheet of paper labelled "connection ribbon" can be used to take care of this connection. Just like in Figure 3.5 this connection sheet is purely metaphorical and it serves only to establish the connectivity. The distance between the two points T is therefore zero. The same applies for the distance between the two points A. (As in topology distances do not matter, the zero distances can be represented by finite distances.) In the physical world, it cannot be avoided that one intersects the first floor in moving from the second floor to the ground floor such that the zero-length connection ribbon will have to "traverse" $\varphi = 2\pi$ even though in the mathematics it is not the intention to have any intersection. (In a finite restriction of the model, this could be avoided by fulfilling the cyclic boundary condition by bending the connection ribbon and establishing the connection in the form of an external loop.) Mathematically, one should not think of the surface as having a cut or a self-intersection. The Riemann surface models the "two-valued function" $f(z) = z^{1/2} = \pm\sqrt{r}e^{-\imath\varphi/2}$, with one sheet representing the solution with the $+$ sign and the other sheet the $-$ sign. Other more complicated Riemann surfaces can be used to represent the n-valued "function" $f(z) = z^{1/n}$. The manifold $\mathbb{M}_N$ used in the text corresponds exactly to the Riemann manifold for the multivalued "function" $f(z) = z^{1/n}$, for $n = N$. Of course, taking the n-th power of f makes the result again single-valued, and it is this gimmick that is used by introducing harmonic polynomials based on the N-th tensor powers $(\xi_0, \xi_1) \otimes (\xi_0, \xi_1) \cdots \otimes (\xi_0, \xi_1)$.

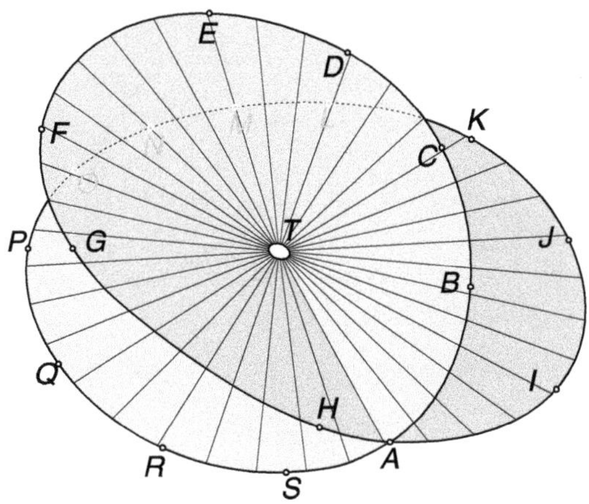

Fig. 6.3 Topology of the Riemann surface for the two-valued square root function $f \in F(\mathbb{C},\mathbb{C}) : z \rightarrow f(z) = z^{1/2}$. The letters A to H correspond to a part of the surface that can be imagined as similar to the surface of a cone with apex T, with a cut along the line AT, because there is no connectivity at A that would permit moving directly from H to B. It has to be imagined that the view is taken from inside the cone in the upward direction towards the apex T of the cone. The other part of the surface can for convenience be imagined as a second cone with apex T and a cut along AT where the view would be down towards the apex T from outside the cone. For the topology, the exact shapes and distances do not matter; only the connectivity is important. The connectivity is between S and B and between H and I, but not between H and B or S and I. The part ATD of the surface can be imagined as being almost vertical, while the part $TGHA$ is almost horizontal. Getting back to A after a first turn the surface "traverses" itself on the line TA and continues along IJK. The part of the surface labeled $LMNO$ is hidden behind the part $ABCDEFG$, becoming visible again in the part $PQRS$. Only after completing a second turn the surface truly connects back to its starting position along the line TA. The surface is not supposed to intersect itself truly along the line TA in the sense that there would be points of the sector STB that would also belong to the sector HTI, because there is no connectivity between H and B, or between S and I. The "intersection" only occurs when one tries to make a paper model of it in the physical world as described in Figure 6.2. This is more clear in the model shown in Figure 6.4.

is reached again. This is illustrated in Figure 6.5. Here again a manifold $\mathbb{M}_N = \mathbb{R}^2 \times \{e^{\imath(k-1)\pi/N}, k \in [1,N] \cap \mathbb{N}\}$ could be used. In the example, $N = 4$. In general, we will have $N = n + \ell$, as it will take n turns around an "ellipse" to become truly periodical, but we will also have gone through n perihelion shifts $2\pi\ell/n$. Here, spinor fields are being used that are functions from the set $F(\mathbb{M}_N, \mathbb{C})$. The orbit on the set $\mathbb{M}_N$ will then no longer intersect itself. It might actually be necessary to introduce a manifold $\mathbb{M}_{2N}$. In

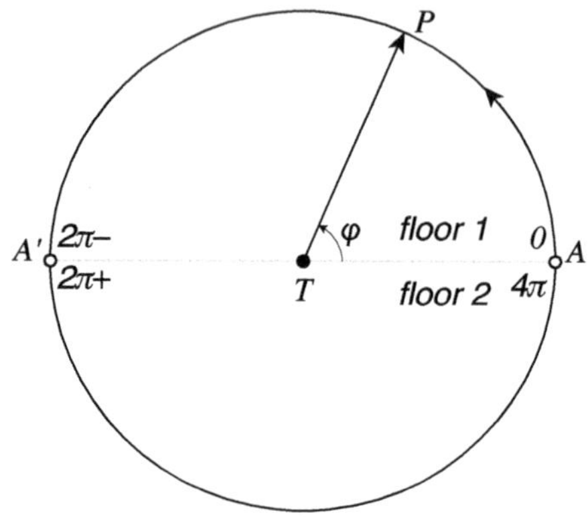

Fig. 6.4 The simplest representation of the Riemann surface for the two-valued square root function $f \in F(\mathbb{C},\mathbb{C}) : z \rightarrow f(z) = z^{1/2}$. It is based on a mapping between the angle φ on the Riemann surface and the angle φ_1 of the Euclidean plane, defined by $\varphi = 2\varphi_1$. This way a 2π turn can be visualized as different from a 4π turn, while 4π can be connected to 0. The points $\varphi \rightarrow 0+$, $\varphi \rightarrow 2\pi-$, $\varphi \rightarrow 2\pi+$, and $\varphi \rightarrow 4\pi-$ in this figure correspond exactly to those in Figure 6.2. It can be appreciated from this figure that there is no self-intersection along $\varphi \rightarrow 2\pi$.

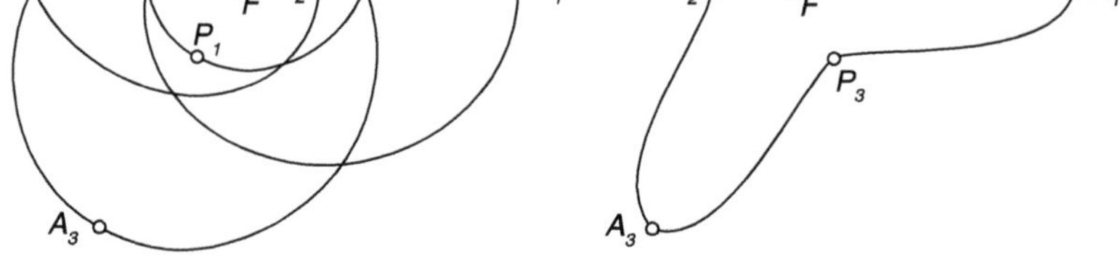

Fig. 6.5 Rosette-like orbit with a perihelion shift of $\frac{2\pi}{3}$ represented in real space (left) and on the Riemann surface $\mathbb{M}_4$ (right). The fixed focus of the rotating ellipse is the point F located at the centre of the drawings. The other six points on the orbit are the three perihelia P_j and the three aphelia A_j. The correspondence between the orbital polar angle ϕ used in the drawing on the left-hand side and the orbital polar angle ϕ' used in the drawing on the right-hand side is given by $\phi' = 4\phi$. A full circle on the drawing on the right-hand side would thus correspond to 8π. Orbital polar angles are noted here by ϕ to distinguish them from spin polar angles φ in the previous figures. On the Riemann manifold $\mathbb{M}_4$ the orbit no longer intersects with itself such that one can define a spinor field on the orbit that is a true function.

the factor 2 it must be remembered that spinors have periodicity 4π rather than 2π in the rotation angle.[1]

6.2.7 *Solution to the paradox — part 2: Equivalence of the Bohr model and the Schrödinger equation in the problem of the hydrogen atom (spin axis along the z-axis)*

The solution to the paradox consists in accepting that the gyroscope has indeed not made exactly an integer number of full turns when the space station has completed a full orbit. In a first, non-relativistic approach it can be assumed that the orbit does not cross Itself. The following simplifying assumptions will be made:

(1) The orbit is just a circle.
(2) The electron rotates uniformly.
(3) The rotation axis of the electron remains fixed in space parallel to the z-axis.

A more general situation will be discussed later. The point is that the SU(2) wave function for a spinning top with rotation axis parallel to the z-axis can be written as $e^{-\imath\omega_0\tau/2}\,[1,0]^{\top}$ such that the spinor will have only one non-zero component ($e^{-\imath\omega_0\tau/2}$). The whole situation can then be described with two rotation angles: one for the orbit and one for the rotation of the electron. As this is a non-relativistic approach, only the physical effect (1) of the list given in Subsection 6.2.5 will play a role.

Let the starting position be called P_1. The wave function is written as $\psi(\mathbf{s}, \varphi)$, whereby the variable $\mathbf{s} = \mathbf{e}_z$ denotes the rotation axis and the

[1]Cartan used a topological argument to prove that it is impossible to define a single-valued spinor field in curved space, by considering a closed loop in the curved space, that he made shrink to a point. It is thus important that this should not be possible. It will be shown below how one can identify the group parameters of the rotation group (i.e. the coordinates of the isotropic vector that codes the triad) with the particle coordinates by using spherical coordinates as discussed in Section 3.10. To prevent a closed loop from being shrunk to a point, one can exclude an open ball around the origin from the space wherein the motion takes place. When in addition the motion is planar, the space will then no longer be simply connected. This idea will be encountered several times in the book. The introduction of the manifold serves then to render the topologies compatible. As noted in the caption of Figure 6.2 the situation is similar to that encountered when one tries to define the n-th root $\sqrt[n]{z}$ of a complex number $z \in \mathbb{C}$ for a natural number $n \in \mathbb{N}$. The "function" $z \to \sqrt[n]{z}$ is also not single-valued. One introduces then exactly the same types of Riemann surfaces to define a single-valued function.

variable $\varphi = \omega t$ describes the rotation angle of the electron spin around this axis. If ever it were necessary to express the values of these quantities in the rest-frame of the electron, the symbols φ_0 and ω_0 would be used for them ($\varphi_0 = \omega_0 \tau$). Even in non-relativistic quantum mechanics, the phase $\omega t - \mathbf{k}\cdot\mathbf{r}$ corresponds to $\omega_0\tau$ after Lorentz transformation, and as such it contains a relativistic element. But otherwise, ω is not considered to vary with the velocity, and it can be almost assumed that the phase can be written as $\omega_0 t - \mathbf{k}\cdot\mathbf{r}$. Of course, this will change in relativistic quantum mechanics.[2]

The variable $\phi = \tilde{\omega} t$ corresponds to the position of the space station along its orbit. As the orientation of the axis of the gyroscope is considered as fixed in space, $\mathbf{s}$ is constant. Note that two different forms $\varphi = \omega t$ and $\phi = \tilde{\omega} t$ of the Greek character phi are used to distinguish the two types of angles. Up to now φ has always been used for a rotation about an axis $\mathbf{s}$, to clearly distinguish it from spherical coordinates (θ, ϕ). As a spinor codes a rotation, its value will only contain the variable φ, but φ can be a function of the position (x, y, z) of the electron, and as such of ϕ.

[2]For an ellipse with perihelion precession, where it takes turning N times around the ellipse to get back to the starting position, it is in reality necessary to consider $2N$ turns rather than N, in order to take into account that the periodicity of spinors is 4π rather than 2π. In what follows, other types of periodicity will be discussed, where it also takes N turns to get back to the initial position, but the global periodicity is $2N$ due to the 4π-periodicity of spinors. The factors 2 and N in the number $2N$ have two very different origins. It is the combination of both that is responsible for the number $2N$. But for the clarity of the discussion, one must be able to discuss both separately. Solutions for the hydrogen atom will be discussed both within the context of the Schrödinger equation and of the Dirac equation. In the context of the Schrödinger equation, single-component wave functions are needed where it is as though the existence of two-component spinors and the need for a factor 2 are ignored. Therefore, the factor of 2 will be ignored within the context of the Schrödinger equation. This corresponds in a sense to a treatment with a wave function expressed in the variables $(x, y, z) \in \mathbb{R}^3$ (in the vector representation based on the isomorphism defined in Section 3.10) rather than (ξ_0, ξ_1), where due to the quadratic dependence of (x, y, z) on (ξ_0, ξ_1), and the possibility of defining polar coordinates in the vector representation, a 2π-turn will already bring the wave function back to its original value. To discuss the wave functions that are expressed in terms of harmonic polynomials rather than spinors (and this way take into account the factor 2) the manifold will be noted as $\mathbb{M}_N^*$. It will be as though we did not know that it takes a rotation angle of 4π rather than 2π to get a wave function back to its initial value. The discussion will be in terms of periods $2\pi N$ rather than $4\pi N$, keeping in mind that ultimately N may need to be doubled to $2N$ to take into account the 4π-periodicity of spinors. When discussing a wave function that recovers its starting value after three turns, it may thus also imply that in reality six turns should be considered to take into account the 4π-periodicity, and when the "wave function" is discussed, it will be in terms of how it would have been if it were not for this factor 2.

At the starting position of the space station $t = 0$ and thus $\psi(P_1) = \psi(\mathbf{s}, 0)$. Imagine that after one period of the space station along its orbit (i.e. when $\phi = 2\pi$) the gyroscope has gone beyond its initial value $\psi(\mathbf{n}, 0)$, because $\omega t \not\equiv 0\,(mod\,2\pi)$, and that it has reached the value $\varphi = 2\pi + 2\pi/3$. (In a case where the gyroscope is slower, we will have for example $\varphi = 2\pi - 2\pi/3$.) The wave function then becomes $\psi(P_2) = e^{\imath 2\pi/3}\psi(\mathbf{n}, 0)$. The space station will have recovered its initial position in physical space, but the triad that codes the gyroscope will not have recovered its initial phase angle, such that the combined motion has not yet come to a full period. In order to address this difference, this combined situation of the space station and the gyroscope will be noted as a "position" P_2. After two turns $\Delta\phi = 2\pi$ (i.e. twice $\Delta\varphi = 2\pi + 2\pi/3$) the space station will get to P_3, and the wave function will be $\psi(P_3) = e^{\imath 4\pi/3}\psi(\mathbf{n}, 0)$, assuming that the rotation is uniform. Finally, after three turns $\Delta\varphi = 2\pi + 2\pi/3$ the wave function will become equal to $\psi(P_1) = \psi(\mathbf{n}, 0)$ again. Hence, the full motion will be truly periodic with an angular period that is 6π in physical space. It is therefore necessary to introduce $\mathbb{M}^*_N$, with $N = 3$. This is also the reason for the introduction of three "positions" P_j. This may at first appear meaningless, as they are identical in physical space $\mathbb{R}^2$, but the point is that they are not identical within $\mathbb{M}^*_N$. The full wave function will recover its starting value after three turns of the type $P_1 \to P_2$. This is illustrated in Figure 6.6.

Also, if the wave function takes an increment of $2\pi + 4\pi/3$ rather than $2\pi + 2\pi/3$, the full motion will have a period that is three times as large as the interval that is necessary to cover the turn $P_1 \to P_2$. In other words, in the situation with an increment $2\pi + 2\pi/3$ as originally discussed, the real period of the combined system of the gyroscope and the space station is 8π in the angular variable φ that describes the rotation of the gyroscope, and 6π in the angular value ϕ that describes the position on the orbit. (When the gyroscope is trailing behind, such that the increment is $2\pi - 2\pi/3$, the period in the angular variable φ will be 4π.) To make sure that the wave function is single-valued, $\mathbb{M}^*_3$ is introduced, because the true period of the system corresponds to three turns in physical space.

Consider now three space stations along the orbit at positions P_{k+1} in $\mathbb{M}^*_3$, with φ-coordinates $\phi + k8\pi/3$, and $k \in [0, 2] \cap \mathbb{Z}$. The copies are purely mathematical constructions, without physical meaning; there is no physical force between them. Their introduction is just a mathematical expedient. The three space stations have wave functions $\psi_k = e^{\imath 2\pi k/3}\,\psi_0$, with $\psi_0 = \psi(\theta, \phi) = \psi(\theta, 0)e^{\imath\varphi}$, where $\theta = \pi/2$. The combined state is then represented by $\Psi = \psi_0 \otimes \psi_1 \otimes \psi_2 = e^{\imath 2\pi}\,[\psi(\theta, 0) \otimes \psi(\theta, 0) \otimes \psi(\theta, 0)]$. When the gyroscope is trailing behind, the phase factor will be $e^{-\imath 2\pi}$ instead of $e^{\imath 2\pi}$.

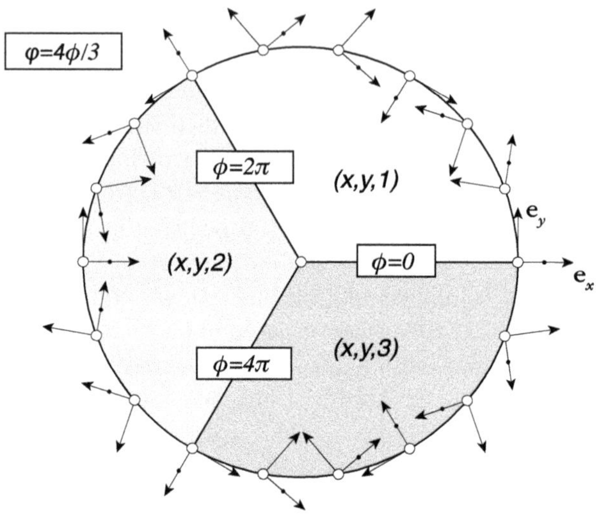

Fig. 6.6 The Riemann manifold $\mathbb{M}_3^*$ with the spinning triad on a circular orbit. The angle of rotation of the spin is φ, while the angle that defines the position of the triad on its orbit is ϕ. The axis of the rotational motion of the spin is the z-axis, which is perpendicular to the plane of the figure. The orbit is circular and the plane of the orbital motion is the Oxy plane, such that the drawing corresponds to the assumptions made in the text. The three copies of $\mathbb{R}^3$ that occur in $\mathbb{M}_3^*$ have been represented by the three 120-degree sectors of the Oxy plane and labelled (x, y, j), where $j \in \{0, 1, 2\}$. Each value of j has been represented by a sector of a different shade. Successive orientations of the triad are shown, but the x- and y-axes have been labelled for only one of them, in order not to overburden the figure. To identify $\mathbf{e}_x$ and $\mathbf{e}_y$ in the frames with unlabelled axes we have marked the $\mathbf{e}_x$ vectors by a dot. In the figure the angle φ of the spin is varying faster than the angle ϕ of the orbital motion. Both motions are uniform and related by $\varphi = 4\phi/3$.

Now, after one turn $\phi = 2\pi$ in physical space, each wave function ψ_k will have been multiplied by $e^{\imath 2\pi/3}$. Each space station will have taken the previous position of its upstream neighbour in $\mathbb{M}_3^*$, and the total state will be indistinguishable from the original state. The combined state Ψ in ϕ-space $\mathbb{M}_N^*$ will then contain the φ-dependence $e^{\imath 3\varphi}$, which is of the type $e^{\imath N\varphi}$ for $N = 3$. The wave function tensor product Ψ belongs to the irreducible representation in terms of harmonic polynomials of degree 3 (when ψ_k is of degree 1). Actually, by introducing the tensor field corresponding to the tensor product of N virtual identical copies of the spinor field (or N identical copies of the spherical-harmonic field of degree 1), a tensor field is defined on $\mathbb{M}_N^*$ that is again periodic with period 2π in the variable φ. It is then no longer necessary to distinguish between $\mathbb{R}^2$ and $\mathbb{M}_N^*$. These tensor fields are the spherical harmonics.

Hence, the original idea of introducing a spinor field failed, as it led to a field that was not single-valued in physical space. We therefore introduced the manifold $\mathbb{M}_N^*$. In a second stage, we constructed an alternative wave function that becomes single-valued again in $\mathbb{R}^2$ by taking N copies of the space station in $\mathbb{M}_N^*$. The states in $\mathbb{M}_N^*$ are constructed by taking products of wave functions of such virtual copies. In the end the distinction between $\mathbb{M}_N^*$ and $\mathbb{R}^2$ can be dropped because in both spaces the period has become 2π. In other words, the introduction of the spherical harmonics is necessary to render it possible to describe the triad (X, Y, Z) as a single-valued function of the orbit parameters (x, y, z), such that we could play the game of "two movies for the price of one", as we called it earlier on. The introduction of the spherical harmonics corrects for the error that the wave function would not be single-valued. As the $1/r$-potential has rotational symmetry, its influence remains correctly expressed in the manifold $\mathbb{M}_N^*$. Hence, the harmonic polynomials of degree N can serve to represent solutions that have a wave function with a ϕ-dependence of the type $e^{\imath\phi/3}$ (with a period 6π in ϕ) rather than $e^{\imath\phi}$ (with a period 2π in ϕ).

The argument can be formalized as follows. The rotation of the gyroscope is described by a spinor (ξ_0, ξ_1) based on the isotropic vector (X, Y, Z) $= \mathbf{e}_X + \imath\mathbf{e}_Y$. Introducing spherical coordinates, and taking into account that the rotations are only around the Z-axis, such that the spherical coordinates reduce to polar coordinates, we have $\mathbf{e}'_X + \imath\mathbf{e}'_Y = e^{\imath\varphi}(\mathbf{e}_X + \imath\mathbf{e}_Y)$. The motion along the circular orbit is also a rotation. This rotation is described by a spinor (η_0, η_1) based on the isotropic vector $\mathbf{e}_x + \imath\mathbf{e}_y$. Taking again into consideration that the rotations are only around the z-axis, and introducing again spherical coordinates that reduce to polar coordinates, we have $\mathbf{e}'_x + \imath\mathbf{e}'_y = e^{\imath\phi}(\mathbf{e}_x + \imath\mathbf{e}_y)$. For this motion, the coordinates of the particle can be used to describe the rotation. It suffices to take $(x, y, z) = r(\cos\phi, \sin\phi, 0)$. Here, r does not really belong to the spherical wave function.

The quantity (X, Y, Z) is thus no longer $(x/r, y/r, z/r)$ itself but a function of $(x/r, y/r, z/r)$. Through the use of polar coordinates, the relationship is expressed through $\varphi = \frac{4}{3}\phi$. By introducing spherical harmonics of degree 3, the period becomes 2π rather than $\frac{2\pi}{3}$.[3] This way it is possible

[3]Taking the n-th tensor power of the spinor defined on a Riemann surface is like solving the problem that $c = \sqrt[n]{z}$ is not a single-valued function of $z \in \mathbb{C}$ by taking the n-th power $c^n = 1$.

to express the rotations of the gyroscope (which are rotations in a space of spherical harmonics in (X, Y, Z)) as rotations in the space of harmonic polynomials of degree 3 in $(x/r, y/r, z/r)$. The present argument will be elaborated for the case that $\mathbf{s} \neq \mathbf{e}_z$ in Subsection 6.2.9, such that it will become more general. In the present approach in terms of Riemann surfaces, the angular-momentum operators actually turn out to be mere degree operators, used to describe a difference between the periods of the spin and of the orbital motion.

(The following two subsections serve to settle a lot of technical details, and are not essential for the understanding of the ideas. The reader may therefore skip them on a first reading and jump to Subsection 6.2.10.)

6.2.8 *Mathematical equivalence between the Bohr model and the solution of the Schrödinger equation for the hydrogen atom — working out the details*

6.2.8.1 *Overview of questions to be treated*

In deriving the form of a spinor in SU(2) it was necessary to remove at a certain stage the parameter r from the formalism by taking the limit $r \to 0$. The justification for this was that r is not a suitable parameter to describe an element of the rotation group. In the same spirit, the space-time coordinates of quantum mechanics are *a priori* not pertinent for a description of the Lorentz group. One of the consequences of this is that an orbit, which is a closed loop in the space part of space-time, does not necessarily correspond to a closed loop on the Lorentz group. This is due to the physical effects (2)–(4) described in Subsection 6.2.5. At least the physical effect (2) will always be present. The quantization conditions express the additional constraints that must be satisfied by a closed orbit to make it a closed loop on the Lorentz group. These constraints ensure that the wave function is a function. The wave function simultaneously defines the action of the group element on the coordinates of space-time that do not belong to the actual world line, opening the way for a probabilistic approach.

That the treatment in terms of spherical harmonics within the context of the Schrödinger equation corresponds to the treatment of the periodicity sketched using the manifold $\mathbb{M}_N^*$ is for the moment an intuition, but it does not yet hold in all the details.

(1) The separation of variables in the differential equation removes expressions of the type ωt from the spatial dependence. As there is a relationship of the type $\phi = \tilde{\omega} t \propto \omega t$ to describe the motion, one may fear that

ϕ will also have disappeared from the spatial dependence. It will have to be explained why the separation of variables does not remove ϕ from the spatial dependence, and what the exact meaning of ϕ is in the Schrödinger equation.

(2) The exact number of modes there are within a representation will also have to be discussed. In the Schrödinger equation this is $2\ell + 1$, which is always an odd number. In the Dirac equation it is $2J + 1$, which is always an even number. As will become obvious, this difference between $2\ell + 1$ and $2J + 1$ cannot be discussed on the basis of purely geometrical arguments. Consideration must also be given to angular momentum "up" and angular momentum "down" states, with respect to a certain axis. This is in accordance with considerations about the energy within a magnetic field and with the fact that in the absence of an external magnetic field, the modes are degenerate. Dirac's solution with the even number $2J + 1$ is necessary to reproduce the correct number of Zeeman sub-states in a magnetic field when the degeneracy is lifted.

(3) The spherical harmonics are components of a tensor. When spherical coordinates are introduced for the calculation of the energy levels of the hydrogen atom, the derivatives must be replaced by covariant derivatives. Moreover, the expression for the covariant derivatives will depend on the rank of the tensor. This has not been done in the solutions for the Schrödinger and Dirac equations for the hydrogen atom. In using only the non-relativistic calculations (i.e. the Bohr model and the Schrödinger equation), it is possible to understand the reason for this, as will be explained below. Contrary to popular belief, the Schrödinger equation does describe the rotation of the electron when it moves along its orbit. The phase of quantum mechanics corresponds to the rotation angle.

6.2.8.2 *Why the variable ϕ does not disappear after the separation of variables*

Let us simplify for the moment the discussion by considering the spin-axis as parallel to the orbital axis. Later on this will require a detailed discussion, to avoid problems of degeneracy within the formalism. The initial form of the spinor contains the (invariant) phase angle $\varphi_0 = \omega_0\tau$. In a moving frame this becomes of the form $\varphi_0 = \omega t - \mathbf{k}\cdot\mathbf{r}$. As discussed in Subsection 6.2.2, here $\mathbf{r}$ and t no longer correspond necessarily to the actual position coordinates of the electron. They can also be hypothetical values.

It is then possible to write $\omega t - \mathbf{k}\cdot\mathbf{r} = \tilde{\omega}t + (\omega - \tilde{\omega})t - \mathbf{k}\cdot\mathbf{r}$, where $\tilde{\omega}$ corresponds to the orbital motion. Here $\tilde{\omega}$ must be considered to be a constant, even for non circular motion, because the energy is constant. This will be discussed in Subsection 6.2.10.2, where it will also be clarified under which circumstances $(\mathbf{r}, t)$ can become hypothetical values. The quantity $(\omega - \tilde{\omega})t - \mathbf{k}\cdot\mathbf{r}$ can be recast into the form $\Phi(\mathbf{r}(t))$. The gimmick is then that $\Phi(\mathbf{r}(t)) = (\omega - \tilde{\omega})t - \mathbf{k}\cdot\mathbf{r}$ can be expressed as a pure function $\Phi(\mathbf{r})$ of $\mathbf{r}$ because the wave function has been rendered a function. Note that for circular orbits, $\mathbf{k}\cdot\mathbf{r}$ will be a constant with time, while for elliptical (or other non-circular) orbits, $\mathbf{k}\cdot\mathbf{r}$ may vary with time.

It is now possible to assume that in the separation of variables it is $\tilde{\omega}$ that should be removed from the total phase angle. After a separation of variables in the differential equation, the variable that has not been removed and that will become an argument of a harmonic polynomial is then Φ. Thus $\omega t - \mathbf{k}\cdot\mathbf{r}$ has been rewritten as $\tilde{\omega}t + \Phi$, expressing that it is both periodic in spin (with angular frequency ω) and in orbit (with angular frequency $\tilde{\omega}$). First, determine the energy from the calculation of an orbit for a point-like particle ($\tilde{\omega}$), and then add a quantization condition by expressing that Φ must be a function of $\mathbf{r}$. This expression takes into account the angular frequency of the spin. The spherical coordinate Φ can thus be identified with the phase of the wave function, even if it no longer exactly corresponds to the value ϕ for the position on the orbit. This settles problem (1) of Subsection 6.2.8.1, and also the problem of on which precise angular variable the wave function truly depends. For uniform circular motion it is possible to introduce a curvilinear coordinate $\varsigma = R\phi$ on the circle and to define $k = 1/R$, such that $\mathbf{k}\cdot\mathbf{r}$ takes the form $k\varsigma = \phi$. The variable $d\varsigma$ occurs then in the instantaneous infinitesimal Lorentz transformation $dt = \gamma(d\tau - v\,d\varsigma/c^2)$ that relates dt to $d\tau$. With a definition $k = 2\pi/\lambda$ for the relation between the wave vector and the wavelength, this corresponds to a wavelength $\lambda = 2\pi R$.

6.2.8.3 *A set of several modes that can be treated on the same manifold* $\mathbb{M}_N^*$

We have N copies of the electron wave function, labelled with the variable $j \in [1, N] \cap \mathbb{N}$. They are placed on positions P_j with coordinates x_j on a circle. This circle symbolizes the manifold $\mathbb{M}_N^*$, if one does not take into account the factor 2 that results from the 4π-periodicity of spinors. Each arc between P_j and P_{j+1} symbolizes a whole orbit. The positions P_j symbolize all the same position P on the orbit in real physical space. The point P_j is reached when the electron reaches the position P in physical space after j orbital periods. The image of another physical position Q on the orbit after going around j times would thus correspond to a position Q_j on the arc between P_j and P_{j+1} in $\mathbb{M}_N^*$. These other positions are thus not represented by the vertices of the regular polygon $P_0P_1 \cdots P_{N-1}$, which only images the position P. In each point of this polygon, there is a difference of phase between the spin of the electron and the phase of the orbital motion. If this difference is δ_1 in P_1, then it will be $\delta_j = j\delta_1$ in P_j. We have thus a manifold $\mathbb{M}_N^*$ pictured as a circle. The angle on the circle is a measure for the phase difference. The phase differences δ_j in P_j take the values $2\pi j/N$, where $j \in [0, N-1] \cap \mathbb{N}$. A uniform motion on the circle representing $\mathbb{M}_N^*$ describes exactly the phase difference between spin and orbit rotation angles. The same circle can be used to describe a situation where the phase difference grows faster, such that it is already $\delta_1^{(k)} = k\delta_1$ in P_1. The situations with the basic phase differences $\delta_1^{(k)} = k2\pi/N$, with $k \in [0, N-1] \cap \mathbb{N}$ can all be described on the same circle. They lead to phase differences $\delta_j^{(k)} = 2\pi jk/N$ in P_j for the "mode" k.

A mode with $k = 2\pi(N-1)/N$ is then indistinguishable from a mode with $k = -2\pi/N$, as the labels of the modes are defined modulo 2π. But physically the modes $k = 2\pi(N-1)/N$ and $k = -2\pi/N$ are not the same thing. In the mode $k = 2\pi(N-1)/N$ the rotation of the electron spin is faster than the rotation along the orbit, while in the mode $k = -2\pi/N$ it is slower. In terms of rotational energy this is not the same thing. Hence, even if $2\pi(N-1)/N$ and $-2\pi/N$ are the same phase, $k = 2\pi(N-1)/N$ and $k = -2\pi/N$ are not the same mode. The slower modes can also be described on the same circle. If the faster modes are described by wave functions Ψ_k it would suffice to take wave functions $\Psi_{-k} = \Psi_k^*$ to describe them. This implies thus that there are $2(N-1)+1 = 2N-1$ modes that are described with the aid of the same manifold. It has been taken into account that for $k = 0$, the value $-k$ does not correspond to a new mode. It can be understood on this basis why in a representation based on harmonic polynomials and the manifold $\mathbb{M}_N^*$ there must always be an odd number of sub-states.

In the Dirac approach this argument no longer applies. The reasoning described above is not correct, as the case $k = 0$ corresponds to a phase *difference* between the rotation angles involved in the electron spin and the orbital motion at the positions P_j on the orbit. But it neglects the fact that the orbital and spin periods can be made to coincide in two different ways: one with the electron spinning in the same direction as the orbital motion, and one with the electron spinning in the opposite direction. If these motions were thought of in terms of current loops, then this would really make a difference within a magnetic field. (It is this point that is taken into account in the even number of states used in the Dirac approach and not in the odd number of states within the Schrödiger approach.)

6.2.8.4 *Analogy with translational invariance: A key to the number of modes*

To treat the temporal translational invariance of the system on the manifold $\mathbb{M}_N^*$, symmetry-adapted functions are introduced. Using only the modes $k \geq 0$, we have a simple paradigm for such symmetry-adapted functions, *viz.* the model of translational invariance in space with cyclic boundary conditions used in solid-state physics and described in Section 2.10. The symmetry-adapted functions are there of the form $\psi_\kappa(j) = e^{\imath\kappa j 2\pi/N}$. As the orbital motion is planar, the two approaches should in principle coincide. This analogy will be developed as follows:

(1) It will be shown that in both systems, the different modes correspond to different degrees of polynomials in some variable. In harmonic polynomials this will be the variable $e^{\imath\phi}$.
(2) The degrees will correspond to the various ratios between the electron and orbital angular velocities. The angular momentum operator $\hat{\mathrm{L}}_z$ projects out the degree in $e^{\imath\phi}$, and the harmonic polynomials of a given representation also have a total degree that one can access by $\hat{\mathrm{L}}^2$.
(3) One should then in principle also be able to understand the number of modes through the analogy. But following the analogy it should also be possible to obtain an even number of modes in the Schrödinger approach, which is not the case.

It is for this reason that consideration must be given to clockwise and anticlockwise de-phasing. The quantities k and κ can be used to describe the various symmetry-adapted functions, when $k \geq 0$. The number of modes is equal to the number of sites. These modes could also be pictured as N equally spaced points on a circle. It would then mean that $k = 2\pi(N-1)/N$ and $k = -2\pi/N$ were considered to be the same mode. This is the case in the solid-state physics model, but not in the electron model. The states would rather have to be on an infinite line, where $k \in \mathbb{Z}$. However, treating $|k| > N - 1$, will have to be done with another manifold than $\mathbb{M}_N^*$. The restricted set of modes with $|k| < N$, can thus be pictured as a line segment without periodic boundary conditions. It can be avoided that the modes $-k$ and k are the same by giving them different pre-factors that contain a supplementary functional dependence. (This happens for example in solid-state physics if there is more than one atom in a unit cell. In solid state physics there is also an expedient to turn the topology of a line segment into that of a circle with periodic boundary conditions by introducing mirror images, which doubles the number of sites.)

6.2.8.5 *Analogy with translational invariance: Use of polynomials of various degrees*

There are thus different functions which describe various rates for the phase difference of the electron. These functions are labelled using k and they can all be written in the form $u^k = e^{2\pi jk/N}$, where $u = e^{2\pi j/N}$. Hence, k is the degree of the function in the variable u, and the function is thus a polynomial in the variable u. Now, the symmetry-adapted functions for rotational symmetry are obtained by taking the tensor product of the spinors. As described in Section 3.9 and Subsection 3.10.5, they are of the form $\xi_0^{2\ell-K}\xi_1^K$, where $K \in [0, 2\ell]$. The variable $|K|$ plays here the role of a degree in z. The degree can vary between 0 and $\ell = N - 1$. When the isomorphism discussed in Section 3.10 is introduced, spherical coordinates can be used, and the degree K gives rise to a degree m in the variable $e^{\imath\phi}$. The values of m can now also be negative, and they vary with integer steps between $-\ell = -(N-1)$ and $\ell = (N-1)$. As discussed in Section 3.9, these steps are integer because combinations $\xi_0\xi_1$ are made. The variable m takes thus $2N - 1 = 2\ell + 1$ different values, if $N = \ell + 1$. There is thus a one-to-one correspondence between the two ways of introducing symmetry-adapted functions, such that the number of modes can really correspond to a number of the type $2\ell + 1$. The point is that the development must start from a model with $N' = 2\ell + 1$ sites. The choice of the number of sites must then not only be motivated by geometrical considerations (such as about the number of perihelia in a closed orbit) but also by energy considerations (in terms of clockwise and anticlockwise de-phasing).

In considering the harmonic polynomials $Y_{\ell,m}(\theta, \phi)$ and restricting the motion to a plane parallel to the Oxy plane such that θ no longer is a variable but becomes fixed, the functional dependence $e^{\imath m\phi}$ of the modes coincides with that for the modes in a model based on translational invariance with cyclic boundary conditions. The modes with indices $-m$ and m are also related as ψ and ψ^*. It follows then that m can really be interpreted as the rate of electron de-phasing, and that the identification made between the spherical harmonics and the various periodicity ratios holds.

In the context of the Dirac equation, the viewpoint will become different. The $2N$ modes will have to be interpreted as N modes with spin up and N modes with spin down, such that the two modes for $k = 0$ will no longer be identical, because they will represent different energies when a magnetic field is switched on.

6.2.8.6 *Spherical harmonics and their degrees*

As discussed in Section 3.9 and Subsection 3.10.5, the spherical harmonics of degree ℓ are of the type $A\xi_0^{2\ell-K}\xi_1^K$, where A is a constant. The total degree is indeed 2ℓ in the spinor quantities, and ℓ in the variables (x, y, z). For a given value of ℓ, there are thus $2\ell + 1$ different polynomials, corresponding to the combinations $\xi_0^{2\ell-K}\xi_1^K$, where $K \in [0, 2\ell] \cap \mathbb{Z}$. Putting $m = \ell - K$, it can be seen that $m \in [-\ell, \ell] \cap \mathbb{Z}$. For $\mathbf{s} = \pm\mathbf{e}_z$, the two matrices $\frac{1}{2}\,[\,e^{-\imath\varphi/2}(\mathbb{1} + \mathbf{s}{\cdot}\boldsymbol{\sigma}) + e^{\imath\varphi/2}(\mathbb{1} - \mathbf{s}{\cdot}\boldsymbol{\sigma})\,]$ become diagonal with one of the two values $e^{-\imath\varphi/2}$ and $e^{\imath\varphi/2}$ on the diagonal elements, and therefore transform $\xi_0 \to \xi_0 e^{-\imath\varphi/2}$, $\xi_1 \to \xi_1 e^{\imath\varphi/2}$. With $\ell - K = m$, we obtain then $\xi_0^{2\ell-K}\xi_1^K \to e^{-\imath m\varphi}\xi_0^{2\ell-K}\xi_1^K$, such that the polynomials are really of degree m in the variable $e^{-\imath\varphi}$, which can be identified with $e^{\imath\kappa j 2\pi/N}$. (This clearly shows that rotations around the z-axis are special in that they do not mix the various polynomials, such that each polynomial builds a one-dimensional representation. This is due to the fact that the rotations around the z-axis build an abelian sub-group, and representations of abelian groups are always one-dimensional. This is simply a matter of choice of basis.) It is then possible to think of using the various polynomials to describe the different ratios between the period of the rotation of the electron and the period of the orbit, by identifying the number $\ell + 1$ (where ℓ is the degree of the polynomial) with the number N in $\mathbb{M}_N^*$ and by identifying the degree m with the number m in the ratio $(N \pm m)/N$ that links ϕ to φ in the relation $\phi = [\,(N \pm m)/N\,]\varphi$.[4]

6.2.8.7 *Degree operators*

As discussed in Subsection 3.10.5, the operators $\hat{\mathrm{L}}_z$ and $\hat{\mathbf{L}}^2$ have a mathematical meaning that precedes any application of the mathematics to physical problems. These degrees have nothing to do with the tilt of a rotation axis, even if this is at variance with what one might have anticipated, based

[4]The number N that is used here to specify $\mathbb{M}_N^*$ does not correspond at all to the principal quantum number, which is traditionally noted as n in textbooks. But the numbers ℓ and m, will correspond to the traditional corresponding quantum numbers.

on the way physics textbooks present $\hat{\mathbf{L}}$ by a vector model. For example, the "coupling" of angular momenta is often pictured as vector summation, while in reality it corresponds to regrouping terms in tensor products, based on the fact that all representations in terms of harmonic polynomials, whatever their degree ℓ, are based on tensor powers 2ℓ of the same spinor $[\xi_0, \xi_1]^\top$ that just differ by the number 2ℓ used in the power.[5] The ratio N/m will be independent from the tilt of the spin axis.[6]

6.2.8.8 *Why we do not use covariant derivatives in the hydrogen problem*

We are dealing here with various polynomials of the same total degree ℓ, and different ratios between the orbital period N and the period of the spin motion described by m. The essential point is that the number $m \in [-\ell, \ell] \cap \mathbb{Z}$. The differences in m (the expectation value of $\hat{\mathrm{L}}_z$) between two different polynomials are always integer. Remember that the formalism deals with calculating products of rotations $R_2 \circ R_1$. For a general rotation R_2 with rotation axis $\mathbf{s}_2$ of a rotation R_1 coded by a general spinor with "spin axis" $\mathbf{s}_1 \neq \mathbf{e}_z$, the whole tensor of $2\ell + 1$ polynomials will be linearly transformed if $\mathbf{s}_2 \neq \mathbf{e}_z$. (The crucial point here is that $\mathbf{s}_2 \neq \mathbf{e}_z$, not that $\mathbf{s}_1 \neq \mathbf{e}_z$.) As this does not enter into consideration for the solution of the Schrödinger equation for the hydrogen atom, this implies that the motion is planar, and that the orbital rotation axis $\mathbf{s}_2$ is fixed.

The representation matrix $\frac{1}{2}\left[e^{-\imath\varphi/2}(\mathbb{1} + \mathbf{s}\cdot\boldsymbol{\sigma}) + e^{\imath\varphi/2}(\mathbb{1} - \mathbf{s}\cdot\boldsymbol{\sigma})\right]$ of a rotation remains diagonal if and only if $\mathbf{s} = \mathbf{e}_z$. Hence, when the

---

[5]The identification of $\hat{\mathrm{L}}_z$ with an angular momentum operator is only valid for polynomials from $F(\mathbb{R}^3, \mathbb{C})$ in the real-vector representation based on the stereographic projection discussed in Section 3.10. This operator truly works on particle coordinates (x, y, z) introduced through the Lorentz transformation $\tau = \gamma(t - \mathbf{v}\cdot\mathbf{r}/c^2)$ of the proper time. Polynomials from $F(\mathbb{C}^3, \mathbb{C})$ have nothing to do with particle coordinates, and the angular-momentum operator $\hat{\mathrm{L}}_z = \frac{\hbar}{\imath}\frac{\partial}{\partial\phi}$, where z would be a particle coordinate, can thus not be defined for them. Only the more general meaning $\hat{\mathrm{L}}_z$ of a degree operator, where $z \in \mathbb{C}$ is the z-component of an isotropic vector, continues to exist for such polynomials, as discussed in Subsection 3.10.5 and Section 12.2.

[6]Note that the symmetries with respect to a variable v are always expressed by an operator $\frac{\partial}{\partial v}$, *viz.* $\frac{\partial}{\partial t}$ for time invariance, $\frac{\partial}{\partial x}$ for translational invariance, and $\frac{\partial}{\partial\phi}$ or $\frac{\partial}{\partial\varphi}$ for rotational invariance. The corresponding symmetry-adapted functions are always of the type $e^{\imath 2\pi k_v v}$ to obtain a periodicity that expresses the invariance. This is the reason why quantum mechanics is "wave mechanics". One of the goals of this book was to elucidate the particle-wave duality by testing the idea that a particle is not a wave, and that the wave functions are just a means to account for the symmetry.

rotations are restricted to the Oxy-plane, the set of rotations considered are restricted to an abelian subgroup. The representations become then one-dimensional, and the group factorizes. The same will then apply to the higher-dimensional representations based on tensor products. The set of spherical harmonics of a representation will thus not transform as a tensor under a rotation around the z-axis. Hence, we are working with the restriction of the three-dimensional rotation group to two-dimensional rotations in the orbital plane. As we are only rotating within the plane, the various components of the tensor remain separate quantities, and can therefore be treated as scalars, such that it is not necessary to consider the covariant derivatives. The whole θ-part of the wave function is therefore superfluous at this level, but the representations of the three-dimensional rotation group can be used nevertheless. The fact that harmonic polynomials are used as scalars proves that the rotations considered are restricted to the plane. This point also settles the apparent contradiction that resides in drawing an analogy between a case with translational symmetry (which corresponds to an abelian group) and the harmonic polynomials which have rotational symmetry (and correspond to a non-abelian group).

6.2.9 *Mathematical equivalence between the Bohr model and the solution of the Schrödinger equation for the hydrogen atom — tilted spin axis*

6.2.9.1 *Mathematical solution*

As discussed, the way the two periods of the physical problem were identified with the numbers N and m needs justification. There are indeed two problems with the approach taken up to now. The first is that when the rotation axis of the electron is made to coincide with the z-axis, then its spinor becomes $[1, 0]$. This implies that all the terms $\xi_0^{2\ell-K}\xi_1^K$ except the one with $K = 0$ become zero.[7] It seems then futile to associate the cases

[7] One may note that this problem disappears if $(0, 1, \imath)$ rather than $(1, \imath, 0)$ is taken for the isotropic vector. This choice permits one to use the particle coordinates of the Oxy plane within the isomorphism defined in Section 3.10. The problem of the wave functions that become zero is actually solved by realizing that it is possible to move on to a two-dimensional abelian description of the rotation group when the rotation axis coincides with the z-axis. The spinors are then replaced by scalars, and the wave functions defined in Subsection 6.2.8.5 can be used. In the three-dimensional formalism the harmonic polynomials are obtained as (tensor) products of the spinor wave functions

where the angular velocities of the electron spin around its axis and the motion of the electron along its orbit are in a rational ratio with a wave function with $m \neq \ell$. The second problem is that it seems arcane to claim that the rotation axis of the electron would be always aligned with the z-axis. These two problems solve each other mutually. It will be shown that the approach also holds when the spin axis is tilted, i.e. when $\mathbf{s}_1 \neq \mathbf{e}_z$ (but keeping the orbital axis $\mathbf{s}_2 = \mathbf{e}_z$). This will solve the problem of lack of generality and also the issue that all components of the tensor except the one for $k = 0$ would be zero. In fact, the case $\mathbf{s}_1 = \mathbf{e}_z$ is degenerate and masks the true φ-dependence. When $\mathbf{s}_1 \neq \mathbf{e}_z$, the $k \neq 0$ components of the tensor will no longer be zero. The treatment of this case is less obvious; the fact that the group is not abelian causes confusion.

When the electron is in orbital motion, the spin axis will co-move because it is a true vector. Here, the revision of the definition of the spin from the quantity $\mathbf{n}$ (spinning-sphere model) to $\mathbf{s}$ (spinning-top model), discussed in Subsection 5.4.2, becomes crucial. Imagine the definition of spin based on $\mathbf{n}$ had been maintained. The general form for a spinor corresponding to a uniform rotation around $\mathbf{n}$ is $\Psi = [\cos(\omega t/2) - \imath n_z \sin(\omega t/2), -\imath(n_x + \imath n_y)\sin(\omega t/2)]^\top$. When simultaneously a rotation of an angle Ωt around the z-axis is applied to this spinor, then the second component will become $e^{\imath\Omega t/2}\,[-\imath(n_x + \imath n_y)\sin(\omega t/2)]$. This kind of reasoning would be dangerous relativistically, but let us consider it non-relativistically. Using spherical coordinates, $-\imath(n_x + \imath n_y)\sin(\omega t/2)$ can be rewritten as $-\imath e^{\imath\phi}\sin\theta\sin(\omega t/2)$, and $e^{\imath\Omega t/2}[-\imath(n_x + \imath n_y)\sin(\omega t/2)]$ as $-\imath e^{\imath(\phi+\Omega t/2)}\sin\theta\sin(\omega t/2)$. The value of the term $e^{\imath(\phi+\Omega t/2)}$ in the latter expression shows that the new spin axis is the initial spin axis rotated around the z-axis by an angle of $\Omega t/2$ rather than Ωt. This nicely illustrates two points: (1) The rotation axis is not co-rotating because of the factor 2 in the angles, showing that $\mathbf{n}$ does not behave as a vector. (2) A

with themselves. In the two-dimensional formalism, the wave functions are also obtained by products of the wave functions with themselves, but these wave functions are now scalars instead of spinors. This shows that the wave functions for the subgroup are not obtained by blindly inserting the rotational parameters into the three-dimensional wave functions. The new procedure used serves precisely to prevent obtaining only zero wave functions. This change of wave functions accounts for the difference observed in the wave functions used in the Schrödinger and the Dirac equations. It is a matter of the dimension of the representation. After restriction of the motion to a plane, the rotation group becomes commutative while the Lorentz group does not. The restriction of the Lorentz group to a plane can thus not have a one-dimensional representation.

spinor has to be rotated over 4π to come back to its starting position. But as we use the notion that the spin axis should be $\mathbf{s} = \mathbf{e}'_z$ rather than $\mathbf{n}$, the spin axis that will be co-rotating as $\mathbf{e}'_z$ is a true vector. This will become even more crucial in the discussion of Thomas precession.

6.2.9.2 *Co-moving frames*

By definition true vectors are co-rotating, and as the spin axis is a true vector it will therefore co-rotate. The ideas can be easily visualized on the rotating Earth. When you turn by 30 degrees around your vertical axis in a room at ten o'clock, it will have the same final effect as when you do it at eleven o'clock. When you rotate at ten o'clock, your rotation will be followed by the rotation of the Earth during the lapse of time between ten and eleven o'clock. When you turn at eleven o'clock, it will be preceded by this rotation of the Earth. This shows why the rotation group is not commuting; the rotations cannot be commuting within a fixed absolute frame because the frame where they do commute is the co-moving frame. The rotations at ten and at eleven o'clock are about an axis that has varied because it looks constant in a relative, co-rotating frame. As this varying co-moving axis is by definition not fixed within the absolute frame, the rotations cannot commute within the absolute frame. If the rotation were made at eleven o'clock about the "true" ten o'clock axis, the one that remains fixed within the absolute frame, and which could clearly be identified by making a fixed mark, the result would not be the same. The co-moving axis of rotation has transformed as a vector as we have materialized it by associating it with the vertical axis of your body, and it is therefore of the type $\mathbf{e}'_z$.

But from all this it can be seen that if the electron has made a 2π turn around the orbit, then the orientation of its axis $\mathbf{s}$ will coincide again with its original orientation, because the relative frame and the absolute frame will coincide again. Therefore, it is still possible to write $\Psi = \psi_0 \otimes \psi_1 \otimes \psi_2 = e^{\imath 2\pi} \left[\psi(\theta,0) \otimes \psi(\theta,0) \otimes \psi(\theta,0)\right]$ at the special positions of the orbit considered in the reasoning of Subsection 6.2.7.

From this proof it is obvious that $\hat{\mathrm{L}}_z$ has nothing to do with the orientation of the spin axis. It has also nothing to do with the projection of the orbital angular momentum onto the z-axis, as the orbital momentum is aligned with the z-axis. It is related to a periodicity. It is a measure for the number of times the period of the spin goes into the true total period of the combined motion, according to Chern's theorem. The quantum number m was defined in terms of a period also in Heisenberg's initial paper. It would not make sense to relate it to a "quantized orientation" of the spin axis. How could such a mysterious rotational discontinuity possibly come out self-consistently of a mathematical formalism that corresponds to the continuous rotation group SO(3)? It would imply that the mathematics are wrong! What can be understood perfectly, is that the degrees N, m of a polynomial would be quantized quantities. That is true by definition.

6.2.9.3 *Degrees of polynomials and the paradoxical "mutual incompatibility" of the operators $\hat{\mathrm{L}}_x$, $\hat{\mathrm{L}}_y$, $\hat{\mathrm{L}}_z$ for the components of the angular momentum*

The claim that the quantities ℓ, m are only degrees of polynomials and that they have nothing to do with the spin axis can be further justified. The harmonic polynomials of total degree ℓ have been catalogued according to their degree m in the variable $e^{\imath\phi_z}$, where ϕ_z is an angle defined within the Oxy-plane. In fact, $e^{\imath m\phi_z}$ is the common ϕ_z-part that can be factorized out for polynomials of the type $\sum_k c_k\,(x+\imath y)^{m+k}(x-\imath y)^k z^{\ell-m-2k} = r^\ell P_{\ell,m}(\cos\theta_z)\,e^{\imath m\phi_z}$. Here (r,θ_z,ϕ_z) are the spherical coordinates. This is useful for orbital motion restricted to the Oxy-plane. If one needed to catalogue harmonic polynomials according to their degree in an analogous variable $e^{\imath\phi_x}$, where ϕ_x is an angle defined within the Oyz-plane, it would be necessary to make first a change of basis. This would require introducing a system of different spherical coordinates (r,θ_x,ϕ_x) wherein $x = \cos\theta_x, y = r\cos\theta_x\cos\phi_x, z = r\cos\theta_x\sin\phi_x$. It would then be possible to consider polynomials of the type $\sum_k c_k\,(y+\imath z)^{m+k}\,(y-\imath z)^k x^{\ell-m-2k} = r^\ell P_{\ell,m}(\cos\theta_x)e^{\imath m\phi_x}$ of degree m in the variable $e^{\imath\phi_x}$. This would then be useful for orbital motions restricted to the Oyz-plane. But in doing this we would have introduced a different set of harmonic polynomials, i.e. a different basis. The wave functions cannot have the same degree m_z in $e^{\imath\phi_z}$ in all their terms and the same degree m_x in $e^{\imath\phi_x}$ in all their terms at the same time. The point is that there are various terms in the polynomial of degree m_z in $e^{\imath\phi_z}$ and total degree ℓ, which contain still different degrees in the variables $x+\imath y$, $x-\imath y$ and z. Therefore, the operators $\hat{\mathrm{L}}_z$ and $\hat{\mathrm{L}}_x$ cannot be used simultaneously. The operator $\hat{\mathrm{L}}_x$ only makes sense as the value of the operator $\hat{\mathrm{L}}_z$ within the (r,θ_x,ϕ_x)-based basis. Within the basis defined with (r,θ_z,ϕ_z) the operator will not be able to project out a meaningful eigenvalue m_x that would be the degree of a term $e^{\imath m_x\phi_x}$ that could be factorized out. In quantum mechanics it is argued that the operators $\hat{\mathrm{L}}_x$ and $\hat{\mathrm{L}}_z$ do not commute and therefore do not have simultaneous eigenvalues. But this does not mean that the component of the orbital angular momentum along the x-axis would not exist. This component does indeed exist and we know its value: zero. The quantities that do not exist simultaneously are the degrees m_z and m_x because they qualify basis vectors that belong to two different bases corresponding to different reference frames. The classical physical components L_x and L_y of the angular momentum are not the expectation values of the operators $\hat{\mathrm{L}}_x$ and $\hat{\mathrm{L}}_y$. They are zero for orbital motion in the Oxy-plane, because the angular momentum is aligned with the z-axis.

6.2.9.4 *Thomas precession and co-moving frames*

Even with circular orbits, one may fear that the previous construction with identical copies may break down in the relativistic case, because when the spin axis $\mathbf{e}'_z$ and the normal to the orbit $\mathbf{e}_z$ are not aligned, there may be doubts that $\mathbf{e}'_z$ could remain fixed in the co-moving frame, due to the Thomas precession (physical effect 3 in Subsection 6.2.5). But an understanding of Thomas precession shows that this fear is not justified. The essence of Thomas precession is that the composition of two non-collinear pure boosts is no longer a pure boost, but contains a rotation. Therefore, four-vectors will also rotated. But of course, in the co-moving frame one does not notice this rotation of the vectors. As the spin vector is a true vector, it will therefore remain fixed in the co-moving frame.

The previous argument of commutativity in the co-moving frame will still hold. This is the principle of relativity, which just corresponds to an argument based on a similarity transformation in group theory; the group looks the same all over.

Of course, the orientation of the spin axis in the lab frame will change. But this can be accounted for by stipulating that what defines a true period of the global system is that the spin axis must come back to its initial position after a true period. It may thus be necessary to use a different Riemann manifold with a different ratio of periodicities to that used in the non-relativistic approach of the same orbit, but the general idea that two periodicities must be adjusted remains the same. In reality, the problem is solved abstractly: a certain set of quantum numbers is chosen, which comes down to selecting a certain ratio. The orbits must then be reshuffled to keep the same ratio, rather than reshuffling the quantum numbers to keep the same orbit. The Lorentz contraction and time dilatation along the orbit (physical effect 2) can then also be treated by the previous argument. The situation will of course remain very obviously simple if $\mathbf{s} = \mathbf{e}_z$.

6.2.9.5 *Degeneracy of the solutions for planar motion*

The orbits treated up to now were thus planar. The use of the spherical harmonics with the full three-dimensional symmetry is, however, instrumental for understanding how they are linked together and for labelling them. To make sure that the symmetry calculations are carried out correctly, the couplings are calculated by constructing $Y_{\ell,m} = Y_{\ell_1+\ell_2,m_1+m_2}$ from the set $\{Y_{\ell_1,m_1} \otimes Y_{\ell_2,m_2}\}_{m=m_1+m_2}^{\ell=\ell_1+\ell_2}$ within a three-dimensional approach. This description can then be "cut" with a two-dimensional plane to obtain the restriction to the two-dimensional subgroup, wherein $L_{m_1+m_2} = L_{m_1} + L_{m_2}$. Coupling means here that a representation of higher dimension $Y_{\ell_1+\ell_2,m_1+m_2}$ is constructed from two representations of lower dimensions Y_{ℓ_1,m_1} and Y_{ℓ_2,m_2}. This is possible as all representations of degree n', (where $2n' \in \mathbb{N}$ can be odd), are constructed from tensor powers $\otimes_1^{2n'}(\xi_0, \xi_1)$, taking $2n'$ identical copies of the basic spinor (ξ_0, ξ_1). This way, a better insight into the relationships between the various polynomials is obtained. An important point is that the labels (i.e. the quantum numbers) used to denote the three-dimensional polynomials with the full spherical symmetry do correspond to their restrictions to the plane, as the wave functions are classified with respect to the values of (L, L_z).

Within the restriction to the two dimensional subgroup, the various components of the tensor are degenerate and all have the same energy. It is only when a magnetic field $\mathbf{B}$ is introduced that (in the traditional interpretation of the formalism) the electron triad (or tetrad) starts to undergo precession around the axis of $\mathbf{B}$ (for which one traditionally takes the z-axis). The degeneracy is then lifted, and it becomes necessary to rotate the whole tensor of $2\ell + 1$ spherical harmonics to take into account that the axis of rotation is no longer along the z-axis. Due to this linear transformation, the various components of the tensor become mixed up; the tensor is no longer diagonal in the frame wherein the rotation axis would be the z-axis. One is then really forced to use spin matrices and to diagonalize them to find a number of states of fixed energy.

6.2.10 *Non-relativistic elliptical orbits: How broken symmetry leads to the Copenhagen interpretation of quantum mechanics*

6.2.10.1 *Orbits without rotational symmetry*

When the orbit is an ellipse, the rotational symmetry is obviously broken. Treating classical elliptical orbits is an intermediary step between the Bohr/Schrödinger and the Bohr-Sommerfeld/Dirac approaches to the hydrogen atom. Because the rotational symmetry is broken, it should not be possible to solve the problem by using symmetry-adapted functions with rotational symmetry. Nevertheless, it can be done, but at a high cost. In the Bohr-Sommerfeld/Dirac approach, the orbits involved are more complicated than just ellipses, because the ellipse will undergo perihelion precession. There will also be an additional complication in that the mass will no longer be constant. It is the latter feature in particular that will present a significant difficulty and requires the introduction of the "spin".

For elliptical motion, the time evolution of the variables $\mathbf{r} = (x, y)$ and $\mathbf{v} = (v_x, v_y)$ will not be uniform. They are thus not very suitable for describing the uniform motion of the electron spin. (Relativistically the electron spin is uniform in the co-moving frame, non-relativistically it can be assumed that it is also uniform in the lab frame.) This is a new situation which is different from the one that prevails on a circular orbit. There it was possible to use the position coordinates (x, y, z) as variables to describe the spin due to a simple geometrical spin-orbit coupling ($\phi = \frac{4}{3}\varphi$ in our example). As on an elliptical orbit $\phi(t)$ does not vary uniformly with time, the polar coordinate ϕ cannot be written as $\tilde{\omega}t$ where $\tilde{\omega}$ would be a constant. Hence, it is difficult to express the uniformly varying angle $\varphi = \omega t$ in terms of $\phi(t)$. As the velocity vector $\mathbf{v} = (v_x, v_y)$ also does not follow a uniform motion as a function of a polar angle, the same applies for the velocity coordinates.

6.2.10.2 *The motion is uniform in terms of covered area, which serves as a generalized angle*

For elliptical motion, the law of the conservation of angular momentum L_z still holds. This conservation law can be restated in the form of *Kepler's area theorem*. The vector product $dS\,\mathbf{e}_z = \frac{1}{2}\mathbf{r} \wedge \mathbf{p}\,dt/m = \frac{1}{2}L_z\mathbf{e}_z dt/m$ describes the area dS covered by the orbit in the lapse of time dt. The variation of S is uniform, i.e. $\exists\Omega \in \mathbb{R} \parallel S = S_0\Omega t/2\pi$, because the angular momentum L_z is a constant of motion. This leads to the idea to consider the area S as a generalized rotation angle that evolves uniformly and could be used to create a uniform link (i.e. a linear relationship) between the orbital motion and the spin rotation of the electron. The generalized rotation angle can then be considered as an angle in angular momentum space, or in area space. In area space, the covered area is described by $S(t) = S_0\Omega t/2\pi$. There is rotational symmetry in area space. The orbit is a "circle" of radius S_0 in area space, and the rotation angle is Ωt. The uniformness of this rotation in area space expressed by $S(t)$ can be linked to the uniformness of the rotation of the electron. In other words, after the necessary arrangements have been taken to ensure that $\phi = \Omega t$ varies between 0 and $k2\pi$ for one turn around the generalized orbit on the manifold $\mathbb{M}_N^*$, it will be possible to recover $\phi = \frac{k}{n}\varphi$ (just as in the Bohr model in the example with $\frac{k}{n} = \frac{4}{3}$). It can thus be expected that $\phi = 2\pi S/S_0$ will have a period $N2\pi$ in $\mathbb{R}^3$, that can be described with the use of a manifold $\mathbb{M}_N^*$. The covered area will be used as a clock that is ticking at a constant rate.

To see what would occur without the area theorem, one can introduce two spinors: $\psi_r(\mathbf{r}, t)$ derived from $\mathbf{e}_x + \imath\mathbf{e}_y$, where the spin is described in terms of the position vector $\mathbf{r}$, and $\psi_v(\mathbf{v}, t)$ derived from $\mathbf{e}_{v_x} + \imath\mathbf{e}_{v_y}$, where the spin is described in terms of the velocity vector $\mathbf{v}$. Introducing coordinates $(x/r, y/r)$, $(v_x/v, v_y/v)$, we will have $\psi_r(\mathbf{r}, t) = \sqrt{(x - \imath y)/2r}$ and $\psi_v(\mathbf{r}, t) = \sqrt{(v_x - \imath v_y)/2v}$. The two spinors are identical in the sense that they describe the same uniform rotation of the electron spin. But as this rotation is described in the variables (x, y) and (v_x, v_y), the rotation angle $\phi_r(t)$ that must be applied to (x, y), and the rotation angle $\phi_v(t)$ that must be applied to (v_x, v_y) are different and they are not linear functions of time.

Here, the orbit coordinates (x, y, z) have been used as parameters to code the triad parameters (X_r, Y_r, Z_r). By a similar procedure it is also possible to use the velocity coordinates (X_v, Y_v, Z_v).

For the electron spin there will be two non-linear functions F and G, such that $(X, Y, Z) = F(X_r, Y_r, Z_r)$, $(X, Y, Z) = G(X_v, Y_v, Z_v)$, or two non-linear functions f and g such that $(\xi_0, \xi_1) = f(\xi_{0r}, \xi_{1r}) = g(\xi_{0v}, \xi_{1v})$ where (ξ_0, ξ_1) will be the first column within a *uniform-rotation* matrix $e^{-\imath(Et - \mathbf{p}\cdot\mathbf{r})}\frac{1}{2}[\mathbb{1} + \mathbf{s}\cdot\boldsymbol{\sigma}] + e^{+\imath(Et - \mathbf{p}\cdot\mathbf{r})}\frac{1}{2}[\mathbb{1} - \mathbf{s}\cdot\boldsymbol{\sigma}]$, i.e. $e^{-\imath(Et - \mathbf{p}\cdot\mathbf{r})}[1, 0]^\top$ if the rotation is around the z-axis. It is thus possible to keep a scalar calculation.

That there could a spinor (ξ_0, ξ_1) in the lab frame that would correspond to a uniform rotation is thus not at all trivial. In fact, (ξ_{0r}, ξ_{1r}) and (ξ_{0v}, ξ_{1v}) do not correspond to a uniform rotation. The rotation coded by (ξ_0, ξ_1) can only be uniform in the generalized angle $\phi(t) = 2\pi S(t)/S_0$, and this is in turn due to the fact that the angular momentum around the z-axis $(xp_y - yp_x)$ is a constant of motion.

6.2.10.3 *The motion is no longer described in terms of true position coordinates*

The Dirac equation was obtained by starting from the Rodrigues formula applied to a rotating electron. The electron was uniformly rotating in its proper time τ. Then the gimmick of multiplying τ with m_0c^2 and putting

$m_0c^2\tau_0$ equal to $\hbar\omega_0\tau_0/2$ was introduced in the function $e^{\imath 2\pi\tau/\tau_0}$. In the end, the phase angle $2\pi\tau/\tau_0$ is then no longer expressed in τ, but in the action variable $Et - \mathbf{p}\cdot\mathbf{r}$. This opened up the way to describe a motion that is no longer uniform and periodic in t, as uniform in a generalized angle $(Et - \mathbf{p}\cdot\mathbf{r})/\hbar$ that is based on the area covered. This area is expressed in the angular momentum, and it is proportional to the proper time of the electron, because the electron rotates uniformly in its own rest frame. It is the expedient of putting $m_0c^2\tau_0 = \hbar\omega_0\tau_0/2$ that permits the crucial change of variables. This constitutes thus a justification *a posteriori* for the *ad hoc* postulate introduced in Subsection 5.1.3 to make e.g. the step from (5.11) to (5.12). The fact that the (x, y) and (v_x, v_y) values must combine into a constant value $(xp_y - yp_x)$ can only be expressed by calculating the expectation value of $\hat{\mathrm{L}}_z$ on $e^{-\imath(Et-\mathbf{p}\cdot\mathbf{r})}[1,0]^\top$. The period of this value must be an integer number of times the period of the electron in the generalized angle $2\pi[\,(m_0c^2\tau)/(m_0c^2\tau_0)\,]$. The new fact in the case of the elliptical orbits is thus that the number m becomes explicitly the expectation value of the angular momentum operator $\hat{\mathrm{L}}_z$. The numbers $N = \ell + 1$ and m are related to the number of turns of a motion that can be described as "uniform", provided a generalized rotation angle is introduced. With this generalized angle, all the arguments can now be repeated to find a common period for the orbital and the spin motion of the electron. The only difference is that it must now be done in terms of the generalized angle. In other words, the motion must not be described as uniform in space, but as uniform in the action variable.

The introduction of the coordinates (x, y) and the corresponding polar coordinates (ρ, ϕ) in the solution of the Schrödinger equation leads to $x + \imath y = re^{\imath\phi}$ for a rotation around the z-axis. But later on, it turns out that the motion in this angular variable ϕ must be uniform. The coordinates (x, y) can therefore not be the actual position coordinates on the elliptical orbit. As ϕ is changing uniformly with time, it must be proportional to the area coordinate. Hence, ϕ corresponds to the position angle on a circular orbit that leads to the same area variable as the motion on the elliptical orbit. The true angular variable on the ellipse would be different from ϕ. The variable might be noted as ϕ_*, but this notation will never be used: it is not proportional to φ. (The variable ϕ remains proportional to φ, as it is uniform.) Similarly, (x, y) are not the coordinates on the elliptical orbit but on the equivalent circular orbit. In fact, by describing the motion in the generalized angle, the true coordinates have been rendered "unknowable" within the formalism. Perhaps the term "not describable" would be more appropriate mathematically, but from the physics point of view the term unknowable is perfectly appropriate due to Heisenberg's uncertainty principle, which addresses the way one can obtain experimental information about a microscopic system.

6.2.10.4 *Solutions of the wave equation as an equivalence class*

Perhaps other types of coordinates and symmetry-adapted functions could be used, such that it would remain possible to describe the true coordinates, but this will not improve the quantum mechanical calculations of the energies. It would be useless. To permit the description of the physics by the group theory it has been necessary to construct a language of areas, generalized angles and fake position and momentum coordinates on virtual circular rather than actual orbits. It is not possible to distinguish the elliptical motion from the circular motion when the generalized angles are used. Therefore, the Bohr model does not talk about elliptical orbits. One might have wondered why they were eliminated from the discussion in the Bohr model. It may have given the impression that they were squeezed out of existence due to some mysterious quantization. Within the context of the Schrödinger equation, the choice not to address the variables (x, y) individually corresponds also to the fact that the wave function only contains the time part of the coordinate transformation in the Lorentz transformation. The coordinates only occur in the combination $Et - \mathbf{p} \cdot \mathbf{r}$ within the phase of the scalar wave function. It is thus possible to vary the orbits and this will not introduce errors, as long as the value of $Et - \mathbf{p} \cdot \mathbf{r}$ is not changed. This means that equivalence classes of orbits defined by the values of $Et - \mathbf{p} \cdot \mathbf{r}$ exist. It is now clear why it was possible to recover circular orbits in Subsection 6.2.1 as solutions of the Klein-Gordon equation for a particle that travels along a straight line in truly uniform motion in free space. The straight line and the circle are orbits that belong to a same equivalence class. Since (x, y) and (p_x, p_y) are no longer described individually, but only in the combination $Et - \mathbf{p} \cdot \mathbf{r}$, it is possible to introduce logical probabilities for the individual parameters. It remains questionable what this truly means, but as the orbital planes can have all orientations in space, rotational symmetry will eventually be recovered, such that the probability calculus may become meaningful.

The angle ϕ corresponds only to the real position parameters (x, y) when the motion is circular. For each value of m such a circular motion exists. Elliptical orbits also exist but they are beyond scrutiny. It is now clear why the action variable is indivisible in quantum mechanics; it must be rendered indivisible in order to permit a description of the orbits with symmetry-adapted functions of spatial rotational symmetry. It does not mean, however, that the true coordinates (x, y) and (p_x, p_y) do not exist. Just as in the case of the eigenvalue equation for the spin operator, it is found here that the Schrödinger equation does not define a single physical motion, but a whole infinite set of them. The set is just defined by a criterion. For spin, it is that the solutions share the z-axis. For the Schrödinger equation, it is that they share the same energy. An equation can be mathematically described as the definition of a set. Here this takes a very unexpected twist, as the sets turn out to be larger than initially may have been imagined.

6.2.10.5 *The Copenhagen interpretation appears*

In the Copenhagen interpretation of quantum mechanics the mathematical formalism has just been *hineininterpretiert*; it describes things in the mathematics, not in the physics. But it does so by giving hints that are not rigorous and plain wrong when taken literally. In a sense the Copenhagen interpretation is a set of prescriptions that allows one to carry out correctly the group-theoretical calculations without understanding the details of the underlying mathematics. It explains what has to be done to conform to the rule to "shut up and calculate" we evoked before. It tries to offer a short-cut to difficult mathematics. But to achieve this, it is necessary to introduce a folk lore of alienating, counterintuitive physical images that are incorrect. They have, however, the virtue that they will guide the physicist through the mathematical jungle unscathed provided he is prepared to give up asking certain questions. This will in any case be a frustrating experience.

The *ad hoc* rules of the Copenhagen interpretation describe the failure of all attempts to force the description of non-circular orbits into a harness of spatial rotational symmetry while this symmetry is very obviously broken. A construction based on a generalized angle can save the rotational symmetry within the mathematical formalism on another level, but at the price of rendering the action variable indivisible. One is then forced to drop the idea of providing a complete description of the orbits, because they do not survive the premise of spatial rotational symmetry, which has (erroneously) been elevated to the status of a touchstone. The probabilities, however, pass the test of the spatial symmetry, such that they become the building bricks for describing the mathematics.

One could muse on the theme of spontaneous symmetry breaking to oppose the broken spatial rotational symmetry of the orbits to the conserved rotational symmetry of the probabilities that occur here, but this would only be a showy description that forgoes the task of explaining that the rotational symmetry has been saved in terms of a generalized angle in covered-area or angular-momentum space. Despite the fact that the set of rules introduced by Bohr has been tremendously useful and successful, it has a drawback, in that it denies us a deeper understanding and undermines confidence in rational thinking. The latter is simply unacceptable. Despite the success of the Copenhagen quantum imagery, the entire realms of Newtonian and relativistic mechanics with their classical imagery remain valid. The working assumption that quantum mechanics could be just the group theory of relativity remains possible.

6.2.10.6 *Elliptical orbits within the equivalence classes of the Bohr model*

In a Bohr model with strictly circular orbits, there is no need for an identification between angular momentum and n, even though Bohr made the link. The various solutions with different expectation values for $\hat{\mathrm{L}}_z$ all have the same energies, but they differ in the ratio between the orbital and the spin periods. With hindsight, we can understand now that the Bohr model already accounted for elliptical orbits, justifying the link Bohr made. One can however not describe the elliptical orbits in detail, and therefore the discussion is limited to a description of equivalence classes of orbits. Each class contains a circular orbit that can be used to represent and label it, and is used to describe the class. In the classical case there is no problem with a varying mass, such that the relationship between the covered area S and L does not require caution. A scalar representation can thus be used and there will be various solutions with the same energy that differ only in the expectation value of $\hat{\mathrm{L}}_z$. Of course, this raises the question about the physical meaning of the operator $\hat{\mathbf{L}}$, because $\hat{\mathbf{L}}$ cannot be a vector. However, the other components $\hat{\mathrm{L}}_x$ and $\hat{\mathrm{L}}_y$ can never be used, except to construct ladder operators. The reason for this is that they belong to a different basis. However $\hat{\mathrm{L}}_x \pm \imath\hat{\mathrm{L}}_y$ makes sense.

In summary, in pursuing the quest to treat elliptical orbits, the heart of quantum mechanics has been reached. The journey has led to the discovery of the origin of the relation $E = \hbar\omega$. It has been discovered why quantization is in terms of angular momentum, why it is necessary to abandon the notion of a classical orbit for symmetry reasons, and why individual parameters like position and momentum become unknowable within the theoretical framework chosen. This is the beginning of a good understanding of quantum mechanics.

It should not be a surprise that rotational symmetry is recovered in the generalized angle that corresponds to the area covered, or to the angular momentum. Noether's theorem says that to each symmetry corresponds a conserved quantity. For rotational symmetry, this is the angular momentum. It is for this reason that the rotational symmetry must be described in angular-momentum space.

6.2.11 *The uncertainty principle and the two-dimensional character of the motion as crucial ingredients*

It has been argued that it does not make sense to determine the position and the velocity of a particle simultaneously in a symmetry-based formalism, because the orbits do not have rotational symmetry; only angular momentum does. The lack of precision that results from this has one significant advantage: if both the position and the velocity in the problem were treated according to classical mechanics, then the whole orbit would be predetermined right from the start. There would then be no further liberty to adjust the orbit to make sure that the wave function was single-valued and it would not be possible to discover the quantization rules. This is also the reason why in the old quantum theory these rules had to be added in an *ad hoc* fashion. Hence, the fact that it has not been tried to describe the orbit completely is crucial.

There is a very important point at stake here. In classical mechanics we are used to a classical paradigm for the calculation of the motion of an object. It is possible to apply a theorem that the translational and rotational degrees of freedom are decoupled. The translational motion is described by applying Newton's law to the motion of the centre of mass, such that the object behaves like a point mass. If a spanner is thrown in free space, its centre of mass will follow a straight line, and the spanner will rotate around this centre of mass. The same remains valid when the centre of mass describes an orbit as illustrated in Figure 6.7. To describe the rotational degrees of freedom, it would be necessary to consider the total angular momentum of the forces, and the moment of inertia of the object. It can be done, but it would be completely irrelevant, because the motions are decoupled.

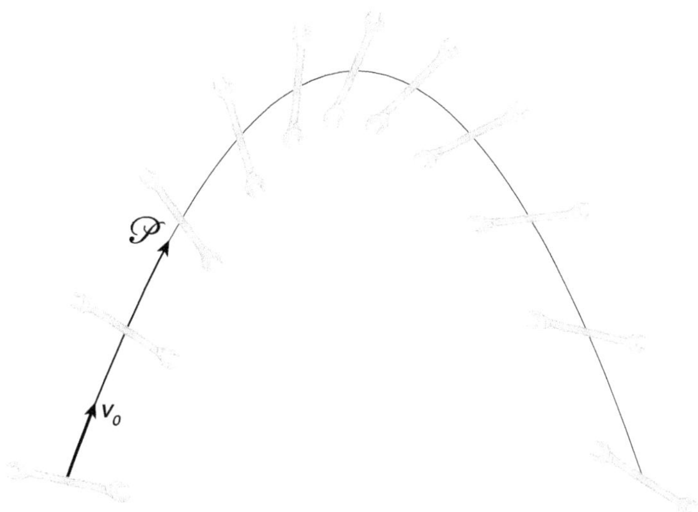

Fig. 6.7 In classical mechanics the motion of the centre of mass of a spanner (the parabola $\mathscr{P}$) and its rotation around its centre of mass are decoupled, as illustrated here in a simulation of a stroboscopic picture that would be taken with a constant frequency of flashes. The initial velocity is $\mathbf{v}_0$. In the drawing, the horizontal component of the translational motion of the centre of mass and the rotational motion of the spinning spanner around its centre of mass are both uniform. The rotational energy of the spanner constitutes a negligible contribution to its mass or total energy. We could therefore have taken any other rotational frequency for the spinning motion and it would not affect the orbit, even for the motion of an electric charge q within, for example, a constant electric field $\mathbf{E}$, where the mass no longer cancels out in the Newtonian equation of motion $m\mathbf{a} = q\mathbf{E}$. (In the analogue of a particle in the gravitational field of the Earth, the mass of the particle cancels out in the equation of motion $m\mathbf{a} = m\mathbf{g}$, where $\mathbf{g}$ is the gravitational constant of the Earth.)

Relativistically, it is more of an issue, as the two types of motion are no longer decoupled. Angular momentum of a body is defined as $\sum_j \mathbf{OP}_j \wedge \mathbf{p}_j$, where $\mathbf{OP}_j$ are the position vectors of the constituent particles. But this definition tacitly assumes a scheme of simultaneity for the vectors $\mathbf{OP}_j$, and in a moving frame, this scheme of simultaneity will be upset, such that angular momentum becomes part of a six-component tensor. The vectors $\mathbf{O'P'}_j$ will no longer be simultaneous, and in order to account for this, the centre of mass must also be considered, the definition of which also depends on the reference frame. The rotational and translational degrees of freedom are thus no longer decoupled. It is well known that angular momentum and the centre of mass cannot be separated in the theory of relativity, and that they make up a tensor, just like the electric and the magnetic field. It is wrong to treat such a tensor as two independent, separated vectors; the angular momentum and the centre of mass must be treated simultaneously, and this can be done by using spinors.

This relativistic remark will have no relevance when the velocity of the centre of mass of the electron is low. Up to now, great care has been taken to calculate the true periodicity of the motion, treating the simultaneous periodicity of the orbit and the spin. According to Newtonian mechanics, there is no need for this, as spin and orbit are decoupled. Relativistically, the decoupled scheme breaks down, illustrating that the decoupling is not a necessity. Certainly, relativity would then justify the need for quantum mechanics, but it cannot justify that quantum mechanics is also needed non-relativistically. The relativistc objections given above would also not apply for a point mass. After all the efforts to carry out the spinor calculations rigorously, the approach has still not revealed why it is so important to describe the periodicity of the combined translational and rotational motion.

It appears paradoxical that one has to consider the true periodicity of a system, taking into account both orbit and spin, when the orbit is already a circle and thus perfectly periodic. What is at stake here is rather the total energy that corresponds to the true period. When a spanner rotates very fast, this implies that it has a lot of rotational energy. This means that its "rest mass" (where "rest" is in translational terms) will be higher. The rotation can thus truly affect the orbit when the interactions are electromagnetic. Conversely, translational motion changes the clock rates, and thus also affects the "speed" of the rotation. Relativistically, spin and orbit are thus coupled. (As already shown, the kind of "spin-orbit coupling" being described here does not correspond to the quantity $\mathbf{L}\cdot\boldsymbol{\sigma}$ used in the Dirac theory.)

For a macroscopic charged object, the rotational motion accounts only for a negligible contribution to that object's total (translational) "rest mass". Therefore, no differences will be observed between orbits of such objects of the "same mass" and different states of rotation. This is then the limit of large quantum numbers. But for a spinning electron, its translational "rest mass" could be entirely rotational energy. The question how many turns the electron makes in a given amount of time, is extremely important for the calculation of the total energy of an orbit, because increasing the speed of the rotational motion will increase the "rest mass" of the electron proportionally. This shows relativity at work, through the relation $E = mc^2$, even when the translational speeds that come into play are not relativistic.

Quantum mechanics is thus relativistic even for non-relativistic velocities. As the phase of the wave function is relativistic, it accounts with great precision for most of the mass effects. It was seen in Section 6.1 that this phase has been wrongly interpreted as related to a wave in space, instead of a wave in proper time. At low velocities, γ within $\gamma(t - vx/c^2)$ can be taken to be 1. By wrongly interpreting the time wave $f(t - vx/c^2)$ as a wave in space through $w = c^2/v$ and $f[(x - c^2t/v)(-v/c^2)] = g(x - wt)$, it has been possible to keep track of the phase. Requiring the wave function to be single-valued necessitates the phase to be taken into account, such that the calculations are carried out correctly. One may have wondered why there is a relativistic phase within the non-relativistic Schrödinger equation. The answer seems to be that it allows us to calculate the mass, and thus the total energy correctly.

As in the hydrogen spectrum only mass differences are measured, the contribution m_0c^2 to mc^2 can be subtracted out. But the phase remains correctly coded to account for what is truly important. It is for this reason that the phase of the wave function is so much more important than the orientation of the spin axis. The orientation of the spin axis will only play a role in magnetic fields (where changes in the orientation of the spin manifest themselves in changes of the total energy), or in electric fields when the translational motion becomes relativistic and one may want to consider the energy of some induced electric dipole within the Coulomb field.

From this analysis the conclusion can be drawn that the message quantum mechanics brings us is that *spin and orbit are also not decoupled "non-relativistically"* (i.e. at low translational velocities). It may indeed never be possible to calculate the true orbit, as the information about the influence of the spin on the orbit may be unaccessible to experimental investigation, as shown by Heisenberg's analysis [Bohm (1951)] (in terms of the uncertainty principle) of how an experiment works. The orbits are then unknowable, and this corresponds exactly to the situation encountered in trying to deal with elliptical orbits. It is only really possible to know the probabilities. As the symmetry theory ignores the details of the non-circular orbits anyway, it is perfectly suited for this situation where the coupling between the orbit and the spin are not known. The only situation that can be treated is the one of a true periodicity. When, after a true period, a particle comes back to the same position, it will also have the same momentum, the same orientation of its spin axis, and the same phase for its spin. It is then certain that this periodic motion can be continued indefinitely, such that the motion must be radiation-less. Some of the properties of the particle are known at the points that correspond to the various periods. These points serve then as beacons on the orbit, while no further details are known about the orbit in between, but this is sufficient to successfully unleash group theory onto the problem.

This is an illustration of the limitations and the power of group theory. Group theory does not contain clues as to the precise underlying mechanism, since wildly different mechanisms can still have the same symmetry. But a given formalism can cover several possible situations simultaneously (see Chapter 11). This is exactly what might have happened here; a number of precautions have been entered into the theory to cover a non-essential relativistic exception (the electric dipole effect), but in doing so we have also covered a non-relativistic exception we were unaware of and that proves to be crucial (the "rest mass effect").

One could imagine a crude picture for the coupling, even if the orbital motion is not relativistic. The fact that the electron spins means that it must have some internal structure and that it is not a true point mass. It is impossible to define some Lorentz transformation for the whole of space-time that would describe a uniformly rotating frame corresponding to the internal motion of an electron. Beyond a given radius, the motion would have to be faster than light. This relativistic argument serves here only to point out that a rotating frame can only be defined for a restricted area in space. Within the most drastic restriction, there could be uniform motion on a circle of one specific well-defined radius, and the rotating frame would only be defined to keep track of a position on this circle. If the motion is not circular, it may be necessary to limit the details of the description to that of a motion which is uniform in terms of a generalized angle (related to angular momentum) as described earlier.

The internal structure of the electron is unknown. The important point is then that there should be true periodicity. In the relativistic model evoked, the separation between spin and orbit could just be a method to split the problem into two smaller problems. What would have to be described could correspond to something that is analogous to a periodicity in the motion of the Moon around the Sun. This could be handled by treating the periods for the motion of the Moon around the Earth (corresponding to the spin), and of the Earth around the Sun (corresponding to the orbit) and trying to fit them into a common scheme.

The heuristics within the hydrogen atom are a search for a manifold that could serve as a phase space for a motion that is periodic both in terms of orbit and spin, with a given commensurate ratio p/q between the two periods, such that there is a true common period for the combined motion. (The feasibility of postulating a true periodicity for the combined motion will be discussed in Subsection 6.2.12.) Describing this periodicity correctly requires a phase space, where the wave function is single-valued and the rotations are described by spinors. To satisfy the requirement that the wave function should be single-valued it is necessary to introduce the manifolds $\mathbb{M}_N^*$ and different representations for each value of N. This way the quantization rules have been discovered unwittingly. That it is impossible to describe the position and the momentum simultaneously within the group theory prevents introducing classical assumptions that may sidetrack the calculations towards unsuitable solutions for the orbit. These solutions would neglect the coupling between the rotational and translational motions of the electron. The two types of motion are indeed coupled and the quantization conditions lead to the correct coupling. This corresponds exactly to Heisenberg's idea of stripping the formalism of all unjustified assumptions.

The quantization rule then becomes equivalent to Chern's topological argument. But it is also connected to Cartan's topological argument. In both cases it seems necessary to avoid a contradiction in the definition of the wave function by the possibility that a loop in space could be shrunk to a point. It is these topological constraints on the wave function that force the quantization. This idea will also show up in the discussion of the double-slit experiment. (Note also a strong similarity with the way Cauchy integrals are prevented from being zero for complex analytic functions.) It therefore seems crucial that the phase space is two-dimensional and not simply connected.

6.2.12 *Solution to the paradox — part 3: Quantization as an emergent property*

The gimmick with the the quantum numbers n, ℓ and m permits treating all cases where the ratio between the periods of the gyroscope and the space station are a rational number p/q, and the true period of the orbit contains n perihelia. One could of course also imagine quasi-periodic solutions, which would give rise to infinite-dimensional representations. However, these give rise to continuum states with high energy levels in the physical theory. This offers another viewpoint on the quantization of the energy. The ratio between the periods of the orbital motion and of the spin explores the full continuum of $\mathbb{R}$, but in working it out, the calculated spectrum turns out the exact discrete experimental values. A fundamental postulate that energy and angular momentum would be quantized is not needed; only the solutions of the equations make it appear that way. Due to the fact that the irrational solutions lead to energies in the continuum, rational numbers for the orbit-to-spin ratio suffice for all practical purposes. In summary, the harmonic polynomials permit the treatment of relativistic effects like perihelion precession or Thomas precession. The present use renders their meaning less abstract. We have a mathematical image of the identical copies that are used to construct them. It is astounding how all these details have been correctly accounted for in the formalism without paying any attention to their underlying geometrical meaning.

It can be appreciated from this that the postulate that the wave function should be a function is not a property that can be derived; the philosophy is different. It is assumed that for any solution there is some representation wherein its symmetry can be described. The "postulate" that the wave function must be a function is rather a kind of heuristics that permits the various representations to be found, and the corresponding energies calculated. The orbits are not quantized and the quantum number n is related to the dimension of the representation rather than to angular momentum. But the infinite number of non-periodic orbits (which would illustrate that there is no real quantization) have energies that lie in the continuum. The energy of the orbits puts them in an order that makes them appear quantized. Of course, the derivation contains the tacit assumption that the orbits should be stable. The philosophy here is also heuristic. The approach based on the assumption that the energy must be constant suggests that a game of give-and-take between the translational and rotational degrees of freedom could be a mechanism to obtain stability. This could be achieved without recourse to radiation, by stipulating that the energy must be constant.

This shows that Bohr's argument that the classical limit corresponds to large quantum numbers is perfectly appropriate. The set of orbits is not a continuum, which becomes evident in describing them correctly by using group theory. For large quantum numbers we get the continuum states and there the set of orbits becomes practically continuous, such that it corresponds to everyday notions of continuity. But at the atomic scale and for small quantum numbers, the set of orbits is not a continuum as it contains gaps. Nevertheless, there is no discontinuity in the group of rotations, and there is no quantization of angular momentum or of the energy in principle.

In a classical analogy, in a certain point $\mathbf{r}$ within a $1/r$-potential, it will be possible to calculate $V(\mathbf{r})$ and $\mathbf{A}(\mathbf{r})$. This enables the magnitude $v(\mathbf{r})$ to be calculated from the total energy, but not the direction of $\mathbf{v}(\mathbf{r})$.

In other words, in a $1/r$-potential the direction of $\mathbf{v}$ will also be a hypothetical quantity in the formalism. In fact, several orbits might exist that take the particle through the same point $\mathbf{r}$. As long as the orbit is not detailed further by specifying initial conditions, all orbits will be possible. From a classical viewpoint it is then not possible to determine these initial conditions in an experiment without disturbing the system very badly through the interaction with the macroscopic measuring device. But the exact argument might be that it is not possible to discover the true laws of motion.

The energies one calculates for the hydrogen atom from the Schrödinger equation and the Dirac equation are identical to those obtained in the old quantum theory from the Bohr and the Bohr-Sommerfeld quantization schemes. The old and the new formulations contain the same essential ingredients, be it that this is hard to detect due to the difficulty of the group-theoretical formulation of the new theory. These essentials are introduced by the quantization condition in the old theory, and by the postulate that $\psi(\mathbf{r})$ is single-valued in the new theory. The new theory offers a way to introduce these quantization conditions implicitly. This will be further discussed in Chapter 8.

6.3 Probabilities and many histories

The coordinates (x, y, z) in the formalism are hypothetical quantities. In fact, the classical derivation of the Dirac equation defines the spinors only on the world line of the particle. The values of the wave function on the rest of space-time are a bonus. They define a time pattern that is imposed on space-time by the instantaneous Lorentz transformations at all points of the world line. For points (x, y, z, t) that are not on the world line, this time pattern determines the clock readings of a particle if it had been in this position (x, y, z) rather than in the actual position (x_0, y_0, z_0) of the particle on the world line, according to the prevailing conditions of the instantaneous Lorentz transformation. Moreover, the free-space wave function in the most simple Schrödinger approach leads to a value for the quantity $\psi^*\psi$ that is constant over all of $\mathbb{R}^3$. As the position of the triad has not been specified, the probability that a particle is in a certain point of $\mathbb{R}^3$ is also a constant. This suggests interpreting $\psi^*\psi$ as a probability density. The presence of a potential can lead to velocity changes that give rise to a modulation of the amplitude of ψ that are uniquely due to kinematic effects.

Both from the Schrödinger equation and the Dirac equation one can derive this way a continuity equation for the suggested probability charge-current density.

To do so it suffices to calculate $\frac{d}{dt}\,\psi^*\psi$ or $\frac{d}{dt}\,\Psi^\dagger\Psi$ by using the equations and their complex conjugate or adjoint equation. For the more fundamental Dirac equation, this teaches us how to define a probability charge-current density with the necessary Lorentz invariant properties. In fact, the charge-current density is a four-vector field. As four-vectors are rank-two tensors in terms of spinors, the charge-current density must be quadratic in terms of spinor quantities, which is the case. Secondly, the probability charge-current density has Lorentz invariance and the part of it that corresponds to the probability density is positive definite as it should be. Finally, it satisfies the appropriate continuity equation.

This is considered to be very important, as it shows in a clear way how at a certain point it is possible to make the leap from a deterministic to a probabilistic formulation by just extending the meaning of the wave function from the actual world line to the whole of space-time. At the same time we no longer describe a single history, but a whole set of histories that are mutually consistent in that they share the same wave function after extension of the definition domain from the actual path to the whole of space-time. The condition that the wave-function must be single-valued works like a filter that aids in finding such sets of mutually consistent histories that fit together into one wave function. Histories that are not compatible this way just belong to different wave functions. The fact that all the different wave functions are mutually orthogonal can then be used to construct a probability formalism for quantum mechanics.

One wave function ψ_1 is seen as describing a set S_1 of possible orbits. This wave function is normalized to 1 by integrating over whole space: e.g. $\int |\psi_1|^2\, d\mathbf{r} = 1$ (in some cases one may prefer $\int |\psi_1|^2\, d\mathbf{r}\, dt = 1$). Such a set S_1 of orbits described by ψ_1 can be given a probability weight $|c_1|^2$. Something analogous can be done for a set S_2 with a wave function ψ_2 and a weight c_2. Imagine that $|c_1|^2 + |c_2|^2 = 1$. Then $\int |\, c_1\psi_1 + c_2\psi_2\,|^2\, d\mathbf{r}$ will treat the probabilities correctly, because ψ_1 and ψ_2 are orthogonal, in the sense $\int \psi_2^*\psi_1\, d\mathbf{r} = 0$. Hence, when two wave functions are orthogonal, then $\int |c_1\psi_1 + c_2\psi_2|^2 d\mathbf{r} = |c_1|^2 + |c_2|^2$. This way the superposition principle can be justified by construction and by definition.

It must be noted that it is not possible to justify the superposition principle in a case where the wave function is not normalized by integration over space. But in cases where the probabilities are obtained by integration over space, it is possible to justify the extension of the spinor formalism to the group ring [Sagan (2001)]. By choosing weights of c_j within the group ring element $\sum_j c_j\psi_j$, statistical weighting factors $|c_j|^2$ can be introduced for the elements of a set $\{\psi_j\}$ that contains all possible histories for a given situation.

Note that it is not possible to justify the superposition principle in hypothetical situations that would not be subject to the definition based on an integration procedure as outlined here. But it may also be noted that the use of the superposition principle in the treatment of the double-slit experiment is improper. This involves using the wave functions ψ_1 and ψ_2 corresponding to two different potentials V_1 and V_2 to find a third wave function ψ for a third potential V. That has nothing to do with the linearity of the three Schrödinger equations and is thus a fake superposition principle that is not even justified by quantum mechanics. The double-slit experiment must be analysed in a way other than by such a fake superposition principle. The correct treatment could for example be based on a proof that Huyghens' principle applies to the solutions of the Schrödinger and Dirac equations. An attempt will be made to develop a different approach in Chapter 10.

6.4 Further support for the "all-histories" approach

Even in the absence of an $1/r$-potential (6.5) has solutions with rotation symmetry. It is important to find out why they were introduced. It may be noted that it also seems difficult to understand how a problem of a rotating frame that travels in uniform motion at a constant velocity can lead to solutions with rotational symmetry around an arbitrary fixed point. This will be explained in a hand-waving manner for circular orbits. In the pristine problem, the wave function $\Psi(\mathbf{r}, t)$ only has a meaning for $\mathbf{r} = \mathbf{r}_0$ where $\mathbf{r}_0$ is the actual position of the particle at time t. For the other values of $\mathbf{r}$ at time t, $\Psi(\mathbf{r}, t)$ is only hypothetical. As in the Lorentz transformation for τ the quantities $(\mathbf{r}, t)$ only occur in the combination $\omega t - \mathbf{k} \cdot \mathbf{r}$, the same argument could have been obtained for another actual position $\mathbf{r}'$ and wave vector $\mathbf{k}'$, with $\mathbf{r}' \cdot \mathbf{k}' = \mathbf{k} \cdot \mathbf{r}$. This does not change the wave function on the new world line. In fact, the actual values of the wave function in the pristine problem are defined on a world line only, not on entire space-time. Inserting the known solution $e^{\imath(\omega t - \mathbf{k} \cdot \mathbf{r})}$ into the Schrödinger equation (6.5) leads to $k_x^2 + k_y^2 + k_z^2 + \omega_0^2/4 = \omega^2$. Hence, when ω is fixed, only k^2 is determined by the equation. This means that $\mathbf{k}$ can be varied at will provided $\mathbf{k} \cdot \mathbf{r}$ is constant.

A motion with $\mathbf{k} \parallel d\mathbf{r}$ and $\mathbf{k} \cdot d\mathbf{r}$ fixed cannot only be obtained on world lines corresponding to uniform motion on a straight line, but also on world lines corresponding to circular uniform motion. The key point is that the circular orbit and the uniform motion generate locally the same temporal pattern $e^{\imath(\omega t - \mathbf{k}\cdot\mathbf{r})}$ on their world lines. This is why these additional solutions for (6.5) are found, with now an inkling as to where they come from. The spinor function is thus only defined on the actual world line. Its extension to $\mathbb{R}^4$ defines then a behaviour for hypothetical values, that is given by the harmonic polynomials and corresponds to other possible world lines that would generate the same pattern. To propose that the same argument of a temporal pattern will apply in the presence of a potential it is necessary to make the substitutions $\frac{\partial}{\partial ct} \to \frac{\partial}{\partial ct} - \frac{\imath q}{\hbar c} V$ and $\boldsymbol{\nabla} \to \boldsymbol{\nabla} - \frac{\imath q}{\hbar c} \mathbf{A}$. It may be noted in this respect that the continuity equation that serves to justify the probabilistic interpretation of quantum mechanics has also only been derived in free space.

6.5 Final remark

The relatively simple picture of a wave equation for the electron in a $1/r$-potential breaks down when one assumes that the electron has a magnetic moment, which is created by its rotation. In the absence of coupling of this magnetic moment to the Coulomb potential, the total angular momentum should remain constant, as the electron is moving around in an 1/r-potential. When also the coupling of such a magnetic moment of the electron to the Coulomb potential is considered, then it becomes from a classical viewpoint very difficult to calculate the detailed orbit of the electron. The moving magnetic dipole will bring about an electric dipole in the lab frame. The additional interactions between this electric dipole and the Coulomb potential will modify both the orbit and the rotation of the triad/tetrad, the change of orbit will change the Thomas precession, and the combined changes in the rotation of the triad will in turn change the magnetic dipole. To find a wave equation for this complicated situation, it would be necessary to proceed as described in Section 5.5, and then find a way to calculate the magnetic moment of the electron in free space.

The induced electric dipole will be responsible for a true spin-orbit interaction, different from the textbook term $\mathbf{L}\cdot\boldsymbol{\sigma}$, which only codes $\hat{\mathbf{L}}$ within

SL(2,$\mathbb{C}$) and has nothing to do with spin-orbit interaction. In fact, the electric dipole will be parallel to $\mathbf{v} \wedge \mathbf{s}$. This electric dipole may not be aligned along the radius $\mathbf{r}$. The question arises then of whether the total force exerted on the electron is still central. A central force is the warrant for two-dimensional motion and for the conservation of total angular momentum. This seems to imply also that the electric dipole may force the electron onto a very complicated non-planar orbit. The point is however that the forces exerted on a dipole resume to a couple exerting a torque. Non-relativistically, one then subdivides the whole problem into two parts. There is a total force exerted on the centre of gravity, which determines the orbit, while the torque determines the rotational motion around the centre of gravity. Following this non-relativistic scenario, the orbit would not be changed by the presence of the electric dipole. Relativistically, however, the rotation induced by the torque changes the mass of the electron, and this in turn will affect the orbit. However, the motion may remain planar even if the electron spin vector $\mathbf{s}$ is no longer perpendicular to the orbit. This is the key point that must be established. If the motion remains planar, the symmetry of the problem will then not be of the type SO(3,1), but SO(2,1).

An electric dipole parallel to $\mathbf{v} \wedge \mathbf{s}$ is in principle not treated correctly in the hydrogen atom. In fact, the Dirac equation does not contain a term in $\mathbf{s}$. When the Dirac equation is "squared", the coupling term $\mathbf{s} \wedge \mathbf{E}$ can thus not enter into the treatment. As noted in the discussion of the anomalous g-factor, treating coupling terms involving $\mathbf{s}$ requires a different minimal substitution then the one for a point charge introduced in Section 5.6. It is not realistic to propose a theory for the charge distribution of the electron by treating it like a point charge without a dipole moment and then hoping that the dipole moment of the charge distribution will just emerge from the calculations by miracle.

Chapter 7

The Hidden-Variables Issue and the Bell Inequalities

7.1 Is our approach a hidden variables theory?

It has been shown that the coordinates (x, y, z) (and in an $1/r$-potential also $\mathbf{v}/v$) are hypothetical values, and that in the Dirac equation the variables $\mathbf{s}$ eventually become hidden inside the spinor. Hence, most of the parameters that conceptually define the spinor have become unknown hidden variables. In this sense, the approach of this book looks like a hidden variable approach, and based on this observation the reader might want to reject it for reasons that will become clear hereafter.

A reason why it may not be possible to know the variables could be the uncertainty principle in the way Heisenberg understood it. For Heisenberg, position and momentum do exist simultaneously but they cannot be measured simultaneously because we interact with them too much with our macroscopic apparatus. The two quantities are just *not knowable* simultaneously. Bohr did not agree with this and instead postulated that position and momentum *do not exist* simultaneously. This is something that is very hard to comprehend, especially since in the formalism the quantities $\mathbf{p}$ and $\mathbf{r}$ appear well defined in the wave function, in contradiction with Bohr's statement. One could try to avoid this contradiction by postulating that the quantities $\mathbf{p}$ and $\mathbf{r}$ have another, unknown, non-classical meaning. But how does one then justify the corresponding definition of the operators for them? Finally, in the derivation of the Dirac equation given here, $\mathbf{p}$ and $\mathbf{r}$ correspond perfectly to their classical analogues.

It may be noted that the angular-momentum operators and their commutation relations are entirely derived from group theory. But in Euclidean geometry there is no place for uncertainty despite the fact that the angular-momentum operators are not commuting. All this creates doubt as to whether the uncertainty principle is really as fundamental as claimed by Bohr.

There are thus two possible objections against the quantum leap that consists in claiming that these concepts are no longer valid on the quantum scale. First of all, compelling evidence must exist that shows that the leap is compulsory, which does not seem to be the case. Secondly, it is like claiming that one can derive a mathematical theory T1 from a set of axioms that leads to another theory T2 that contradicts the theory T1 that one wants to derive. Bohr's assumption is much stronger than Heisenberg's assumption. In the light of the principle of Occam's razor, the additional content in Bohr's assumption appears *ad hoc*, unless experimental evidence existed that would render it compelling.

The meaning of the uncertainty principle has not only led to discussions between Bohr and Heisenberg. The idea that position and momentum could exist simultaneously despite the uncertain relationship was the motive behind the famous EPR paradox introduced by Einstein, Podolsky, and Rosen [Einstein *et al.* (1935)].

The history of the ensuing development is well known and will be described below. Bell set out to investigate if a scheme could exist that could be used to design a decisive experiment that would differentiate between the viewpoints of Bohr and Einstein. He formulated his famous inequalities [Bell (1965)] and applying these to the results of the experiments of Aspect *et al.* [Aspect *et al.* (1982)] refuted the hidden variables assumption. Actually, the equalities that were used in the experiment of Aspect *et al.* were a variant of the inequalities of Bell, designed by Clauser, Horne, Shimony, and Holt (CHSH) [Clauser *et al.* (1978)].

Heisenberg's interpretation appears to validate the approach of this book. But it seems as though this approach would not fit into Bohr's vision. This would then imply that it is in principle in contradiction with the conclusions that can be drawn from the experimental results of Aspect *et al.* In order to avoid a high-handed rejection of the ideas expressed here with a lapidary comment based on a remark of this type, it must be proved that there is something wrong with these inequalities, which is what will be done in this chapter.

7.2 The Bell inequalities

The essentials about the variant of the Bell-type inequalities designed by CHSH have been expounded in great clarity by Shimony [Shimony (1983)].

His presentation of the derivation of the inequalities corresponds largely to the argument presented by Clauser and Horne [Clauser *et al.* (1978)]. In the experiments of Aspect *et al.*, one considers a composite system of two particles. The first particle impinges upon an apparatus d_1 with an adjustable parameter A (a device that can take two orientations A_1 and A_2). Two complementary outcomes, labelled $\oplus$ (where the particle is transmitted by the device and registered in a detector behind the device) and $\ominus$ (where the opposite is true) are possible for each device setting. In a completely analogous way, the second particle travels to another apparatus d_2. The adjustable parameter is now called B (also a device that can take two orientations B_1 and B_2), and again two complementary results $\oplus$ and $\ominus$ are possible for each device setting. In the following the term "detector" will also be used to refer to an apparatus. In the real experiments, an apparatus consists of a polarizer and an associated detector, and ancillary equipment.

The probabilities of the $\oplus$ readings are noted by $p(A)$ and $p(B)$. The probabilities for the $\ominus$ results are then $1 - p(A)$ and $1 - p(B)$ respectively. These probabilities are compared with a very general Bell-type inequality. Shimony [Shimony (1983)] explains how one can derive such a very general inequality for probabilities that the particles have properties A_1, A_2, B_1, and B_2. The derivation of the inequality starts from a generally valid algebraic inequality for any four numbers $(r_1, r_2, s_1, s_2) \in [0,1]^4$ that can be easily checked on a Venn diagram in set theory:

$$-1 \leq r_2 s_2 + r_2 s_1 + r_1 s_2 - r_1 s_1 - r_2 - s_2 \leq 0. \tag{7.1}$$

From this one derives for probabilities $(p(a_1), p(a_2), p(b_1), p(b_2)) \in [0,1]^4$:

$$-1 \leq p_\mu(a_2 \cap b_2) + p_\mu(a_2 \cap b_1) + p_\mu(a_1 \cap b_2) - p_\mu(a_1 \cap b_1) - p_\mu(a_2) - p_\mu(b_2) \leq 0 \tag{7.2}$$

where $p_\mu(a \cap b)$ denotes the probability for the joint outcome of a and b. The index μ is used to indicate that the probabilities might depend on a number of parameters, over which one still has to integrate using a distribution $\rho(\mu)\, d\mu$, which yields the final result:

$$-1 \leq p(A_2 \cap B_2) + p(A_2 \cap B_1) + p(A_1 \cap B_2) - p(A_1 \cap B_1) - p(A_2) - p(B_2) \leq 0 \tag{7.3}$$

where $p(A \cap B)$ denotes now the probability for the joint outcome of A and B. This is the CHSH inequality, in the version presented by Clauser and Horne. It is this inequality that was reported to be violated by quantum mechanics.

7.3 Several types of "independence"

It is often claimed that the derivation of the Bell inequalities is so simple that it is hard to imagine what could go wrong with it. It is therefore important to outline the assumptions that are made to obtain the derivation. The derivation of the inequality is based on an assumption of statistical independence. In the derivation of (7.2) from (7.1) statistical independence is translated by $p_\mu(a \cap b) = p_\mu(a)\, p_\mu(b)$. It is this assumption that renders it possible to derive (7.2) from (7.1). In fact, the real expression should read $p_\mu(a \cap b) = p_\mu(a)\, p_\mu(b \,|\, a)$, where the "conditional" probability $p_\mu(b \,|\, a)$ for the occurrence of b provided a has occurred (where for mnemonics $|$ could be read as "provided that") is identified with $p_\mu(b)$, based on the independence conditions.

By making the distance between the detectors large enough, one makes sure that the experimental results are Einstein-independent. The outcome of an experiment in detector d_1 cannot influence the outcome of an experiment in detector d_2. The information obtained at detector d_1 would have to travel faster than light to be available at detector d_2 when the measurement is carried out at detector d_2. This expresses the principle of locality, which is a basic ingredient of realism. It implies that the experiments are carried out in a double-blind fashion. The term "Einstein independence" is used for locality, in order to be able to confront it with statistical independence.

CHSH have used the assumption of statistical independence as a self-evident consequence of Einstein independence. Both types of independence express that what goes on in detector d_1 must be completely independent of what goes on in detector d_2. The assumption of CHSH can be summarized as the logical implication $\mathfrak{I}_1$: Einstein independence $\Rightarrow$ statistical independence. As already stated, it is this statistical independence that is used in the derivation of the inequalities and needed to make the step from (7.1) to (7.2). It is the Einstein independence that is warranted by the design of the experiments. Based on the logical implication $\mathfrak{I}_1$ one can then argue that the protocol of the experimental set-up demands that statistical independence applies to the probabilities that intervene in the experiments. The statistical independence implies in turn that the Bell-type inequality applies to the probabilities measured. This second implication can be noted as $\mathfrak{I}_2$: statistical independence $\Rightarrow$ Bell-type inequality. Combining $\mathfrak{I}_1$ and $\mathfrak{I}_2$ we have then the logical implication: Einstein independence $\Rightarrow$ Bell-type inequality.

From the experimental violation of the Bell-type inequality, i.e. $\neg$(Bell-type inequality) it follows then that $\neg$(Einstein independence), as in general, for logical propositions P_j, we have $(P_1 \Rightarrow P_2) \Leftrightarrow (\neg P_2 \Rightarrow \neg P_1)$. The experiment will then function as a proof by a *reductio ex absurdo* that nature is non-local. This is truly a major concern because it raises the issue of the compatibility of quantum mechanics and relativity. In fact, the experiments show then that there is a correlation between the results in the two detectors, despite the fact that they are Einstein-independent and despite the fact that we cannot explain this correlation by a hidden variables *ansatz*. Ghirardi *et al.* [Ghirardi *et al.* (1980)] argued that there could be a signal that travels faster than light between detector d_1 and d_2 that permits them to match their results such that they are in agreement with the observed correlation law, but that this would not be in disagreement with relativity because an observer is not able to extract information from the signal, and therefore the signal does not enclose information. In other words, one can violate the law if nobody sees it. This is really not convincing as it is an *ad hoc* redefinition of the concept of information. The signal does contain information because the detectors use it to tune their answers so as to comply with the correlation law, and the consequences of this tuning can be observed. The argument of Ghirardi *et al.* would be like claiming that there is no information in a letter that a person has written to another person if a third person has not opened the letter and also read it, or using Einstein's methaphore that the Moon is not there when you are not looking at it. This kind of argument is not able to assuage many concerns, even if admittedly these worries reflect a world vision based on realism. It is difficult to believe that detectors exchange signals, even if this may be an aspect of realism and the experiments seem to indicate that realism should be discarded. The problem with relativity remains acute, not in the least because relativity and quantum mechanics blend together into relativistic quantum mechanics, which has been extremely successful. How can relativistic quantum mechanics be right if relativity could have a problem? It is the concern about this issue with relativity that has motivated the work presented in this chapter.

The implication $\mathfrak{I}_1$ can be dismissed by showing that it is not necessarily true. The consequence of this would then be that the principle of locality remains unscathed. The best way to show that an implication is not generally valid is to give a counter-example. This is based on the fact that $\neg(\forall x : p(x)\ \textit{true}) \Leftrightarrow (\exists x : p(x)\ \textit{not true})$. For formulating such counter-examples *Gedankenexperiments* will be used. The point of such

Gedankenexperiments is not to describe a true possible experiment, but to provide a logical counter-example that should confirm that the implication $\mathfrak{I}_1$ is not generally valid. In fact, the best examples will be those that do not rely on any physical argument, but on pure logic that even somebody without training in physics could understand.

The *Gedankenexperiments* also have another role. There is often a debate over whether the Sapir–Whorf hypothesis in cognitive science — that language is necessary for conscious thought — is correct. But within the context of the subtleties in arguments about probabilities there are numerous examples where the thoughts were already there a long time before words are found to express them, which clearly illustrates that the strong version of this assumption, claiming it would be universally valid, does not hold up. Sometimes one finds out that one has verbalized a correct idea with completely wrong words. There is often a genuine communication barrier in discussing subtleties in probability calculus, even internally. Common language just does not provide the tools needed for such a discussion; it is not rich enough to permit certain distinctions to be expressed in a way that is clear and not laborious, lengthy, and cumbersome. With respect to such fine distinctions, common language is just ambiguous. The word "independence" is a case in point: many very different issues can go under the same wording of independence. The examples are helpful in overcoming and bringing down this communication barrier. They permit one to seize the idea immediately without getting into lengthy formulations. The difficulties mentioned here already indicate that the argument that leads to the Bell-type inequalities may not be as simple as is being claimed.

The main question to be answered is what kind of conclusions can be drawn from the violation of the inequalities. Could there be a loophole in the CHSH formulation of the Bell inequalities? The answer to the question can only be yes or no, but providing the correct answer will prove to be an extremely delicate and subtle task. Several times in the following developments the answer will seem to change sides. It will be argued that in certain situations the numbers $p(A_j \cap B_k)$ in Eq. 7.3 can only be obtained from those in 7.2 by an integration over *different* probability densities $\rho_{jk}(\mu)d\mu$ rather than by integration over a *unique common* distribution $\rho(\mu)d\mu$.

7.4 *Gedankenexperiment* with snooker balls

The experiments described measure probabilities for four properties A_1, A_2, B_1, and B_2 of particles. For each of the four properties only the values

$\oplus$ or $\ominus$ can be obtained. These values can equivalently be noted as 1 and 0. There are thus 16 possible values the four joint properties $(A_1, A_2, B_1, B_2) \in ([0,1] \cap \mathbb{N})^2$ can take. Let us therefore consider a large set S of snooker balls. There are 16 types of balls, labelled by the integer numbers from 0 to 15, where it is assumed that all white balls are labelled with the number zero. For each label n there is an even number of balls, such that for each label we can constitute complete sets of pairs of balls with that label. The probability distribution law for the labels n in the set S is given by $p(n) = n/120$. This distribution law is normalized to 1 as $\sum_{n=0}^{15} n = 120$. The labels n of the balls are written in binary notation on the balls as $\underline{j_4 j_3 j_2 j_1}$. This means that it is possible to define that the value for A_1 will be j_1, for A_2 will be j_2, for B_1 will be j_3, and for B_2 will be j_4. This is summarized in Table 7.1.[1]

From this table it can be read that the probability for $j_3 = 1$ is $(4 + 5 + 6 + 7 + 12 + 13 + 14 + 15)/120 = \frac{76}{120}$. Similarly, the probability that $j_1 = 1$ is $\frac{64}{120}$, and the probability that $j_1 = 1 \,\&\, j_3 = 1$ is $\frac{1}{3}$, which is easily checked. We have thus $p(j_1 = 1, j_3 = 1) = \frac{1}{3} \neq p(j_1 = 1)p(j_3 = 1) = \frac{76}{225}$ such that the probabilities $p(j_k)$ are not statistically independent. This can be checked here on an example but in general it is always possible to invent a distribution that does not satisfy the independence conditions.

Let us now measure these probabilities experimentally with a dedicated set-up. We have two copies d_1 and d_2 of a type of device that can be adjusted such that it will be able to read automatically one of the four binary digits. Let us assume that device d_1 can read the digits j_1 or j_2, but only one at a time, for example by using a switch to select which one of the two will be measured. Let us further assume that it is possible to switch between the two options very quickly in a random fashion. For a given choice for the position of the switch, the device will thus read 0 or 1 on a ball. The second device will be used for reading the digits j_3 or j_4, again only one at a time. The apparatus is identical to apparatus d_1, but now it is programmed to switch between the possibilities of reading j_3 or j_4, also very quickly and in a random fashion. Suppose one wants to measure the probabilities (for example $p(j_1 = 1 \,\&\, j_3 = 1)$, $p(j_1 = 1 \,\&\, j_4 = 1)$, $p(j_2 = 1 \,\&\, j_3 = 1)$, $p(j_2 = 1 \,\&\, j_4 = 1)$) in the whole set of snooker balls. The problem is that with device d_1 it is only possible to read j_1 or j_2 at a time, not both, while with device d_2 the same applies with j_3 or j_4.

[1]The following natural conventions have been used: $q_1 = \sum_{n=0}^{15} p(n)$, $<j_k> = \sum_{n=0}^{15} p(n) j_k(n)$, $<j_k j_m> = \sum_{n=0}^{15} p(n) j_k(n) j_m(n)$, and $Q = -j_1 j_3 + j_1 j_4 + j_2 j_3 + j_2 j_4 - j_2 - j_4 \in \{-1, 0\}$ as explained in the text.

Table 7.1 Example of probability distribution for a large set of snooker balls

n	$p(n)$	j_4	j_3	j_2	j_1	j_1j_3	j_1j_4	j_2j_3	j_2j_4	Q	W
0	0	0	0	0	0	0	0	0	0	0	
1	1/120	0	0	0	1	0	0	0	0	−1	
2	2/120	0	0	1	0	0	0	0	0	0	x
3	3/120	0	0	1	1	0	0	0	0	−1	x
4	4/120	0	1	0	0	0	0	0	0	−1	
5	5/120	0	1	0	1	1	0	0	0	−1	x
6	6/120	0	1	1	0	0	0	1	0	0	x
7	7/120	0	1	1	1	1	0	1	0	0	x
8	8/120	1	0	0	0	0	0	0	0	0	x
9	9/120	1	0	0	1	0	1	0	0	0	x
10	10/120	1	0	1	0	0	0	0	1	−1	x
11	11/120	1	0	1	1	0	1	0	1	−1	x
12	12/120	1	1	0	0	0	0	0	0	−1	x
13	13/120	1	1	0	1	1	1	0	0	0	x
14	14/120	1	1	1	0	0	0	1	1	−1	x
15	15/120	1	1	1	1	1	1	1	1	0	x
quantity	q_1	$< j_4 >$	$< j_3 >$	$< j_2 >$	$< j_1 >$	$< j_1j_3 >$	$< j_1j_4 >$	$< j_2j_3 >$	$< j_2j_4 >$		
value	1	92/120	76/120	68/120	64/120	40/120	48/120	42/120	50/120		
quantity						$< j_1 >< j_3 >$	$< j_1 >< j_4 >$	$< j_2 >< j_3 >$	$< j_2 >< j_4 >$		
value						76/225	92/225	741/1800	34/75		

As for each label there is an even amount of balls, we will constitute pairs of balls with the same label $\underline{j_4 j_3 j_2 j_1}$. Each ball of a pair is placed in an identical box, and each pair labelled by labelling each of the two boxes of the pair with the same serial number s. The pairs of boxes are then placed in a larger box that is also labelled with the same serial number s. Imagine that $\#S$ is a multiple $2m14400$ of 14400, such that the realizations of the probabilities will correspond to integer numbers of boxes and balls; the probabilities will then not require consideration of fractional balls or fractional boxes. The serial number s will then run from 1 to $m14400$. A machine is now constructed that can be fed with the large boxes, and will open them and fire one of the smaller boxes to the device d_1 and the other to the device d_2. The devices are able to receive the little boxes and open them. They can choose randomly if one will read digit 1 or digit 2 for device d_1 or digit 3 or 4 for device d_2. They can read the instantaneous randomly chosen digit, and then write the serial number and that digit in a computer file. Finally, they are able to put the balls back into their little boxes and put these boxes into two containers (c_1 and c_2, or c_3 and c_4, according to the choice of which digit has been read). When the whole experiment is over, it will be possible to redistribute the boxes over containers c_{13}, c_{23}, c_{14}, and c_{24} of reunited pairs of boxes. The container S_{13} will receive the pairs of boxes with the same serial number s where one box of the pair came from container c_1 at device d_1, and the other box from container c_3 at device d_2.

In order to improve the analogy with the real experiments, it will be assumed that during the experiment no information has leaked to the technician in charge of device d_1 about what happens at device d_2 and *vice versa*. This way everything will be double-blind and independent. Of course, this could be done by making a building with three adjacent rooms. The middle room contains the apparatus that opens the large boxes and fires the small boxes into two opposite tubes which traverse the two walls between the middle room and the two outer rooms where the technicians are situated with their measuring devices d_j. But as the fear remains that they might have radio contact, the two tubes must be very long, for instance several light years. There would not be the slightest possibility that the technicians could exchange information as signals sent from one side to the other would only arrive many years after the whole measurement process had been finished. This way the locality condition and Einstein independence are satisfied.

Now there is problem of fair sampling. It is important to make sure that when $p(j_1 = 1, j_3 = 1)$ is measured in the subset $S_{13} \subset S$ of pairs of balls selected with the switching devices d_1 and d_2, the outcome will be the same as in a measurement on the whole set. This subset S_{13} is nothing other than the contents of container c_{13}. If the random choices have been arranged correctly then the probability $\tilde{p}(1)$ that $p(j_1)$ is measured will be the same as the probability $\tilde{p}(2)$ that $p(j_2)$ is measured, such that $\tilde{p}(1) = \tilde{p}(2) = \frac{1}{2}$. Note that these probabilities $\tilde{p}(k)$ are not probabilities for the readings $p(j_k)$ of the digits but for the choices of the digit j_k that will be read. The same will apply for the probabilities $\tilde{p}(3)$ and $\tilde{p}(4)$ that $\tilde{p}(j_3)$ and $\tilde{p}(j_4)$ respectively are measured, such that $\tilde{p}(3) = \tilde{p}(4) = \frac{1}{2}$. When the choices are completely random, i.e. statistically independent, the probability $\tilde{p}(1 \cap 3)$ that $p(j_1 = 1 \,\&\, j_3 = 1)$ is measured will be $\tilde{p}(1)\tilde{p}(3) = \frac{1}{4}$. This reasoning will hold for all combinations $\tilde{p}(k \cap l)$. This confirms fair sampling, provided the feeding process of the large boxes is entirely random. Fair sampling can be achieved in principle without requiring Einstein independence; it suffices to have a good random generator. In fact, the real question is whether the random generator will be of impeccable quality or otherwise. But if it is not of exacting quality, then there is nothing that Einstein independence can do to improve the situation. The fair sampling will imply that the result of the measurement of $p(j_1 = 1 \,\&\, j_3 = 1)$ using the set $S_{13} \subset S$ will be the same as for the measurement of $p(j_1 = 1 \,\&\, j_3 = 1)$ by using the whole set S. In other words, the value obtained as a result of this measurement will be $p(j_1 = 1 \,\&\, j_3 = 1) = \frac{1}{3} \neq p(j_1)p(j_3 = 1) = 76/225$.

The statistical independence that can be invoked is thus the one of the probabilities $\tilde{p}(k \cap l) = \tilde{p}(k)\tilde{p}(l)$, not the one of the probabilities $p(j_k = 1 \,\&\, j_l = 1)$ that could be noted as $p(j_k \cap j_l)$. The latter probabilities are not statistically independent, by construction. And again, Einstein independence cannot change anything with respect to this situation. The choice of which digit is read on one ball is independent of the analogous choice for the other ball, but the readings themselves are not independent. The choices of the measurements are independent, but the outcomes of these measurements do not need to be independent. But in the CHSH derivation the independence of the former is used as would-be evidence for the independence of the latter. The reader can check that this experiment corresponds exactly to the description by Shimony that was reproduced above. Hence, the logic used in the derivation of the Bell inequalities is flawed, as it assumes that Einstein independence would always imply statistical independence, while in this example this would mean $p(j_1 = 1 \,\&\, j_3 = 1) = p(j_1 = 1)p(j_3 = 1)$, which is wrong. In fact, there are two types of statistical independence at stake here: statistical independence for probabilities $\tilde{p}$, and statistical independence for probabilities $p(j_j = 1 \,\&\, j_k = 1)$. Einstein independence does not intervene in any aspect of the experiment. This illustrates that at least one set of probabilities exists where Einstein independence does not imply statistical independence, while in the work of CHSH the statistical independence is presented as a trivial consequence of locality in a way that is universally valid.

Statistical independence can thus not be represented as a necessary consequence of locality (or Einstein independence) that would be absolutely indispensable in deriving the inequalities. In the *Gedankenexperiment* described here we have thus $\neg$(Einstein independence $\Rightarrow$ statistical independence), such that the assumption that CHSH made about the implication $\mathfrak{I}_1$ is not generally valid. It is thus not possible that a violation of the Bell inequalities would have an impact on any issues about locality. It is pointless to quibble if our *Gedankenexperiment* could describe the experiments of Aspect *et al.* or otherwise, because what we want to point out is a *logical error.* The Bell inequalities are built only on logic, without any input regarding physical issues. Precisely because they did not address physical issues, the Bell inequalities have been claimed to be completely general. In the derivation of the CHSH inequality it is thus possible to identify an error within the very same axiomatic framework as the one on which the derivation has been built: viz. a framework of pure logic that is devoid of any physical issues. It is done by giving a counter example showing that the assumption used to derive it is not a logical necessity. The *Gedankenexperiment* may not correspond to a real physical experiment, but it could be used to make a Monte Carlo simulation that could serve as an example for a case wherein the CHSH assumption is not true, such that the assumption is not a logical necessity. As a consequence of the results of the experiments, the principle of locality has come under fire, but this is only due to an incorrect transcription of the principle into the formalism using $\mathfrak{I}_1$.

7.5 The Bell inequalities without any assumption about locality

What has been shown now is that the proof of the Bell inequalities contains a flaw. But it is not because somebody gives an incorrect proof for a proposition that we can conclude that the proposition itself is wrong; it is thus not because the introduction of the implication $\mathfrak{I}_1$ was not justified that one can conclude that the measurements are not always statistically independent. In fact, there is a way to approach the snooker ball experiment from a different angle. The probability $p(j_1 = 1 \,\&\, j_3 = 1)$ corresponds to $< j_1 j_3 >$ in Table 7.1, and this probability is calculated by multiplying the numbers $j_1(n) j_3(n)$ (which are all 0 or 1) by $p(n)$ and summing the results. This fact turns any attempt to claim that the Bell inequality would be wrong for the snooker ball experiment into a fiasco. It can in fact be considered that n plays here the role of μ in the presentation of Shimony. The quantities $p(n)$ are then the counterparts of $\rho(\mu)d\mu$. The numbers 0 and 1 entered into $j_1(n)$ and $j_3(n)$ are in reality not probabilities but the actual numbers measured. When trying to calculate the probability that j_k takes the value 1 it is necessary to calculate the $\sum_n j_k(n)p(n)$. Here, $j_k(n)$ is the weighting factor for the probability $p(n)$. This weighting factor can be used to select the events where the answer to the question: "Is k equal to 1?" is yes. Their number $\sum_n j_k(n)p(n)$ will be compared with the total number of events $\sum_n p(n) = 1$. The quantity $j_k(n)$ functions here thus as a weighting factor $w_n(k)$ (or expectation value) rather than as a probability, and it is a binary value. Let us now look at the Boolean algebra of two binary variables w_1 and w_2:

$$\begin{array}{|c|cc|} \hline w_1 w_2 & 0 & 1 \\ \hline 0 & 0 & 0 \\ 1 & 0 & 1 \\ \hline \end{array} \quad \leftarrow w_1 \qquad . \tag{7.4}$$
$$\uparrow \; w_2$$

Here, $(w_1 = 1 \,\&\, w_2 = 1) \Leftrightarrow w_1 w_2 = 1$. In other words, $w(j_1(n) = 1 \,\&\, j_3(n) = 1)$ can be written as $w(j_1(n) j_3(n) = 1) = j_1(n) j_3(n)$. The latter can be rewritten as $w(j_1(n))w(j_2(n)) = w(j_1(n) = 1)w(j_3(n) = 1)$.

This derivation does not rely conceptually on statistical independence. This is all the better, as it has been illustrated that "independence" is a poorly defined concept that leads to confusion. Defining the concept unambiguously requires introducing subtle distinctions for which suitable terminology is not readily available. The various possible concepts of "independence" are part of that very subtle side of probability calculus where everyday language falls short of being an adequate vehicle of expression. The proof given in the preceding lines relies conceptually on the fact that all information can be coded in a binary way and on the multiplication table of the numbers 0 and 1. It relies on linear independence (which means that the probabilities factorize). Whereas statistical independence implies linear independence, the converse is not true. Probabilities may factorize even when they are not statistically independent. The multiplication table codes the "and" function in Boolean algebra. Because the answers of an experiment can only be 0 or 1 they can then be identified with the weighting factors for the probability that the answer is 1. Now, there are four measurements with 0 or 1 as the only possible outcomes, and all possible outcomes of the four experiments can now be considered together. This leads to 16 possibilities, which can be coded with numbers j_1, j_2, j_3, and j_4, meaning that for each possibility we will have $Q = -j_1j_3+j_1j_4+j_2j_3+j_2j_4-j_2-j_4 \in \{-1, 0\}$, as shown in Table 7.1, such that the Bell inequality is always verified for the weighting factors. Combining these weighting factors with the probabilities for the combinations will then lead to their exact probabilities if fair sampling is assumed. Integration over the positive-valued distribution $\rho(\mu)d\mu$ will preserve the inequality.

This illustrates that there are several different criteria for independence in this analysis and it is cumbersome to phrase their exact meaning clearly. In fact, the proof of the Bell inequality is not well defined, because it does not state how the probabilities it talks about must be defined for a given example. It is then not obvious how to be sure that all possibilities have been covered. This could lead, for instance, to making the wrong choice of probabilities for which the factorization rule must be established. It could be stated that the weights w are the independent probabilities under consideration, because they obey a multiplication law, *viz.* a multiplication law for Boolean variables. These weights are linearly independent probabilities, but it cannot be argued that they are statistically independent, because there exist correlations in the set of snooker balls. This demonstrates that there are probabilities that intervene in the description of the *Gedankenexperiment* which are "independent", while there are other probabilities that also intervene which are not "independent", and that despite that fact, everything is locally defined. This illustrates once more that the "independence" is not easily expressed in clear terms, and that it has (at least *a priori*) nothing to do with locality.

The inequalities are even more general than anticipated, because they do not depend on a locality condition. It is now possible to make a list of all possible configurations, with all possible variables for the particles and with all possible variables for the atoms that belong to the devices d_j. For each of these configurations there will be one of the 16 possibilities for the joint outcome of the four different measurements, and the Bell inequalities will be verified. This result can be obtained even without assuming any cause for the results at all. One forgets about probability distributions that would be responsible for the answers obtained and just counts the numbers. These numbers must then satisfy the Bell inequalities. The only remaining issue in justifying the inequalities is the fair sampling assumption. What is measured in S_{13}, for example, can be reasonably extrapolated to what would have been measured in S, provided that were possible.

The procedure to formulate the proof would run as follows. The first step consists mapping out the complete configuration space of the particles and the mesuring device. The next step consists in dividing the configuration into bins such that histograms can be made. Finally, a probability is attributed to each bin. This way a probability distribution is defined over whole configuration space. If the bins are small enough, it can be assumed that an answer w can be defined that is 0 or 1 for that bin. The Bell inequalities would then be automatically correct.

It is perhaps interesting to ask at this stage what the conclusion should be if the inequality is violated. As the implication $\mathfrak{I}_1$ is no longer used in the derivation, the violation cannot be used to claim that the assumption of locality has been shown to be wrong. Certainly, it is concluded in the literature that the assumption that there would be a probability distribution is wrong. The reason for this could be that there are no hidden variables. If there are no hidden variables, then there is no distribution. Eventually, it will be necessary to conclude that there is no *common* probability distribution. But it will not imply that there are no hidden variables. The conclusion that there would be no hidden variables has an inherent problem: if there are no hidden variables, how can it then be that in the experiments the conservation of angular momentum is respected? This problems seems to relaunch the issue of non-locality, while it just has been settled.

There are two problems with the *Gedankenexperiment* that must be acknowledged for. First of all, in the example with the snooker balls, not all probabilities $p(n)$ take part in the measurement of Q. Whether they participate or not is shown in column W in Table 7.1. For the n-values ($n \in \{0, 1, 4\}$) that do not participate, arbitrary counting rates must be invented without any impact on the counting rates for the values of n that contribute to Q. But changing these counting rates changes the normalization of the probabilities. The probabilities are thus not well defined, but for each possible choice of definition for the probabilities, the counting rates will be in agreement with the Bell inequalities. This is due to the fact that the choice of the set S is arbitrary; there is no physical law that imposes a choice of S. It can be hoped that the choice of sets in physics is less arbitrary, such that the probabilities are defined.

A second problem is that it must be established where the scheme, which seems to be general, could have gone wrong. What the outcome of the experiments seems to suggest is that it cannot be assumed that it is possible to define a distribution of all possible configurations and then decide for each configuration if the answer is 1 or 0. It will be shown now that one can also understand this based on classical reasoning with a second *Gedankenexperiment.*

The idea of the second experiment is to describe the interaction of the particles with the measuring device. This is not a case of simply registering properties of particles as with the snooker balls, but properties of the interactions of a particle with a measuring device. The Bell inequalities can never be dismissed just by considering the properties of particles in splendid isolation without interactions. But without making the particles interact with a measuring device, it will never be possible to find out what these properties are.

7.6 Second *Gedankenexperiment*

7.6.1 *Purpose*

It will be argued now that there cannot be a *common* distribution function that describes all possible configurations, as assumed in the proof of the inequality, by introducing $\rho(\mu)\, d\mu$. In fact, while there might be a distribution of all relevant parameters describing all particles in a measuring device d_j, the problem is that it is necessary to rotate this measuring device as a whole. In fact, in the experiments of Aspect *et al.* the configurations that correspond to the probabilities $p(A_j \cap B_k)$ are obtained by rotating the polarizers. This changes the whole distribution of orientations of the molecules. It would be possible to consider a configuration space that accounts for all possible orientations[2] $(<\Phi_A>, <\Phi_B>) \in [0, 2\pi] \times [0, 2\pi]$ of the two polarizers, such that a single probability distribution could be defined over the configuration space $[0, 2\pi] \times [0, 2\pi]$, but then it would be necessary to reason in terms of probability densities rather then probabilities. The probabilities for a single configuration of the settings of the two polarizers would then be infinitesimal. In probability densities $p(x)dx$, the function p can take values outside $[0, 1]$. Elaborating this would require some effort, and thus a simpler approach with a smaller configuration space will be taken, that will allow us to focus on the essence of the problem.

[2]The notation will become clear in Section 7.6.2.

This second *Gedankenexperiment* will be a kind of model for the experiments of Aspect *et al.*, but this should not be taken too literally. The properties ascribed to the particles and the polarizers in this fictitious experiment may differ from the true properties of photons and polarizers. The purpose is simply to identify logical possibilities to investigate if the inequalities are really cogent on purely logical grounds.

7.6.2 *Description*

Imagine pairs of particles that are characterized by an angle φ, where φ parameterizes the direction of the spin vector or the polarization vector of the particles. In each pair the angle φ is the same for the two particles. The angles have a distribution $g(\varphi)d\varphi$. Imagine that the device consists of molecules that are also characterized by an angle Φ. It could then be possible to idealize the polarizer and describe it as a geometric plane. The molecules of the polarizer (with zero thickness) could be little beads that are all situated within this plane, whereby Φ characterizes the orientation of the beads within the plane. Both φ and Φ could be defined with respect to a reference frame in the laboratory. This model takes inspiration from the way one usually explains how a polarizer works. The oscillating electric-field vector of a photon can jiggle electrons within a linear molecule if the orientation of this molecule is parallel to the electric-field vector. The motion of the electrons corresponds to a current. Due to collisions, the electrons will be scattered and dissipate energy. The energy of the photon is then absorbed. If the orientation of the molecule is perpendicular to the electric-field vector of the photon, then the electron current will be confined to a much smaller region such that dissipation of energy is much harder. The energy of the photon is then not absorbed and the photon is transmitted. Presumably, the correct formulation should be conceived mentally as a two-step process. First, the photon is absorbed, creating a plasmon excitation. Then the system can return to its ground state by re-emitting the photon or otherwise (by dissipating energy through scattering). Re-emission would then correspond to transmission. Imagine then that the particles interact with only one bead, *viz.* the bead that is the closest to the point where the trajectory of the particle intersects the plane of the beads. Within a polarizer A, the beads have a narrow distribution $q(\Phi)d\Phi$ around a mean value $< \Phi_A >$. It can be imagined that there is a function w that describes completely what will happen to the particle. It could, for example only depend on the angle $\Phi - \varphi$ and only take the values 0 and

1. (The value 1 corresponds to transmission, the value 0 corresponds to absorption.)

The expectation value for the result that both answers are 1 can then be written as:

$$W(A_j \cap B_k) = \int_{<A_j>-\delta}^{<A_j>+\delta} \int_{<B_k>-\delta}^{<B_k>+\delta} \int_{\varphi_1}^{\varphi_2} w(\varphi - \Phi_{A_j}) q(\Phi_{A_j}) d\Phi_{A_j} \times w(\varphi - \Phi_{B_k}) q'(\Phi_{B_k}) d\Phi_{B_k} g(\varphi) d\varphi. \quad (7.5)$$

7.6.3 *Absence of a common distribution of hidden variables*

The numbers w are thus integrated over a joint distribution for the variables Φ_{A_j}, Φ_{B_k}, and φ. To derive the Bell inequalities it is then necessary to write an inequality with six integrals of this type. Now the problem is that the distributions that are intervening in the six integrals are not all the same, because $q(\Phi_{A_1})d\Phi_{A_1}$ is not the same distribution as $q(\Phi_{A_2})d\Phi_{A_2}$. The distribution $q(\Phi_{A_2})d\Phi_{A_2}$ is the distribution $q(\Phi_{A_1})d\Phi_{A_1}$ rotated over the angle $< \Phi_{A_2} > - < \Phi_{A_1} >$.

This is visualized in Figure 7.1 by showing two imaginary distributions for the beads corresponding to two settings A_1 and A_2. It is now possible to use the expedient to bodily rotate the distributions g, w, and the polarizer jointly back so as to recover the initial polarizer position. This rotation is illustrated in Figure 7.2 for a general distribution function p. The function that must be rotated as a whole is gw. But p can be used as a general notation, to illustrate symbolically g, w, or gw. This presentation of the state of affairs is actually an extreme simplification, as the positions of the particles will in general not coincide as in Figure 7.2, and the beads will also not coincide as a different portion of a polarizer might be exposed to the beam after a rotation. The presentation is also highly artificial as it treats distributions for angles $\Phi_{A_1}(m)$ and $\Phi_{A_2}(m)$ of molecules m in the polarizer in terms of distributions for angles φ for the particles. This simplification will only function as an exercise serving to illustrate the difficulties on a one-particle distribution, rather than on a real multi-particle distribution, as would be the case in a correct description of the molecules in the polarizers. By using this expedient, it is possible to make the calculation for the probabilities that intervene in the description of the situation with the polarizer setting A_2 with the same distribution for the beads as for the

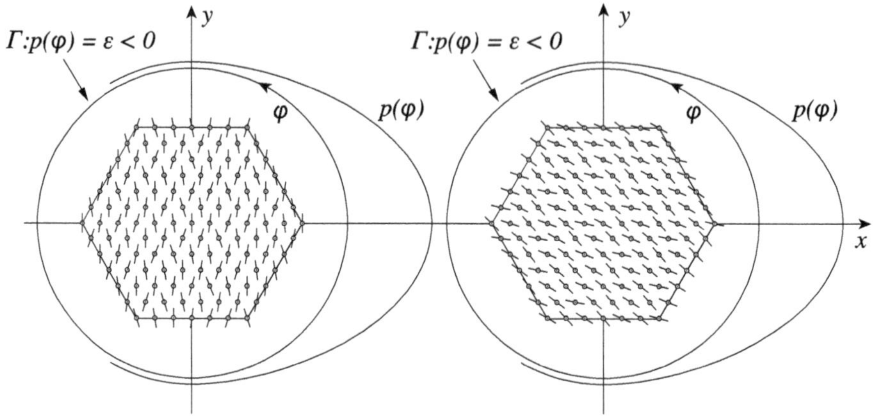

A_1: Initial orientation of the polarizer A *A_2: Polarizer A after a rotation $\Phi_A = \pi/3$*

Fig. 7.1 Two orientations of a same polarizer A are illustrated to show that the distribution functions for the hidden variables are not the same. This is illustrated by a hexagonal patch of the polarizer A before and after an anticlockwise rotation over an angle $\Phi_A = \pi/3$. The small circles note molecular positions. The beads through these small circles note the orientations of the molecules. The drawings do not pretend to represent the reality. They merely serve to illustrate the ideas that underpin the argument. In general, after a rotation around the origin of the reference frame Oxy, positions (x, y) for molecules in the hexagonal area will no longer match. For simplicity, in the case illustrated here the positions do match. In reality the distributions for the molecular positions will also be different. An imaginary non-uniform angular distribution $p(\varphi)$ for the angle φ describing some vector quantity associated with the particle is also shown. The circle Γ corresponds to the base line for the angular distribution $p(\varphi)$. It has an offset that renders it possible to clearly see the curve $p(\varphi)$ vs φ, when $p(\varphi) = 0$. In other words, Γ displays a constant function $p(\varphi) = \epsilon < 0$, where the offset ϵ is a small number.

setting A_1. It simply suffices to rotate the values $g(\varphi)w(\varphi)$. The value of $p'(\varphi')$ for the rotated position φ' is not equal to the value of $p(\varphi)$ for $\varphi = \varphi'$, as illustrated in Figure 7.2. This would present a problem when trying to define a joint probability distribution for the configurations A_1 and A_2 by just defining it for the configuration A_1. Fortunately, in an experiment, only one of the two configurations occurs at a time such that it will not be necessary to consider a configuration space that contains two copies of A. Simply using two copies of $g(\varphi)$ and $w(\varphi)$ will suffice.

However, a similar but not identical remark applies to the relative positions of the polarizers A and B. The remark is not strictly identical as the distribution of the beads within polarizer B is only the same as the one within polarizer A on average, not in its microscopic details. In a first

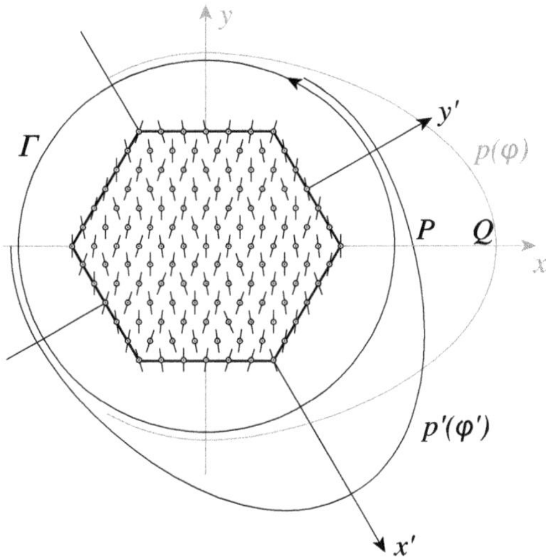

Fig. 7.2 As presumably only the relative position between the distribution $p(\varphi)$ and the polarizer plays a role, the two are rotated together to find the initial orientation of the polarizer back. This way two settings A_1 and A_2 can be described for polarizer A with a same distribution $q(\Phi_A)$, but with two different functions p and p' (which is a rotated version of p). This in turn could allow one to describe the configuration space for the two polarizers as a Cartesian product of the configuration spaces for the two initial settings A_1 and B_1, and the complete configuration space as the Cartesian product of the configuration spaces for the two initial polarizer settings and a configuration space for the particle pairs. But the fact that the points P and Q at $\varphi = 0$ on the figure do not coincide illustrates that the functions $p(\varphi)$ for the setting A_1 and $p'(\varphi')$ for the setting A_2 (where $p'(\varphi')$ is a rotated version of $p(\varphi)$) for the particle are different. The same applies for the two positions of the polarizers A and B, such there seems to be a conflict between what is chosen for p with respect to A and what is chosen for p with respect to B. As discussed in the text, this shows that there really is no common distribution of hidden variables for the two experiments, even if the hidden variables and distributions for them do exist. A different description with a different configuration space should thus be sought which accounts for the differences observed here and would simultaneously be able to save the derivation of the inequalities.

approach this can be neglected. When the orientations of A and B are different, then with respect to the polarizer A it is necessary to rotate $g(\varphi)$ over a different angle than for the polarizer B. Two different copies of $g(\varphi)$ will then be required. This is in principle contradictory, as one must integrate only once over some distribution $g(\varphi)d\varphi$. We are then forced nevertheless to consider a configuration space that contains four copies of the configuration space for the polarizer settings A_1 and B_1, and define for each copy some distribution $g(\varphi)d\varphi$. But what should be taken for this distribution? Based on intuition one could take the geometric mean $\sqrt{g(\varphi)g'(\varphi')}$. This implies calculating with probability amplitudes $\sqrt{g(\varphi)}$ rather than with probabilities. But while using intuition can be good, it is advisable to seek a mathematical argument to justify the use of the rule that has been guessed here.

Such an argument exists, as in reality, there is not one single particle but two. They do have the same distributions for φ. Let us consider photons that are not part of a correlated pair. They could have a distribution f, such that probability densities are expressed as $f(\varphi)\,d\varphi$. The joint distribution for two photons would then be $f(\varphi)f(\varphi')\,d\varphi\,d\varphi'$. When the two photons are correlated, the configuration space must be restricted to $\varphi = \varphi'$. Physicists would write the correlation condition as $\delta(\varphi - \varphi')$, treating it like a function in integrating over the density $f(\varphi)f(\varphi')\,d\varphi\,d\varphi'$, such that the integrand will contain $f(\varphi)f(\varphi')\delta(\varphi - \varphi')\,d\varphi\,d\varphi'$. Of course, this is only a shorthand for the correct formulation in terms of distributions, because delta functions are mathematical nonsense, but it summarizes the idea symbolically in a concise way. Making the integration over the variable φ' will lead to $[\,f(\varphi)\,]^2\,d\varphi$. Hence, $g(\varphi)$ can be identified with $f^2(\varphi)$. In principle, the distributions of f and g may not be normalized simultaneously to 1. In the problem with two polarizers, two different rotated versions of f will be needed. The correlation condition will then no longer be $\delta(\varphi - \varphi')$ but take another expression, such as $\delta(\varphi - \varphi' - (< \Phi_A > - < \Phi_B >))$, because the two distribution functions have been rotated. There is now no longer a conflict between the two rotations because they are not applied to a single distribution function. This results in an expression of the type $\sqrt{g(\varphi)g'(\varphi')}$, rather than $g(\varphi)$, or $\sqrt{p(\varphi)p'(\varphi')}$ rather than $p(\varphi)$ (for $p = gw$). But this stresses again that there is no common probability distribution for the whole configuration space, because the value of $\sqrt{g(\varphi)g'(\varphi')}$ will be different in the four configurations, such that the six quantities in an inequality use different probability distributions. This shows that (7.5) was not correct; $g(\varphi)d\varphi$ should be replaced by something that expresses the idea $f(\varphi)f(\varphi')\delta(\varphi-\varphi')\,d\varphi\,d\varphi'$.

There is a snag in the expression for this constraint: not only can the correlation condition not be expressed by a function, but the procedure to define it also involves the use of numbers that do not belong to $[0, 1]$. The operation of a singular distribution δ on a continuous test function f can be defined by a limit procedure:

$$f(0) = \lim_{\epsilon \to +0} \int_{-\pi}^{\pi} f(x)\delta_\epsilon(x)\,dx,$$

$$\text{where: } \delta_\epsilon(x) = \begin{cases} \dfrac{1}{2\epsilon} & \forall x \in [-\epsilon, \epsilon] \\ 0 & \forall x \in [-\pi, \pi] \backslash [-\epsilon, \epsilon]. \end{cases} \tag{7.6}$$

The order of the procedures of taking the limit and performing the integration can not be reversed, and the limit $\lim_{\epsilon \to +0} \delta_\epsilon$ does not exist. One will have to fiddle a bit with this definition to adapt it to the integration interval used for φ, but this is a mathematical technicality of little interest. There is a problem if, for example, φ and φ' are defined over $[0, 2\pi]$, because then $[-\epsilon, \epsilon]$ is not a subset of $[0, 2\pi]$. The definition:

$$f(0+) = \lim_{\epsilon \to +0} \int_{0}^{2\pi} f(x)\delta_\epsilon(x)\,dx,$$

$$\text{where: } \delta_\epsilon(x) = \begin{cases} \dfrac{1}{\epsilon} & \forall x \in [0, \epsilon] \\ 0 & \forall x \in [0, 2\pi] \backslash [0, \epsilon] \end{cases} \tag{7.7}$$

can then be considered. This also has the advantage of avoiding a continuity problem when δ_ϵ is translated from 0 to another point φ' in $[0, 2\pi]$, *viz.* that the function w might present a jump from 0 to 1 in φ'. The integral in (7.6) would then yield $\frac{1}{2}(f(0-) + f(0+))$. This is all rather technical and of minor physical interest. But the assumption that the probabilities take values in $[0, 1]$ is essential in the derivation of the inequalities. There are thus two options to present the calculation:

(1) Use a common distribution function for φ and φ', accepting that the correlation condition leads to pseudo-probabilities $\delta_\epsilon(x)$ that can be larger than 1. The consequence of this will be that (7.1) can no longer be used as a starting point for the derivation.
(2) Take the distribution function $\sqrt{gg'}$ for φ that is obtained after the integration of the correlation condition over φ' as a true probability distribution, that takes again values in $[0, 1]$ (and must be combined with $\sqrt{ww'}$), but then it will no longer be possible to make the further calculations of the six integrals with the same probability distributions. This renders then the step from (7.2) to (7.3) invalid.

It is not necessary to carry out the difficult procedure of introducing the expressions for the correlations to obtain the conclusion presented as option

(2). It was clear at the outset that four different distributions for (Φ_A, Φ_B) would have to be used rather then a single common distribution. The molecular orientations in the two polarizers depend on *four differently correlated distributions*, one for each of the four different relative angles between these polarizers. It has been possible to translate this here into a difference between distributions for $\sqrt{gg'}$ due to the introduction of an expedient, that only became possible after an extreme simplification of the real case. But the general conclusion will nevertheless be that there are two options: either trying to describe the whole configuration space with one single common distribution, and accept option (1), or abandoning any hope of finding a common distribution, and take option (2). Of course, these two presentations are equivalent; the correct treatment of option (1) must yield option (2). The issue that f and g may not be normalized to 1 simultaneously is also relevant to this problem.

The simplification used here may not be the best one with which to illustrate the difficulties, because one could argue that all versions of $\sqrt{gg'}$ are the same when the photon distribution is uniform. For the sake of generality, no assumptions have been made about g, which means that it can have a non-uniform distribution, and the reasoning is built on this general assumption, since the intention was to discuss logical possibilities. In reality, the true problem resides in the correlations between the Φ_A and Φ_B distributions. One could start then by keeping p fixed and considering distribution functions for (Φ_{A_j}, Φ_{B_k}). It would then be necessary to introduce a correlation condition for Φ_{A_j} and Φ_{B_k}. But Φ_{A_j} and Φ_{B_k} have many-particle rather than single-particle distributions, such that formulating the constraints would be much more involved. Imagine again for simplicity that the filters A and B are strictly identical rather than identical on average. A molecule m will have position $\mathbf{r}_m$ and its orientation will be $\Phi_{A_1}(\mathbf{r}_m)$. After a rotation R over an angle $< \Phi_{A_2} > - < \Phi_{A_1} >$, the new position will be $\mathbf{r}_{m'} = R(\mathbf{r}_m)$, and the new orientation will be $\Phi_{A_2}(\mathbf{r}_{m'}) = \Phi_{A_1}(\mathbf{r}_m) + < \Phi_{A_2} > - < \Phi_{A_1} >$. It is thus necessary to consider the couples (m, m') that are defined by a correlation condition written in shorthand as $\delta(\mathbf{r}_{m'} - R(\mathbf{r}_m))$. For each couple (m, m') that satisfies this condition, there must be a correlation that could be noted symbolically as: $\delta(\Phi_m + < \Phi_{B_k} > - < \Phi_{A_j} > - \Phi'_{m'})$. The correlation condition is thus of a type that can be noted symbolically as: $\prod_m \delta(\Phi_m + < \Phi_{B_k} > - < \Phi_{A_j} > - \Phi'_{m'})\,\delta(\mathbf{r}_{m'} - R(\mathbf{r}_m))$. The number of factors in the product remains finite, such that there is no need to consider infinite products. This approach leads to the same conclusion that there are two options, even if this might be very elaborate to write down in an equation. The simplification made is instrumental in avoiding this complexity and showing how this conclusion can be reached, by using an example where it is only necessary to consider correlations between two variables φ and φ'. An alternative approach that illustrates the ideas is given in Figure 7.3.

When the polarizers are no longer strictly identical, but only equal on average, it will be even more tedious to translate the correlation into mathematics. In fact, the complication that the polarizers are not strictly identical should also be considered in the simplified discussion of the correlation condition in terms of φ and φ' (while in the treatment given above, it has been ignored). To be loophole-free, a correct proof of the Bell inequalities must be able to cover all possible correlations between the polarizers, even those that are tedious or impossible to describe in a mathematically rigorous way. This will require discussions with intricate detail and depend on the experiment described. It may well be impossible to develop a proof without introducing simplifying assumptions. The derivation of the inequalities is thus not as simple and general as has been claimed.

What is noticeable here is that there is not the slightest problem in calculating any of the six integrals individually, but that a problem emerges when attempting to treat these integrals collectively with the aim of setting up an inequality between them. This is subtle, but most painfully illustrates that the issue of an existence proof cannot always be belittled as unpractical mathematical hair-splitting. If option (2) is adopted to describe the situation, then it is indeed wrong to take it for granted that a common probability density distribution exists with which the six integrals could be calculated simultaneously. The reason for this is that the distributions of the molecules in the two polarizers must be correlated and that for each integral a differently correlated distribution must be used. The failure of the Bell inequalities then only confirms that it is indeed not possible to define a unique probability distribution for the configuration space that contains all six experiments, and that the answers cannot be obtained from the type of probability calculus used in (7.5). Probability amplitude calculus must be used instead. The probability amplitudes are here probabilities in their own right, of a more elementary form. One should, however, refrain from extrapolating this result too casually to a general rule without providing any further proof for it. Not all physical problems where probability amplitudes are used contain two correlated particles. One might of course feel that there are plenty of other possible approaches that could help to generalize the idea on a case-by-case basis, but such ideas must be pursued

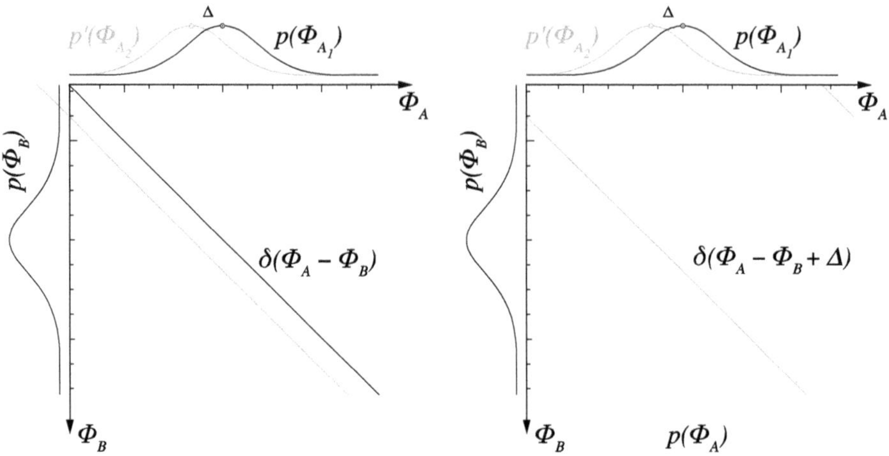

Fig. 7.3 The probabilities must be integrated over the common probability density $p(\Phi_A, \Phi_B)d\Phi_A d\Phi_B$. We may consider (Φ_A, Φ_B) as expressed with respect to a fixed reference frame in the laboratory. The probability distribution p is defined for $(\Phi_A, \Phi_B) \in [-\pi, \pi] \times [-\pi, \pi]$. This set $[-\pi, \pi] \times [-\pi, \pi]$, which is the square region illustrated in the two figures, is thus the definition domain of the common probability distribution. The probability densities $p(\Phi_A)$ and $p(\Phi_B)$ have been displayed on their corresponding axes. They will be the same when the polarizers are identical on average. In the figure it is shown what happens when it is assumed that the polarizers are strictly identical, such that a photon encounters an angle $\Phi_B = \Phi_A$ when its companion photon encounters the angle Φ_A. When the two polarizers are oriented the same way, this must be expressed by using the correlation constraint $\delta(\Phi_A - \Phi_B)$. In the part on the left of the figure, the curves drawn in black illustrate this situation. The curves drawn in grey illustrate the distribution $p(\Phi_{A_2})$ when A has been turned by an angle Δ from the original setting A_1 to a setting A_2. The corresponding new constraint is also shown. The constraint can now be expressed by $\delta(\Phi_A - \Phi_B + \Delta)$. These changes come down to rotating the probability distribution for the orientations of the beads by an angle Δ from p to p'. On the right-hand side the effect of this change of variables $\Phi_{A_2} = \Phi_{A_1} - \Delta$ is shown, whereby the angles are reduced modulo 2π to make them fit again into the interval $[-\pi, \pi]$. The effect of this reshuffling operation on the new constraint $\delta(\Phi_A - \Phi_B + \Delta)$ is shown. This clearly shows that we integrate over $(\Phi_{A_1}, \Phi_B) \in [-\pi, \pi] \times [-\pi, \pi]$ and $(\Phi_{A_2}, \Phi_B) \in [-\pi, \pi] \times [-\pi, \pi]$ with different correlation constraints. Alternatively, it can be said that we must integrate over a different integration domain after the rotation of a polarizer. The constraint $\Phi_B = \Phi_A$ is perhaps not realistic, but its treatment is instrumental in clearly showing what happens in general to a constraint for a probability density $p(\Phi_A, \Phi_B)d\Phi_A d\Phi_B$ due to the rotation of a polarizer. Imagine, for example, that $\Phi_B = \Phi_A$ tags a ridge of local maxima of the correlated distribution function. By following the fate of this line under the rotation of a polarizer, two different polarizer configurations will correspond to differently correlated probability densities, unless the original distribution density $p(\Phi_A, \Phi_B)d\Phi_A d\Phi_B$ were to be uniform, which it certainly is not. We may note that we have an integration of a product of a probability density with a response function. Option (1) discussed in the text corresponds to stipulating that the probability density

Fig. 7.3 (Figure on facing page) remains fixed and the response function is changed by rotating a polarizer, while option (2) corresponds to stipulating that the distribution densities are changed by the rotation and the response function remains fixed. The two descriptions are equivalent. In option (2), there will be four different constructions of the probability distribution like in the figure, corresponding to four combinations of angles.

with appropriate rigour. In fact, the question of whether to use probability amplitudes or traditional probabilities is tied up with the particle-wave duality: It can therefore not be expected that it should be possible to obtain from the approach some stunning conclusion for free.

Hence, it is not too difficult to imagine a case where it is impossible to define a joint probability distribution that would apply to the six measurements simultaneously. The Bell-type inequality can then not be derived, even though hidden variables could exist. One may feel inclined to dismiss this conclusion, but it is exactly the same one as was used to claim that there are no hidden variables, *viz.* that it cannot be taken for granted that a (common) probability distribution exists. The conceptual difference with the traditional approach intervenes only later on when it must be explained why there is no common probability distribution. The traditional approach consists in stipulating that there is no such common distribution as hidden variables simply do not exist; it is meaningless to stipulate the existence of a probability distribution for things that do not exist. One could propose a different reading for the observation that a common probability distribution does not exist. It is actually less mysterious than the traditional rationale, because it is really hard to understand how nature would manage without hidden variables to comply with the conservation of angular momentum over large separation distances without entering in conflict with relativity.

7.7 Conclusion

This chapter will conclude with a comparison between two options to explain the violation of Bell-type inequalities:

- No probability distribution exists for the experiments. This is due to the fact that hidden variables simply do not exist. Quantum mechanics is correct, but it is unclear how nature manages to respect the law of conservation of angular momentum over large distances without using hidden variables. It is conceded that realism and relativity might have a problem.

- No *common* probability distribution exists for the experiments. Nevertheless hidden variables do exist. Both quantum mechanics and relativity are correct.

It is obvious which of the two choices is preferable. It is based on a geometrical loophole. There are not only correlations between the photons in the experiments of Aspect *et al.*; there are also geometrical correlations between the hidden-variable distributions within the polarizers. These geometrical correlations are in a sense non-local, but this is trivial, as Euclidean geometry is based on a tacit assumption of universal simultaneity and therefore defined in a non-local way. This does not imply any stunning spooky action at a distance. The angle between the simultaneous settings of two polarizers that are separated by light years of distance is a non-local quantity, to which it will only be possible to obtain access after having brought the information about these simultaneous positions together at some data collection and treatment centre. If the polarizers are light years apart, then it will take years for the information to make its way to this data centre. This remark also applies to the definition of a Lorentz frame. Synchronizing clocks in a Lorentz frame would require a one-way signal that travels at infinite velocity. Einstein synchronization is not a non-local one-way but a local two-way procedure, based on sending light rays to and fro between two observers. The definition of simultaneity that tacitly underlies the use of Euclidean geometry is responsible for the faster-than-light velocities of the de Broglie waves as discussed in Section 6.1, because the wave function is defined at an instant t simultaneously over the whole of $\mathbb{R}^3$, just like we do for the definition of a Lorentz frame, while this is in principle not possible. The definitions of a Lorentz frame or of a wave function are in this sense non-local, and it is this which creates the EPR paradox. This way of defining the wave function at an instant t over the whole of $\mathbb{R}^3$ by using geometry will be used again in Chapter 8. In the construction of the wave function it is assumed that the potential is simultaneously defined over the whole of $\mathbb{R}^3$.

In our understanding, the discussions between Einstein and Bohr cannot be locked up into the self-delusion of a *tertium non datur*. Even if the viewpoints of Einstein and Bohr appear very hard to reconcile, they are not mutually exclusive. On the contrary, they are complementary in a beautiful way. Einstein brought in the element of the hidden variables, Bohr the point of the interaction with the measuring device. It might in view of all this be recommended not to draw too quickly one's conclusions from the violations of the Bell-type inequalities observed.

An objection in principle against the derivation of Bell-type inequalities, is that they treat probability densities systematically as scalars. In a relativistic context this violates their four-vector symmetry. Quantum mechanics accounts automatically for all possible symmetries, as illustrated by the way multi-component spinors lead to a four-vector density $(\Psi^\dagger\Psi, \Psi^\dagger\boldsymbol{\alpha}\Psi)$ in the Dirac theory. In contrast with a simple Bell-type argument, quantum mechanics contains a built-in safety belt against the possibility of a subtle unknown relativistic probability paradox. The crucial question is thus if something in the formalism could catch a "mystery axiom" that would set quantum mechanics apart from a group-theoretical treatment of relativistic probability calculus. The possibility to derive the Dirac equation classically suggests that such an axiom could only lie hidden in the *self-consistent* extrapolation of the definition of the wave function from the orbit to entire space-time (see also the discussion at the very end of Chapter 10).

Chapter 8

Equivalence of the Bohr-Sommerfeld and Dirac Theories for the Hydrogen Atom

8.1 Introduction

Both the quantization of the angular momentum in the hydrogen atom (see Chapter 6) and the interference pattern in the double-slit experiment (see Chapter 10) can be understood as following from the postulate that the "wave function" must be a function. To make it a function it may be necessary to define it on a manifold rather than on $\mathbb{R}^3$, such as a Riemann surface. The postulate itself can be understood as expressing the fact that the rest mass of the electron is its rotational energy.

8.2 Quantization of the Coulomb problem — first approach

8.2.1 *General formulas in relativistic dynamics*

Relativistically, the Newtonian equation of motion can be preserved, provided it is taken under the form $\mathbf{F} = \frac{d\mathbf{p}}{dt}$, where $\mathbf{p} = m_0\gamma\mathbf{v}$, such that the mass $m = \gamma m_0$ can vary.[1] Noting $\frac{d\mathbf{v}}{dt} = \mathbf{a}$, we have then:

$$\mathbf{F} = m_0\frac{d\gamma}{dt}\mathbf{v} + m_0\gamma\mathbf{a}. \tag{8.1}$$

[1]It would perhaps be better to formulate this differently. The quantities $(\gamma, \gamma\boldsymbol{\beta})$ and $(\gamma c, \gamma\mathbf{v})$ are four-vectors. $(\gamma c, \gamma\mathbf{v})$ could be called a velocity four-vector, because it is the generalization of the concept of velocity that renders it covariant. When $(\gamma, \gamma\boldsymbol{\beta})$ is multiplied with the constant m_0c^2 we obtain the energy-momentum four-vector $(E, c\mathbf{p})$. It is thus more logical to state that the mass does not change and that it is the velocity that changes. Saying that the mass changes is just a means to carry out the calculations correctly.

To calculate $\frac{d\gamma}{dt}$, γ is written under the form $\gamma = (1 - \frac{\mathbf{v}\cdot\mathbf{v}}{c^2})^{-\frac{1}{2}}$. This gives then:

$$\frac{d\gamma}{dt} = -\frac{1}{2}\gamma^3(-2\frac{\mathbf{v}\cdot\mathbf{a}}{c^2}) = \frac{\mathbf{v}\cdot\mathbf{a}}{c^2}\gamma^3. \tag{8.2}$$

Introducing the notations $\mathbf{a}_{\|}$ and $\mathbf{a}_{\perp}$ for the components of $\mathbf{a}$ that are parallel and perpendicular to $\mathbf{v}$, we have then $\mathbf{a} = \mathbf{a}_{\|} + \mathbf{a}_{\perp}$. This way the first term on the right-hand side of (8.1) becomes:

$$m_0\frac{d\gamma}{dt}\mathbf{v} = m_0\frac{\mathbf{v}\cdot\mathbf{a}}{c^2}\gamma^3\mathbf{v} = m_0\frac{v^2}{c^2}\gamma^3\,\mathbf{a}_{\|}. \tag{8.3}$$

Substituting (8.3) into (8.1) we obtain:

$$\mathbf{F} = m_0\frac{v^2}{c^2}\gamma^3\,\mathbf{a}_{\|} + m_0\gamma\mathbf{a}_{\|} + m_0\gamma\mathbf{a}_{\perp}. \tag{8.4}$$

Using $1 + \frac{v^2}{c^2}\gamma^2 = \gamma^2$ then produces the well-known result:

$$\mathbf{F} = m_0\gamma\mathbf{a}_{\perp} + m_0\gamma^3\mathbf{a}_{\|}. \tag{8.5}$$

Here, $m_0\gamma$ and $m_0\gamma^3$ are called the parallel and perpendicular inertia. The quantity $m_0\gamma$ is the relativistic mass. By multiplying with $\mathbf{v}$ this leads to:

$$\mathbf{v}\cdot\mathbf{F} = m_0\gamma^3\mathbf{a}\cdot\mathbf{v}, \tag{8.6}$$

such that according to (8.2):

$$m_0\frac{d\gamma}{dt}\mathbf{v} = m_0\frac{\mathbf{v}\cdot\mathbf{a}}{c^2}\gamma^3\mathbf{v} = \frac{\mathbf{v}\cdot\mathbf{F}}{c^2}\mathbf{v}. \tag{8.7}$$

Substituting this into (8.1) we obtain:

$$\mathbf{F} = \frac{\mathbf{v}\cdot\mathbf{F}}{c^2}\mathbf{v} + m_0\gamma\mathbf{a}. \tag{8.8}$$

From this it follows that:

$$\mathbf{v}\wedge\mathbf{F} = m_0\gamma\mathbf{v}\wedge\mathbf{a}, \tag{8.9}$$

and:

$$\mathbf{v}\wedge\mathbf{F}dt = m_0\gamma\mathbf{v}\wedge d\mathbf{v}. \tag{8.10}$$

8.2.2 *The Coulomb field*

The Coulomb force is given by:

$$\mathbf{F} = \frac{1}{4\pi\epsilon_0}\frac{q_1 q_2 \mathbf{r}}{r^3}. \tag{8.11}$$

For an electron with charge $-|e|$ within the potential of a nucleus with charge $Z|e|$ the potential energy is given by:

$$U(r) = \frac{k}{r} = -\frac{Ze^2}{4\pi\epsilon_0}\frac{1}{r}, \tag{8.12}$$

following the introduction of the notation:

$$k = -\frac{Ze^2}{4\pi\epsilon_0} < 0. \tag{8.13}$$

In the following, a detailed account of Sommerfeld's approximation to the calculation of the orbit is given, with the aim of rendering the reader familiar with the methods of calculation. The law of conservation of energy is in many textbooks written as:

$$m_0c^2(\gamma - 1) + U(r) = E. \tag{8.14}$$

It is also what Sommerfeld used. The conservation law:

$$m_0c^2\gamma + U(r) = E \tag{8.15}$$

is much more logical. This equation results from the idea that within a potential the rest mass changes, due to the potential energy, from m_0 to m_* given by:

$$E = m_*c^2 = m_0c^2 + U(r). \tag{8.16}$$

Rewriting this as $E - U(r) = m_0c^2$ and expressing this in a moving frame using covariance we obtain then:

$$(E - qV)^2 - c^2(\mathbf{p} - q\mathbf{A})^2 = m_0^2c^4, \tag{8.17}$$

which is equivalent to the Dirac equation:

$$(E - qV)\gamma_{ct} + (c\mathbf{p} - q\mathbf{A})\cdot\boldsymbol{\gamma} = m_0c^2\mathbb{1}. \tag{8.18}$$

In (8.17) and (8.18), $(V, \mathbf{A})$ is the four-potential. It is also possible to derive (8.15) from (8.17). The potential energy U is related to the potential V through $qV = U$. In the absence of a magnetic field $\mathbf{A} = \mathbf{0}$, such that:

$$(E - U)^2 - c^2p^2 = m_0^2c^4. \tag{8.19}$$

Using $1 + \gamma^2\frac{v^2}{c^2} = \gamma^2$ it follows from this that:

$$E - U = \sqrt{m_0^2c^4 + c^2\gamma^2m_0^2v^2} = m_0c^2\sqrt{1 + \gamma^2\frac{v^2}{c^2}} = \gamma m_0c^2, \tag{8.20}$$

in conformity with (8.15). In a sense, as the potential energy is determined up to a constant, the difference between (8.14) and (8.15) may not look important. (8.14) can be rewritten as:

$$\gamma m_0 c^2 + U(r) = E_S + m_0 c^2. \tag{8.21}$$

This equation can be identified with (8.15) by putting $E = E_S + m_0 c^2$. It does not make any difference for the calculations of radiative transitions whether the calculations are made with E or with E_S, because $E_{S,1} - E_{S,2} = E_1 - E_2$. The difference between E_S and E may thus indeed not matter. In Sommerfeld's approach $\lim_{r\to\infty} E_S = 0$, while in the approach described here $\lim_{r\to\infty} E = m_0 c^2$. In Sommerfeld's approach $\lim_{r\to\infty} E_S = 0$ corresponds then to the kinetic energy at infinite distance, and as the potential energy is also zero, this sets the value for the total energy to zero. The energy $E(r \to \infty) - E(r_0) > 0$ that is radiated away when the electron is initially at rest at $r \to \infty$ and settles on an orbit that contains the point r_0 is positive. From this it follows that the energy $E(r_0)$ should be negative. This is also confirmed by Sommerfeld's final formula for the energy levels. Therefore E must correspond to the (rescaled) total energy.

8.2.3 *Sommerfeld's approach*

Let us carry out the calculation in Sommerfeld's approach. From (8.14) it is possible to calculate:

$$\gamma = 1 + \frac{E - U(r)}{m_0 c^2}, \tag{8.22}$$

and:

$$\frac{v^2}{c^2} = \frac{\gamma^2 - 1}{\gamma^2} = \frac{[\,E - U(r)\,]\,[\,E - U(r) + 2m_0 c^2\,]}{[\,m_0 c^2 + E - U(r)\,]^2}. \tag{8.23}$$

In the following calculations the quantity $m_0^2 \gamma^2 v^2$ will often be needed. This will be:

$$m_0^2 \gamma^2 v^2 = m_0^2 c^2 (\gamma^2 - 1) = \frac{[\,E - U(r)\,]\,[\,E - U(r) + 2m_0 c^2\,]}{c^2}. \tag{8.24}$$

8.2.4 *Equation of the orbit in Sommerfeld's approach*

The orbit can be calculated by first introducing the auxiliary variable $u = \frac{1}{r}$ and using the expression for the velocity in polar coordinates:

$$v_\phi = r\frac{d\phi}{dt}, \quad v_r = \frac{dr}{dt}, \tag{8.25}$$

such that:

$$v^2 = \left(\frac{dr}{dt}\right)^2 + r^2\left(\frac{d\phi}{dt}\right)^2. \tag{8.26}$$

The equation $\ell = mr^2\frac{d\phi}{dt} = \gamma m_0 r^2\frac{d\phi}{dt}$ for the angular momentum leads to the two identities:

$$\frac{dr}{dt} = -\frac{\ell}{\gamma m_0}\frac{du}{d\phi} \tag{8.27}$$

and

$$\frac{d\phi}{dt} = \frac{\ell}{m_0\gamma r^2}, \tag{8.28}$$

such that:

$$v^2 = \left[\frac{\ell}{m_0\gamma}\frac{du}{d\phi}\right]^2 + \left[\frac{\ell}{m_0\gamma}\frac{1}{r}\right]^2 \tag{8.29}$$

or:

$$\frac{m_0^2\gamma^2 v^2}{\ell^2} = \left(\frac{du}{d\phi}\right)^2 + u^2. \tag{8.30}$$

(8.24) can now be used for $m_0^2\gamma^2 v^2$. We obtain then:

$$[\,E - ku\,]\,[\,(E - ku + 2m_0c^2\,] = \ell^2c^2\left[\left(\frac{du}{d\phi}\right)^2 + u^2\right]. \tag{8.31}$$

By rearranging the terms we obtain the following so-called Binet equation:

$$(k^2 - \ell^2c^2)u^2 - 2k(E + m_0c^2)\,u + E(E + 2m_0c^2) = \ell^2c^2\left(\frac{du}{d\phi}\right)^2. \tag{8.32}$$

By differentiating both sides with respect to ϕ and "dividing out" the common factor $2\,\frac{du}{d\phi}$ we obtain the differential equation for the orbit $u(\phi)$:

$$(k^2 - \ell^2c^2)u - k(E + m_0c^2) = \ell^2c^2\,\frac{d^2u}{d\phi^2}. \tag{8.33}$$

This can be written in the form:

$$-\left(1-\frac{k^2}{\ell^2c^2}\right)u-\frac{k}{c^2\ell^2}(E+m_0c^2)=\frac{d^2u}{d\phi^2}. \tag{8.34}$$

The solution of this differential equation for the orbit is of the form $u = K+A\cos\omega_1\phi$, where $\omega_1^2 = 1-\frac{k^2}{\ell^2c^2} = 1-\frac{Z^2e^4}{16\pi^2\epsilon_0^2\ell^2c^2}$, $K = -\frac{k}{\omega_1^2c^2\ell^2}(E+m_0c^2)$, which can be rewritten as $K = \frac{Ze^2}{4\pi\epsilon_0\omega_1^2c^2\ell^2}(E+m_0c^2) > 0$, and A is arbitrary. Such an orbit does not need to be closed, but if it is intended that the wave function is a function belonging to a representation that is finite-dimensional then the orbit must be closed.[2] Such a closed orbit is illustrated in Figure 8.1. The two important parameters that define the orbit in terms of the constants of motion E and ℓ are:

$$\omega_1=\sqrt{1-\frac{k^2}{\ell^2c^2}} \tag{8.35}$$

and

$$K=-\frac{k}{\omega_1^2\ell^2c^2}(E+m_0c^2). \tag{8.36}$$

At first sight it appears as though the parameter A is free, but in fact it is determined by K and ω_1, such that there are only two free parameters. An alternative set of free parameters is (E,ℓ). Two quantization conditions will arise due to the demand that the wave function be a true function (i.e. a single-valued function of position space). One is that the orbit must be closed. The orbit may intersect itself and then the "wave function" could take different values at such an intersection point. But this can be taken care of by replacing the ordinary position space $\mathbb{R}^3$ by a Riemann surface. The second condition is that the period of the spin and of the orbit must be commensurate such that it is possible to visualize the truly periodic motion also on a Riemann surface. The wave function is then single-valued over the whole orbit in the position space defined by such a Riemann surface. Note that in the first quantization condition, the number of copies of $\mathbb{R}^3$ the Riemann surface contains will correspond to the dimension of the representation of the rotation group used in quantum mechanics. The requirement

[2]Exactly the same condition occurs also in the quantum mechanical treatment of the wave equations. The solutions are obtained in the form of a power series, and at a certain point it is desirable that this power series contains only a finite number of terms. When the quantum numbers increase, so too does the energy of the calculated levels. We enter then into a continuum regime.

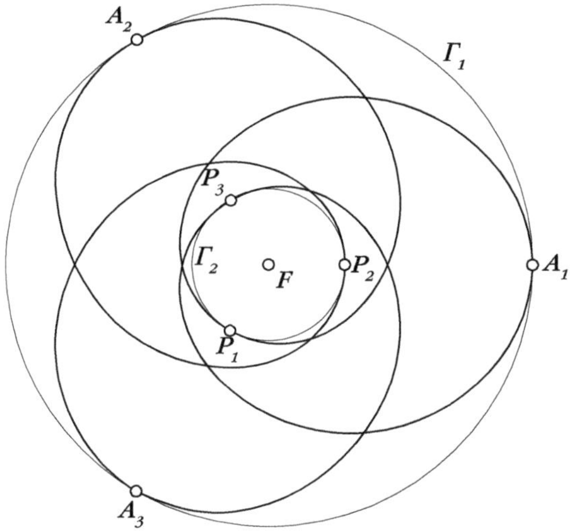

Fig. 8.1 Rosette-like orbit with a perihelion shift of $\frac{2\pi}{3}$. The fixed focus of the rotating ellipse is the point F located at the centre of the drawing. The other six points on the orbit are the three perihelia P_j and the three aphelia A_j. These six points define two "kissing" circles Γ_1 and Γ_2, as discussed in the text. These circles are also shown.

that the orbit be closed corresponds to the requirement that the dimension of the representation should be finite-dimensional. In quantum mechanics the same condition is obtained by ensuring that the power series expansion for the radial part of the wave function is not an infinite series. The way to find these quantization conditions does not depend on some special quantization of nature. It rather corresponds to going through all representations and calculating the energies that one obtains from them. We find then that the energies are discrete and increasing when the dimension of the representation increases, as explained in Footnote 2. The infinite-dimensional representations also lead to viable results, but these will not be seen.

From $u = K + A\cos\omega_1\phi$ it must be obvious that $K > |A|$ must be satisfied, otherwise r will become infinite when $K + A\cos\omega_1\phi$ becomes zero. Only when $K > |A|$ can $K + A\cos\omega_1\phi$ never reach the value zero. The solution:

$$r(\phi) = \frac{1}{K + A\cos\omega_1\phi} \tag{8.37}$$

can be compared with the standard equation for an ellipse of eccentricity w with one of its foci at the origin:

$$r(\phi) = \frac{a(1-w^2)}{1+w\cos\phi}. \tag{8.38}$$

The great axis of the ellipse corresponds to $\phi = 0$. For $\phi = 0$, we will have $r = a(1-w)$, while for $\phi = \pi$ we have $r = a(1+w)$. The equation of the orbit is periodic in ϕ with period 2π as $r(2\pi) = r(0)$. To compare (8.37) and (8.38) better we put $w = A/K$, and $1/K = a(1 - A^2/K^2)$. It follows then that $a = \frac{K}{K^2-A^2}$. The would-be perihelion distance of the pseudo-ellipse will thus be $a(1-w) = \frac{K}{A(A+K)}$, while the would-be aphelion distance will be $a(1+w) = \frac{K}{A(A-K)}$. These are not the correct values; they are only obtained by an identification of the rosette with an ellipse.

It is clear from this comparison that the ratio K/A is related to the eccentricity of the pseudo-conic defined by the orbit. When $A = K$ the orbit will become a pseudo-parabola. When $A > K$ the orbit will become a pseudo-hyperbola. Only a limited range of ϕ-values will then lead to positive values of r^2, i.e. real values of r. The limiting values $\phi \in \{-\phi_0, \phi_0\}$ that are the solutions of the equation $K = -A\cos(\omega\phi)$ will define the asymptotes of the pseudo-hyperbola. These two asymptotes can be described together by the equation $\phi^2 - \phi_0^2 = 0$. The minimum value of r will then be reached for $\phi = \phi_0$ and this corresponds to the perihelion of the pseudo-hyperbola. The requirement that $|A| < K$ should be satisfied is thus dictated by the desire to study bound states.

8.2.5 *The angular momentum in Sommerfeld's approach*

The orbital rosette has a circumscribed circle Γ_1 that contains all the aphelia of the orbit, and an inscribed circle Γ_2 that contains all the perihelia. These circles "kiss" the rosette at all the common points at which the tangent to the circle and to the orbit coincide. From this it follows that $\mathbf{r} \perp \mathbf{v}$ at these special points, such that it is possible to calculate $\boldsymbol{\ell}$ by using $\boldsymbol{\ell} = rm_0\gamma v\mathbf{e}_z$. Now, in the perihelia we have $u = K + A$, while in the aphelia $u = K - A$. We have $U(r) = k(K + A\cos\omega_1\phi)$, γ is given by (8.22), while v can be obtained from (8.23). It is now possible to calculate ℓ in an aphelion and in a perihelion, and this should yield the same value, as ℓ is a constant of the motion. Equating the two values yields an equation in A. As the expressions in v contain square roots, it is preferable to write the equation by equating the values of $\ell^2 = r^2m_0^2\gamma^2v^2$ at the two points. Now,

$U(r) = ku$. Hence at these special points $U = k(K \pm A)$, such that using (8.24) for $m_0^2\gamma^2 v^2$ produces:

$$\ell_+^2 = \frac{(E - k(K+A))(E - k(K+A) + 2m_0c^2)}{c^2(K+A)^2}, \tag{8.39}$$

in the perihelion, and:

$$\ell_-^2 = \frac{(E - k(K-A))(E - k(K-A) + 2m_0c^2)}{c^2(K-A)^2} \tag{8.40}$$

in the aphelion. We can now equate $\ell_+^2 = \ell_-^2$. For a given orbit with well-defined values of ℓ and E, this is an equation in A that permits one to obtain the equation of the orbit that corresponds to these values of ℓ and E. It is easy to check that $A = 0$ is a solution, because it renders the left-hand side ℓ_1^2 and the right-hand side ℓ_2^2 equal. Moreover, when $A = A_0$ is a solution, then $A = -A_0$ must also be a solution, because the substitution $A_0| - A_0$ just interchanges the left- and right-hand sides. The solutions are thus of the type $0, 0, A_0, -A_0$. Performing the algebra, one obtains indeed the equation for A:

$$A^2 = K^2\left(1 - \frac{E(E + 2m_0^2)}{E + m_0c^2}\,\frac{1}{kK}\right). \tag{8.41}$$

Now, $k < 0$ as identified. It follows then that $A^2 > K^2$, which seems to contradict the requirement that $K > |A| > 0$ in order to prevent $r(\phi)$ from reaching infinite values. But the contradiction is avoided because $E < 0$ in Sommerfeld's approach as already discussed. The same derivation can be made using (8.15).

8.2.6 *The perihelion shift*

Let us now consider (8.37) and assume $A > 0$ and $K > 0$. Then $r(\phi)$ reaches its minimum $\frac{1}{K+A}$ for $\phi = 0$, and it will return to this minimum value when ϕ reaches the value:

$$\phi = \frac{2\pi}{\sqrt{1 - \frac{k^2}{\ell^2c^2}}} > 2\pi. \tag{8.42}$$

For small values of $\frac{k}{\ell c}$, the orbit will closely resemble an ellipse, but after running through the ellipse once, the perihelion will shift by the amount:

$$\phi_s = 2\pi\left(\frac{1}{\sqrt{1 - \frac{k^2}{\ell^2c^2}}} - 1\right). \tag{8.43}$$

Therefore, ϕ_s can be called the perihelion shift and these notions can be adopted also for the case that $\frac{k}{\ell c}$ is not small. Note that the ellipse has been oriented in such a way that the aphelion is to the left of the perihelion by assuming $A > 0$, because it is this condition that implies that $\frac{1}{K+A}$ is a minimum for $r(\phi)$. The orbit will look like a rosette. This rosette-like orbit will close if ω_1 is a rational number $l/n \in \mathbb{Q}$, since for $\phi = 2\pi n$, we have $\omega_1 \phi = 2\pi l$. When $\phi = 2\pi n$ we will have travelled l times around the pseudo-ellipse. Such a closed orbit is illustrated in Figure 8.1.

8.2.7 *Towards quantization*

Let us now try to express the condition that the wave function should be a function. For an electron whose spin is around the z-axis the wave function in the rest frame is of the form:

$$\Psi(\tau) = \frac{1}{2}\{\,[\,\mathbb{1} + \mathbf{e}_z{\cdot}\boldsymbol{\sigma}\,]\,e^{-\imath\omega_0\tau/2} + [\,\mathbb{1} - \mathbf{e}_z{\cdot}\boldsymbol{\sigma}\,]\,e^{+\imath\omega_0\tau/2}\,\}, \tag{8.44}$$

which is also:

$$\Psi(\tau) = \frac{1}{2}\{\,[\,\mathbb{1} + \mathbf{e}_z{\cdot}\boldsymbol{\sigma}\,]\,e^{-\imath m_0 c^2\tau/\hbar} + [\,\mathbb{1} - \mathbf{e}_z{\cdot}\boldsymbol{\sigma}\,]\,e^{+\imath m_0 c^2\tau/\hbar}\,\}. \tag{8.45}$$

The first column of $\Psi(\tau)$ is:

$$\psi_-(\tau) = \left(\begin{array}{c} 1 \\ 0 \end{array}\right) e^{-\imath m_0 c^2 \tau/\hbar}. \tag{8.46}$$

In the lab frame, the phase $m_0 c^2 \tau/\hbar$ will become $(Et - \mathbf{p}\cdot\mathbf{r})/\hbar$. With the notations of (4.9), the spinor in the lab frame will be:

$$\psi_-(\mathbf{r}, t) = \left(\begin{array}{c} a \\ c \end{array}\right) e^{-\imath(Et - \mathbf{p}\cdot\mathbf{r})/\hbar}, \tag{8.47}$$

where a and c contain (partial) information on $(s_{ct}, \mathbf{s})$ as:

$$\left(\begin{array}{c} a \\ c \end{array}\right)(a^* \;\; c^*) = \frac{1}{2}\mathbf{L}\,(\mathbb{1} + \mathbf{e}_z{\cdot}\boldsymbol{\sigma})\,\mathbf{L}^\dagger. \tag{8.48}$$

Let us now search for periodic orbits. This means that we have a period T, such that $\mathbf{r}(t+T) = \mathbf{r}(t)$ and $\psi(\,\mathbf{r}(t), t\,) = \psi(\,\mathbf{r}(t+T), t+T\,)$. There is something confusing about the part $[a \;\; c]^\top$ of the spinor. It is obtained after applying a general Lorentz transformation. In SU(2) one can associate the fact that $c \neq 0$ with a spin vector $\mathbf{s}$ that is no longer parallel to $\mathbf{e}_z$. But this is no longer true in SL(2,$\mathbb{C}$). The point is actually that vectors transform quadratically, while spinors transform linearly under Lorentz transformations. Consider the special case where the orbit takes place completely within the Oxy plane while at some time, e.g. $t = 0$, we have $\mathbf{s} = \mathbf{e}_z$. Then this will not be modified by the subsequent orbital motion, despite the fact that $[a \;\; c]^\top$ is no longer proportional to $[1 \;\; 0]^\top$. For a boost along $\mathbf{u} = (u_x, u_y) = (\cos\alpha, \sin\alpha)$ in the Oxy plane we will have $(a, c) = (\sqrt{\frac{\gamma+1}{2}}, -\sqrt{\frac{\gamma-1}{2}}e^{\imath\alpha})$, such that $c \neq 0$. But a calculation shows that such a boost leaves $\mathbf{e}_z$ invariant. As far as boosts are concerned, the phase of the entry on the first line of the spinor is entirely determined by $(Et - \mathbf{p}\cdot\mathbf{r})/\hbar$ as $a = \sqrt{\frac{\gamma+1}{2}}$ is real. The entry on the second line of the spinor could have a different phase due to the presence of $e^{\imath\alpha}$. A succession of boosts in the Oxy plane will result in a boost within the Oxy plane multiplied by a rotation around the z-axis. These rotations are of the type $(a, c) = (e^{-\imath\chi/2}, e^{+\imath\chi/2})$, such that they can produce a contribution to the phase that does not come from $S/\hbar = (Et - \mathbf{p}\cdot\mathbf{r})/\hbar$. The matrix $[a \;\; c]^\top$ will be completely defined by $\mathbf{L}(\mathbf{r}, t)$ such that it is completely defined by the position of the electron on the Riemann manifold, of which an example is shown in Figure 6.5. In fact, the position on the Riemann manifold also defines the rotational parameters, because that was the reason for introducing Riemann surfaces in the first place. The geometrical information contained in $[a \;\; c]^\top$ will be completely determined by the boost vector and the rotational parameters. The boost vector is unambiguously defined by the position of the electron on the orbit. The rotational parameters are unambiguously defined by the fact that $\mathbf{s} = \mathbf{e}_z$, and by a rotation angle. This angle is different from the angle described in the phase $S/\hbar = (Et - \mathbf{p}\cdot\mathbf{r})/\hbar$; it corresponds to the angle χ and a possible effect of Thomas precession. The orientation of the spin vector does not change: it is a constant that can be left out of the description as its value will always be known. (For example, in SU(2) this would imply that the spinning top is described by a scalar wave function describing just the phase.)

In this chapter the results for the hydrogen atom obtained from the Dirac equation will be compared with those obtained from the Sommerfeld approach. It is now very important to realize that in the derivation of the minimal substitution developed in this book, the angles α and χ have not been treated, only the boost part of the Lorentz transformation has been treated. It was anticipated in Footnote 30 of Chapter 5 that this might perhaps be naive and that it would be necessary to return to this issue in the present chapter. As the minimal substitution corresponds to what has been used in the Dirac approach to the hydrogen atom, the angles α and χ must also be ignored here. This may be inexact. Imagine that χ varies with time.

It could then redefine the true rotation angle of the electron frame from $\omega_0\tau$ to $\omega_0\tau \pm \chi$, and the energy from $\frac{\hbar\omega_0}{2}$ to $\frac{\hbar\omega_0}{2} \pm \frac{\hbar}{2}\frac{d\chi}{d\tau}$. The corresponding equivalent change in mass may modify the orbit. A correct description would require the derivation of a self-consistent description of the orbit that takes into account the small changes in mass due to the Thomas precession. But such corrections are now clearly seen to be beyond the Dirac approach. This discussion raises the interesting question of whether these corrections could correspond to some corrections obtained from quantum electrodynamics.

The only difference that can be introduced by taking into account the spin by a two-component description in SL(2,$\mathbb{C}$) must thus reside within the angular effects that are being neglected by using the minimal substitution. This is the reason why one may expect to obtain the same results from the Dirac and the Bohr–Sommerfeld approaches. Imagine in a first approach that $\mathbf{e}'_x$ co-rotates with the boost vector $\mathbf{v}$, such that $\mathbf{e}'_x = \mathbf{u} = \mathbf{v}/v$. Thomas precession implies that this is not exact. The description of the Thomas description is hidden within $[a \;\; c]^\top$ rather than in $(Et - \mathbf{p}\cdot\mathbf{r})/\hbar$. In neglecting the angular effects, all attention can be focussed on the phase $S/\hbar = (Et - \mathbf{p}\cdot\mathbf{r})/\hbar$ of the wave function along the orbit, and treating it as though it is the phase of the first line in the spinor (8.47). The only error then being made is neglecting the Thomas precession.

In [Greiner (1990)] it is shown that the wave functions for the Coulomb problem are of the type:

$$\Psi = \imath g(r) \begin{pmatrix} \sqrt{\frac{j+m}{2j}}\, Y_{\ell,m-\frac{1}{2}} \\ \sqrt{\frac{j-m}{2j}}\, Y_{\ell,m+\frac{1}{2}} \end{pmatrix}, \tag{8.49}$$

where $j = \ell + \frac{1}{2}$, and $m \pm \frac{1}{2} \in \mathbb{Z}$. Hence, each entry is a harmonic polynomial. Such a harmonic polynomial corresponds to a construction that permits one to represent a wave function that is periodic on a Riemann surface $\mathbb{M}_\ell$ with a period $2\pi\ell$ as a wave function with period 2π on $\mathbb{R}^2$ as discussed in Subsection 6.2.7. On this Riemann surface the wave function is then $A\, e^{-\imath(Et-\mathbf{p}\cdot\mathbf{r})/\hbar}$, where A does not contain θ and ϕ. This type of solution illustrates that there can indeed be a phase difference between the two entries of (8.49) due to a pure boost. As discussed above, the first line of these solutions may perhaps fail to include a varying contribution $e^{-\imath\chi(\mathbf{r})/2}$ to the phase from (a, c), such that it should perhaps not have been assumed that A does not depend on ϕ. But in the present approach this is emphatically neglected.

On going from one point $\mathbf{r}_1$ of the orbit Γ to another point $\mathbf{r}_2$, this phase $S/\hbar$ will increase by $\frac{1}{\hbar}\int_{\mathbf{r}_1}^{\mathbf{r}_2} Edt - \mathbf{p}{\cdot}d\mathbf{r} = \frac{1}{\hbar}\int_{\mathbf{r}_1}^{\mathbf{r}_2} Edt - p_\phi d\varsigma_\phi - p_r d\varsigma_r$. Here, $d\mathbf{r} = (d\varsigma_\phi, d\varsigma_r)$ is the infinitesimal displacement vector along the orbit, and p_ϕ and p_r are the components of the momentum $\mathbf{p}$. This notation should not be confused with the notation for the canonical momenta $\tilde{p}_r$ and $\tilde{p}_\phi$ from the Hamiltonian formalism that is often used in the literature. These canonical momenta are $\tilde{p}_\phi = \ell$ and $\tilde{p}_r = \gamma m_0 \frac{dr}{dt}$, and they are *a priori* different from the momenta p_r and p_ϕ that are used here. According to (8.25) we have here: $p_r = \gamma m_0 v_r = \gamma m_0 \frac{dr}{dt}$ and $p_\phi = \gamma m_0 v_\phi = \gamma m_0 r \frac{d\phi}{dt}$. With $\frac{d\phi}{dt} = \frac{\ell}{m_0 \gamma r^2}$ we have then $p_\phi = \frac{\ell}{r} = \frac{\tilde{p}_\phi}{r}$. We have thus $p_r = \tilde{p}_r$ and $p_\phi \neq \tilde{p}_\phi$. Now $d\varsigma_\phi = r d\phi$ and $d\varsigma_r = dr$, such that $p_\phi d\varsigma_\phi = \ell d\phi = \tilde{p}_\phi d\phi$ and $p_r d\varsigma_r = \gamma m_0 \frac{dr}{dt} dr = \tilde{p} dr$. We have thus:

$$\Delta S = \int_{\mathbf{r}_1}^{\mathbf{r}_2} Edt - \tilde{p}_\phi\, d\phi - \tilde{p}_r\, dr. \tag{8.50}$$

Over a full orbital period, this phase factor must be an integer number of times 2π. This is the true quantization condition. Via Eq. 8.49 one introduces in quantum mechanics the assumption that $ae^{-\imath(Et-\mathbf{p}\cdot\mathbf{r})/\hbar}$ factorizes as $ae^{-\imath(Et-\mathbf{p}\cdot\mathbf{r})/\hbar} = F(t)\Phi(\phi)R(r)$, with a similar expression for the other component of ψ. This corresponds to the procedure of separating the variables within the partial differential equation. This implies that R is a function of r only and that Φ a function of ϕ only. The phase in R is then only $-\frac{1}{\hbar}\int_{\mathbf{r}_1}^{\mathbf{r}_2} \tilde{p}_r dr$, and the phase in Φ is only $-\frac{1}{\hbar}\int_{\mathbf{r}_1}^{\mathbf{r}_2} \tilde{p}_\phi d\phi$. For R and Φ to be functions we must thus have $\frac{1}{\hbar}\oint_\Gamma \tilde{p}_\phi d\phi = 2\pi n$ and $\frac{1}{\hbar}\oint_\Gamma \tilde{p}_r dr = 2\pi l$, where Γ represents the full closed orbit that corresponds to the true period. These are exactly Sommerfeld's quantization conditions.

At first sight, this contains a tacit assumption that r and ϕ are not correlated. This is true for a circular orbit, but it will not be true for a rosette-like orbit. In reality, this assumption is related to the way the wave function is defined by extending its definition from the orbit to the whole space $\mathbb{R}^3$. Originally, $\psi(\mathbf{r}, t)$ is defined only for $\mathbf{r} \in \Gamma$, i.e. for points on the orbit. A partial differential equation is then defined for ψ. Hence, it is only necessary to solve the differential equation for points on the orbit. But by solving the differential equation by standard methods, a wave-function will be found that is defined on the whole of space-time. This represents a natural extension of the wave function from the orbit to the whole of space-time. This extension of the domain of definition has a physical counterpart. In making a Lorentz transformation of τ, the quantities $\mathbf{r}$ and t enter into the phase of ψ. The position of the particle must therefore be specified, while originally in the co-moving frame of the electron this was not necessary. The phase must only be determined on the exact orbit, but as it is necessary to introduce $(\mathbf{r}, t)$ into the expressions, it is also obtained at other points of space-time. At these other points, the phase of the wave function tells us what the clock readings on the spin clock of the particle would have been if the particle had been there. The wave function is thus a natural extension of the wave function from the orbit to whole space-time. This extension of the wave function produces the phase that would have been found if the electron were on another orbit that passes through other points. The other orbit must be compatible with the initial orbit, in the sense that it produces a compatible value for the phase, i.e. the clock readings on the spin clock.

The wave function for all $\mathbf{r} \in \mathbb{R}^3$ is then defined simply by considering the whole set of orbits Γ_ϕ where Γ_ϕ is Γ rotated over an angle ϕ. Together with these orbits new points of $\mathbb{R}^3$ that do not belong to Γ enter into consideration, which leads to an extention of the definition domain of the wave function. Let the rotation that rotates Γ to Γ_ϕ be called R_ϕ. At the point $\mathbf{r}_\phi = R_\phi(\mathbf{r})$ we define then $\psi(\mathbf{r}_\phi) = e^{-\imath\ell\phi/\hbar}\,\psi(\mathbf{r})$. Note that this definition which links *two points on different orbits* is consistent with the rule $-\frac{1}{\hbar}\int_{\mathbf{r}_1}^{\mathbf{r}_2} \ell d\phi$ for the ϕ-part of the phase change between two points on a same orbit. As discussed below, this is the only way to make a self-consistent definition. For two points on an orbit with the same value of r, the spatial part of the phase difference will then be exactly $-\frac{1}{\hbar}\int_{\mathbf{r}_1}^{\mathbf{r}_2} \ell d\phi$. By defining the wave function this way it acquires rotational invariance. It must be very clear that the orbit Γ itself does not have rotational invariance. By rotating Γ to Γ_ϕ with the rotation R_ϕ, all points $R_\phi(\mathbf{r})$ with the same value for r will obtain the same value $R(r)$.

It must still be proved that this definition is self-consistent. The difficulty is namely that there are $2n$ points $\mathbf{r}_j$ with $|\mathbf{r}_j| = r_0$ on the orbit Γ. The rule $\Delta S/\hbar = -\frac{1}{\hbar}\int_{\mathbf{r}_1}^{\mathbf{r}_j} \ell d\phi$ permits the extension of the definition of ψ from $\mathbf{r}_1$ to $\mathbf{r}_j$. But the value of the phase S of ψ at $\mathbf{r}_j$ does not need to be defined, because $\mathbf{r}_j$ belongs to the orbit and the value of the phase of the wave function is already defined by the motion of the electron on this orbit. It must thus be ensured that the value S' that might be attributed to this phase by using the definition of the extension $dS/\hbar = -\frac{1}{\hbar}p_\phi d\phi$ complies with S. This can only be achieved by using the same rule as is used on the orbit. The Φ-part of the definition is thus in agreement with (8.50).

It is easy to show by symmetry that the points $\mathbf{r}_j$ all have the same value of $R(r)$. In fact, consider an aphelion A and calculate the phase in two adjacent points P_1 and P_2 on a circle of radius r using (8.50). At one point P_1 the electron will be moving outwards, while at the other point P_2 the electron will be moving inwards (or *vice versa*). From an inspection of the orbit on the Riemann surface it is easy to see that $\Delta S_{AP_1} = \Delta S_{AP_2}$. In fact the paths AP_1 and AP_2 are mirror-symmetrical with respect to OA. In two symmetrical points $Q_1 \in AP_1$ and $Q_2 \in AP_2$ the velocities are the same because the total energy E is a constant and the potential energies are equal because $|OQ_1| = |OQ_2|$. Therefore, $\mathbf{p}(Q_1)$ will be the mirror image of $-\mathbf{p}(Q_2)$, while $d\mathbf{r}(Q_1)$ will be the mirror image of $-d\mathbf{r}(Q_2)$. From this it follows that $\mathbf{p}(Q_1)\cdot d\mathbf{r}(Q_1) = \mathbf{p}(Q_2)\cdot d\mathbf{r}(Q_2)$ which completes the proof.

Similarly, there might be several points $\mathbf{r}'_j(r_j, \phi_j)$ on Γ with the same value for ϕ_0 for ϕ_j and different values r_j. To obtain self-consistency, the same rule must be used for the extension as was used on the orbit, i.e. $dS/\hbar = -\frac{1}{\hbar}p_r dr$. By proceeding this way for all variables, we obtain a natural extension of the wave function to the whole of $\mathbb{R}^4$. By construction, this wave function will have rotational invariance and it will automatically be a function that allows for a separation of variables as postulated in the solution of the Dirac equation. The ωdt part of dS will lead to the equation $E = \hbar\omega$. It is perhaps useful to remind that $\varphi = \omega t$ refers to a spin angle, while ϕ refers to an orbital angle that characterizes the position on the orbit. They are thus two different variables that should not be added, even if both are polar coordinates.

After going around the orbit the phase may not be a multiple of 2π, such that it will become only a multiple of 2π after going several times around the orbit. Therefore, the definition domain must be extended from $\mathbb{R}^3$ to a Riemann surface. For this natural extension of the wave function from the orbit to the whole Riemann surface to be a true function on the Riemann surface, the orbit must then satisfy the two quantization conditions $\frac{1}{\hbar}\oint_\Gamma p_\phi d\phi = 2\pi n$ and $\frac{1}{\hbar}\oint_\Gamma p_r dr = 2\pi l$. These conditions lead then automatically to wave functions $R(r)Y_{l,m}(\theta, \phi)$ wherein the variables have been separated. The first steps in the solution of the Dirac equation lead also to the finding that the wave function must be of the form $R(r)Y_{l,m}(\theta, \phi)$. In other words, the only generalization of the wave function from the orbit to the whole of space-time is that which satisfies Sommerfeld's quantization conditions. By adopting these conditions, navigating within the wave function will become independent from the path taken in space-time. Note that this Dirac equation simply expresses the same physics of Lorentz transforming (8.45) by another approach.

This shows that making the *ansatz* of a separation of the variables as used in the solution of the Dirac equation is equivalent to the Sommerfeld quantization conditions. As discussed, it may be necessary to introduce a Riemann surface to make the construction possible. The wave function will then not be a function on $\mathbb{R}^3$, but on the Riemann surface. For a non-circular orbit we will then still have factorizations $R(r)\Theta(\theta)\Phi(\phi)$ even though the phase changes along the orbit.

From this discussion it is clear that it is thus normal that the Dirac equation yields the same result as the Bohr-Sommerfeld approach. Both approaches are equivalent to using the minimal substitution in order to calculate the relativistic dynamics and adding a constraint to make sure that the wave function is a true function. The postulate that the wave function must be a function could express the fact that the rest mass of the electron corresponds to its rotational energy.

Finally, note that the construction of the wave function presented in this chapter is geometrical and therefore non-local as discussed in Section 7.7. Due to its construction the wave function is simultaneously defined over the whole of $\mathbb{R}^3$.

Chapter 9

The Problem of the Electron Spin within a Magnetic Field

9.1 Landau levels in a magnetic field

9.1.1 *A calculation without harmonic oscillators*

The Dirac equation for the problem of an electron in a magnetic field is:

$$\left[\left(c\hat{\mathrm{p}}_x + \frac{qBy}{2}\right)\gamma_x + \left(c\hat{\mathrm{p}}_y - \frac{qBx}{2}\right)\gamma_y + \hat{\mathrm{E}}\gamma_{ct}\right]\psi = m_0c^2\Psi. \tag{9.1}$$

Here a constant magnetic field $\mathbf{B} = B\mathbf{e}_z$ can be defined by the vector potential $\mathbf{A} = \frac{1}{2}\mathbf{B}\wedge\mathbf{r}$. (9.1), follows then by choosing a reference frame with its z-axis aligned with $\mathbf{B}$, such that $\mathbf{B} = B\mathbf{e}_z$. The value $\frac{1}{2}\mathbf{B}\wedge\mathbf{r}$ taken here for $\mathbf{A}$ and which leads to $\mathbf{A} = B(\frac{1}{2}y\mathbf{e}_x - \frac{1}{2}x\mathbf{e}_y)$ in this reference frame is not the only possible solution for the equation $\mathbf{B} = \boldsymbol{\nabla}\wedge\mathbf{A}$. Landau made another choice, *viz.* $\mathbf{A} = Bx\mathbf{e}_y$. This is somewhat surprising as it very obviously breaks the cylindrical symmetry around the z-axis that characterizes the physical situation. The choice in (9.1) was motived by the desire to respect this cylindrical symmetry. The fact that several solutions exist for $\mathbf{A}$ is related to gauge invariance. An electric potential V is determined up to a constant. For a magnetic potential there is even more liberty. When $(V, \mathbf{A})$ are changed simultaneously, the changes allowed are defined by the Lorentz gauge $\frac{1}{c}\frac{dV}{dt} + \boldsymbol{\nabla}\cdot\mathbf{A} = 0$.

It is assumed here that the motion takes place in the Oxy plane. The matrix equation consists of four coupled scalar equations. To decouple the equations the Dirac operator must be applied to itself, leading to an expression:

$$\left[-\left(c\hat{\mathrm{p}}_x+\frac{qBy}{2}\right)^2-\left(c\hat{\mathrm{p}}_y-\frac{qBx}{2}\right)^2+\hat{\mathrm{E}}^2\right]\mathbb{1}$$
$$=\left((\hat{\mathrm{E}}^2-c^2\hat{\mathbf{p}}^2)+cqB\hat{\mathrm{L}}_z-\frac{q^2B^2r^2}{4}\right)\mathbb{1} \tag{9.2}$$

for the diagonal term, and:

$$\left[\left(c\hat{\mathrm{p}}_x+\frac{qBy}{2}\right)\left(c\hat{\mathrm{p}}_y-\frac{qBx}{2}\right)-\left(c\hat{\mathrm{p}}_y-\frac{qBx}{2}\right)\left(c\hat{\mathrm{p}}_x+\frac{qBy}{2}\right)\right]\gamma_x\gamma_y$$
$$=-\frac{cqB\hbar}{\imath}\gamma_x\gamma_y,$$
$$\left[\left(c\hat{\mathrm{p}}_x+\frac{qBy}{2}\right)\hat{\mathrm{E}}-\hat{\mathrm{E}}\left(c\hat{\mathrm{p}}_x+\frac{qBy}{2}\right)\right]\gamma_x\gamma_{ct}=0,$$
$$\left[\left(c\hat{\mathrm{p}}_y-\frac{qBx}{2}\right)\hat{\mathrm{E}}-\hat{\mathrm{E}}\left(c\hat{\mathrm{p}}_y-\frac{qBx}{2}\right)\right]\gamma_y\gamma_{ct}=0, \tag{9.3}$$

for the other terms. The terms with $\gamma_x\gamma_t$ and $\gamma_y\gamma_t$ reduce to zero, because $\frac{\partial x}{\partial t}=0$ and $\frac{\partial y}{\partial t}=0$. We obtain then:

$$(\hat{\mathrm{E}}^2-c^2\hat{\mathbf{p}}^2)\Psi=\left[(m_0c^2)^2+\frac{q^2B^2r^2}{4}-cqB\hat{\mathrm{L}}_z\mathbb{1}+\frac{cqB\hbar}{\imath}\gamma_x\gamma_y\right]\Psi. \tag{9.4}$$

Here, $\frac{q^2B^2r^2}{4}=q^2A^2$. After dividing both sides by m_0c^2 we recollect the following term:

$$\frac{cqB\hbar}{\imath m_0c^2}\gamma_x\gamma_y=-\frac{\hbar q}{m_0c}\begin{pmatrix}\mathbf{B}\cdot\boldsymbol{\sigma} & \\ & \mathbf{B}\cdot\boldsymbol{\sigma}\end{pmatrix}. \tag{9.5}$$

Here, $\mathbf{B}\cdot\boldsymbol{\sigma}$ corresponds to the term discussed previously in Section 5.7, and which has been over-interpreted in terms of the potential energy of the magnetic dipole moment of the electron within the magnetic field. By applying the Dirac operator to itself, it has thus indeed been decoupled into two Pauli-like equations in SL(2,$\mathbb{C}$). In fact, in all remaining 4×4 matrices, the two 2×2 blocks on the diagonal are now equal, such that we have two identical equations. We have then:

$$\frac{1}{m_0c^2}(\hat{\mathrm{E}}^2-c^2\hat{\mathbf{p}}^2)\Psi=m_0c^2\Psi-\frac{qB}{m_0c}\hat{\mathrm{L}}_z\Psi-\frac{\hbar q}{m_0c}[\,\mathbf{B}\cdot\boldsymbol{\sigma}\,]\Psi. \tag{9.6}$$

It is important to notice that the operator $\hat{\text{L}}_z$ here stands for $\hat{\text{L}}_z\,\mathbb{1}$, which means that the two entries on the diagonal of its matrix representation are equal. This stands in marked contrast to what happens inside the term $\mathbf{B}\cdot\boldsymbol{\sigma}$ where the two entries have opposite signs. For these reasons, one cannot really consider $\hat{\text{L}}_z\,\mathbb{1}$ as an operator that could embody angular momentum around the z-axis, because it does not have the required vector symmetry. A true operator that could be identified with angular momentum around the z-axis would have to be something like $\hat{\text{L}}_z\,\sigma_z$. The true operator for angular momentum around an axis $\mathbf{s}$ in its most general form would have to be something like $s_x\hat{\text{L}}_x\sigma_x + s_y\hat{\text{L}}_y\sigma_y + s_z\hat{\text{L}}_z\sigma_z$. It is at this point that Dirac introduces the g-factor:

$$\frac{1}{m_0c^2}\,(\hat{\text{E}}^2 - c^2\hat{\mathbf{p}}^2)\Psi = m_0c^2\Psi - \frac{qB}{m_0c}(\hat{\text{L}}_z\mathbb{1} + g\hat{s}_z)\Psi, \tag{9.7}$$

whereby $g = 2$. This was discussed in Section 5.7 and it was explained that this is based on a misinterpretation of the term $\mathbf{B}\cdot\boldsymbol{\sigma}$ as a scalar product between the magnetic field and the spin (in reality it is only a coding of $\mathbf{B}$ in the special language of the group theory). The arcane character of Dirac's substitution can be avoided by using the correct meaning of the term $\mathbf{B}\cdot\boldsymbol{\sigma}$ and using the fact that $\mathbf{B}\cdot\boldsymbol{\sigma} = B\sigma_z$. We obtain then:

$$\frac{1}{m_0c^2}\,(\hat{\text{E}}^2 - c^2\hat{\mathbf{p}}^2)\Psi = m_0c^2\Psi - \frac{qB}{m_0c}(\hat{\text{L}}_z\mathbb{1} + \hbar\sigma_z)\Psi. \tag{9.8}$$

This is equivalent to:

$$\frac{1}{2m_0c^2}\,\hat{\text{E}}^2\,\Psi = \left(\frac{m_0c^2}{2} - \frac{\hbar^2}{2m_0}\Delta - \frac{qB}{2m_0c}(\hat{\text{L}}_z\mathbb{1} + \hbar\sigma_z)\right)\,\Psi. \tag{9.9}$$

9.1.2 *The Pauli equation*

Here, both sides have been divided by 2 for a comparison that will follow below. One does not need to use the Dirac equation to obtain this result. We have $E = mc^2 = \gamma m_0c^2$. A Taylor expansion of γ with respect to v/c yields then $E = m_0c^2 + \frac{1}{2}m_0v^2 + \cdots$ such that to first order $E \approx m_0c^2 + \frac{\mathbf{p}^2}{2m_0}$, where p is given by its non-relativistic value $p = m_0v$. The non-relativistic Hamiltonian is then obtained by dropping the rest mass and is therefore just $\frac{\hat{\mathbf{p}}^2}{2m_0}$. With the minimal substitution $\hat{\mathbf{p}} \to \hat{\mathbf{p}} - e\mathbf{A}\cdot\boldsymbol{\sigma}/c$ in a gauge that respects the cylindrical symmetry, we obtain then the Schrödinger equation:

$$\hat{\text{E}} = \frac{1}{2m_0}\,(\hat{\mathbf{p}} - q\mathbf{A}/c)^2 = \frac{1}{2m_0}\left[\left(\frac{\hbar}{\imath}\frac{\partial}{\partial x} + \frac{qBy}{2c}\right)^2 + \left(\frac{\hbar}{\imath}\frac{\partial}{\partial y} - \frac{qBx}{2c}\right)^2\right], \tag{9.10}$$

or:

$$\hat{\mathrm{E}}\psi = -\frac{\hbar^2}{2m_0}\Delta\psi + \frac{q^2B^2r^2}{8m_0c^2}\psi - \frac{qB}{2m_0c}\frac{\hbar}{\imath}\left[x\frac{\partial}{\partial y} - y\frac{\partial}{\partial x}\right]\psi. \tag{9.11}$$

This contains the same terms as (9.9). Even the term $\frac{q^2B^2r^2}{8m_0c^2}$ from the Dirac approach is reproduced (as we have divided by $2m_0c^2$). The term $-\frac{qB}{2m_0c}\hat{\mathrm{L}}_z\mathbb{1}$ is reproduced, but the spin term is lost. However, it can be recovered by taking a more serious approach to the non-relativistic limit of the Dirac equation, *viz.* the Pauli equation:

$$\hat{\mathrm{E}}\Psi = \left[\underbrace{\frac{1}{2m_0}(\hat{\mathbf{p}} - q\mathbf{A}/c)^2 + qV}_{\substack{(9.11)\\ \text{with } qV \text{ added}}} \quad \underbrace{-\frac{q\hbar}{2m_0c}\mathbf{B}\cdot\boldsymbol{\sigma}}_{\substack{\text{spin term}\\ (9.9)}}\right]\Psi. \tag{9.12}$$

It can be obtained by performing the minimal substitution before taking the non-relativistic limit. (9.10) is obtained by performing the minimal substitution *after* taking the non-relativistic limit.

To be more precise and to remain within the same spirit of what was done before, $\hat{\mathrm{E}}\mathbb{1}$, $(\hat{\mathbf{p}} - q\mathbf{A}/c)^2\mathbb{1}$ and $qV\mathbb{1}$ should be written in this matrix equation. The Landau gauge breaks the cylindrical symmetry, and in this gauge the term $\frac{q^2B^2r^2}{8m_0c^2}$ that one obtains in the symmetry-respecting approach has to be replaced by $\frac{q^2B^2y^2}{2m_0c^2}$. In both approaches the term reduces to $\frac{q^2A^2}{2m_0c^2}$. The reason that there are several energy levels is the presence of the operator $\hat{\mathrm{L}}_z\mathbb{1}$, rather than the presence of some harmonic oscillator. The introduction of the harmonic oscillator is thus just a formal, mathematical expedient.

9.2 Traditional description of the spin in a magnetic field

9.2.1 *Current loops associated with orbital angular momentum*

In the traditional definition of the operators, the term $\frac{qB}{2m_0c}\hat{\mathrm{L}}_z$ is nothing other then the operator corresponding to $\frac{q}{2m_0c}\mathbf{B}\cdot\mathbf{L} = \mathbf{B}\cdot(\frac{q}{2m_0c}\mathbf{r}\wedge m_0\mathbf{v})$, where $m = m_0$ has been taken in the non-relativistic limit. This becomes $\frac{1}{c}\mathbf{B}\cdot(\frac{1}{2}\mathbf{r}\wedge q\mathbf{v}) = \frac{1}{c}\mathbf{B}\cdot\boldsymbol{\mu}$, where $\boldsymbol{\mu}$ is just the magnetic moment as one would obtain it from a calculation for a circular current loop. The current loop has exactly the same value as that produced by the orbital motion of the electron, such that the classical picture is recovered.

9.2.2 *The traditional algebra*

Traditionally it is stated that the "energy of the electron in a magnetic field" is described by:

$$\hat{H} = \frac{\hbar q}{2m_0 c}\mathbf{B}\cdot\boldsymbol{\sigma}. \tag{9.13}$$

The term $\frac{\hbar q}{2m_0 c}\mathbf{B}\cdot\boldsymbol{\sigma}$ comes straight out of the Dirac equation for an electron in a magnetic field as shown in the preceding lines. It suffices to assume that the electron is at rest in (9.12). Traditionally, the term $\mathbf{B}\cdot\boldsymbol{\sigma}$ is interpreted as the scalar product of $\mathbf{B}$ with a quantity $-\boldsymbol{\mu} = \frac{q}{m_0 c}\mathbf{S}$:

$$\hat{H} = -\boldsymbol{\mu}\cdot\mathbf{B}, \tag{9.14}$$

where $\mathbf{S}$ is supposed to be the spin $\frac{\hbar}{2}\boldsymbol{\sigma} = \frac{\hbar}{2}(\sigma_x, \sigma_y, \sigma_z)$. This takes inspiration from the physical image of a current loop in classical electromagnetism, as explained in Subsection 9.2.1. With such a current loop, one can associate a magnetic dipole $\boldsymbol{\mu}$. Such a dipole has a potential energy $-\boldsymbol{\mu}\cdot\mathbf{B}$ in the magnetic field.

As already discussed in Section 5.7, this interpretation is incorrect as $\mathbf{B}\cdot\boldsymbol{\sigma}$ only codes the magnetic field within the formalism of SU(2) and $\frac{\hbar}{2}(\sigma_x, \sigma_y, \sigma_z)$ is not the spin. The spin would be coded by $\frac{\hbar}{2}\mathbf{s}\cdot\boldsymbol{\sigma}$. There is thus no scalar product of $\mathbf{B}$ with a hypothetical quantity $(\sigma_x, \sigma_y, \sigma_z)$. The vector quantity $-\frac{\hbar q}{2m_0 c}\mathbf{B}$ is not an energy term as it is not a scalar. At the very best it could be a momentum term.

The term $\frac{q\hbar}{2m_0 c}\mathbf{B}\cdot\boldsymbol{\sigma}$ does not code any interaction between the spin and the magnetic field. It associates the magnetic field with some scalar multiplication constant that makes sure that the term that contains $\mathbf{B}\cdot\boldsymbol{\sigma}$ ends up with the same dimensions as the other terms in the equation. This scalar constant could therefore be considered as a kind of conversion factor.

Nevertheless, Dirac's calculation comes very close to reproducing exactly the experimental result. In fact, the over-interpretation leads to the same mathematics as the correct interpretation, because the "scalar product" of $\mathbf{B} = B\mathbf{e}_z$ with the presumed spin operator $\mathbf{S} = \frac{\hbar}{2}(\sigma_x, \sigma_y, \sigma_z)$ evidently leads to:

$$\hat{H} = \frac{qB}{m_0 c}\hat{S}_z, \tag{9.15}$$

while in the correct interpretation $\mathbf{B} = B\mathbf{e}_z$ automatically leads to the equivalent result $\mathbf{B}{\cdot}\boldsymbol{\sigma} = B\sigma_z$. The eigenvectors of σ_z are then eigenvectors of the energy operator because they contain only one non-zero component, and thus give rise to stationary states. In the context of an electron within a magnetic field one can define the cyclotron frequency $\omega_c = \frac{qB}{m_0 c}$. It is a quantity that naturally appears in the equations.

9.2.3 *Can the anomalous magnetic moment of the electron be calculated?*

In the naive model of a current loop that generates the magnetic dipole moment $\boldsymbol{\mu}$, it appeared enigmatic how it could ever be possible to calculate the dipole moment that corresponds to the spin of the electron. Should we find a model for some current loop that could exist within the electron and creates this magnetic dipole? How could one manage to describe such a loop without knowing anything about the internal structure of the electron in terms of charge and current distributions? What would be the shape, orientation and diameter of such a current loop? Should we guess some three-dimensional charge-current distribution?

One has absolutely not the slightest idea about a possible starting point from which one could try to develop a reasoning about such charge-current distributions. In the present stage of knowledge, there is no experimental value for some finite size of the electron. If the electron has a finite size, it is too small to be measured with the presently available technologies. How can one then possibly make assumptions about an internal current loop?

Another approach to the problem of the magnetic moment consists in exploring the gravitational analogue of the macroscopic spinning top. The equation of motion of a spinning top is based on $\frac{d\mathbf{L}}{dt} = \mathbf{r} \wedge \mathbf{F}$. Here we have $L = I\omega$, where I is the moment of inertia, and ω the angular frequency of the spinning top. Again, a calculation of the moment of inertia requires knowledge of the mass distribution of the electron, and among other elements the radius of the electron. Again it is difficult to know where to start from to obtain this information.

But it is nevertheless claimed that $\boldsymbol{\mu}$ (or g) is calculated in quantum electrodynamics with extraordinary precision. What kind of insight about the charge-current and/or mass distributions is it that has permitted the people who developed quantum electrodynamics to perform this calculation? Feynman's written legacy reveals how he was a most fantastic teacher. How come he has not transmitted that dearly needed insight to us? The problem looks so impenetrable that one feels stale. Having to be able to find a way to calculate $\boldsymbol{\mu}$ appears to be an insurmountable problem.

The idea of postulating a term of the type $\boldsymbol{\mu}{\cdot}\mathbf{B}$ is based on intuition. This term cannot materialize directly within the equations through $\mathbf{B}{\cdot}\boldsymbol{\sigma}$, but only through $\mathbf{B}\cdot\mathbf{s}$. The physics that lead to the expression $\boldsymbol{\mu}{\cdot}\mathbf{B}$ look like the most intuitive thing of the whole procedure, and therefore it is the last thing one might suspect to be questioned in the attempt to make sense of the problem. But on purely mathematical grounds, casting the contribution to g that comes from $\mathbf{B}{\cdot}\boldsymbol{\sigma}$ in the form $\boldsymbol{\mu}{\cdot}\mathbf{B}$ is simply incorrect.

To end this section a very important remark must be made. The minimal substitution has been derived by considering the energy and momentum of a point-like particle characterized by only one parameter, *viz.* its charge. No other parameters have been considered that could further characterize a particle that would also possess some magnetic dipole moment $\boldsymbol{\mu}$. It can therefore reasonably asked how one can possibly expect that a potential-energy term for a magnetic dipole moment $\boldsymbol{\mu}$ will come out of the equations. As noted in Section 5.7, with a God-given Dirac equation it might be believed that the equation captures a mystery axiom that does the trick. But the derivation of the Dirac equation developed in this book from a well-defined set of assumptions spells the end of such hopes. What comes out as a result from an equation must have been put into it as part of the assumptions that are necessary to derive it. When an equation is derived for a particle by simply specifying its charge and saying that it spins, then it will not be possible to obtain a result that would correspond to the intuition for a magnetic dipole moment $\boldsymbol{\mu}$. That over-interpreting the results of the calculations in terms of $\boldsymbol{\mu}$ is *a priori* gratuitous is confirmed by the experimental values for the g-factors of the proton and the neutron which do not agree at all with those calculations. That it all works out so well for the electron must thus be considered as a coincidence, and raises the interesting issue of the reasons for this success.

The intuition about a dipole moment associated with the spin could well be correct in principle, but it can *a priori* not be validated by calculations based on an equation derived for a point charge. There is no obligation to believe that the part of g that comes out of the Dirac equation corresponds to the energy of a dipole.

9.3 Problems with the traditional treatment of spin dynamics

9.3.1 *Criterion for conservation of energy*

There is an important result that comes out of the analysis of the structure of the Pauli equation. It expresses that when the electron is at rest, the only solutions that lead to fixed energy levels are those where the electron spin is aligned with the axis of the magnetic field, because this is the only way to have only one frequency within the spinor "eigenvector". The energy operator $-\frac{\hbar}{\imath}\frac{\partial}{\partial t}\mathbb{1}$ can only project out an eigenvalue from a spinor when the two entries of the spinor contain only one identical frequency.[1] But a vector operator $\mathbf{B}{\cdot}\boldsymbol{\sigma}$, has in principle the effect of pulling out two different frequencies from the two spinor entries. The only way to escape from this conclusion and to pull out only one frequency is to make sure that one of the two spinor entries becomes zero by aligning the spin axis of the spinor with the axis of the magnetic field, such that the frequency associated with this entry stays mute.

This well-known alignment condition is a baffling non-intuitive result. Classically, it would be expected that the spin could have any orientation with respect to the magnetic field. One may ask at which point in the derivations this non-intuitive feature has been introduced. In the traditional textbook approach to treating the possibility of a spin that is not aligned with the magnetic-field axis, one finds that the energy cannot be a constant. The minimal substitution is an expedient that allows the Lorentz transformations to be written in terms of $(E, c\mathbf{p})$ despite the fact that the rest mass in a potential changes. This rest mass at a point $\mathbf{r}$ can be calculated by considering a number of radiative processes that bring a particle at rest from infinity to a situation where it is at rest in $\mathbf{r}$. In the situations to be calculated here, the particle is often not at rest, but we still need

[1] These conclusions are based on using the definition $\hat{\mathrm{E}} = -\frac{\hbar}{\imath}\frac{\partial}{\partial t}\mathbb{1}$ for the energy operator. The idea that $\hat{\mathrm{E}} = -\frac{\hbar}{\imath}\frac{\partial}{\partial t}\mathbb{1}$ will be challenged later on in this chapter.

its rest mass. The radiation that is considered in the minimal substitution is therefore virtual. The minimal substitution provides some information about energy-momentum within an electromagnetic potential in terms of emitted virtual radiation. Due to the virtual radiation, the rest mass of the electron in an electric potential is changed. In a magnetic field it is slightly different, because the magnetic field is related to a vector potential. A possible interpretation could then be that to come to rest, the electron must already have travelled in the magnetic field and radiated away virtual momentum. This a directional effect, as it works with $\mathbf{A}$, not with V. One could speculate that this directional information could be registered within the electromagnetic potential. A specific Lorentz transformation of V defines one specific value of $\mathbf{A}$, but $\mathbf{A}$ has a certain gauge freedom, which implies that its direction is not pre-established. It is therefore difficult to see how this speculation could be further developed.

9.3.2 *The problem with the description of the states that are not eigenstates*

In (9.13) a Hamiltonian term has been derived, that renders it possible to study the behaviour of the electron spin within a magnetic field. But severe conceptual problems arise when questioning what might happen to the spin if it is initially not aligned with the z-axis, such that it is not an eigenvector of σ_z. *A priori* this looks like a perfectly valid initial condition for the problem.

The problem of the misaligned spin could be considered as a special case of a more general problem, *viz.* what happens in any physical model that is described by stationary states when the initial state is not one of the stationary states. From the Bohr model it is known that the electron radiates to settle on a stationary state. One may then expect something similar for the spin of the electron in a magnetic field, but it is not obvious how this should be described. The equations have not been set up to account for emission of radiation, and it is not obvious how they could be changed to do so.

Let us now address the conceptual problems that arise in trying to describe what happens with a misaligned spin. It was shown in Subsection 5.1.3 that the eigenvector of the matrix $\mathbf{s}{\cdot}\boldsymbol{\sigma}$ that corresponds to the eigenvalue $+1$ is $e^{-\imath\varphi/2}[\,s_z+1, s_x+\imath s_y\,]^\top$. For $\mathbf{s} = (\sin\theta\cos\phi,\, \sin\theta\sin\phi,\, \cos\theta)$ the value of $[\,s_z+1, s_x+\imath s_y\,]^\top$ can be rewritten in the form: $2\cos(\theta/2)$ $e^{-\imath\varphi/2}[\,\cos(\theta/2), \sin(\theta/2)e^{\imath\phi}\,]^\top$, where the constant $2\cos(\theta/2)e^{-\imath\varphi/2}$ can be dropped in order to normalize the eigenvector. Here, the two symbols ϕ and φ have a different meaning. The choice of the contribution $e^{-\imath\varphi/2}$ to this term is arbitrary as the eigenvector is determined up to a phase constant. As stated earlier, the eigenvector is not a spinor. The quantity $[\cos(\theta/2), \sin(\theta/2)e^{\imath\phi}]^\top$ referred to as the spin is an eigenvector for a vector quantity $\mathbf{s}{\cdot}\boldsymbol{\sigma}$ that may be called the vector of the spin axis, and therefore a set of spinors. It is a set of spinors that corresponds to the concept of a vector quantity in the group theory.

Such a set is not a spinor quantity, even if within the expression of the eigenvector the arbitrary phase factor $e^{-\imath\varphi/2}$ is included. The spin is the set of all spinors (i.e. rotations) that share the same rotation axis $\mathbf{s} = \mathbf{e}'_z$. This set is the ray $\zeta = \{\psi_\varphi, \varphi \in \mathbb{R} \parallel \psi_\varphi = e^{-\imath\varphi/2}[\cos(\theta/2), \sin(\theta/2)e^{\imath\phi}]^\top\}$. Here, ψ_φ is not a spinor but a sum of two spinors. The phase factor $e^{-\imath\varphi/2}$ has a precise meaning for the spinors that belong to the ray, and can be used to label the spinors. But it has no importance for the ray itself, apart from the fact that it is used to label the elements of the ray ζ. Hence, the ray can also be represented by $\zeta(0)$. The dynamics can be introduced by describing the function $\psi \in F(\mathbb{R}, \mathscr{I}) : \forall\tau \in \mathbb{R}, \psi(\tau) = e^{-\imath\omega_0\tau/2}[\cos(\theta/2), \sin(\theta/2)e^{\imath\phi}]^\top = e^{-\imath\omega_0\tau/2}\zeta(0)$. In the dynamics, $\psi(\tau)$ runs thus periodically through the whole ray. The ray has been noted as ζ rather than ψ to make it very obvious that it is not a single sum of two spinors, but a set of such sums. Note also that for a wave equation in free space, a ray contains only one frequency.

The time evolution $\zeta(\tau)$ within the ray can be described by using a precise value of $\zeta(0)$ as an initial value. That value can be taken to correspond to an eigenstate of an arbitrary spin vector $\mathbf{s}$. For the sake of generality it is assumed then that it is *a priori* not aligned with $\mathbf{e}_z$. The initial-value condition to be imposed on ζ is then: $\zeta(0) = c_1[1,0]^\top + c_2[0,1]^\top = c_1|\uparrow> +c_2|\downarrow>$, with $c_1 = \cos(\theta/2)$, $c_2 = \sin(\theta/2)e^{\imath\phi}$, and $c_1c_1^* + c_2c_2^* = 1$.

In many textbooks one introduces an evolution operator that is derived from (9.15). With this evolution operator and the initial values $\zeta(0)$ one can then calculate the time dependence of the ray $\zeta(\tau)$ within a magnetic field. It is shown then that the solution for $\zeta(\tau)$ is a mixed state that contains two frequencies: $\zeta(\tau) = c_1e^{-\imath(\omega_0-\omega_c)\tau/2}[1,0]^\top + c_2e^{+\imath(\omega_0+\omega_c)\tau/2}[0,1]^\top$, where ω_c is the cyclotron frequency.[2] It is further shown that this time dependence

[2]This can also be derived from (9.12) by noting that $V = 0$ and that $\frac{1}{2m_0}(\hat{\mathbf{p}} - q\mathbf{A})^2$ leads to the three terms on the right-hand side of (9.11). The first and the third of these terms are zero for a particle at rest and the second term is ignored. Only the term

$\zeta(\tau)$ describes a state wherein the spin axis undergoes precession, which is a "mixed state" $\zeta(\tau)$. The puzzling thing about this mixed state is that it does not seem to have a constant energy, as can be seen by applying the operator $-\frac{\hbar}{\imath}\frac{\partial}{\partial t}\mathbb{1}$ to it. One wonders if this does not violate the law of conservation of energy-momentum.

Traditional quantum mechanics explains the fact that the energy would not be constant away by invoking an uncertainty in the energy due to Heisenberg's uncertainty principle. The time evolution is interpreted within a probability framework. It is then claimed that the conservation of energy is respected for the probability distribution, not for the single events. This has far-reaching philosophical implications. Two strongholds of classical physics are severely challenged by this explanation, and the equations demand the introduction of two revolutionary ideas. The law of conservation of energy is declared to be no longer always valid and classical determinism must be abandoned to allow for probabilistic behaviour. These are things Einstein would not like: *"Der gute Gott würfelt nicht"*.

9.3.3 *Is there an Ariadne's thread that can lead us out of these conceptual problems?*

It is thus obvious that mixed-state solutions are fraught with conceptual problems, because it looks as though they do not have a fixed energy. It is preferable to consider that these conceptual problems are due to an improper, messy use of the mathematics. After simplifying a mathematically exact equation by introducing approximations, it will no longer be possible to make sense of the exact algebraic solutions of the resulting simplified equation by exact reasoning. The exact algebraic result will just be refractory to analysis.

One can then run in circles in trying to make sense of the solutions of the simplified equation but one is deemed to fail as they no longer comply with the exact logic. The only way out of this impasse is recognizing that one has trapped oneself within a fake problem. Most of the time making a diagnosis of this type can be awfully difficult. Could the present situation also require such a diagnosis?

It will nevertheless be attempted first to find solutions for the paradox within the logic of the equation obtained.

$-\frac{q\hbar}{2m_0c}\mathbf{B}\cdot\boldsymbol{\sigma}$ remains. From the solution of the ensuing equation for a particle with an arbitrary spin that is not aligned with $\mathbf{B}$ we obtain then the same result.

9.3.4 *Solution of the conceptual problems with mixed states in terms of sets?*

Two ways will be considered to solve the paradox that the energy of the mixed state would not be constant (within the context of the traditional Dirac equation). One solution is based on a description of sets (of spinors); the other is based on an analysis of the energy operator applied to the time evolution of a single particle.

The initial value $\zeta(0)$ is an assumed initial condition. It is supposed to be a value for the spin at a moment $\tau = 0$ in the time evolution of a hypothetical stationary situation wherein the spin axis would not be aligned with $\mathbf{B}$. As discussed in Subsections 5.4.1 and 9.3.2, the value of the ray $\zeta(0)$ corresponds to the definition of a set of spinors. For some special sets of spinors wherein all spinors share the same rotation axis it is possible to define a spin vector. In free space, all spinors of such a set have the same energy. But it is quite conceivable that sets of spinors could exist for which it is not possible to define a spin vector or a single energy.

In fact, it is in general only possible to assume initial values $\psi(0)$ for spinors, not for sets. Claiming that a unique initial value exists for $\zeta(0)$ consists in assuming that there is a point $\tau = 0$ in the time evolution where all instantaneous motions $\psi(\tau)$ turn out to have the same instantaneous rotation axis. This assumption defines a set. But if at other times the instantaneous motions no longer share a common axis or a common energy, then the set is somewhat artificial. It is nevertheless possible to introduce such an assumption and make calculations based on it. The calculations of the time evolutions of the spinors $\psi(\tau)$ could then be made for all possible values $\psi(0)$ using the wave equation. For each solution it can then be verified that the spinor belongs to $\zeta(\tau)$ thereby checking that the same result is obtained as with the evolution operator. From $\zeta(\tau)$ a value can be calculated for $\mathbf{s}(\tau)$ and for the energy that apparently makes sense. But the question is what those values for $\mathbf{s}(\tau)$ and the energy are supposed to mean.

It has been checked on certain examples that a superposition $\zeta_1 + \zeta_2$ of two rays ζ_1 and ζ_2 can describe a union of disjoint sets. Even the set that corresponds to a well-defined spin in free space contains two subsets, *viz.* of left-handed and right-handed frames. Pushing this idea further, the result of the calculation tells us that the set of spinors is a disjoint union of two subsets. For each subset a spin vector can be defined that is aligned with $\mathbf{B}$. One subset is spin-up and one subset spin-down. Each subset has a time evolution that does not violate the conservation of energy and where $e^{-\imath(\omega_0 \pm \imath\omega_c)\tau}$ applies to all spinors of the subset. The phase factor $e^{\pm\imath\omega_c\tau}$ can therefore be meaningfully applied to the ray defined by the subset. It is then possible to reconstruct a fictive weighted value $\zeta(\tau) = c_\downarrow\zeta_\downarrow(\tau) + c_\uparrow\zeta_\uparrow(\tau)$ from the well-defined values $\zeta_\downarrow(\tau)$ and $\zeta_\uparrow(\tau)$, even if it is not appropriate to define a unique spin and a unique energy for the whole set. It is only possible to define unique spins for each of the two subsets. Each subset will have a well-defined energy that remains constant in time, but average spins and energies may be calculated. As the weights $|c_\downarrow|^2$ and $|c_\uparrow|^2$ of the subsets and their frequencies do not change with time, the average energy will be constant and the sum $|c_\downarrow|^2 E_\downarrow + |c_\uparrow|^2 E_\uparrow$ of the energies contained in the individual subsets.

The average value for s_z will be given by $\lambda_\uparrow|c_\uparrow|^2 + \lambda_\downarrow|c_\downarrow|^2$, where the eigenvalues are given by $\lambda_\uparrow = 1$ and $\lambda_\downarrow = -1$. This yields $\cos^2\frac{\theta}{2} - \sin^2\frac{\theta}{2} = \cos\theta$, and corresponds to the classical idea of a spin component. The average energy will be $|c_\uparrow|^2 E_\uparrow + |c_\downarrow|^2 E_\downarrow$ and this is constant in time. *This value of the energy cannot be calculated by bluntly applying the energy operator* $\hat{\mathrm{E}}$ *of quantum mechanics to the mixed state!* It should be clear that in this analysis based on sets the original idea that $\zeta(0)$ would describe a well-defined unique spin for a single particle has been abandoned in favour of an interpretation of $\zeta(0)$ in terms of an average quantity calculated over a set.

From a certain point of view, it could be claimed that this analysis is a cheat, because the idea was to describe a single particle. But this is not true. The probabilistic approach is forced upon us by the way the calculus is undertaken. By treating the spinors like vectors, sets are introduced, and to continue obtaining meaningful results probabilities must then be used. In fact, in SU(2) the exact equation it all started from is $\frac{d}{dt}\psi(t) = -\imath\,[\,\mathbf{s}\cdot\boldsymbol{\sigma}\,]\,\frac{\omega}{2}\psi(t)$. In order to solve it the eigenvectors of $[\,\mathbf{s}\cdot\boldsymbol{\sigma}\,]$ are calculated. The eigenvectors are no longer spinors, but an algebraic sum of two spinors that can be identified with a set that defines the spin vector. These vector states satisfy a simplified equation $\frac{d}{dt}\psi(t) = \pm\imath\,\frac{\omega}{2}\psi(t)$. Without introducing the assumption that the wave functions describe sets, the simplified equation cannot be obtained. The simplified equations describe spin-up and spin-down vector states. As mentioned many times, spinors cannot be added like vectors. In the process of defining spin it has been specified what it means to add two spinors that contain the *same frequency*: Their sum can be interpreted as the definition of a set. This is why the equations developed here do not describe single particles but sets.

But when a linear combination of spin-up and spin-down states is made with coefficients $c_\downarrow$ and $c_\uparrow$, again a procedure is introduced that is *a priori* not mathematically defined, because the two contributions contain now *two different frequencies*. It must thus be established what kind of meaning can be given to this kind of operation in terms of sets. The analysis above shows then that it could be interpreted as a set with a fraction $|c_\downarrow|^2$ of spin-down states and a fraction $|c_\uparrow|^2$ of spin-up states. The quantities $|c_\downarrow|^2$ and $|c_\uparrow|^2$ are then indeed probabilities, and the traditional interpretation of the formalism used in quantum mechanics is recovered. In other words, this describes a *statistical ensemble*. It is thus meaningless within the formalism to identify the linear combination with a single spinor state that would describe an electron with a tilted spin axis, because the formalism can only be derived by introducing sets. In order to seek only single-particle solutions, it is necessary to adhere to the equation $\frac{d}{dt}\psi(t) = -\imath\,[\,\mathbf{s}{\cdot}\boldsymbol{\sigma}\,]\,\frac{\omega}{2}\psi(t)$ and search for a way to solve it meaningfully without introducing sets or without mixing frequencies. In Section 9.5 it will be shown how to formulate such a single-particle approach. It will be necessary to derive first some equations for a single particle undergoing precession before one can address this issue. More will be derived than may be needed, but the results and the methodology are useful for possible further research. It is claimed in textbooks that the mixed state leads to a notion of a spin axis that undergoes precession. It is for this reason that precession will now be studied in its own right. It will then be possible to use these results to analyse the problem of the misaligned spin for a single particle.

9.4 Describing precession of a single particle

9.4.1 *Describing spin with an arbitrary orientation of the spin axis and without precession*

Let us start by describing a spinning top whose physical rotation axis is given by $\mathbf{e}'_z$, where the triad of the top is defined by the spinor $[\xi_0, \xi_1]^\top$. The vector $\mathbf{e}'_z$ is then, according to (3.19), coded by:

$$\mathbf{e}'_z{\cdot}\boldsymbol{\sigma} = \begin{pmatrix} \xi_0\xi_0^* - \xi_1\xi_1^* & 2\xi_0\xi_1^* \\ 2\xi_0^*\xi_1 & \xi_1\xi_1^* - \xi_0\xi_0^* \end{pmatrix}. \tag{9.16}$$

It is easy to check that $[\,\mathbf{e}'_z{\cdot}\boldsymbol{\sigma}\,]\,[\xi_0, \xi_1]^\top = [\xi_0, \xi_1]^\top$. Let us introduce spherical coordinates (θ, ϕ) to describe $\mathbf{e}'_z = (\sin\theta\cos\phi, \sin\theta\sin\phi, \cos\theta)$.

We obtain then:

$$\mathbf{e}'_z\cdot\boldsymbol{\sigma} = \begin{pmatrix} \cos\theta & \sin\theta\, e^{-\imath\phi} \\ \sin\theta\, e^{+\imath\phi} & -\cos\theta \end{pmatrix}. \tag{9.17}$$

Hence, $\cos\theta = \xi_0\xi_0^* - \xi_1\xi_1^*$. As $\xi_0\xi_0^* + \xi_1\xi_1^* = 1$, it follows that $\cos(\theta/2) = \sqrt{\xi_0\xi_0^*}$ and $\sin(\theta/2) = \sqrt{\xi_1\xi_1^*}$. Further calculation yields then $\xi_0 = \cos(\theta/2)e^{-\imath\phi/2}$ and $\xi_1 = \sin(\theta/2)e^{+\imath\phi/2}$ as a possible solution.

It was demonstrated in Section 3.8 that $\mathbf{e}'_z$ defines a spinor only up to a rotation angle χ. In fact, (9.16) is invariant under the substitutions $\xi_0 \to \xi_0 e^{\imath\chi}$, $\xi_1 \to \xi_1 e^{\imath\chi}$. Therefore, the values $\xi_0 = \cos(\theta/2)e^{-\imath\phi/2}$ and $\xi_1 = \sin(\theta/2)e^{+\imath\phi/2}$ are only one possible choice of solution. It is advisable to adhere to this single choice, as the phase factor will come out in yet another way from our calculations. Note that it is also possible to find the values of ξ_0 and ξ_1 by searching for the eigenvectors of the matrix in (9.17). This illustrates then the spin concept, as we find all rotations (defined by their spinor coordinates ξ_0 and ξ_1) that share the same z'-axis.

The vector product $\mathbf{e}_z \wedge \mathbf{e}'_z = (\sin\theta)\,\mathbf{m}$ can be calculated from the identity: $[\,\mathbf{e}_z\cdot\boldsymbol{\sigma}\,][\,\mathbf{e}'_z\cdot\boldsymbol{\sigma}\,] - [\,\mathbf{e}'_z\cdot\boldsymbol{\sigma}\,][\,\mathbf{e}_z\cdot\boldsymbol{\sigma}\,] = 2\imath\sin\theta\,[\,\mathbf{m}\cdot\boldsymbol{\sigma}\,] = 4\imath\sin(\theta/2)$ $\cos(\theta/2)\,[\,\mathbf{m}\cdot\boldsymbol{\sigma}\,]$. From this we find:

$$\imath\sin(\theta/2)\,[\,\mathbf{m}\cdot\boldsymbol{\sigma}\,] = \frac{1}{\sqrt{\xi_0\xi_0^*}}\begin{pmatrix} 0 & \zeta_0\zeta_1^* \\ -\xi_0^*\xi_1 & 0 \end{pmatrix}. \tag{9.18}$$

Let us now calculate the anticlockwise rotation (i.e. around $-\mathbf{m}$) that maps $\mathbf{e}'_z$ onto $\mathbf{e}_z$. If θ is considered as a non-algebraic quantity, then the angle of that rotation is θ. The rotation is therefore given by:

$$\mathbf{R}_0 = \cos(\theta/2)\mathbb{1} + \imath\sin(\theta/2)\,[\,\mathbf{m}\cdot\boldsymbol{\sigma}\,] = \frac{1}{\sqrt{\xi_0\xi_0^*}}\begin{pmatrix} \xi_0\xi_0^* & \xi_0\xi_1^* \\ -\xi_0^*\xi_1 & \xi_0\xi_0^* \end{pmatrix}. \tag{9.19}$$

The determinant of the matrix that occurs here is 1 as it should be. The inverse matrix is thus:

$$\mathbf{R}_0^{-1} = \frac{1}{\sqrt{\xi_0\xi_0^*}}\begin{pmatrix} \xi_0\xi_0^* & -\xi_0\xi_1^* \\ \xi_0^*\xi_1 & \xi_0\xi_0^* \end{pmatrix}. \tag{9.20}$$

A rotation over an angle χ around $\mathbf{e}'_z$ of the triad defined by the spinor $[\xi_0, \xi_1]^\top$ will yield a new triad. The spinor that defines this new triad can be calculated by multiplying the spinor $[\xi_0, \xi_1]^\top$ with the rotation matrix:

$$\mathbf{R} = \mathbf{R}_0^{-1}\begin{pmatrix} e^{-\imath\chi/2} & 0 \\ 0 & e^{\imath\chi/2} \end{pmatrix}\mathbf{R}_0. \tag{9.21}$$

The idea here is to rotate $\mathbf{e}'_z$ to $\mathbf{e}''_z = \mathbf{e}_z$, perform the rotation over an angle χ around $\mathbf{e}_z$, and then transform $\mathbf{e}''_z$ back to recover the original orientation of the vector $\mathbf{e}'_z$. The rotation around $\mathbf{e}'_z$ is calculated thus by a similarity transformation. In order to describe a triad in a uniform spinning motion around $\mathbf{e}'_z$ it suffices now to put $\chi = \omega_0\tau$. Here τ is used to clearly indicate that we are considering an electron that is translationally at rest. The rotation matrices $\mathbf{R}$ corresponding to the uniform motion are then given by:

$$\mathbf{R} = \begin{pmatrix} e^{-\imath\omega_0\tau/2}\xi_0\xi_0^* + e^{\imath\omega_0\tau/2}\xi_1\xi_1^* & \xi_0\xi_1^*(e^{-\imath\omega_0\tau/2} - e^{\imath\omega_0\tau/2}) \\ \xi_0^*\xi_1(e^{-\imath\omega_0\tau/2} - e^{\imath\omega_0\tau/2}) & e^{\imath\omega_0\tau/2}\xi_0\xi_0^* + e^{-\imath\omega_0\tau/2}\xi_1\xi_1^* \end{pmatrix}. \tag{9.22}$$

$\mathbf{R}$ can be written as a linear combination $\mathbf{R}_+e^{\imath\omega_0\tau/2} + \mathbf{R}_-e^{-\imath\omega_0\tau/2}$. In comparing the matrices $\mathbf{R}_+$ and $\mathbf{R}_-$ with the matrix given by (5.25), it can be seen that $\mathbf{R} = \frac{1}{2}\left[\,(\mathbb{1} + \mathbf{e}'_z{\cdot}\boldsymbol{\sigma})e^{-\imath\omega_0\tau/2} + (\mathbb{1} - \mathbf{e}'_z{\cdot}\boldsymbol{\sigma})e^{+\imath\omega_0\tau/2}\,\right]$. This is exactly what was obtained previously for the Rodrigues formula, with $\mathbf{n}$ replaced by $\mathbf{e}'_z$, i.e. $\cos(\omega_0\tau/2)\mathbb{1} - \imath\sin(\omega_0\tau/2)\,[\,\mathbf{e}'_z{\cdot}\boldsymbol{\sigma}\,]$. This should not be a surprise as $\mathbf{R}$ is exactly a rotation with axis $\mathbf{n} = \mathbf{e}'_z$. The difference is that we have now:

$$\psi = \begin{pmatrix} \xi'_0 \\ \xi'_1 \end{pmatrix} = \mathbf{R}\begin{pmatrix} \xi_0 \\ \xi_1 \end{pmatrix} = \mathbf{R}\psi_0, \tag{9.23}$$

such that the new "Rodrigues equation" becomes now $\psi = (\cos(\omega_0\tau/2)\mathbb{1} - \imath\sin(\omega_0\tau/2)\,[\,\mathbf{e}'_z{\cdot}\boldsymbol{\sigma}\,])\;\psi_0$ rather than $\psi = (\cos(\omega_0\tau/2)\mathbb{1}\; - \imath\sin(\omega_0\tau/2)\,[\,\mathbf{e}'_z{\cdot}\boldsymbol{\sigma}])\;[1,0]^\top$. Note that if at $\tau = 0$ the triad just corresponds to an anti-clockwise rotation by an angle θ around $\mathbf{e}_y$, (such that $\mathbf{e}'_y = \mathbf{e}_y$, $\mathbf{e}'_z = \sin\theta\,\mathbf{e}_x + \cos\theta\,\mathbf{e}_z$ and $\mathbf{e}'_x = \cos\theta\,\mathbf{e}_x - \sin\theta\,\mathbf{e}_z$), then ψ_0 corresponds to the first column of the rotation matrix $\cos(\theta/2)\,\mathbb{1} - \imath\sin(\theta/2)\,[\,\mathbf{e}_y{\cdot}\boldsymbol{\sigma}\,]$.

From this we can derive:

$$\frac{d}{d\tau}\psi = -\imath\frac{\omega_0}{2}\,[\,\mathbf{e}'_z{\cdot}\boldsymbol{\sigma}\,]\,\psi, \tag{9.24}$$

such that the differential equation (5.10) derived earlier remains valid, provided $\mathbf{n}$ is replaced by $\mathbf{e}'_z$. Here, $\mathbf{e}'_z$ is the true rotation axis of the spinning top, and the equation is now covariant. When $\mathbf{e}'_z = \mathbf{e}_z$, we will, after simplification, obtain the classical Dirac-like equation $\frac{d}{d\tau}\psi = -\imath\frac{\omega_0}{2}\psi$. But this form cannot be used as a starting point for a calculation that would generalize it to an equation for a rotation with an arbitrary axis. For such a generalization it is necessary to keep the unsimplified equation (9.24).

ξ'_0 and ξ'_1 can also be calculated directly from (9.22). In fact, carrying out the calculations in (9.23) yields:

$$\psi = \begin{pmatrix} \xi'_0 \\ \xi'_1 \end{pmatrix} = \begin{pmatrix} e^{-\imath\omega_0\tau/2} & 0 \\ 0 & e^{-\imath\omega_0\tau/2} \end{pmatrix} \begin{pmatrix} \xi_0 \\ \xi_1 \end{pmatrix}. \tag{9.25}$$

Hence, $\xi'_0 = e^{-\imath\omega_0\tau/2}\xi_0$, $\xi'_1 = e^{-\imath\omega_0\tau/2}\xi_1$, such that $\mathbf{e}'_z$ only defines a spinor up to a proportionality factor $e^{\imath\chi}$. (That ξ_0 and ξ_1 are multiplied by the same factor is due to the fact that the physical spin axis has been defined as $\mathbf{e}'_z$ rather than as $\mathbf{n}$. As for $\mathbf{e}'_z = \mathbf{e}_z$ the spinors have only one non-zero component, they contain only one sign of the frequency ω_0, and this will remain true when $\mathbf{e}_z$ is moved to $\mathbf{e}'_z \neq \mathbf{e}_z$ by a similarity transformation.) Using these values, the analogue of (5.25) can be calculated. As ξ'_0 and ξ'_1 are multiplied by the same factor, we find that $\mathbf{e}'_z$ has not changed. This is as expected due to the fact that the rotation is around $\mathbf{e}'_z$. With a rotation axis $\mathbf{n}$ different from $\mathbf{e}'_z$ the result would have been less simple: the value for the new axis $\mathbf{e}'_z$ would have varied with time.

Hence, there is a single frequency within the system, even if the spin axis is not the z-axis. There is another axis that functions as the easy axis. This axis is related to the z-axis through a similarity transformation. Hence, when there is a single axis in a problem, it can always be considered as an easy axis that gives rise to only one frequency. It is in a sense just a matter of making a change of reference frame. The difficulties will only arise when there are two different physically relevant axes within the system.

9.4.2 *A Rodrigues-like equation for an electron whose spin axis undergoes precession*

9.4.2.1 *Arbitrary initial orientation of the spin axis with precession around the z-axis*

Spin with an arbitrary orientation of the spin axis was discussed in Subsection 9.4.1. Let us now investigate what will happen in the case that there is also precession of the spin axis. When the rotation axis undergoes simultaneously a precession by an angle $\Omega\tau$ around the z-axis, then the global rotation matrix is given by:

$$\begin{aligned} \mathbf{M} &= \begin{pmatrix} e^{-\imath\Omega\tau/2} & 0 \\ 0 & e^{\imath\Omega\tau/2} \end{pmatrix} \mathbf{R}_0^{-1} \begin{pmatrix} e^{-\imath\omega_0\tau/2} & 0 \\ 0 & e^{\imath\omega_0\tau/2} \end{pmatrix} \mathbf{R}_0 \\ &= \mathbf{D}\mathbf{R}_0^{-1}\mathbf{C}\mathbf{R}_0. \end{aligned} \tag{9.26}$$

The last part of this equation introduces and defines the diagonal matrices $\mathbf{D}$ and $\mathbf{C}$.

There is a worry here in that the rotation group is not abelian, such that it is necessary to justify the order in which the operations are performed. However, it is possible to choose the order in which the two rotations are treated, provided it is done correctly. Here, we consider $\mathbf{M}\psi$ as $\mathbf{D}\mathbf{R}_0^{-1}\mathbf{C}\mathbf{R}_0\mathbf{D}^{-1}$ working on $\mathbf{D}\psi$. The position of the $\mathbf{e}'_z$-axis in $\mathbf{D}\psi$ after a given amount of time is known with certainty, as is the number of rotations around this axis that has taken place within the same amount of time. These can be calculated from the starting position. To calculate the effect of the rotations $\mathbf{C}$ around $\mathbf{e}'_z$, it was necessary to rotate $\mathbf{e}'_z$ back to $\mathbf{e}_z$ with $\mathbf{R}_0$, perform the rotations and then rotate back to the original position of $\mathbf{e}'_z$ with $\mathbf{R}_0^{-1}$. This is the way to move the effect of the rotations around the instantaneous axis $\mathbf{e}'_z$ from one orientation of $\mathbf{e}'_z$ to another. In calculating the effect of the rotations around $\mathbf{e}_z$ there is no need to make the distinction between $\mathbf{n}$ and the local value of $\mathbf{e}'_z$, because in this special case they coincide. Moving such effects around is thus performed with a similarity transformation based on the rotation $\mathbf{R}_0$ that carries the move. To calculate the effect of the rotations around the instantaneous axis $\mathbf{e}'_z$ while $\mathbf{e}'_z$ is turning, a similar similarity transformation based on $\mathbf{D}^{-1}$ is used. If part of the rotation of the axis $\mathbf{e}'_z$ around $\mathbf{e}_z$ is undertaken before considering the rotations around $\mathbf{e}'_z$, the axis of the rotation needed to bring $\mathbf{e}'_z$ to $\mathbf{e}_z$ would no longer be $\mathbf{m}$, and the rotation would no longer be $\mathbf{R}_0$. In other words, it would be necessary to make the similarity transformation with a rotation other than $\mathbf{R}_0$. That would theoretically be possible but much more complicated. The way the problem was treated in (9.26) is simple in that it permits the use of the fixed value of $\mathbf{m}$ and it is correct, as it uses similarity transformations to move around in the group. The explicit calculation of rotation matrix $\mathbf{M}$ yields:

$$\mathbf{M} = \begin{pmatrix} \xi_0\xi_0^* e^{-\imath(\omega_0+\Omega)\tau/2} + \xi_1\xi_1^* e^{\imath(\omega_0-\Omega)\tau/2} & 2\xi_0\xi_1^*(e^{+\imath(\omega_0-\Omega)\tau/2} - e^{-\imath(\omega_0+\Omega)\tau/2}) \\ 2\xi_1\xi_0^*(e^{+\imath(\omega_0+\Omega)\tau/2} - e^{-\imath(\omega_0-\Omega)\tau/2}) & \xi_0\xi_0^* e^{\imath(\omega_0+\Omega)\tau/2} + \xi_1\xi_1^* e^{-\imath(\omega_0-\Omega)\tau/2} \end{pmatrix}. \tag{9.27}$$

We have thus:

$$\psi = \begin{pmatrix} \xi'_0 \\ \xi'_1 \end{pmatrix} = \mathbf{M} \begin{pmatrix} \xi_0 \\ \xi_1 \end{pmatrix} = \mathbf{M}\psi_0. \tag{9.28}$$

Here, ψ_0 and $\mathbf{R}_0$ are constant matrices, while ψ and the matrices $\mathbf{D}$, $\mathbf{C}$ that occur within $\mathbf{M}$ are varying. Calculating $\mathbf{M}\psi_0$ from (9.27) looks an imposing task, but the effect of $\mathbf{R}_0^{-1}\mathbf{C}\mathbf{R}_0$ on ψ_0 has already been calculated, only the effect of $\mathbf{D}$ needs to be added. Therefore:

$$\psi = \begin{pmatrix} \xi_0'' \\ \xi_1'' \end{pmatrix} = \begin{pmatrix} e^{-\imath(\omega_0+\Omega)\tau/2} & 0 \\ 0 & e^{-\imath(\omega_0-\Omega)\tau/2} \end{pmatrix} \begin{pmatrix} \xi_0 \\ \xi_1 \end{pmatrix}, \tag{9.29}$$

from which the new value $\mathbf{e}_z''$ of $\mathbf{e}_z'$ can now be calculated. This will now be:

$$\begin{aligned} [\,\mathbf{e}_z''(\tau)\cdot\boldsymbol{\sigma}\,] &= \begin{pmatrix} \xi_0\xi_0^* - \xi_1\xi_1^* & 2\xi_0\xi_1^*\, e^{-\imath\Omega\tau} \\ 2\xi_0^*\xi_1 e^{\imath\Omega\tau} & \xi_1\xi_1^* - \xi_0\xi_0^* \end{pmatrix} \\ &= \begin{pmatrix} \cos\theta & \sin\theta\, e^{-\imath(\phi+\Omega\tau)} \\ \sin\theta\, e^{\imath(\phi+\Omega\tau)} & -\cos\theta \end{pmatrix}. \end{aligned} \tag{9.30}$$

This is all pure geometry and could have been written from scratch with a little thought about the components of the rotating vector $\mathbf{e}_z''$. Its z-component $\cos\theta$ should remain fixed and the other components should rotate with an angular frequency Ω. The combined motion can be written as $\psi(\tau) = [\,\cos(\omega\tau/2)\mathbb{1} - \imath\,[\,\mathbf{e}_z''(\tau)\cdot\boldsymbol{\sigma}\,]\,\sin(\omega\tau/2)\,]\,\psi_0$. By assuming that $\phi = 0$ corresponds to $\tau = 0$, the quantity ϕ drops out of the equation, such that it becomes simplified. This can also all be derived by examining what happens to the differential equation. From $\frac{d}{d\tau}\mathbf{D} = -\imath\frac{\Omega}{2}\sigma_z\mathbf{D}$ and $\frac{d}{d\tau}\mathbf{C} = -\imath\frac{\omega_0}{2}\sigma_z\mathbf{C}$ it follows that:

$$\frac{d}{dt}\psi = -\imath\frac{\Omega}{2}\sigma_z\psi - \imath\frac{\omega_0}{2}\mathbf{D}\mathbf{R}_0^{-1}\sigma_z\mathbf{C}\mathbf{R}_0\psi_0. \tag{9.31}$$

To recover ψ again, $\mathbf{R}_0\mathbf{D}^{-1}\mathbf{D}\mathbf{R}_0^{-1}$ is introduced between σ_z and $\mathbf{C}$, yielding:

$$\frac{d}{d\tau}\psi = -\frac{\imath}{2}\,\left[\,[\,\boldsymbol{\Omega}\cdot\boldsymbol{\sigma}\,] + \mathbf{D}\mathbf{R}_0^{-1}\,[\boldsymbol{\omega}_0\cdot\boldsymbol{\sigma}\,]\,\mathbf{R}_0\mathbf{D}^{-1}\,\right]\,\psi, \tag{9.32}$$

with the definitions $\boldsymbol{\Omega} = \Omega\mathbf{e}_z$ and $\boldsymbol{\omega}_0 = \omega_0\mathbf{e}_z$. Furthermore, $\mathbf{D}\mathbf{R}_0^{-1}\,[\boldsymbol{\omega}_0\cdot\boldsymbol{\sigma}\,]\,\mathbf{R}_0\mathbf{D}^{-1}$ just codes the new instantaneous rotation axis $\boldsymbol{\omega}_0''\cdot\boldsymbol{\sigma} = \omega_0\,[\,\mathbf{e}_z''\cdot\boldsymbol{\sigma}\,]$ after the precession, as $\mathbf{e}_z''\cdot\boldsymbol{\sigma} = \mathbf{D}[\,\mathbf{e}_z'\cdot\boldsymbol{\sigma}\,]\mathbf{D}^{-1}$ and $\boldsymbol{\omega}_0' = \omega_0\mathbf{e}_z' = \mathbf{R}_0^{-1}\,[\boldsymbol{\omega}_0\cdot\boldsymbol{\sigma}\,]\,\mathbf{R}_0$. This can also be shown by straightforward calculation of $\mathbf{D}\mathbf{R}_0^{-1}\,\sigma_z\,\mathbf{R}_0\mathbf{D}^{-1}$ using (9.19) and (9.20) and the definition of $\mathbf{D}$ as given by (9.26). The same value is obtained then for $\mathbf{e}_z''(\tau)\cdot\boldsymbol{\sigma}$ as given by (9.30). Hence, we have:

$$\frac{d}{d\tau}\psi = -\frac{\imath}{2}\,[\Omega\,[\,\mathbf{e}_z\cdot\boldsymbol{\sigma}\,] + \omega_0\,[\,\mathbf{e}_z''(\tau)\cdot\boldsymbol{\sigma}\,]]\,\psi = -\frac{\imath}{2}\,[\,\boldsymbol{\Omega} + \boldsymbol{\omega}_0''(\tau)\,]\,\psi. \tag{9.33}$$

The rotation vectors thus add up, according to:

$$\boldsymbol{\omega}(\tau) = \boldsymbol{\omega}_0''(\tau) + \boldsymbol{\Omega}. \tag{9.34}$$

Putting $\hbar\omega_0 = 2m_0c^2$ as before, we have then: $mc^2\mathbf{e}_\omega = m_0c^2(\mathbf{e}''_z(\tau) + \frac{\hbar\Omega}{2m_0c^2}\mathbf{e}_z)$. Hence, the global rotation axis is not aligned with the precession axis, but slightly tilted with respect to it, and it is not fixed. The value $\omega = |\boldsymbol{\omega}|$, however, is fixed.

This will now be written somewhat differently. The equation that accounts for the precession can be written as:

$$\psi(\tau) = \left(\cos\frac{\Omega\tau}{2}\,\mathbb{1} - \imath\sin\frac{\Omega\tau}{2}\,[\,\mathbf{e}_z\cdot\boldsymbol{\sigma}\,]\right)\left(\cos\frac{\omega_0\tau}{2}\,\mathbb{1} - \imath\sin\frac{\omega_0\tau}{2}\,[\mathbf{e}'_z\cdot\boldsymbol{\sigma}]\right)\psi_0. \tag{9.35}$$

With $\mathbf{D} = \cos\frac{\Omega\tau}{2}\,\mathbb{1} - \imath\sin\frac{\Omega\tau}{2}\,[\,\mathbf{e}_z\cdot\boldsymbol{\sigma}\,]$, and $\mathbf{R} = \cos\frac{\omega_0\tau}{2}\,\mathbb{1} - \imath\sin\frac{\omega_0\tau}{2}\,[\,\mathbf{e}'_z\cdot\boldsymbol{\sigma}\,]$, this can be rewritten as $\psi = \mathbf{D}\mathbf{R}\psi_0$, whereby $\mathbf{R}\psi_0 = e^{-\imath\omega_0\tau/2}\psi_0$. The equation $\psi = \mathbf{D}\mathbf{R}\psi_0$ is intuitively clear. We rotate around $\mathbf{e}'_z$ and rotate the result around $\mathbf{e}_z$. This is thus a rotation with an angle $\Omega\tau$ after a rotation with an angle $\omega_0\tau$. From all this it follows that $\frac{d}{d\tau}\psi = (\frac{d\mathbf{D}}{d\tau}\mathbf{R} + \mathbf{D}\frac{d\mathbf{R}}{d\tau})\,\psi_0$. As $\frac{d\mathbf{D}}{d\tau} = -\imath\frac{\Omega}{2}\,[\,\mathbf{e}_z\cdot\boldsymbol{\sigma}\,]\mathbf{D}$, the first term yields $-\imath\frac{\Omega}{2}\,[\,\mathbf{e}_z\cdot\boldsymbol{\sigma}\,]\psi$. As $\frac{d\mathbf{R}}{d\tau} = -\imath\frac{\omega_0}{2}\,[\,\mathbf{e}'_z\cdot\boldsymbol{\sigma}\,]\mathbf{R}$, the second term yields $-\imath\frac{\omega_0}{2}\mathbf{D}\,[\,\mathbf{e}'_z\cdot\boldsymbol{\sigma}\,]\,\mathbf{R}\psi_0$. This is also $-\imath\frac{\omega_0}{2}\mathbf{D}\,[\,\mathbf{e}'_z\cdot\boldsymbol{\sigma}\,]\,\mathbf{D}^{-1}\,\mathbf{D}\mathbf{R}\psi_0$ or $-\imath\frac{\omega_0}{2}\,[\,(\mathbf{e}''_z(\tau))\cdot\boldsymbol{\sigma}\,]\,\psi$, such that the same result is obtained. The precession itself can be written as $\psi = \mathbf{D}\mathbf{R}\mathbf{D}^{-1}\,\mathbf{D}\psi_0$, which yields:

$$\psi(\tau) = \left[\cos\frac{\omega_0\tau}{2}\,\mathbb{1} - \imath\sin\frac{\omega_0\tau}{2}\,[\,(\mathbf{s}(\tau))\cdot\boldsymbol{\sigma}\,]\right]\,\mathbf{D}\psi_0, \tag{9.36}$$

where $\mathbf{s}(\tau) = \mathbf{e}''_z(\tau)$ is given by:

$$\mathbf{s}(\tau) = \cos\theta\,\mathbf{e}_z + \sin\theta\,\cos\Omega\tau\,\mathbf{e}_x + \sin\theta\,\sin\Omega\tau\,\mathbf{e}_y. \tag{9.37}$$

Within SU(2), this vector is expressed by:

$$\mathbf{s}(\tau)\cdot\boldsymbol{\sigma} = \begin{pmatrix} \cos\theta & \sin\theta e^{-\imath\Omega\tau} \\ \sin\theta e^{\imath\Omega\tau} & -\cos\theta \end{pmatrix}. \tag{9.38}$$

At $\tau = 0$ this corresponds then indeed to $\cos\theta\,\mathbf{e}_z + \sin\theta\,\mathbf{e}_x$. The term between brackets in (9.36) is a rotation around $\mathbf{s}(\tau)$. It works on the spinor of a triad that has been rotated around $\mathbf{e}_z$ such that its $\mathbf{e}'_z$ vector coincides with $\mathbf{s}$. This is also intuitively clear. Here, a rotation is performed with an angle $\omega_0\tau$ after a rotation with an angle $\Omega\tau$. This confirms thus what was stated in the beginning, *viz.* that it is possible to choose the order in which the two rotations are treated, provided it is done correctly.

9.4.2.2 *Tilting the precession axis*

The most general possible motion of the electron spin implies that the precession axis itself might undergo precession around a secondary axis, etc.... The description must then be based on a hierarchical series expansion in terms of precession axes of increasing orders.

A precession around an axis $\mathbf{m}$ that is not the z-axis will now be considered. For $\mathbf{D}$ one must then take $\cos\frac{\Omega\tau}{2}\mathbb{1} - \imath\sin\frac{\Omega\tau}{2}[\,\mathbf{m}\cdot\boldsymbol{\sigma}\,]$ instead of $\cos\frac{\Omega\tau}{2}\mathbb{1} - \imath\sin\frac{\Omega\tau}{2}[\,\mathbf{e}_z\cdot\boldsymbol{\sigma}\,]$. From $\mathbf{M} = \mathbf{D}\mathbf{R}_1$, where $\mathbf{R}_1 = \mathbf{R}_0^{-1}\mathbf{C}\mathbf{R}_0$, it follows then that $\frac{d}{d\tau}\psi = (\frac{d}{d\tau}\mathbf{M})\mathbf{R}_1\psi(0) + \mathbf{M}(\frac{d}{d\tau}\mathbf{R}_1)\psi(0)$. Using the same methods as before, we find then that $\frac{d}{d\tau}\psi = -\frac{\imath}{2}(\boldsymbol{\Omega} + \boldsymbol{\omega}_0(\tau))\,\psi$. The quantities $\omega_j\mathbf{e}_j(\tau)$ add up like vectors, but it must be taken into account in the vector additions that the lower-hierarchy rotation axes are co-moving.

Let us calculate the effect of rotating about $\mathbf{m}$. The result of rotating $\mathbf{s}\cdot\boldsymbol{\sigma}$ will then be:

$$\mathbf{s}'(\tau)\cdot\boldsymbol{\sigma} = \left[\cos\frac{\Omega\tau}{2}\mathbb{1} - \imath\sin\frac{\Omega\tau}{2}[\mathbf{m}\cdot\boldsymbol{\sigma}]\right][\mathbf{s}\cdot\boldsymbol{\sigma}]\left[\cos\frac{\Omega\tau}{2}\mathbb{1} + \imath\sin\frac{\Omega\tau}{2}\,[\,\mathbf{m}\cdot\boldsymbol{\sigma}]\right]. \tag{9.39}$$

The calculation yields:

$$\mathbf{s}'(\tau)\cdot\boldsymbol{\sigma} = \cos\Omega\tau\,[\,\mathbf{s}_\perp\cdot\boldsymbol{\sigma}\,] + [\,\mathbf{s}_\parallel\cdot\boldsymbol{\sigma}\,] + \sin\Omega\tau\,[\,(\mathbf{m}\wedge\mathbf{s})\cdot\boldsymbol{\sigma}\,]. \tag{9.40}$$

Here, $\mathbf{m}\wedge\mathbf{s} = \mathbf{m}\wedge\mathbf{s}_\perp$, which is an understandable result. The part $\mathbf{s}_\parallel$ of $\mathbf{s}$ that is parallel to the rotation axis $\mathbf{m}$ remains unaltered. The part that is perpendicular to $\mathbf{m}$ will turn in the plane perpendicular to $\mathbf{m}$. The vectors $\mathbf{s}_\perp$ and $\mathbf{m}\wedge\mathbf{s}$ can be used as a basis for this plane, and the pre-factors $\cos\Omega\tau$ and $\sin\Omega\tau$ are then expressing the rotation in a very natural way. We have $\mathbf{s}'_\perp(\tau) = \cos\Omega\tau\,[\,\mathbf{s}_\perp\cdot\boldsymbol{\sigma}\,] + \sin\Omega\tau\,[\,(\mathbf{m}\wedge\mathbf{s}_\perp)\cdot\boldsymbol{\sigma}\,]$.

9.5 Solution to the conceptual problems associated with misaligned spin

9.5.1 *Preamble*

There is a whole set of arguments to show that the conceptual problems that arise in the traditional approach to the physics of a single particle with a misaligned spin axis in a magnetic field are simply due to improper use of the mathematics.

It has already been noted that the equations of quantum mechanics are obtained from equations for single particles by replacing the single-particle wave functions by rays. Without the replacement, it is impossible to derive the equations of quantum mechanics. The equations for single-particle behaviour and for sets are therefore fundamentally different, and the problem of a single particle with a misaligned spin cannot be treated using the traditional equations. This point will be further stressed below.

The calculations from Section 9.4 now also show that an equation that ultimately will describe precession for a single particle cannot be derived by starting from the equation $\frac{d}{d\tau}\psi = -\imath\frac{\omega_0}{2}\,[\,\mathbf{s}{\cdot}\boldsymbol{\sigma}\,]\,\psi$, because it is easy to verify that the latter equation is built on the *ansatz* $\frac{d\mathbf{s}}{d\tau} = 0$, which emphatically assumes that there is no precession.

It is necessary to introduce here a remark about the minimal substitution. The minimal substitution can also be applied to the single-particle equation. But it was noted in Chapter 8 that the minimal substitution introduced is based on simplifications. The angles α and χ were ignored and it was assumed that $\mathbf{s}$ is perpendicular to the orbital plane, such that it is parallel to $\mathbf{e}_z$. If ever the electron made a small motion within some orbital plane, then this plane would have to be perpendicular to $\mathbf{B}$. The assumption $\mathbf{s} \not\parallel \mathbf{e}_z$ used above would then be at variance with the simplifying assumptions introduced to obtain the minimal substitution, such that the derivation of the minimal substitution would have to be revised. The discussion in Chapter 8 showed that the true wave function must be of the form:

$$\begin{aligned}\psi(\mathbf{r},t) &= \mathbf{C}_R(\mathbf{r},t)\phi_B(\mathbf{r},t)\\ &= \begin{pmatrix}\sqrt{\frac{\gamma+1}{2}}\,e^{-\imath\chi/2} & \\ & -\sqrt{\frac{\gamma-1}{2}}\,e^{\imath\alpha}\,e^{\imath\chi/2}\end{pmatrix}\phi_B(\mathbf{r},t),\end{aligned} \tag{9.41}$$

where it is only the part ϕ_B that satisfies the equation with the minimal substitution. The quantities γ, α, and χ can here be space-time dependent. The correction factor $\mathbf{C}_R$ is purely kinematic and defined by the orbit. But it will lead to a correction of the mass, such that the orbit must be calculated again. This is an iterative procedure. If it converges, a self-consistent orbit will be obtained. This line of approach will not be developed any further here because it leads to very complicated equations. It can be argued that these complicated equations can be avoided by just considering a particle at rest.

In the following approach to the problem of the energy for a single particle, the analysis will be based on a description of precession in its own right, without invoking the presence of a magnetic field to explain why the precession is there. The hope behind this approach is to find a description whereby states that would not have a constant energy can be avoided. For a single particle this is not the only riddle to be solved; it must also be understood why the spin states are quantized.

It will be shown that precession in its own right does not violate the law of conservation of energy. This can be shown by studying precession in the absence of a magnetic field. In such a description that accounts for precession: (a) the energy levels will no longer be quantized, and (b) the law of conservation of energy will not be violated, provided the operator $-\frac{\hbar}{\imath}\frac{\partial}{\partial t}\mathbb{1}$ is interpreted correctly.

The point is then that precession can lead to a well-defined energy in free space. It is not possible at this juncture to make a statement about what happens within a constant magnetic field, because this requires the use of different, more complicated equations.

9.5.2 *Energy operators*

Let us go back to (5.38). This equation is correct, but the difference with the true Dirac equation is that the solutions the Dirac equation defines are sums of spinors that define states which correspond to the idea of spin. In other words, the Dirac equation describes sets, while (5.38) describes single-particle states. Our understanding of the way one can derive the Dirac equation from (5.38) follows thus the same lines as the analysis in Subsection 9.3.4 of the mixed states in terms of sets. The vector states defined in order to derive the Dirac equation from (5.38) are combining different states than those used to define spin in SU(2). They correspond to left- and right-handed representations, whereas in SU(2) they correspond to spin-up and spin-down solutions within the same representation, but the idea remains the same, *viz.* that spinor states are combined to define vector states.

To describe single-particle rather than set behaviour it is thus necessary to return to (5.38). With this modified Dirac-like equation in a rest frame, it becomes imediately obvious in SL(2,$\mathbb{C}$) that the operator $-\frac{\hbar}{\imath}\frac{\partial}{\partial t}\mathbb{1}$ should not be interpreted as $\hat{\mathrm{E}}$, as it pulls out a 2×2 matrix $m_0c^2[\, s_t(t)\mathbb{1} + \mathbf{s}(t)\cdot\boldsymbol{\sigma}\,]$ in front of ψ. It is rather $-\frac{\hbar}{\imath}[\, s_t(t)\mathbb{1} + \mathbf{s}(t)\cdot\boldsymbol{\sigma}\,]\frac{\partial}{\partial t}$ that would yield m_0c^2 as an eigenvalue. And if the length of the time-varying vector $s_t(t)\mathbb{1} + \mathbf{s}(t)\cdot\boldsymbol{\sigma}$ does not vary with time, then $-\frac{\hbar}{\imath}[\, s_t(t)\mathbb{1}+\mathbf{s}(t)\cdot\boldsymbol{\sigma}\,]\frac{\partial}{\partial t}$ will still extract a fixed energy value from ψ. In a correct single-particle formalism, the meaning of $-\frac{\hbar}{\imath}\frac{\partial}{\partial t}\mathbb{1}$ is thus not that of an energy operator.

All this is related to the weird way the physical operators are introduced in quantum mechanics. They have been obtained from a reasoning applied to the very simple situation of a plane wave within the context of the Schrödinger equation. Consequently it is assumed that it is possible to extrapolate this result to all other possible situations. It was noted that this is a questionable procedure when the definition of the angular momentum operators was discussed in Subsection 3.10.5.5. For the energy operator it is also a questionable procedure to extrapolate the definitions derived within the context of the scalar one-component Schrödinger equation to the context of the non-scalar multi-component Pauli or Dirac equations. It is now clear that the extrapolation required depends on the context wherein it is intended to be used.

This is already obvious from the Rodrigues-like equation for the spin in SU(2), which is the simplest possible equation for spin in a formalism where the eigenfunctions are no longer scalar quantities but spinors. If this equation is used as a starting point for these extrapolations, then the operator $-\frac{\hbar}{\imath}\frac{\partial}{\partial t}\mathbb{1}$ does not project out ω_0. It rather pulls out the matrix $\omega_0\,[\,\mathbf{s}\cdot\boldsymbol{\sigma}\,]$ in front of ψ. In more complicated situations, it will define even more complicated matrices, as is illustrated by (9.34) for the case of precession. It has then to be redetermined what kind of condition will ensure that there is only one frequency in the state.[3] It is after choosing to describe vector states rather than spinor states that the equation can be simplified and the operator $-\frac{\hbar}{\imath}\frac{\partial}{\partial t}\mathbb{1}$ used instead of $-\frac{\hbar}{\imath}\,[\,\mathbf{s}\cdot\boldsymbol{\sigma}\,]\frac{d}{dt}$.

9.5.3 *With a correct energy operator the law of conservation of energy is not violated*

As in (9.34) the angle θ between $\mathbf{e}_z''(\tau)$ and $\mathbf{e}_z$ remains constant, a single frequency $(\Omega^2+\omega_0^2-2\omega_0\Omega\cos\theta)^{\frac{1}{2}} = |\boldsymbol{\Omega}+\boldsymbol{\omega}_0(\tau)|$ can actually be defined, where $\cos\theta = \mathbf{e}_z\cdot\mathbf{e}_z''(\tau)$, $\boldsymbol{\Omega}=\Omega\mathbf{e}_z$, and $\boldsymbol{\omega}_0=\omega_0\mathbf{e}_z''$. When in this expression ω_0 is factorized out, the second-order term Ω^2/ω_0^2 ignored, and a Taylor series expansion $(1+x)^{\frac{1}{2}}\approx 1+\frac{1}{2}x$ in $\omega_0(1-2\frac{\Omega}{\omega_0}\cos\theta)^{\frac{1}{2}}$ made, we find then $\omega_0-\Omega\cos\theta$, wherein ω_0 corresponds to the rest energy of the electron. This result has been derived for a motion in free space, in the absence of any

[3]In the simple case of spin without precession, one should thus rather use $-\frac{\hbar}{\imath}\,[\,\mathbf{s}\cdot\boldsymbol{\sigma}\,]\frac{d}{dt}$, as the energy operator to start from in SU(2). This then also solves the problem of the negative energies in the Dirac equation. As it may take some algebra to find out what the spin axis $\mathbf{s}$ is, one can proceed by searching for the eigenvalues of $-\frac{\hbar}{\imath}\frac{\partial}{\partial t}\mathbb{1}$. The energies will then just be the absolute values of these eigenvalues, and this way it is possible to take a shortcut to the tedious algebra of calculating $\mathbf{s}$. There is thus no need for a Dirac sea or an infinite-dimensional Majorana representation to solve the problem of negative-energy states. Note that the total energy of an electron-positron pair is zero in the anti-particle interpretation of the negative-energy states, while in reality it is twice 511 keV, which is completely consistent with the approach described here.

potential. The problem that the energy would not be conserved when the axis of the spinning top is precessing because it is not constant does now no longer subsist. Concomitantly, a whole continuum of possible spin states will be found rather than just the two quantum-mechanical spin states corresponding to $\theta = k\pi$, with $k \in \mathbb{Z}$ in the absence of precession.

9.5.4 *The difference between sets and single-particle states*

Let us now discuss the use of $-\frac{\hbar}{\imath}\frac{\partial}{\partial t}\mathbb{1}$ as an energy operator in quantum mechanics for the stationary states and why it leads then to the correct results. It is possible to propose here that $-\frac{\hbar}{\imath}\frac{\partial}{\partial t}\mathbb{1}$ is a operator defined for sets, while the correct energy operator for single-particle states would rather be $-\frac{\hbar}{\imath}\,[\,\mathbf{s}(t)\cdot\boldsymbol{\sigma}\,]\,\frac{\partial}{\partial t}$. In the single-particle equation, the vector part $\mathbf{s}(t)\cdot\boldsymbol{\sigma}$ within $-\frac{\hbar}{\imath}\,[\,\mathbf{s}(t)\cdot\boldsymbol{\sigma}\,]\,\frac{\partial}{\partial t}$ can be removed by squaring.[4] The operator $-\hbar^2\frac{\partial^2}{\partial t^2}\mathbb{1}$ corresponds thus to $\hat{\mathrm{E}}^2\mathbb{1}$. The whole issue is introduced in the approximation made in going from (9.9) to (9.12). Here, $\hat{\mathrm{E}}^2\mathbb{1} = -\hbar^2\frac{\partial^2}{\partial t^2}\mathbb{1} = m_0^2c^4[\,s_t(t)\mathbb{1} + \mathbf{s}(t)\cdot\boldsymbol{\sigma}\,]^2$, which is entirely correct, is replaced by $-\frac{\hbar}{\imath}\frac{\partial}{\partial t}\mathbb{1} = \hat{\mathrm{E}}\mathbb{1}$, while $-\frac{\hbar}{\imath}\frac{\partial}{\partial t}\mathbb{1} = m_0c^2[\,s_t(t)\mathbb{1} + \mathbf{s}(t)\cdot\boldsymbol{\sigma}\,] \neq \hat{\mathrm{E}}\mathbb{1}$.

If after squaring a Dirac-like equation with a vector potential, the additional terms that combine $\mathbf{s}\cdot\boldsymbol{\sigma}$ and $\mathbf{A}\cdot\boldsymbol{\sigma}$ are ignored, then the spin axis must be forcedly aligned with the magnetic field due to the diagonal character of $\hat{\mathrm{E}}^2\mathbb{1}$ and the presence of $\mathbf{B}\cdot\boldsymbol{\sigma}$ in (9.9). But if they are not ignored, it might be necessary to consider an equation where $\mathbf{s}$ is no longer constant with time. It is perhaps from such an approach that one could obtain more insight about a situation where the rotation axis of the single-particle state is not aligned with the magnetic field.

9.5.5 *Important conclusion about the problem of a particle with a misaligned spin*

The situation can be summarized as follows:

- In the absence of a field, the state describing precession for a single particle makes sense. The law of conservation of energy is not violated and the

[4] It is assumed here that the electron is at rest and that there is no electromagnetic potential; the phenomenon of precession is considered in its own right. It is then possible to write $[\,\mathbf{s}(t)\cdot\boldsymbol{\sigma}\,]\frac{\partial}{\partial t}\psi(t) = -\imath\omega\psi(t)$. The term $[\,\mathbf{s}(t)\cdot\boldsymbol{\sigma}\,]\,[\,\frac{\partial}{\partial t}\mathbf{s}(t)\cdot\boldsymbol{\sigma}\,]\,\psi$ in the squared equation reduces to zero as $[\,\mathbf{s}(t)\cdot\boldsymbol{\sigma}\,]^2 = \mathbb{1}$. When the electromagnetic potential is introduced, there will be additional terms with respect to the traditional approach due to the presence of $[\,\mathbf{s}(t)\cdot\boldsymbol{\sigma}\,]$ in the equations.

energy contains a term $\cos\theta$ as classically expected. It simply requires using the correct energy operator to reach that conclusion.
- For a single particle in a magnetic field, the resulting mixed state does not make sense. More complex equations must be introduced to obtain a correct single-particle description.
- But for a set of particles in a magnetic field the mixed state makes sense again. The law of conservation of energy is then again not violated. For calculating the energy of such a set there is no simple energy operator; the calculations must be made using the probabilities as weighting factors in order to obtain meaningful results.

In these three different cases one must thus use three different procedures to calculate the energy. Just as for other operators, the procedure of defining the energy operator in quantum mechanics, by generalizing a result derived from the simple case of a plane wave in the context of the Schrödinger equation, is simply not justified.

Traditional treatments use a classical description with a $\cos\theta$ term for the spin of a particle in a magnetic field. This is of course incorrect, as it is in conflict with quantum mechanics. For a set of particles, one could claim that the $\cos\theta$ term is an effective quantity introduced to translate the correct description of the set as a distribution of the particle spins over the spin-up and the spin-down states in terms of an angle.

9.6 What is the coupling of the spin to the magnetic field?

It must be admitted that the expression for the interaction of the spin with the magnetic field is unknown. It was always assumed that it was given by the term that contains $\mathbf{B}\cdot\boldsymbol{\sigma}$, but this has been shown to be incorrect. *A priori*, the expression for $\mathbf{A}_{rot}$ in a constant magnetic field $\mathbf{B}$ does not need to be $\mathbf{A} = \frac{1}{2}\mathbf{B} \wedge \mathbf{r}$, because that kind of expression corresponds to a dependence on position $\mathbf{r}$. When $\mathbf{A} = \frac{1}{2}\mathbf{B} \wedge \mathbf{r}$ is used in a minimal substitution, it will be possible to describe translational (i.e. orbital) kinetics. But it could be that an entirely different type of expression $\mathbf{A}_{rot}(\mathbf{s})$ is needed for the potential in order to be able to describe the internal rotational kinetics of the spin of an electron that is translationally at rest.

Imagine that (independently from our findings in Subsection 9.3.4) the orientation of the electron spin in the magnetic field does matter and that the energy of the electron is higher when the spin is not aligned with the magnetic field. To go from the misaligned state to the aligned state, the electron must then shed energy-momentum, and this will lead to an extra term in the minimal substitution, as discussed in Section 5.6.

As there is no translational motion there is no need for a dependence of $\mathbf{A}_{rot}$ on $\mathbf{r}$, whereby it would be necessary to have $\boldsymbol{\nabla} \wedge \mathbf{A}_{rot}(\mathbf{r}) = \mathbf{B}(\mathbf{r})$. A dependence of the type $\mathbf{A}_{rot}(\mathbf{s})$ is needed, whereby $\mathbf{s}$ is a spin variable, and the relation between $\mathbf{A}_{rot}$ and $\mathbf{B}$ could be based on $\boldsymbol{\nabla}_{\mathbf{s}} = (\frac{\partial}{\partial s_x}, \frac{\partial}{\partial s_y}, \frac{\partial}{\partial s_z})$, for example, instead of $\boldsymbol{\nabla}_{\mathbf{r}} = (\frac{\partial}{\partial x}, \frac{\partial}{\partial y}, \frac{\partial}{\partial z})$. It is not even certain that this should be $\boldsymbol{\nabla}_{\mathbf{s}} \wedge \mathbf{A}_{rot}(\mathbf{s}) = \mathbf{B}_{rot}(\mathbf{s})$. All these problems transpire in a painful way when one tries to apply the expression for the magnetic potential $\mathbf{A}(\mathbf{r}) = \frac{1}{2}\mathbf{B} \wedge \mathbf{r}$ for a constant magnetic field to the spin dynamics. It is then a mystery what should be taken for $\mathbf{r}$. It is also a mystery where the orientation of the spin vector $\mathbf{s}$ comes in.

In fact, the expression $\mathbf{A} = \frac{1}{2}\mathbf{B} \wedge \mathbf{r}$ applies to the translational motion of a "point charge" within an electromagnetic field; it cannot apply to the spin of an electron at rest within an electromagnetic field. The expression $\mathbf{B}(\mathbf{r})$ express what kind of relevant field values $\mathbf{B}$ an electric point charge experiences at various positions $\mathbf{r}$ in the magnetic field. "Relevant" here means that the field values $\mathbf{B}(\mathbf{r})$ can be used to calculate a force, and a corresponding potential, which results from the coupling of the charge to the magnetic field. To express the coupling of the spin to the field, the functional dependence $\mathbf{B}(\mathbf{r})$ is not relevant; an expression $\mathbf{B}(\mathbf{s})$ is needed that describes what kind of relevant field values a spin might experience for various orientations $\mathbf{s}$ within the magnetic field.

The functional dependences $\mathbf{B}(\mathbf{r})$ and $\mathbf{B}(\mathbf{s})$ correspond to completely unrelated questions. Nevertheless, the traditional textbook approach tries to relate them, mainly based on a confusion. It is thought that the algebra of the Dirac equation with a minimal substitution designed for translational dynamics gives rise to a coupling between the spin and the magnetic field through the term $\mathbf{B}{\cdot}\boldsymbol{\sigma}$. If that were true, this would indeed describe how the spin interacts with a magnetic field and allow the rotational dynamics to be addressed. But $\mathbf{B}{\cdot}\boldsymbol{\sigma}$ is not a coupling term between the magnetic field and the spin. Such a coupling term for rotational dynamics cannot come about in the equations if it has not been entered. It cannot come about by putting the system into translational motion. It must thus be entered on the basis of a different minimal substitution that describes the coupling between the spin and the electromagnetic field. This coupling is just unknown. It must be guessed and then validated by checking that it reproduces the experimental results.

A magnetic field cannot do work on an electron during displacements $d\mathbf{r}$ in physical space, because the Lorentz force $\mathbf{F} = q\mathbf{v} \wedge \mathbf{B}$ it exerts on the electron is always perpendicular to $d\mathbf{r} = \mathbf{v}dt$. Hence, for any path Γ we have $-\int_{\Gamma} \mathbf{F}{\cdot}d\mathbf{r} = -\int_{\Gamma} q(\mathbf{v} \wedge \mathbf{B}){\cdot}\mathbf{v}dt = 0$. If there is work to consider, then it has nothing to do with displacements $d\mathbf{r} = \mathbf{v}dt$. It would be the work $-\int_{\Gamma'} \mathbf{F}(\mathbf{s}){\cdot}d\mathbf{s}$ during displacements $d\mathbf{s}$ on a path Γ' in spin space. But the key point is that it creates confusion to search for a magnetic potential in terms of a scalar (i.e. an energy term), as the magnetic potential is a vector (i.e. a momentum term). Trying to impose a scalar violates the symmetry and creates a logical contradiction. The final conclusion of this is that we must have a magnetic potential $\mathbf{A}_{rot}(\mathbf{s})$ that depends on $\mathbf{s}$ rather than a potential $\mathbf{A}(\mathbf{r})$ that would depend on $\mathbf{r}$ and that it must be a vector quantity.

As already mentioned, any vector operator in the wave equation will force the spin into alignment. This follows from the calculation of the eigenvalues of a vector operator as explained in Subsection 9.3.1, but it can also be seen as follows. Let us consider a vector $\mathbf{w}$ that gives rise to a term $[\,\mathbf{w}{\cdot}\boldsymbol{\sigma}\,]\,\psi$ in the equation. For a particle at rest, the spin dynamics can be described within SU(2) such that $[\,\mathbf{s}{\cdot}\boldsymbol{\sigma}\,]\psi = \psi$. When the term $[\,\mathbf{w}{\cdot}\boldsymbol{\sigma}\,]\,\psi$ is replaced by $[\,\mathbf{w}{\cdot}\boldsymbol{\sigma}\,]\,[\,\mathbf{s}{\cdot}\boldsymbol{\sigma}\,]\psi$, a term $[\,\mathbf{w}\cdot\mathbf{s}\,]\,\psi$ will be obtained, but also an imaginary term $\imath[\,(\mathbf{w} \wedge \mathbf{s}){\cdot}\boldsymbol{\sigma}\,]\,\psi$. It is only possible to dismiss this imaginary term by assuming $\mathbf{s} \parallel \mathbf{w}$ or by introducing another term that compensates for it. This is just another way to come to the same conclusion that the only meaningful stationary states are those where the spin is aligned with the vector $\mathbf{w}$.

This presentation opens up the possibility for speculations about the case $\mathbf{w} = \mathbf{B}$. Imagine more complicated equations than the Dirac equation were drawn into the considerations, for instance to account for the possibility $\frac{d\mathbf{s}}{dt} \neq 0$. Could one of the additional terms that would occur in the equations then perhaps be used to compensate for the imaginary term $\imath[\,(\mathbf{B} \wedge \mathbf{s}){\cdot}\boldsymbol{\sigma}\,]\,\psi$? Would it then be possible to recover the intuition that a term $\mathbf{B}{\cdot}\boldsymbol{\mu}$ based on $\mathbf{B}\cdot\mathbf{s}$ corresponds to the potential energy for the spin in a magnetic field and that the electron has a magnetic dipole moment $\boldsymbol{\mu}$ associated with its spin $\mathbf{s}$? Only performing the calculations can tell. But a potential positive outcome for these questions will nevertheless have to be assessed in the light of the remark about the minimal substitution at the end of Section 9.2.3. Such a positive outcome would imply new insight, because it seems at face value very difficult to validate the intuition about the magnetic dipole moment of the electron by a rigorous approach based on an equation derived for a point charge. It may in this respect be mentioned how Heisenberg's exchange mechanism is able to explain ferromagnetism purely in terms of the Coulomb interaction and wave function overlap.

Chapter 10

The Double-Slit Experiment and the Superposition Principle

10.1 An approach inspired by the treatment of the Aharonov-Bohm effect

The superposition principle is a genuine problem, as adding up spinors does not make sense. In reality it is part of a second construction that is layered on top of group theory. In fact, the Dirac or Schrödinger equation will, in certain cases, yield more than one solution. Due to the linearity of the equations, these solutions build a vector space, which has a basis. The basis functions ψ_j are orthogonal according to a definition of a scalar product. If these basis functions are given a weight of $c_j = \sqrt{p_j}$, where p_j corresponds to the probability of this basis function, one can build a self-consistent calculus on the mixed states $\sum_j c_j\psi_j$. These mixed states do not have any physical meaning in group theory, but they make sense in the probability calculus that we add on top of it. In other words, quantum mechanics is a combination of two formalisms: the first is group theory, while the second uses the spinors that come out of group theory as elements of an abstract vector space which allow a probability calculus to be performed on sets. The linearity of the Dirac equation is thus not used to prove the superposition principle. It is rather that the linearity permits the introduction of the second construction. This was clearly illustrated in the discussions regarding the spin within a magnetic field **B**, when trying to assume that the spin axis **s** is no longer parallel with **B**. The probability calculus is further justified by the continuity equations that can be derived from the Dirac and Schrödinger equations.

To make this argument plausible, it will now be shown that there might be a way to come to the same conclusions as obtained with the superposition principle, but without invoking it. It will be attempted to illustrate this on the example of the famous double-slit experiment. Here again it will be shown that the counterintuitive result can be obtained by postulating that the wave function should be a function. This does not correspond to a real property of a wave function, but rather to a method of cataloguing them. Such an approach, which does not rely on the superposition principle, is also important within a strictly traditional quantum mechanical context because (as noted in Section 6.3) the textbook treatment of the double-slit experiment uses the superposition principle in an awkward way. Perhaps a correct treatment should rather be based on invoking a Huyghens' principle for the solutions of the wave equations. The double-slit experiment reflects a property of the global structure of a single-valued wave function defined by a wave equation rather than a consequence of the superposition principle.

Imagine that we have a wave function for the double-slit experiment. Particles that travel across the slits are for most of the time travelling in free space. An interaction with the matter of the slits might bend their paths. Let us assume that this only changes the direction of their momentum $\mathbf{p}$, but not its absolute value p, nor their total energy E. The description is simplified as follows. Two points are considered in the plane of the planar device that contains the two slits: point S_1 with position vector $\mathbf{r}_1$ in slit 1, and point S_2 with position vector $\mathbf{r}_2$ in slit 2. The source is at a point A. There is also a point B with position vector $\mathbf{r}$ far behind the double-slit. This is shown in Figure 10.1.

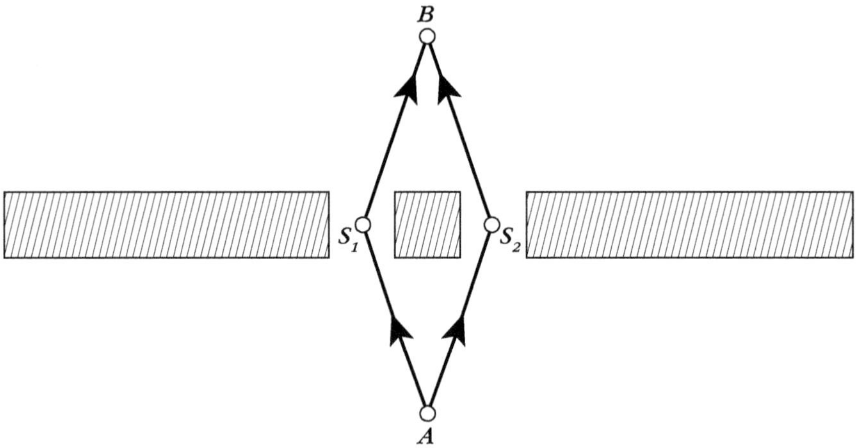

Fig. 10.1 Drawing of the double-slit experiment, showing the notations used in the text.

It is considered that the possible interactions are just an instantaneous kick at time t_0 that changes the momentum from the initial momentum $\mathbf{p}_0$ to $\mathbf{p}_1$ if the particle passes through slit 1, or an an instantaneous kick at time t_0 that changes the momentum from the initial momentum $\mathbf{p}_0$ to $\mathbf{p}_2$ if the particle passes through slit 2. As this addresses the possible interactions in $\mathbf{r}_1$ and $\mathbf{r}_2$, it can now be considered that in all other points of space the particles are travelling in free space. That means that if the particle arrives at $\mathbf{r}$, coming from $\mathbf{r}_1$, its momentum $\mathbf{p}_1$ will have been directed along $\mathbf{r}-\mathbf{r}_1$. Similarly, the particle's momentum momentum $\mathbf{p}_2$ will be directed along $\mathbf{r}-\mathbf{r}_2$ in the case that the particle passes through slit 2.

Consider now the simplified description of the wave function where the whole coding of the tetrad is reduced to a single phase factor, $e^{\frac{\imath}{\hbar}(Et-\mathbf{p}\cdot\mathbf{r})}$. The particle thus has a phase, *viz.* the orientation of its tetrad, while it is not necessary to invoke the presence of a wave behaviour.

As explained by Feynman, the solutions of the two-slit experiment become different when trying to establish through which slit the particle has passed, for example by shining light on the particle at the level of a slit. If the wave function (that corresponds to the true two-slit experiment without a measuring device that tries to establish through which slit the particle has passed) describes the probabilities correctly, it must be acknowledged that it is really not known through which one of the two slits the particle has passed. The probabilities must reflect this lack of knowledge. That means that the wave function must be such that it is not knowable through which slit the particle has passed. This implies that the hypothetical values obtained from the wave function generated by the particle if it actually goes through slit 1 must be such that they are identical to those that would be generated by the particle if it went through slit 2.

Before the slits, this leads to a condition $c_1 e^{\frac{\imath}{\hbar}(\mathbf{p}_0\cdot\mathbf{r}_1-Et_0)} = c_2 e^{\frac{\imath}{\hbar}(\mathbf{p}_0\cdot\mathbf{r}_2-Et_0)}$. Let us simplify by assuming that $c_1 = c_2$. One can indeed make the idealizing assumption that the particle will arrive at the same time at $\mathbf{r}_1$ as it would arrive at $\mathbf{r}_2$. When the source is far away, the two alternative incoming paths can be considered as parallel (see Figure 10.2). It is thus considered that they start from different points A_1 and A_2 in a plane that is parallel to the plane of the slits, such that the vectors $\mathbf{A}_1\mathbf{S}_1$ and $\mathbf{A}_2\mathbf{S}_2$ are equipollent and perpendicular to the plane of the slits. Hence, the real starting point A, has been replaced by two starting points A_1 and A_2. The second condition becomes then:

$$e^{\frac{\imath}{\hbar}[\mathbf{p}_1\cdot(\mathbf{r}-\mathbf{r}_1)-E(t-t_0)]} = e^{\frac{\imath}{\hbar}[\mathbf{p}_2\cdot(\mathbf{r}-\mathbf{r}_2)-E(t-t_0)]}, \tag{10.1}$$

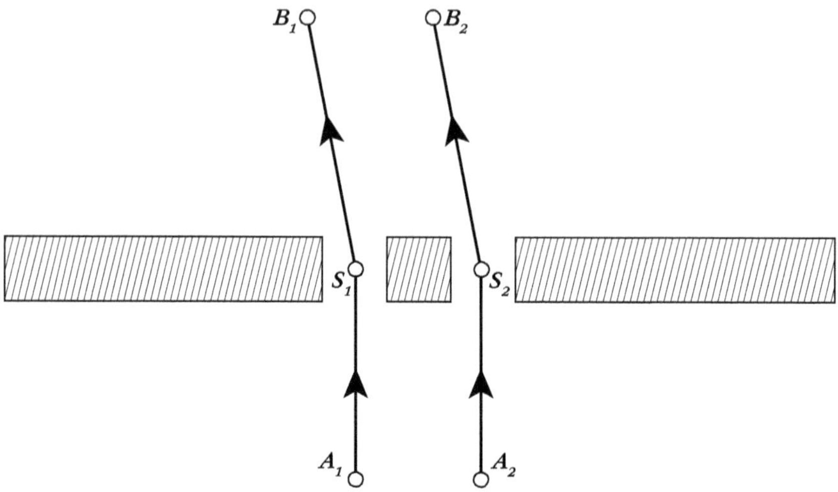

Fig. 10.2 Drawing of the double-slit experiment with the points A and B very far from the slits, such that $AS_1 \parallel AS_2$ and $S_1B \parallel S_2B$. The point A is replaced by two points A_1 and A_2, and the same procedure is applied *mutatits mutandis* to B.

which explicitly expresses that the wave function must be single-valued. The condition leads to:

$$\sin((\mathbf{p}_1\cdot(\mathbf{r}-\mathbf{r}_1)-\mathbf{p}_2\cdot(\mathbf{r}-\mathbf{r}_2))/2\hbar)=0. \qquad (10.2)$$

If $\mathbf{r}$ is taken very far behind the slits, it is again possible to idealize and assume that $\mathbf{p}_1=\mathbf{p}_2$. The end point B can then also be replaced by two end points B_1 and B_2 in the plane perpendicular to $\mathbf{p}_1=\mathbf{p}_2$. Taking all this into account, we find $p(|S_1B_1|-|S_2B_2|)=nh$, where the quantity between the parentheses is the difference in path length between the two possible paths. This leads to a quantization of the directions $\mathbf{p}_1=\mathbf{p}_2$, in agreement with the result of the two-slit experiment. This is all not very rigorous, but it gives an idea of what could be happening. The derivation is rather similar to the one used in the superposition principle in terms of constructive interference. However, the argument does not allow the calculation of the detailed probability distribution in the part of space between the slits and the detector.

The argument could also be reformulated using the path integral:

$$\oint(\mathbf{p}\cdot d\mathbf{r}-Edt)=nh \qquad (10.3)$$

on the closed loop $A_1\ S_1\ B_1\ B_2\ S_2\ A_2\ A_1$, where $\mathbf{p}$ can always be considered as parallel to $d\mathbf{r}$, and the loop as a path through the wave function for which $p = |\mathbf{p}|$ is constant, and $\oint E dt = 0$. The latter is due to the fact that one half of the loop takes us backwards in time. The loop is thus a purely mathematical construction. But on such a closed loop the argument that the wave function must be single-valued can be linked to the one used in the hydrogen atom: to get the equation of motion right, the right rest mass must be used. If one varies the rotational frequency of the particle, its rest mass will change. The time evolution of the phase of the wave function (which keeps track of the frequency of the internal motion) must therefore be coupled to the dynamics of the orbital motion to get the equation of motion right. If one lets the particle spin at will in a decoupled way within the calculations, these will get its mass wrong. The system is described in terms of a ratio r between periodicities on a closed loop. Every ratio corresponds to a dimension of the corresponding representation in group theory. Only within a given well-chosen representation will the corresponding wave function truly be a function. Poorly commensurate ratios, i.e. ratios $r = n/m \in \mathbb{Q}$, where n and m are large, correspond to the limit of large quantum numbers. It will then seem paradoxical to postulate that the wave function is a function, as it is not. In the limit of large quantum numbers, the familiar continuum states will be obtained. But for small quantum numbers it will emerge that there are gaps in the set of orbits. It will now be shown that in the double-slit experiment, the analogue of the gaps will produce a diffraction pattern.

In a single-slit experiment one might also consider such a loop, and also a value of n. If the idea is now generalized to the point that the loop does not need to consist of straight lines, A_1 and A_2 are merged back into the original position A, and B_1 and B_2 are merged back into the original position B, then we obtain a condition: $\oint pdr = nh$. By considering another possible end point B' and invoking the postulate that the wave function is single-valued, we will find $\oint pdr = n'h$, where it is conceivable that $n' \neq n$. However, using a point B' that is infinitesimally close to B, the continuity of the wave function can be used to prove that $n' = n$. By adding up a huge quantity of such infinitesimals, the loop can be shrunken gradually to zero to show that $n = 0$. In the double-slit, however, it is not possible to shrink the loop to zero when its two branches go through different slits.

In this argument, the hypothetical loophole that the amplitude of the wave function could become zero has been ignored, but with functions of the type $ce^{\frac{\imath}{\hbar}(Et - \mathbf{p}\cdot\mathbf{r})}$, this will only happen if $c = 0$. This suggests thus that in analogy with the problem of the hydrogen atom, there is not one wave function in the double-slit experiment, but several wave functions with different values of n. Each interference fringe corresponds to one wave function. The grounds for the quantization are also topological, *viz.* that the space available to the particle is not simply connected. There seems to be some quantization of angular momentum at work in the double-slit experiment as well. The angular momentum is introduced by the kicks given to the particle by the potential. The reader will have noticed that this approach merely copies the idea behind the Aharonov-Bohm experiment.[1] This is perhaps a way to explain the double-slit experiment without using the superposition principle. The difficulty of the superposition principle is not only that it is hard to understand physically, but also that it is mathematically ill-defined, as in general spinors cannot be added.

Before the slits, the wave function should be a free-space wave function, with different clock readings in different places when $\mathbf{v} \neq \mathbf{0}$, as discussed in Section 6.1. After the slits, this should be the case as well. To obtain a wave function for the double-slit experiment, these "before" and "after" solutions should be glued together in a self-consistent way. This means that the phase, which represents the clock reading on the internal watch of the particle should not make any jump at any point of space, rendering the wave-function two-valued at some point, because such a jump would correspond to inelastic scattering, a channel that might not be available to a particle with low kinetic energy or that might have a low probability. This implies that not all solutions after the slits can be glued to a solution before the slit. Only solutions after the slits that represent the correct phase relation between the clock readings in the points S_1 and S_2 corresponding to the slits can be glued, such that this leads to a kind of quantization of momentum for the solutions after the slits. This way the argument becomes one about a global property of the solution of a wave equation. It is in the procedure of extending the wave-function to the whole of space-time that one must compel it to be single-valued. The extension can then be interpreted in terms of a probability calculus that is self-consistent over the whole of space-time, and obeys a continuity equation. Even if this approach may not solve all the problems, this appears to be a better way to think about the double-slit experiment than in terms of interference, which has revealed itself historically to be a conceptual stalemate. The issue in the

[1] Note that the Aharonov-Bohm effect is in reality not generally valid. The true quantization condition is that $\oint V \cdot cdt - \mathbf{A} \cdot d\mathbf{r}$ should be quantized. The Aharonov-Bohm effect is only a limiting case of this condition. The Aharonov-Casher experiment is the other limiting case of the same condition.

gluing procedure is really that the wave function should be a function. The heuristic nature of such an *"ansatz"* was discussed in Subsection 6.2.12.

The procedure described here has eliminated the mathematical difficulty that spinors cannot be added. This mathematical difficulty goes hand in hand with the conceptual difficulty of the superposition principle, which leads to a wave-particle duality. Moreover, both difficulties have been reduced to a single issue, *viz.* that the wave function must be single-valued, a point that is understandable now. This is also the key point in the argument here, such that the phase changes accumulated over two different paths through the wave function must be identical. This condition rather belongs to a kind of heuristics that enables one to constitute the whole catalogue of different wave functions that belong to different representation formalisms, rather than being a true property. Here, the heuristics facilitate the discovery of several different wave functions with a different topology. In the hydrogen atom, the heuristic argument that the wave function must be single-valued renders a number of classical orbits impossible, which causes the energy levels to be discrete. A similar elimination of certain orbits could take place here.

By using the superposition principle wave functions are constructed that have the left-right symmetry of the experiment (which is the only symmetry present in the problem). This is mathematically not rigorous but it might be good enough an approximation to the real wave function to reproduce the essentials of the physics. The argument of constructive and destructive interference in the superposition principle in the double-slit experiment reproduces the very same argument that the number of turns the electron makes along the path difference must be an integer, in asking that the wave function should be single-valued.

In certain cases, the superposition principle can be rendered more rigorous by the orthogonality between the wave functions. In fact, when one introduces the group ring for a group, the group elements are considered to be orthogonal in some vector space. The wave functions of quantum mechanics are in general orthogonal with respect to a scalar product based on an integration over space. This scalar product could be taken as the scalar product for the group ring that warrants the orthogonality of the group elements. With a scalar product, $< \sum_j c_j\psi_j, \sum_j c_j\psi_j >$ would be equal to $\sum_j |c_j|^2 < \psi_j, \psi_j >$, due to the orthogonality, i.e. the cross terms would automatically reduce to zero, and there would be no real interference. This would certainly work for a basis of plane waves in free space. The problem is that the scalar product corresponds to an integration over space. It can thus be used when the desired result is an integral of this type. The integration can then also be used to determine the weighting factors. But the integration over space is not satisfactory for explaining interference patterns, as these still have a spatial pattern. The superposition principle is then rather a kind of shortcut that reproduces the same results as the method that does not invoke it.

There is also another very important lesson to be learned from this approach. The physical problem must topologically correspond to a space with a hole to become "quantized". This also plays a role in the "quantization" of the energy levels in the hydrogen atom. If the orbits were not planar, the nucleus would be a singularity that could be avoided if one wanted to reduce the orbit continuously to a point. But when the motion is planar, the singularity cannot be circumvented. Something similar might also be the case in the theory of superconductivity.

It was proposed within the frame box of Section 5.2 that the quantum effects are introduced in the formalism by unwittingly transgressing the limits of validity of the classical framework wherein the Dirac equation is derived in this book. Here are a few examples, where exact results are obtained by brute-force calculations without caring about what they mean and what these limits are.

(1) The wave function is defined over whole space-time rather than on a classical orbit. Sometimes this extends the definition domain to classically forbidden regions, e.g. in quantum tunnelling.
(2) Contrary to classical intuition, it is stipulated that the wave function must be a function. This leads e.g. to the quantization of the energy levels in the hydrogen atom and the interferences in the double-slit experiment. It is not obvious that the rationale proposed to justify this assumption applies to all possible situations.
(3) The superposition principle is not justified geometrically. Introducing it leads e.g. to the paradox of Schrödinger's cat. Superposition can only be understood by assuming it describes probability distributions for statistical ensembles of particles.
(4) Negative energies have been interpreted in terms of anti-particles. They could have been interpreted differently, but the interpretation accounts for the existence of anti-particles.

Obtaining the correct answers without also providing a justification for the transgressions could be considered as just a fluke. Points (1)–(3) seem to be related to the way one has to define (relativistically and non-locally) probability charge-current densities in a self-consistent way. Here the conceptual difficulties arise if one assumes that the waves are describing single particles rather than probability distributions for particles. Experiments are *always* measuring probability distributions since data are *always* obtained by statistical sampling. Point (4) is further discussed in Chapter 11.

One point not touched upon until now in this book is second quantization. As the operators of quantum mechanics correspond to vector quantities and vectors are rank-two tensors in spinor quantities, it is natural to try to express these operators as rank-two quantities of what turn out to be creation and annihilation operators. Second quantization is just a spinorization of the operators that correspond to vector quantities.

Chapter 11

A Caveat About the Limitations of Group Theory

11.1 A same group-theoretical formalism can represent several, very different physical mechanisms

It has been seen how a treatment in terms of symmetry is a powerful aid in physical calculations, but there are some limitations inherent to this method. First in all, it can be illustrated from a simple example in solid-state physics that symmetry does not allow addressing questions about the underlying physical mechanism that might be at work. This will be done using two models described in Figures 2.6 and 2.7 of Chapter 2. It was seen there that in the two different problems the same matrix $\mathbf{M}$ occurs. This is because the symmetry of the two problems is the same. In both cases there is translational symmetry with cyclic boundary conditions, and the number of sites is also the same. The eigenvectors are therefore also identical. The fact that the eigenvectors are wave-like and of the type $(\mathbf{V}^{(k)})_j = e^{\imath\frac{2\pi(j-1)(k-1)}{n}}$ corresponds to the so-called Bloch theorem in solid-state physics, which is based on the fact that a crystal lattice has translational symmetry.[1]

[1]The use of cyclic boundary conditions does not correspond exactly to the physical reality. That this is not a genuine obstacle can be seen from the way the jump problem on a straight line without cyclic boundary conditions can be solved. Only in positions 1 and n will the equations for the probabilities q_j change: $\frac{dp_1}{dt} = -p_1 + p_2$ and $\frac{dp_n}{dt} = +p_{n-1} - p_n$. The problem can be solved by solving the problem with $2n$ sites and cyclic boundary conditions. Note the intervening probabilities q_j with $j \in [1, 2n] \cap \mathbb{N}$. The sites P_{2n+1-j} can thereby be considered as the mirror sites of the sites P_j. It is then easily verified that $p_j = q_j + q_{2n+1-j}$ solves the problem with n sites on a straight line without cyclic boundary conditions. This approach leads to the same phonon dispersion curves as the solution for n sites with cyclic boundary conditions.

Waves are the eigenvectors of problems with translational symmetry. (As already noted, for de Broglie waves the symmetry is with respect to translations in time.) But the underlying physical mechanism in the two problems, *viz.* lattice diffusion and lattice vibrations, is entirely different. Hence, the symmetry treatment cannot offer a clue as to the underlying mechanism, this information is not present in the wave function. The wave function contains only information about the symmetry. In as far as quantum mechanics is purely based on group theory and has been proved to be able to predict the outcome of all experimental results, it seems that it will forever hide the underlying mechanism. At least, it is impossible to search for this mechanism within the formalism. Einstein was right to say that quantum mechanics is incomplete in the sense that the symmetry description is not exhaustive and does not specify what the underlying mechanism could be. But the experimental results never offer a clue as to an underlying mechanism, and quantum mechanics is thus complete in as far as it boils down to reproducing all experimental results. The Copenhagen interpretation is however an absurd, irrational way of defending the latter viewpoint.

Two situations where the mechanism remains elusive are the double-slit experiment and tunnelling in solid-state physics. In both situations the apparent paradox may come from an over-simplified description of the problem in terms of of an idealized potential. For example, in the case of the double-slit experiment, an incoming electron or photon may have an electro-magnetic interaction with the electrons of the material of the double-slit and the details of this are not incorporated into the symmetry argument. A polarization of the material of the slit could indeed imply that an electron detects whether there is a second slit or otherwise, because it could lead to different induced charge distributions. In the case of electron tunnelling in solid-state physics, the electron will certainly not see a truly flat potential. These details are not necessary to describe the symmetry, but they can be essential to understanding the mechanism. The mechanism does not have to be universal and might require a case by case discussion.

There is another point in quantum mechanics where this absence of certain details in the description can be seen at work. It has often been stressed in this book that one should not over-interpret the presence of negative energies in the formalism. Dirac did not originally predict anti-particles on the basis of his equation. It was realized later on that the Dirac theory could be used to incorporate the description of anti-particles, provided certain conventions were accepted, *viz.* that the substitution $\omega | -\omega$ was applied to the wave function that describes the particle in order to obtain the wave function for the anti-particle. This leads to the notion that positrons have negative frequencies ω and negative energies $\hbar\omega$. Dirac called it the biggest error of his life not to have predicted the existence of the positron in his first paper. It was not an error. Based on the derivation of the Dirac equation proposed in this book, it cannot be claimed to predict the existence of anti-particles. The Dirac equation is however able to *accommodate* for the existence of anti-particles, provided they are mapped mathematically to the reversed-frequency solutions. This is just a mapping and should not be taken too seriously. The Dirac formalisms for particles and for anti-particles can be considered as two different physical models for the same symmetry theory, just as the phonon and the diffusion models for the matrix $\mathbf{M}$ in (2.21) and (2.24) are two different realizations of translational symmetry. Problems arise, however, if two such different models are used simultaneously, and the phenomena are not kept separate. Diffusion and phonons are definitely two different phenomena. Such problems also arise in the Dirac theory if it is applied to particles and anti-particles simultaneously. As the sum of the energies of an electron and a positron is not zero but twice 511 keV, it becomes a real problem. Hence, describing anti-particles as particles with reversed frequencies will cause problems if a calculation has to be made that involves both particles and anti-particles at the same time. It will also cause confusion because a wave obtained by a substitution $\omega| -\omega$ that describes a clockwise rotating particle can then be confused with an anti-particle. It is also easy to mistakenly interpret negative-frequency waves as "advanced" waves, as the frequency ω and the time t always occur in the combination ωt. When the product is negative, one can draw the wrong conclusion that it is the time which is negative and that the wave is "advanced". In the end, this leads to the picture described by Feynman that anti-particles are travelling backwards in time.

This should in no way be seen as an attempt to belittle Dirac's towering contribution to physics, which has been in all aspects crucial. It is merely the intention to show how due to the absence of the details about the mechanism, the same formalism can cover several different realities. With respect to the problem of particles and anti-particles, the formalism is able to cover the two different realizations. One realization is the particles, the other is the anti-particles. In a sense, the absence of details about the precise mechanism can be a blessing, because it does not confine us to a single theory. This has permitted us to detect mechanisms that do not belong to a single framework. Full-fledged quantum mechanics contains applications that transcend the framework of the initial derivation of the

Dirac equation based on the Rodrigues equation for a rotating electron. The absence of a complete understanding of the mathematics has enabled the transgression of the logical borderlines into new fields with new applications that would have been very hard to predict. It has encouraged unjustified extrapolations that nevertheless worked. It has also facilitated the discovery of anti-particles and tunnelling, because these phenomena can be described by the same symmetry arguments within an extended setting. The theory of symmetry is therefore an ideal tool for scientific research, as it can fool us into making leaps that are logical errors in one given model but not in another. On the basis of these remarks, it seems thus that the history of quantum mechanics has rather followed the description of science given by Paul Feyerabend [Feyerabend (1975)] than the one given by Thomas Kuhn [Kuhn (1996)].

Chapter 12

Spin and Angular Momentum as Vector and Bi-Vector Concepts

12.1 Lorentz covariance of bi-vectors

The example of the energy operator in Chapter 9 showed that the definition of operators in quantum mechanics by straightforward extrapolation should be considered with circumspection. A similar remark was formulated for the angular-momentum operators in Chapter 3. Returning to this issue, it will be noted that relativistic angular momentum is no longer a vector but a bi-vector.

A typical example of a bi-vector is the electro-magnetic field. Lorentz symmetry does not only apply to spinors and vectors, but also to bi-vectors, tri-vectors, and so on. The same is in fact true for rotational symmetry. However, due to the fact that the vector product of two vectors is again a three-component quantity in the rotation group, it is easy to confuse bi-vectors in the rotation group with vectors. Such a confusion will not occur within the Lorentz group. In fact, vectors have four components, while bi-vectors are six-component tensors. With some algebra, one can check from the transformation properties of the electromagnetic field bi-vector $(\mathbf{E}, \mathbf{B})$ that it transforms according to the rule that $E_x^2 + E_y^2 + E_z^2 - c^2 B_x^2 - c^2 B_y^2 - c^2 B_z^2$ must remain a constant. These symmetry arguments are valid for any bi-vector $(b_t \mathbb{1} - \mathbf{b}{\cdot}\boldsymbol{\sigma})\,(a_t \mathbb{1} + \mathbf{a}{\cdot}\boldsymbol{\sigma})$. In this respect, the bi-vector character of the electro-magnetic field corresponds to a rather special combination of four-vectors, *viz.* the four-gradient and the four-potential $(\frac{\partial}{\partial ct}\mathbb{1} - \boldsymbol{\sigma}{\cdot}\boldsymbol{\nabla})\,(V\mathbb{1} + \mathbf{A}{\cdot}\boldsymbol{\sigma})$ as explained in Appendix C.

The same logic could equally be applied to the bi-vector combining space-time and energy-momentum $(ct\mathbb{1} - \mathbf{r}{\cdot}\boldsymbol{\sigma})\,(E\mathbb{1} + c\mathbf{p}{\cdot}\boldsymbol{\sigma}) = (ctE - c^2\mathbf{p}\cdot\mathbf{r})\mathbb{1} + (c^2t\mathbf{p} - E\mathbf{r}){\cdot}\boldsymbol{\sigma} - \imath(\mathbf{r}\wedge c\mathbf{p}){\cdot}\boldsymbol{\sigma}$. Here, $(ct\mathbf{p} - E\mathbf{r}){\cdot}(\mathbf{r}\wedge c\mathbf{p}) = 0$ has been used, because the vector product is perpendicular to both its constituants.

The scalar part is proportional to the relativistic invariant $(Et - \mathbf{p}\cdot\mathbf{r})$. The length of the vector part is also a relativistic invariant. In fact, it can be calculated that $(c^2t\mathbf{p}-E\mathbf{r})^2-(\mathbf{r}\wedge c\mathbf{p})^2 = (Ect-c\mathbf{p}\cdot\mathbf{r})^2-(E^2-c^2p^2)(c^2t^2-r^2)$.

After even more algebra, something analogous can be derived for the operator:

$$(ct\mathbb{1} - \mathbf{r}\cdot\boldsymbol{\sigma})\left(\frac{\partial}{\partial ct} + \boldsymbol{\sigma}\cdot\boldsymbol{\nabla}\right)$$
$$= \left(ct\frac{\partial}{\partial ct} - \mathbf{r}\cdot\boldsymbol{\nabla}\right)\mathbb{1} + \left(ct\boldsymbol{\nabla} - \mathbf{r}\frac{\partial}{\partial ct}\right)\cdot\boldsymbol{\sigma} - \imath(\mathbf{r}\wedge\boldsymbol{\nabla})\cdot\boldsymbol{\sigma}. \qquad (12.1)$$

If one applies the operator $(ct\frac{\partial}{\partial ct} - \mathbf{r}\cdot\boldsymbol{\nabla})$ to the phase factor $e^{\imath(\omega t-\mathbf{k}\cdot\mathbf{r})}$, one obtains $\imath(\omega t - \mathbf{k}\cdot\mathbf{r})\,e^{\imath(\omega t-\mathbf{k}\cdot\mathbf{r})}$, such that it could be considered as the phase operator. The vector part is proportional to a generalized angular momentum operator. The three conventional angular momentum operators $\hat{\mathrm{L}}_{yz}$ that up to a normalization factor correspond to $y\frac{\partial}{\partial z} - z\frac{\partial}{\partial y}$ are complemented by three new operators of the type $\hat{\mathrm{L}}_{xt}$ that (up to the same normalization factor) correspond to $x\frac{\partial}{\partial ct} - ct\frac{\partial}{\partial x}$. By exerting this operator on $e^{\imath(\omega t-\mathbf{k}\cdot\mathbf{r})}$, this will indeed produce a generalization of the three-dimensional angular momentum.

These interpretations of these operators in terms of angular momentum have been derived within the non-relativistic framework of the Schrödinger equation. The representations of the rotation group used in this framework are harmonic polynomials in (x, y, z) derived from mononomials $\xi_0^{\ell-k}\xi_1^k$ that occur in tensor products of the spinors. The interpretation of the variables (x, y, z) in these harmonic polynomials as coordinates is only valid within an extrapolation of the formalism from the isotropic cone $\mathscr{I}$ in $\mathbb{C}^3$ to $\mathbb{R}^3$, and can therefore not be fundamental. What the meaning of these operators becomes in the Lorentz group, where the mononomials contain four variables (a, b, c, d) (or two dotted and two non-dotted components), is even less clear. To figure it out, one should first prove that it is possible to make a similar extrapolation from the light cone $\mathscr{C}$ in $\mathbb{C}^4$ to $\mathbb{R}^4$. Hence, the fundamental meaning of the "angular momentum operators" is not in terms of angular momentum. Angular momentum is not a basic concept of group theory. The relationship can only be established within a derived formalism obtained by extrapolation. The square $\hat{\mathrm{L}}_{tx}^2 + \hat{\mathrm{L}}_{ty}^2 + \hat{\mathrm{L}}_{zt}^2 - \hat{\mathrm{L}}_{xy}^2 - \hat{\mathrm{L}}_{yz}^2 - \hat{\mathrm{L}}_{zx}^2$ of the "angular-momentum operator" reduces to:

$$\hat{\mathbf{L}}^2 = (r^2 - c^2t^2)\left(\Delta - \frac{\partial^2}{c^2\partial t^2}\right) - \left(ct\frac{\partial}{\partial ct} - \mathbf{r}\cdot\boldsymbol{\nabla}\right)^2 - 4ct\frac{\partial}{\partial ct}. \qquad (12.2)$$

Hence, if it is imagined (by analogy with the rotation group), that there could exist sets of hyper-spherical harmonics $\Psi(x, y, z, t)$ of global degree ℓ in the variables x, y, z, t, and of constant degree M in the variable t (that would satisfy $\Box\Psi = 0$ and build a representation of the Lorentz group), then one would find an eigenvalue equation $\hat{\mathbf{L}}^2\Psi = (\ell^2 - 4M)\,\Psi$ for these polynomials. The numbers ℓ and M could then be used as labels for the polynomials. As in principle six variables are needed to define a spinor, the description in terms of the four coordinates is perhaps not the most appropriate one, and another coordinate system with six coordinates could perhaps lead to more elegant expressions. The use of six coordinates, however, is certainly not an absolute necessity, as in the four-dimensional rotation group, the analogous expression based on four coordinates is perfectly elegant in its own right:

$$\hat{\mathbf{L}}^2 = (r^2 + u^2)\left(\Delta + \frac{\partial^2}{\partial u^2}\right) - \left(u\frac{\partial}{\partial u} + \mathbf{r}\cdot\boldsymbol{\nabla}\right)^2 - 2\left(u\frac{\partial}{\partial u} + \mathbf{r}\cdot\boldsymbol{\nabla}\right), \quad (12.3)$$

and it leads to an eigenvalue equation $\hat{\mathbf{L}}^2\Psi = -\ell(\ell+2)\,\Psi$, where it is not necessary to consider the degree of the polynomial in the fourth variable u seperatly in more detail. It is thus the metric of the Lorentz group that renders the expressions less elegant. This kind of argument can be generalized to higher-dimensional groups. The basic quantity in the group theory is thus the degree of the polynomial, not the angular momentum. This is self-evident, as the mathematics of the group theory precede their application in a physical context.

As discussed in Section 5.4, the spin corresponds to an axial-vector concept. It is related to reflections with respect to a three-dimensional hyper-plane that is orthogonal to the spin vector; (there are four such fundamental hyper-planes). Orbital angular momentum is not a vector but a bi-vector concept. Rotations leave a two-dimensional vector space of rotation axes invariant. The changes induced then take place in the two-dimensional vector space that is orthogonal to the vector space of the axes. It is thus by reflection with respect to such two-dimensional planes that one can try to obtain meaningful eigenvectors for rotations. There are six fundamental planes that can serve as reflection planes. Each type is spanned by one of the six possible combinations $\mathbf{e}_\mu, \mathbf{e}_\nu$. However, only three of these are good combinations. Reflecting with respect to a plane spanned by two space-like components, e.g. $\mathbf{e}_x, \mathbf{e}_y$, means that the rotation axes are $\mathbf{e}_z, \mathbf{e}_{ct}$, such that the turn takes place in the $\mathbf{e}_x, \mathbf{e}_y$ plane, which is exact. But, taking another combination like $\mathbf{e}_z, \mathbf{e}_{ct}$ will mean that the turn takes place in the plane spanned by $\mathbf{e}_z, \mathbf{e}_{ct}$, such that this corresponds to a boost, rather than to a real rotation. In fact, it is well known that relativistically the conservation of angular momentum entails also considerations about the motion of the centre of mass. This is due to the fact that the generalized angular momentum operator contains three components that correspond to boosts.

Within the context of the Lorentz group it is thus meaningless to combine spin and angular momentum by summing. Sums of spin and angular momentum only make sense within the restrained non-relativistic context of the rotation group, where bi-vectors can be identified with vectors, because they have the same number of components. It is also not possible to consider a relativistic concept of spin that would have only three components, as there are no bilinear covariants with that number of components in the Lorentz group, as explained in Subsection 5.10.1.5.

12.2 An inconvenient truth

Within this context of the rotation group, it is possible to define the operator $\hat{\mathbf{L}}^2 = (x\frac{\partial}{\partial y} - y\frac{\partial}{\partial x})^2 + (y\frac{\partial}{\partial z} - z\frac{\partial}{\partial y})^2 + (z\frac{\partial}{\partial x} - x\frac{\partial}{\partial z})^2$. Calculation shows that it corresponds to $(x^2+y^2+z^2)\Delta - (x\frac{\partial}{\partial x} + y\frac{\partial}{\partial y} + z\frac{\partial}{\partial z})^2 - (x\frac{\partial}{\partial x} + y\frac{\partial}{\partial y} + z\frac{\partial}{\partial z})$. In the same spirit as for (12.1)–(12.3), one can show that for a harmonic polynomial Ψ in some representation of degree ℓ it leads to $\hat{\mathbf{L}}^2\Psi = -\ell(\ell+1)\,\Psi$. Here, the minus sign is due to the fact that $\frac{\hbar}{\imath}$ has not been included within the operator $\hat{\mathbf{L}}$.[1] The value of $\ell(\ell+1)$ has been interpreted as the expectation value of the square of the angular momentum vector. This interpretation introduces the very intimidating and puzzling notion that the square of the norm of a vector $\mathbf{L}$ would not be ℓ^2, but $\ell(\ell+1)$. This can of course not possibly be true. The whole formalism is derived from the pure geometry of the rotation group. The very starting point of these mathematics is that for any vector $\mathbf{r}$ we have $\mathbf{r}^2 = r^2$, and rotations are defined on the basis of this definition as those linear transformations that leave such squares of vectors invariant. The result $\ell(\ell+1)$ does not imply that there would be some

[1] *Mutatis mutandis* this result is also true when the polynomials are of half-integer degree. In that case the polynomials are of the type $\Psi = \xi_0^{n-j}\xi_1^j \simeq (\frac{x-\imath y}{2})^{\frac{n-j}{2}}(\frac{-x-\imath y}{2})^{\frac{j}{2}}$, where in the expression in terms of x and y the pre-factors have been dropped that could be generated by the signs $\pm$ and $\mp$ in (3.10). The calculation of $(x\frac{\partial}{\partial x} + y\frac{\partial}{\partial y} + z\frac{\partial}{\partial z})\,\Psi$ yields then $\frac{n}{2}\Psi$, which is the exact counterpart of the result $\ell\Psi$ in the case when the degree is integer.

internal contradictions in the mathematics. This is simply a rather cruel example of an *ad hoc* "physical" interpretation of a purely mathematical result that has been misunderstood. The difficulty that the mathematics are not understood is eluded by explaining it away. One assumes that the solution has to be searched for within the domain of physics rather than in the mathematics. One adds then the postulate that physics on the atomic scale is so inscrutable that it is beyond human understanding and imagination. Generically the quantity ℓ corresponds to the total degree of a polynomial, not to some angular momentum, and the value $\ell(\ell+1)$ is a purely algebraic result. It reflects how the Casimir operator $\hat{\mathbf{L}}^2$ can be expressed in terms of the more elementary Casimir operator $(x\frac{\partial}{\partial x} + y\frac{\partial}{\partial y} + z\frac{\partial}{\partial z})$. These operators are Casimir operators as they multiply all polynomials of a given representation by the same constant.

This of course exposes here a very inconvenient truth, but the inconvenience cannot justify hushing it up. The whole point is that the interpretation in terms of particle coordinates of the variables (x, y, z) within the harmonic polynomials is not generic. We can call it a miracle that we can use what we have called the "two movies" simultaneously. It is this miracle that renders quantum mechanics possible. It is somewhat surprising, as the particle coordinates (x, y, z) and r are not pertinent parameters for the coding of a group element. Through the dependence on the the proper time of the spinor, the coordinates (x, y, z, t) become parameters of the wave function. And by the extrapolation of the formalism from the isotropic cone $\mathscr{I}$ to $\mathbb{R}^3$ it becomes obvious that (x, y, z) can be used as the variables wherein one expresses the harmonic polynomials. One pays a price for this as it entails a loss of information about the full *Vielbein*. Such extrapolations create the illusion that spinors would be vectors and that the playground of the calculations would be the four-dimensional manifold of space-time rather than the six-dimensional manifold of the Lorentz group.

It could be dangerous to claim that it should be possible to link every physical quantity to some operator. Such a correspondence between an operator and a physical quantity is ill-defined. As explained in textbooks such as [Messiah (1965)], for more complex physical quantities that are products, it leads to problems about the order of the corresponding operators in the products. In certain cases it is ambiguous how the operator should be defined, and then it takes a detailed investigation to find out which definition will yield the correct results. The crucial test is that the operator should yield the correct "expectation value". The expectation value for $\hat{\mathbf{L}}^2$

does actually not pass this crucial test, and for this reason the corresponding operator should have been rejected.[2]

In the context of quantum mechanics we very often find statements detailing how some result must be "interpreted". But *in claro non interpretatur.* A mathematical result is a mathematical result that is rational in its own right. It is *a priori* not subject to some parallel rationalization or additional interpretation that one could draw in from outside by invoking "physics", and the meaning of a mathematical result cannot always be somebody's first guess.

[2]That $\hat{\mathbf{L}}^2$ does not yield the expectation value ℓ^2 is due to the fact that its basic definition also contains operators other than $x\frac{\partial}{\partial x}$, $y\frac{\partial}{\partial y}$, and $z\frac{\partial}{\partial z}$. When a contribution of the type $x\frac{\partial}{\partial x}$ operates on $x^{k_x}y^{k_y}z^{k_z}$, it yields $k_x x^{k_x}y^{k_y}z^{k_z}$, where the degree of the polynomial is lowered by the differentiation, but restored by the subsequent multiplication by x. Consequently, $(x\frac{\partial}{\partial x})^n$ will indeed yield an expectation value $(k_x)^n$. But with an operator $x\frac{\partial}{\partial y}$, the original degree of the polynomial $x^{k_x}y^{k_y}z^{k_z}$ is not restored and the polynomial is not even an eigenfunction. These effects eventually add up to the additional term $(x\frac{\partial}{\partial x}+y\frac{\partial}{\partial y}+z\frac{\partial}{\partial z})$ within $(x\frac{\partial}{\partial x}+y\frac{\partial}{\partial y}+z\frac{\partial}{\partial z})^2+(x\frac{\partial}{\partial x}+y\frac{\partial}{\partial y}+z\frac{\partial}{\partial z})$, such that one obtains $\ell^2+\ell$ instead of ℓ^2.

Appendix A

Discovering SL(2,$\mathbb{C}$) Starting from the Cartan Representation

For the simplicity of the calculations it is tempting to try the Cartan representation. This leads then to the discovery of SL(2,$\mathbb{C}$). Using this Cartan representation, we discover namely that $\mathbf{V}$ and $\mathbf{V}^{\star}$ are each other's inverse when they represent unit vectors. We also find that $\mathbf{V}\mathbf{V}^{\star} = \mathbf{v}\cdot\mathbf{v}\mathbb{1} = \det(\mathbf{V})\mathbb{1}$ for any four-vector of arbitrary length. Hence, the key to obtaining $\mathbf{v}\cdot\mathbf{v}\mathbb{1}$ is no longer taking $\mathbf{V}^2$, as in SU(2), but taking $\det(\mathbf{V})$.

We can then also discover that it is possible to obtain the Lorentz group without the reflections by using the two-dimensional matrices. In fact, in reflections one swaps between the two SL(2,$\mathbb{C}$) representations, such that for an even number of reflections one stays within the same SL(2,$\mathbb{C}$) representation. This manifests itself in the Cartan γ-matrices by the fact that one always stays on the secondary diagonal. This implies that the blocks always stay separated and are thus decoupled, which really means that one block alone is already building a representation in its own right.

SL(2,$\mathbb{C}$) can then be reconsidered as a representation for the Lorentz group without reflections. This has of course to be worked out. We then find out that the Lorentz transformation is forcedly of the type $\mathbf{V} \rightarrow \mathbf{L}\mathbf{V}\mathbf{L}^{\dagger}$, as previously explained. The definition that a Lorentz transformation conserves the norm $\sqrt{\det(\mathbf{V})}$ of a vector $\mathbf{v}$, leads then to the condition $\det(\mathbf{L}) = 1$. At least, this is the simplest solution.[1] For $\mathbf{z} \in \mathbb{C}$, the condition

[1] In fact, the true condition expressing that the square of a four-vector is a conserved quantity is $\det(\mathbf{L})\det(\mathbf{L}^{\dagger}) = 1$. It would have been possible to take thus also $e^{\imath\chi}\mathbf{L}$, with $\chi \neq 0$, but this gives us a degree of freedom more than really needed, as the value of χ is not conserved in a product $\mathbf{L}_2\mathbf{L}_1$, and becomes $\chi_1 + \chi_2$. It lives thus its own life when compared to the choice $\chi = 0$. This also raises the question of how an overall consistent choice can be defined for χ that will ensure the matrices fulfill the requirement of building a representation. Obviously $\chi = 0$ is a consistent choice. All other options are at least much less evident.

$\mathbf{z} = 1$ corresponds to two real constraints, *viz.* $\Re(\mathbf{z}) = 1 \,\&\, \Im(\mathbf{z}) = 0$. The constraints reduce the 8 real parameters that are initially present in the four complex matrix entries, to 6 free real parameters. This way it can then be established that SL(2,$\mathbb{C}$) is a representation of the Lorentz group, whereby vectors are coded in the way identified.

Appendix B

Differences in the Spinor Formalisms Between SO(3) and SL(2,ℂ)

Within the context of the rotation group, a spinor decomposition for the isotropic vector $\mathbf{e}_x + \imath\mathbf{e}_y$ is given in (3.24). But this decomposition is of no use within the context of the Lorentz group. In fact, the coordinates $(ct, x, y, z) = (0, 1, \imath, 0)$ of this isotropic vector $\mathbf{e}_x + \imath\mathbf{e}_y$ are not real, such that this matrix is not Hermitian. The same will apply to its images under Lorentz transformations. It is thus not possible to express the matrix in terms of ξ_0 and ξ_1 in a way whereby the complex conjugate quantities ξ_0^* and ξ_1^* would also present themselves as useful quantities in the formalism. On the contrary, the matrix:

$$\begin{aligned}\begin{pmatrix} ct+z & x-\imath y \\ x+\imath y & ct-z \end{pmatrix} &= \begin{pmatrix} \eta_0\eta_0^* & \eta_0\eta_1^* \\ \eta_1\eta_0^* & \eta_1\eta_1^* \end{pmatrix} \\ &= \begin{pmatrix} \eta_0 \\ \eta_1 \end{pmatrix} \otimes (\eta_0^*, \eta_1^*)\end{aligned} \tag{B.1}$$

that corresponds to a light ray type of vector $\mathbf{e}_{ct} + \mathbf{e}_z$ is Hermitian as its coordinates $(1, 0, 0, 1)$ are real. Successive Lorentz transformations will preserve this Hermitian character. Hence, here the complex conjugate quantities η_0^* and η_1^* are an unavoidable part of the formalism. There is thus a flagrant difference between $\boldsymbol{\xi}$ and $\boldsymbol{\eta}$: for $\boldsymbol{\eta}$, the presence of its conjugate counterpart $\boldsymbol{\eta}^\dagger$ is imperative, while for $\boldsymbol{\xi}$, it is the absence of the conjugate counterpart $\boldsymbol{\xi}^\dagger$ that is imperative. The question is thus whether the whole can still be combined self-consistently into a single description of the entire tetrad, that is if, for example, the two spinors can be combined into a single general zero-length vector. This general zero-length vector could for instance be as follows:

$$\begin{aligned}\begin{pmatrix} ct+z & x-\imath y \\ x+\imath y & ct-z \end{pmatrix} &= \begin{pmatrix} -\eta_0\xi_1 & \eta_0\xi_0 \\ -\eta_1\xi_1 & \eta_1\xi_0 \end{pmatrix} \\ &= \begin{pmatrix} \eta_0 \\ \eta_1 \end{pmatrix} \otimes (-\xi_1, \xi_0). \end{aligned} \tag{B.2}$$

This uses the expressions for the spinors as given by (3.24) and (B.1). It is here that it becomes apparent that the decomposition as given in (3.24) is misleading. It is the structure of the spinor corresponding to the isotropic vector $\mathbf{e}_x + \imath\mathbf{e}_y$ that is problematic. In fact, when a Lorentz transformation transforms the column spinor $\boldsymbol{\eta}^\top$ into $\mathbf{L}\boldsymbol{\eta}^\top$, then automatically we have $(\boldsymbol{\eta}^\top)^\dagger = \boldsymbol{\eta}^* \to \boldsymbol{\eta}^*\mathbf{L}^\dagger$, such that the formalism for a vector $\mathbf{V} \to \mathbf{RVR}^\dagger$ is consistent with these two transformations combined with the decomposition of $\mathbf{V}$ according to (B.1). This is not the case for the column spinor $\boldsymbol{\xi}$. It could be noted that this problem with the definition of $\boldsymbol{\xi}$ already exists within the rotation group. In fact, in the rotation group a reflection with normal $\mathbf{n}_1$ transforms a vector $\mathbf{V}$ into $-\mathbf{N}_1\mathbf{V}\mathbf{N}_1$, with $\mathbf{N}_1 = \mathbf{n}_1{\cdot}\boldsymbol{\sigma}$. As the Pauli matrices are Hermitian, this can also be written as $\mathbf{V} \to -\mathbf{N}_1\mathbf{V}\mathbf{N}_1^\dagger$. As a general rotation $\mathbf{R}$ is the product of two reflections, this leads to $\mathbf{V} \to \mathbf{RVR}^\dagger$. That success is ultimately achieved within the rotation group is due to the fact that from $(\xi_0, \xi_1)^\top \to \mathbf{N}(\xi_0, \xi_1)^\top$ it follows that $(-\xi_1, \xi_0) \to (-\xi_1, \xi_0)\mathbf{N}^\star$, with $\mathbf{N}^\star = -\mathbf{N} = -\mathbf{N}^{-1} = -\mathbf{N}^\dagger$. With a sequence of two reflections $\mathbf{R} = \mathbf{N}_2\mathbf{N}_1$ this becomes: $(-\xi_1, \xi_0) \to (-\xi_1, \xi_0)\mathbf{N}_1^\star\mathbf{N}_2^\star = (-\xi_1, \xi_0)\mathbf{R}^{-1} = (-\xi_1, \xi_0)\mathbf{R}^\dagger$. From this we can see that decomposing $\mathbf{V} = (\xi_0, \xi_1)^\top \otimes (-\xi_1, \xi_0)$ is compatible with $\mathbf{V} \to \mathbf{RVR}^\dagger$. In the Lorentz group, we have also $(-\xi_1, \xi_0) \to (-\xi_1, \xi_0)\mathbf{N}^\star$. But now $\mathbf{N}^\star = \mathbf{N}^{-1}$. With a sequence of two reflections $\mathbf{L} = \mathbf{N}_2\mathbf{N}_1$, this becomes also $(-\xi_1, \xi_0) \to (-\xi_1, \xi_0)\mathbf{L}^{-1}$, but now $\mathbf{L}^{-1} \neq \mathbf{L}^\dagger$, such that compatibility is lost. Note that in the rotation group, any isotropic vector $\mathbf{e}_j + \imath\mathbf{e}_k$ leads to a matrix $\sigma_j + \imath\sigma_k$, where the part σ_j is Hermitian while the part $\imath\sigma_k$ is anti-Hermitian. In the Lorentz group, there is also such an anti-Hermitian part, but this leads to an impasse because $\mathbf{L}^{-1} \neq \mathbf{L}^\dagger$. The whole problem with $\boldsymbol{\xi}^\top$ is that the quantity $\boldsymbol{\xi}^*$ just does not have a natural place in the formalism, as for the isotropic vector $\mathbf{e}_x + \imath\mathbf{e}_y$ the 2×2 matrix in (3.24) is not Hermitian. The natural quantity is just based on $\boldsymbol{\xi}$ again. Of course, $\boldsymbol{\xi}^\dagger$ could also be used, rather than $\boldsymbol{\xi}$, but then it would be $\boldsymbol{\xi}^\top$ that loses its legitimacy.[1]

[1]To remove one more real parameter the spinors could, for example, be normalized to 1, i.e. $\xi_0\xi_0^* + \xi_1\xi_1^* = 1$. But normalizing spinors in the Lorentz group has no meaning.

The fact that $\mathbf{L}^{-1} \neq \mathbf{L}^\dagger$ is also the reason why the diagonalization approach to find the expressions for the spinors as applied to the rotation group does not apply to the Lorentz group. In fact, in the rotation group, the matrix $\mathbf{V}$ corresponding to a vector $\mathbf{v}$ was written as $\mathbf{V} = \mathbf{SWS}^{-1}$. As the rotations $\mathbf{R}$ work on such a vector as $\mathbf{RVR}^{-1}$, it makes sense to attempt to make an expression of the type $\mathbf{V} = \mathbf{SWS}^{-1}$ for vectors isomorphic to the formalism $\mathbf{RVR}^{-1}$ that applies to rotations by identifying a rotation with an isotropic vector. The way $\mathbf{S} \to \mathbf{RS}$ the rotations work on the left-hand side of $\mathbf{V} = \mathbf{SWS}^{-1}$ will then be equivalent with the way $\mathbf{S}^{-1} \to \mathbf{S}^{-1}\mathbf{R}^{-1}$ they work on the right-hand side. But, as in the Lorentz group $\mathbf{L}$ does not work on a vector as $\mathbf{LVL}^{-1}$ but as $\mathbf{LVL}^\dagger$, there can be no hope that it would be possible to render the structure of the transformation $\mathbf{RVR}^{-1}$ isomorphic with the structure of transformations $\mathbf{LVL}^\dagger$ within the formalism for Lorentz transformations. Working with $\mathbf{L}^\dagger$ on $\mathbf{S}^{-1}$ is not equivalent to working with $\mathbf{L}$ on $\mathbf{S}$, as $\mathbf{L}^\dagger \neq \mathbf{L}^{-1}$. From this, it seems necessary to write a zero-length vector rather as something that has a structure $\boldsymbol{\xi} \cdots \boldsymbol{\xi}^\dagger$, where the central dots represent an unknown expression. What counts here is that the expression starts with $\boldsymbol{\xi}$ and ends by $\boldsymbol{\xi}^\dagger$, such that working with $\mathbf{L}$ on $\boldsymbol{\xi}$ and working with $\mathbf{L}^\dagger$ on $\boldsymbol{\xi}^\dagger$ are equivalent. The coding $\boldsymbol{\xi} \otimes \boldsymbol{\xi}^\dagger$ for light rays in (4.13) and (4.17) in Chapter 4 has exactly this structure. In [Coddens (2008)] this was not understood. The fact that the diagonalisation approach does not work for the Lorentz group might create the impression that the principles (of homogeneity and removing any mention of the length of a vector from the formalism) used to derive the expressions for spinors in the rotation group would not be universal. But the present remark proves how they indeed are generally valid.

As the left-hand side of the condition $\xi_0\xi_0^* + \xi_1\xi_1^* = 1$ is forcedly real and positive, it is indeed a condition on only one parameter. But a Lorentz transformation $\mathbf{L}$ transforms $(\xi_0, \xi_1)^\top$ into $\mathbf{L}(\xi_0, \xi_1)^\top$, and (by Hermitian conjugation) $(\xi_0, \xi_1)^*$ into $(\xi_0, \xi_1)^*\mathbf{L}^\dagger$. Hence, $(\xi_0, \xi_1)^*(\xi_0, \xi_1)^\top$ transforms to $(\xi_0, \xi_1)^*\mathbf{L}^\dagger\mathbf{L}(\xi_0, \xi_1)^\top$, and as $\mathbf{L}^{-1} \neq \mathbf{L}^\dagger$, this cannot be pushed further to show that $\xi_0\xi_0^* + \xi_1\xi_1^*$ would be an invariant, despite the fact that $\det \mathbf{L} = 1$. Hence, to preserve the normalization of a spinor, the formalism should have been able to code a reflection $\mathbf{N}$ into something like $\mathbf{V} \to -\mathbf{NVN}^{-1}$. This is the case in the rotation group where we have $\mathbf{N}^{-1} = \mathbf{N} = \mathbf{N}^\dagger$. In the rotation group we therefore have $\mathbf{R}^{-1} = \mathbf{R}^\dagger$, such that normalization of spinors is conserved. In the rotation group the norm is expressed in terms of the square of a matrix. The normalization problem in the Lorentz group comes from the minus signs in the metric, which demands that the norm is expressed in terms of a determinant rather than in terms of the square of a matrix. We then no longer have $\mathbf{N}^{-1} = \mathbf{N}^\dagger$ (while we still have $\mathbf{N} = \mathbf{N}^\dagger$).

Appendix C

Additional Results on the Lorentz Group

C.1 General form of an element of the Lorentz group

To be able to follow the argument, the reader must be familiar with the notion that a general Lorentz transformation is the composition of a boost and a rotation. A good description of this idea can be found in [Rhodes and Semon (2004)]. In fact, most introductory textbooks treat Lorentz transformations solely on the basis of the equations for a general boost along the x-axis, without discussing what happens when one composes boosts that are not collinear. A set of collinear boosts generates a subgroup of the Lorentz group that does not contain a single rotation (apart from the trivial identity element). Consequently, the rotational part of a general Lorentz transformation and the corresponding concept of Thomas precession are much less familiar to the average reader. The idea is that a general boost $\mathbf{B}(\mathbf{v})$ with boost vector $\mathbf{v} = v\mathbf{u}$, where $\mathbf{u}$ is the unit vector that is parallel with $\mathbf{v}$, is obtained by performing a sequence of three operations. First, make a change of coordinates in the form of a rotation $\mathbf{R}(\mathbf{n}_1, \varphi_1)$ around $\mathbf{n}_1$ (defined by $\mathbf{n}_1 \sin\varphi_1 = \mathbf{e}_x \wedge \mathbf{u} = (0, -u_z, u_y))$ over an angle φ_1 (defined by $\cos\varphi_1 = \mathbf{u}\cdot\mathbf{e}_x$). After this change, the new coordinates for $\mathbf{u}$ are $(1,0,0) = \mathbf{e}'_x$, such that $\mathbf{u}$ is now aligned with the new x'-axis. Within this aligned geometry the boost $\mathbf{B}(\mathbf{v}) = \mathbf{B}(v\mathbf{e}'_x)$ is performed along the x'-axis, and finally applying the inverse rotation $(\mathbf{R}(\mathbf{n}_1, \varphi_1))^{-1}$ we obtain the expression of the boost in the original coordinates. After this boost $\mathbf{B}(\mathbf{v})$, one can still make a rotation $\mathbf{R}(\mathbf{n}, \varphi)$ of the frame around an axis defined by the space-like unit vector $\mathbf{n}$ over an angle φ. The total expression for a general Lorentz transformation is then $\mathbf{R}(\mathbf{n}, \varphi)(\mathbf{R}(\mathbf{n}_1, \varphi_1))^{-1}\mathbf{B}(v\mathbf{e}'_x)\mathbf{R}(\mathbf{n}_1, \varphi_1)$.

C.2 Coding of a general Lorentz transformation: Boost and rotation parameters

It will now be determined how within SL(2, $\mathbb{C}$) the six real independent parameters contained in $\mathbf{v}, \mathbf{n}, \varphi$ are coded into the 2×2 representation matrix of a general Lorentz transformation. Before proceeding it must be certain that the definitions of the 2×2 matrices that are representing the vectors in the rotation group and in the Lorentz group coincide such that the representation of the rotation group becomes embedded into the representation of the Lorentz group. That is, the following definitions must be used together:

$$(ct, x, y, z) \leftrightarrow \begin{pmatrix} ct + z & x - \imath y \\ x + \imath y & ct - z \end{pmatrix},$$
$$(0, x, y, z) \leftrightarrow \begin{pmatrix} z & x - \imath y \\ x + \imath y & -z \end{pmatrix}. \tag{C.1}$$

With this underlying convention for the Pauli matrices, a general rotation $\mathbf{R}(\mathbf{n}, \varphi)$ corresponds then to $\mathbf{L} = \mathbb{1} \cos\frac{\varphi}{2} - \imath \boldsymbol{\sigma} \cdot \boldsymbol{n} \sin\frac{\varphi}{2}$. This is the well-known Rodrigues formula. More specifically, the rotation $\mathbf{R}(\mathbf{n}_1, \varphi_1)$ defined in the previous section becomes:

$$\mathbf{R}(\mathbf{n}_1, \varphi_1) = \frac{1}{\sqrt{2v(v + v_x)}} \begin{pmatrix} v + v_x - \imath v_y & v_z \\ -v_z & v + v_x + \imath v_y \end{pmatrix}, \tag{C.2}$$

while the boost $\mathbf{B}(v\mathbf{e}'_x)$ becomes:

$$\mathbf{B}(v\mathbf{e}'_x) = \begin{pmatrix} \sqrt{\frac{\gamma+1}{2}} & -\sqrt{\frac{\gamma-1}{2}} \\ -\sqrt{\frac{\gamma-1}{2}} & \sqrt{\frac{\gamma+1}{2}} \end{pmatrix}, \tag{C.3}$$

as is easily checked by calculating $\mathbf{BVB}^\dagger$ and verifying that $\det \mathbf{B} = 1$. (This extra check is necessary as $\mathbf{BVB}^\dagger$ defines $\mathbf{B}$ only up to a phase factor.) The total transformation is thus given by $\mathbf{V} \to \mathbf{LVL}^\dagger$ with:

$$\mathbf{L} = \frac{1}{2v(v + v_x)} \begin{pmatrix} \cos\frac{\varphi}{2} - \imath\, n_z \sin\frac{\varphi}{2} & -(\imath\, n_x + n_y) \sin\frac{\varphi}{2} \\ (-\imath\, n_x + n_y) \sin\frac{\varphi}{2} & \cos\frac{\varphi}{2} + \imath\, n_z \sin\frac{\varphi}{2} \end{pmatrix}$$
$$\times \begin{pmatrix} v + v_x + \imath\, v_y & -v_z \\ v_z & v + v_x - \imath\, v_y \end{pmatrix}$$

$$\times \begin{pmatrix} \sqrt{\frac{\gamma+1}{2}} & -\sqrt{\frac{\gamma-1}{2}} \\ -\sqrt{\frac{\gamma-1}{2}} & \sqrt{\frac{\gamma+1}{2}} \end{pmatrix}$$

$$\times \begin{pmatrix} v + v_x - \imath\, v_y & v_z \\ -v_z & v + v_x + \imath\, v_y \end{pmatrix}. \tag{C.4}$$

First, calculate the pure-boost part. It can be written as $[(v+v_x)\mathbb{1}+\imath(v_z\sigma_y - v_y\sigma_z)]\,[\sqrt{\frac{\gamma+1}{2}}\mathbb{1} - \sqrt{\frac{\gamma-1}{2}}\sigma_x]\,[(v+v_x)\mathbb{1} - \imath(v_z\sigma_y - v_y\sigma_z)]/[2v(v+v_x)]$. After somewhat lengthy algebra it is found by using $\sigma_x\sigma_y = \imath\sigma_z(cyclic)$ that it is equal to:

$$\mathbf{B}(\mathbf{v}) = \sqrt{\frac{\gamma+1}{2}}\,\mathbb{1} - \sqrt{\frac{\gamma-1}{2}}\;\mathbf{u}\cdot\boldsymbol{\sigma}, \tag{C.5}$$

where $\mathbf{u} = \mathbf{v}/v$. This leads to the total result:

$$\mathbf{L}(\mathbf{u}, v, \mathbf{n}, \varphi) = \left[\cos\frac{\varphi}{2}\mathbb{1} - \imath\sin\frac{\varphi}{2}\mathbf{n}\cdot\boldsymbol{\sigma}\right]\left[\sqrt{\frac{\gamma+1}{2}}\,\mathbb{1} - \sqrt{\frac{\gamma-1}{2}}\;\mathbf{u}\cdot\boldsymbol{\sigma}\right], \tag{C.6}$$

which using the identity $[\mathbf{a}\cdot\boldsymbol{\sigma}]\,[\mathbf{b}\cdot\boldsymbol{\sigma}] = \mathbf{a}\cdot\mathbf{b} + \imath(\mathbf{a}\wedge\mathbf{b})\cdot\boldsymbol{\sigma}$ becomes:

$$\begin{aligned} \mathbf{L}(\mathbf{u}, v, \mathbf{n}, \varphi) = &\left(\cos\frac{\varphi}{2}\sqrt{\frac{\gamma+1}{2}} + \imath\sin\frac{\varphi}{2}\sqrt{\frac{\gamma-1}{2}}\,(\mathbf{n}\cdot\mathbf{u})\right)\mathbb{1} \\ &-\imath\sin\frac{\varphi}{2}\sqrt{\frac{\gamma+1}{2}}\,[\,\mathbf{n}\cdot\boldsymbol{\sigma}\,] - \cos\frac{\varphi}{2}\sqrt{\frac{\gamma-1}{2}}\,[\,\mathbf{u}\cdot\boldsymbol{\sigma}\,] \\ &-\sin\frac{\varphi}{2}\sqrt{\frac{\gamma-1}{2}}\,[\,(\mathbf{n}\wedge\mathbf{u})\cdot\boldsymbol{\sigma}\,]. \end{aligned} \tag{C.7}$$

This equation is the analogue for the Lorentz group of what the Rodrigues formula is for the rotation group. However, it is much more cumbersome to decode the physical parameters $\mathbf{u}, v, \mathbf{n}, \varphi$ from a 2×2 matrix of SL(2,**C**) than to decode a rotation axis $\mathbf{n}$ and rotation angle φ from a rotation matrix of SU(2).

C.3 Decoding of a general Lorentz transformation: Boost and rotation parameters

To determine the parameters $\mathbf{u}, v, \mathbf{n}, \varphi$, from the general form of a Lorentz transformation in SL(2,$\mathbb{C}$) given by (4.9), with $ad - bc = 1$, it is necessary to solve $\mathbf{u}, v, \mathbf{n}, \varphi$ from the matrix equation $\mathbf{L}(\mathbf{u}, v, \mathbf{n}, \varphi) = \mathbf{L}(a, b, c, d)$. The quantities (a, b, c, d) are nothing other than the spinors of the Lorentz group. This matrix equation can be solved as follows. In order to render the intermediate notations less cumbersome it is necessary to introduce:

$$-\mathbf{g} = \cos\frac{\varphi}{2}\sqrt{\frac{\gamma-1}{2}}\,\mathbf{u} + \sin\frac{\varphi}{2}\sqrt{\frac{\gamma-1}{2}}\,(\mathbf{n}\wedge\mathbf{u}) + \imath\sin\frac{\varphi}{2}\sqrt{\frac{\gamma+1}{2}}\,\mathbf{n}. \tag{C.8}$$

Then:

$$\mathbf{L}(\mathbf{u}, v, \mathbf{n}, \varphi) = \left(\cos\frac{\varphi}{2}\sqrt{\frac{\gamma+1}{2}} + \imath\sin\frac{\varphi}{2}\sqrt{\frac{\gamma-1}{2}}\,(\mathbf{n}\cdot\mathbf{u})\right)\mathbb{1} + \mathbf{g}\cdot\boldsymbol{\sigma} \tag{C.9}$$

From this we obtain $(a+d)/2 = \cos\frac{\varphi}{2}\sqrt{\frac{\gamma+1}{2}} + \imath\sin\frac{\varphi}{2}\sqrt{\frac{\gamma-1}{2}}\,(\mathbf{n}\cdot\mathbf{u})$, $g_x = (c+b)/2$, $g_y = (c-b)/2\imath$, $g_z = (a-d)/2$. We also have: $\mathbf{g}^* - \mathbf{g} = 2\imath\sin\frac{\varphi}{2}\sqrt{\frac{\gamma+1}{2}}\,\mathbf{n}$, and $2\cos\frac{\varphi}{2}\sqrt{\frac{\gamma+1}{2}} = (a+a^*+d+d^*)/2$, such that:

$$\begin{aligned}\imath\left(\tan\frac{\varphi}{2}\right)\mathbf{n} = (c^*+b^*-c-b, \imath c^* - \imath b^* + \imath c - \imath b, a^* - a - d^* + d)/\\ (a+a^*+d+d^*),\end{aligned} \tag{C.10}$$

from which $\frac{\varphi}{2}$ and $\mathbf{n}$ can be determined, using the fact that $\mathbf{n}$ is a unit vector. One can also make the combination $(a+a^*+d+d^*)^2 - 4(\mathbf{g}^*-\mathbf{g})^2 = 8(\gamma+1)$ which leads to:

$$\begin{aligned}8(\gamma+1) &= (a+a^*+d+d^*)^2 - (c^*+b^*-c-b)^2\\ &\quad + (c^*-b^*+c-b)^2 - (a^*-a-d^*+d)^2,\end{aligned} \tag{C.11}$$

from which it follows that

$$\begin{aligned}\gamma &= \frac{aa^*+bb^*+cc^*+dd^*}{2},\\ \cos\frac{\varphi}{2} &= \frac{a+a^*+d+d^*}{2\sqrt{aa^*+bb^*+cc^*+dd^*+2}}.\end{aligned} \tag{C.12}$$

From (C.10) we obtain then:

$$\imath\left(\sin\frac{\varphi}{2}\right)\mathbf{n} = \frac{(c^*+b^*-c-b, \imath c^*-\imath b^*+\imath c-\imath b, a^*-a-d^*+d)}{2\sqrt{aa^*+bb^*+cc^*+dd^*+2}}. \tag{C.13}$$

It follows then that:

$$\cos\frac{\varphi}{2}\mathbb{1} + \imath\sin\frac{\varphi}{2}\mathbf{n}\cdot\boldsymbol{\sigma} = \frac{1}{\sqrt{aa^*+bb^*+cc^*+dd^*+2}}\begin{pmatrix} a^*+d & c^*-b\\ b^*-c & a+d^*\end{pmatrix}. \tag{C.14}$$

Using (C.6), (4.9) and $\mathbf{L}(\mathbf{u}, v, \mathbf{n}, \varphi) = \mathbf{L}(a,b,c,d)$ we obtain $[\cos\frac{\varphi}{2}\mathbb{1} + \imath\sin\frac{\varphi}{2}\mathbf{n}\cdot\boldsymbol{\sigma}]\,\mathbf{L}(a,b,c,d) = [\sqrt{\frac{\gamma+1}{2}}\,\mathbb{1} - \sqrt{\frac{\gamma-1}{2}}\,\mathbf{u}\cdot\boldsymbol{\sigma}]$, where the left-hand side is completely determined in terms of (a,b,c,d):

$$\frac{1}{\sqrt{aa^* + bb^* + cc^* + dd^* + 2}} \begin{pmatrix} aa^* + cc^* + 1 & a^*b + dc^* \\ ab^* + d^*c & bb^* + dd^* + 1 \end{pmatrix}$$
$$= \left[\sqrt{\frac{\gamma+1}{2}}\,\mathbb{1} - \sqrt{\frac{\gamma-1}{2}}\,\mathbf{u}\cdot\boldsymbol{\sigma}\right]. \tag{C.15}$$

From this the expression for $\mathbf{u}$ in terms of (a,b,c,d) can also be determined. We obtain:

$$\mathbf{u} = \frac{1}{\sqrt{(aa^* + bb^* + cc^* + dd^* + 2)}}\,\frac{1}{\sqrt{(aa^* + bb^* + cc^* + dd^* - 2)}}$$
$$(-a^*b - ab^* - dc^* - d^*c,\ \imath ab^* + \imath d^*c - \imath a^*b - \imath dc^*,$$
$$bb^* + dd^* - aa^* - cc^*). \tag{C.16}$$

It serves no purpose to normalize the vector $\sin\frac{\varphi}{2}\,\mathbf{n}$ in the expression $\imath\sin\frac{\varphi}{2}\,\mathbf{n}\cdot\boldsymbol{\sigma}$ by trying to remove $\sin\frac{\varphi}{2}$ from it. The general expressions for the term $\sin\frac{\varphi}{2}$ are cumbersome and $\sin\frac{\varphi}{2}\,\mathbf{n}$ already defines the rotation axis, which is all that counts for the physical and geometrical contents. The normalization of a vector along an axis can be performed *ad hoc* by using the numerical values of its coordinates, without deriving general expression that may underly it due to the fact that it fits into the logic of the Lorentz group. Furthermore, the quantity needed for possible ulterior calculations is $\imath\sin\frac{\varphi}{2}\,\mathbf{n}\cdot\boldsymbol{\sigma}$ anyway.

C.4 Decoding of a general Lorentz transformation: Coordinates of the tetrad

The coordinates of $\mathbf{V}'_2$, which codes $\mathbf{e}'_x + \imath\mathbf{e}'_y$, are $(ab^* + cd^*, cb^* + ad^*, cb^* - ad^*, ab^* - cd^*)$. The real parts of these are $(ct_1, x_1, y_1, z_1) = \frac{1}{2}\,(ab^* + cd^* + a^*b + c^*d, cb^* + ad^* + c^*b + a^*d, cb^* - ad^* + c^*b - a^*d, ab^* - cd^* + a^*b - c^*d)$, while the imaginary parts are $(ct_2, x_2, y_2, z_2) = \frac{1}{2\imath}\,(ab^* + cd^* - a^*b - c^*d, cb^* + ad^* - c^*b - a^*d, cb^* - ad^* - c^*b + a^*d, ab^* - cd^* - a^*b + c^*d)$. We obtain similarly the coordinates $(ct_0 + ct_3, x_0 + x_3, y_0 + y_3, z_0 + z_3) = (aa^* + cc^*, ca^* + ac^*, ca^* - ac^*, aa^* - cc^*)$ from $\mathbf{V}'_1$. The vector $(ct_0 - ct_3, x_0 - x_3, y_0 - y_3, z_0 - z_3)$ must still be determined. The easiest way to do this is to observe that $\mathbf{e}'_{ct} - \mathbf{e}'_x$

is coded by $\mathbf{V}_4' = \mathbf{L}\mathbf{V}_4\mathbf{L}^\dagger$. Hence, the coordinates $(ct_0 - ct_3, x_0 - x_3, y_0 - y_3, z_0 - z_3)$ are given by $(bb^* + dd^*, db^* + bd^*, db^* - bd^*, bb^* - dd^*)$. Thus, $(ct_0, x_0, y_0, z_0) = \frac{1}{2}(aa^* + cc^* + bb^* + dd^*, ca^* + ac^* + db^* + bd^*, ca^* - ac^* + db^* - bd^*, aa^* - cc^* + bb^* - dd^*)$, and $(ct_3, x_3, y_3, z_3) = \frac{1}{2}(aa^* + cc^* - bb^* - dd^*, ca^* + ac^* - db^* - bd^*, ca^* - ac^* - db^* + bd^*, aa^* - cc^* - bb^* + dd^*)$. Hence, we have:

$$\begin{pmatrix} \mathbf{e}_{ct}' \\ \mathbf{e}_x' \\ \mathbf{e}_y' \\ \mathbf{e}_z' \end{pmatrix} = \begin{pmatrix} \frac{1}{2} & 0 & 0 & 0 \\ 0 & \frac{1}{2} & 0 & 0 \\ 0 & 0 & \frac{1}{2\imath} & 0 \\ 0 & 0 & 0 & \frac{1}{2} \end{pmatrix}$$
$$\times \begin{pmatrix} aa^* + cc^* + bb^* + dd^* & ca^* + ac^* + db^* + bd^* \\ ab^* + cd^* + a^*b + c^*d & cb^* + ad^* + c^*b + a^*d \\ ab^* + cd^* - a^*b - c^*d & cb^* + ad^* - c^*b - a^*d \\ aa^* + cc^* - bb^* - dd^* & ca^* + ac^* - db^* - bd^* \end{pmatrix} \cdots$$
$$\begin{matrix} ca^* - ac^* + db^* - bd^* & aa^* - cc^* + bb^* - dd^* \\ cb^* - ad^* + c^*b - a^*d & ab^* - cd^* + a^*b - c^*d \\ cb^* - ad^* - c^*b + a^*d & ab^* - cd^* - a^*b + c^*d \\ ca^* - ac^* - db^* + bd^* & aa^* - cc^* - bb^* + dd^* \end{matrix} \Bigg) \begin{pmatrix} \mathbf{e}_{ct} \\ \mathbf{e}_x \\ \mathbf{e}_y \\ \mathbf{e}_z \end{pmatrix}. \tag{C.17}$$

It can be seen from this that the relationship between $\frac{d}{dx_\mu}$ and $\frac{d}{d\tau}$ in SL(2,$\mathbb{C}$) is not simple as there are no simple expressions for reflection operators in SL(2,$\mathbb{C}$). But on the other hand, the relationship between spinors and their transforms is extremely simple, e.g. $(\xi_0', \xi_1')^\top = \mathbf{L}(\xi_0, \xi_1)^\top$. These difficulties are overcome by making the transition from the two-dimensional to the four-dimensional representation, as explained in [Coddens (2008)].

C.5 The electromagnetic field and angular momentum as a tensor quantities

The aim of this section is not to prove the well-known fact that the electromagnetic field and angular momentum are tensors; it is more to provide the additional elements for the discussion of the generalized Dirac equation. It illustrates how the structure of the matrices automatically imposes the symmetry and that the whole structure of the equations follows from the symmetry. In the following, Gaussian units are used such that the results are immediately comparable with those in [Jackson (1963)]. Let us calculate the product of two matrices that both code four-vectors:

$$\left(\frac{1}{c}\frac{\partial}{\partial t}\mathbb{1} - \boldsymbol{\sigma}\cdot\boldsymbol{\nabla}\right)(V\mathbb{1} - \mathbf{A}\cdot\boldsymbol{\sigma}) = \left(\frac{1}{c}\frac{\partial}{\partial t}V + \boldsymbol{\nabla}\cdot\mathbf{A}\right)\mathbb{1} + \left(-\frac{1}{c}\frac{\partial}{\partial t}\mathbf{A} - \boldsymbol{\nabla}V\right)\cdot\boldsymbol{\sigma} + \imath(\boldsymbol{\nabla}\wedge\mathbf{A})\cdot\boldsymbol{\sigma}. \tag{C.18}$$

Then the first term on the right-hand side is zero due to the Lorenz condition $(\frac{1}{c}\frac{\partial}{\partial t}V + \boldsymbol{\nabla}\cdot\mathbf{A}) = 0$, such that we obtain:

$$\left(\frac{1}{c}\frac{\partial}{\partial t}\mathbb{1} - \boldsymbol{\sigma}\cdot\boldsymbol{\nabla}\right)(V\mathbb{1} - \mathbf{A}\cdot\boldsymbol{\sigma}) \equiv (\mathbf{E} + \imath\mathbf{B})\cdot\boldsymbol{\sigma}. \tag{C.19}$$

As now:

$$\left(\frac{1}{c}\frac{\partial}{\partial t}\mathbb{1} + \boldsymbol{\sigma}\cdot\boldsymbol{\nabla}\right)\left(\frac{1}{c}\frac{\partial}{\partial t}\mathbb{1} - \boldsymbol{\sigma}\cdot\boldsymbol{\nabla}\right) \equiv \Box^2\mathbb{1}, \tag{C.20}$$

we have:

$$\left(\frac{1}{c}\frac{\partial}{\partial t}\mathbb{1} + \boldsymbol{\sigma}\cdot\boldsymbol{\nabla}\right)\left(\frac{1}{c}\frac{\partial}{\partial t}\mathbb{1} - \boldsymbol{\sigma}\cdot\boldsymbol{\nabla}\right)(V\mathbb{1} - \mathbf{A}\cdot\boldsymbol{\sigma}) = -\frac{4\pi}{c}(c\rho\,\mathbb{1} + \mathbf{j}\cdot\boldsymbol{\sigma}), \tag{C.21}$$

or:

$$\left(\frac{1}{c}\frac{\partial}{\partial t}\mathbb{1} + \boldsymbol{\sigma}\cdot\boldsymbol{\nabla}\right)(\mathbf{E} + \imath\mathbf{B})\cdot\boldsymbol{\sigma} = -\frac{4\pi}{c}(c\rho\,\mathbb{1} + \mathbf{j}\cdot\boldsymbol{\sigma}). \tag{C.22}$$

By separating the real and imaginary parts we obtain Maxwell's equations in vacuum:

$$\begin{aligned} \boldsymbol{\nabla}\cdot\mathbf{B} &= 0, \qquad \boldsymbol{\nabla}\wedge\mathbf{E} + \frac{1}{c}\frac{\partial}{\partial t}\mathbf{B} = \mathbf{0}, \\ \boldsymbol{\nabla}\cdot\mathbf{E} &= 4\pi\rho, \qquad \boldsymbol{\nabla}\wedge\mathbf{B} - \frac{1}{c}\frac{\partial}{\partial t}\mathbf{E} = \frac{4\pi}{c}\mathbf{j}. \end{aligned} \tag{C.23}$$

We would expect to have $V\mathbb{1}+\mathbf{A}\cdot\boldsymbol{\sigma}$ rather than $V\mathbb{1}-\mathbf{A}\cdot\boldsymbol{\sigma}$ to start from, but this is only a matter of conventions. One can equally satisfy $\boldsymbol{\nabla}\cdot\mathbf{B} = 0$ by putting $\mathbf{B} = -\boldsymbol{\nabla}\wedge\mathbf{A}$. Starting from $(\frac{1}{c}\frac{\partial}{\partial t}\mathbb{1} - \boldsymbol{\sigma}\cdot\boldsymbol{\nabla})(V\mathbb{1}+\mathbf{A}\cdot\boldsymbol{\sigma})$, it is possible to define: $(\frac{1}{c}\frac{\partial}{\partial t}V - \boldsymbol{\nabla}\cdot\mathbf{A}) = 0$ for the Lorenz condition, $\mathbf{E} = \frac{1}{c}\frac{\partial}{\partial t}\mathbf{A} - \boldsymbol{\nabla}V$, and $\mathbf{B} = -\boldsymbol{\nabla}\wedge\mathbf{A}$. Ultimately, the Maxwell equations will also be obtained. The quantities $(V, \mathbf{A})$ are not measured quantities: they have only been defined to replace the Maxwell equations, and with the alternative choice the wave equation is also obtained. In fact, these issues depend only on the structure of (C.21) and (C.22). It is obvious that at the time the vector potential was defined by $\mathbf{B} = \boldsymbol{\nabla} \wedge \mathbf{A}$, it would have seemed pedantic to define it rather as $\mathbf{B} = -\boldsymbol{\nabla} \wedge \mathbf{A}$. The advantage of the alternative formulation based on $V\mathbb{1} + \mathbf{A}{\cdot}\boldsymbol{\sigma}$ is, however, that it highlights better the four-vector character of the four-potential, and that also the Lorenz condition follows the overall symmetry. This, however, still, has a significant consequence for the so-called minimal substitution. The situation is somewhat confusing, and has been summarized in Tables C.1 and C.2.

From (C.19) it is seen that the electromagnetic field is a bi-vector. It has six rather than three components and due to (C.19) it must transform bi-linearly rather than linearly under a Lorentz transformation. Hence (C.19) exhibits the tensor structure of the electromagnetic field. Note that the left-hand side of (C.21) has the structure of a tri-vector, and therefore contains a quantity with four components.

To illustrate the power of the machinery further we calculate:

$$\begin{aligned}\left[q\mathbb{1} - \frac{q}{c}\mathbf{v}{\cdot}\boldsymbol{\sigma}\right]&[\,(\mathbf{E} + \imath\mathbf{B}){\cdot}\boldsymbol{\sigma}\,]\\ &= \left[-\frac{q}{c}(\mathbf{v}\cdot\mathbf{E})\,\mathbb{1} + \left(q\mathbf{E} + \frac{q}{c}\mathbf{v}\wedge\mathbf{B}\right){\cdot}\boldsymbol{\sigma}\right]\\ &\quad + \imath\left[-\frac{q}{c}(\mathbf{v}\cdot\mathbf{B})\,\mathbb{1} + \left(q\mathbf{B} - \frac{q}{c}\mathbf{v}\wedge\mathbf{E}\right){\cdot}\boldsymbol{\sigma}\right].\end{aligned} \tag{C.24}$$

In the real part it is possible to recognize the four-vector of the Lorentz force. The expression for the work completes the four-vector in its time-like component. The imaginary part contains the anomalous Zeeman term $q\,[\,\mathbf{B}{\cdot}\boldsymbol{\sigma}\,]$ and the spin-orbit coupling term $\frac{q}{c}\,[\,(\mathbf{v}\wedge\mathbf{E})\,{\cdot}\boldsymbol{\sigma}\,]$.

Hence, with the aid of the representation matrices one can easily find all the correct equations. Along the same line of thought, it may also be observed that the angular momentum $\mathbf{r}\wedge m\mathbf{v}$ is a bi-vector in $\mathbb{R}^4$ rather than a vector. It has six rather than three components. The difference between vectors and bi-vectors does not manifest itself in $\mathbb{R}^3$ where they have both three components. Three-dimensional intuition tends thus to hide the fact that angular momentum is a rank-2 tensor (i.e. a bi-vector) rather than a vector. This tensor character becomes obvious by writing:

$$\begin{aligned}(ct\mathbb{1} &- \mathbf{r}{\cdot}\boldsymbol{\sigma})\,(E\mathbb{1} + c\mathbf{p}{\cdot}\boldsymbol{\sigma})/c\\ &= (Et - \mathbf{p}\cdot\mathbf{r})\,\mathbb{1} - (E\mathbf{r} - c^2\mathbf{p}t){\cdot}\boldsymbol{\sigma}/c - \imath(\mathbf{r}\wedge\mathbf{p}){\cdot}\boldsymbol{\sigma}.\end{aligned} \tag{C.25}$$

The three-dimensional concept of angular momentum thereby becomes part of a six-component quantity that can be cast into the form $\mathbf{L}_{ct,\mathbf{r}} + \imath\mathbf{L}_{\mathbf{r},\mathbf{r}}$. The

Table C.1 Conventions for the vector potential $\mathbf{A}$

Lorenz Gauge equation: $\frac{\partial V}{\partial t} - \boldsymbol{\nabla}\cdot\mathbf{A} = 0$	Direct space (covariant)	Dual space (contravariant)
Defining equations for the potentials	$\mathbf{E} = \frac{\partial}{\partial ct}\mathbf{A} - \boldsymbol{\nabla}V$, $\mathbf{B} = \boldsymbol{\nabla}\wedge\mathbf{A}$	
Potentials	$(V, \mathbf{A})$	$(V, -\mathbf{A})$
Derivation with respect to proper time $\frac{d}{dc\tau}$		$(\frac{\partial}{\partial ct}, -\boldsymbol{\nabla})$
$\frac{\hbar}{\imath}\frac{d}{dc\tau}$		$(\frac{\hbar}{\imath}\frac{\partial}{\partial ct}, -\frac{\hbar}{\imath}\boldsymbol{\nabla})$
$\Downarrow$		$\Downarrow$
Operator definitions $(\hat{\mathrm{E}}, c\hat{\mathbf{p}})$		$\hat{\mathrm{E}} = \frac{\hbar c}{\imath}\frac{\partial}{\partial ct}$, $c\hat{\mathbf{p}} = -\frac{\hbar c}{\imath}\boldsymbol{\nabla}$
Minimal substitution $(E, c\mathbf{p}) - q(V, \mathbf{A})$	$(E, c\mathbf{p}) - q(V, \mathbf{A})$	$(\frac{\hbar c}{\imath}\frac{\partial}{\partial ct}, -\frac{\hbar c}{\imath}\boldsymbol{\nabla}) - q(V, -\mathbf{A})$
$\Downarrow$		$\Downarrow$
Operator definitions $(\hat{\mathrm{E}} - q\hat{V}, c\hat{\mathbf{p}} - \hat{\mathbf{A}})$		$\hat{V} = V$, $\hat{\mathbf{A}} = -\mathbf{A}$

Table C.2 Conventions for the vector potential $-\mathbf{A}$

Lorenz Gauge equation: $\frac{\partial V}{\partial t} + \boldsymbol{\nabla}\cdot\mathbf{A} = 0$	Direct space (covariant)	Dual space (contravariant)
Defining equations for the potentials	$\mathbf{E} = -\frac{\partial}{\partial ct}\mathbf{A} - \boldsymbol{\nabla}V$, $\mathbf{B} = -\boldsymbol{\nabla}\wedge\mathbf{A}$	
Potentials	$(V, -\mathbf{A})$	$(V, \mathbf{A})$
Derivation with respect to proper time $\frac{d}{dc\tau}$		$(\frac{\partial}{\partial ct}, -\boldsymbol{\nabla})$
$\frac{\hbar}{\imath}\frac{d}{dc\tau}$		$(\frac{\hbar}{\imath}\frac{\partial}{\partial ct}, -\frac{\hbar}{\imath}\boldsymbol{\nabla})$
$\Downarrow$		$\Downarrow$
Operator definitions $(\hat{\mathrm{E}}, c\hat{\mathbf{p}})$		$\hat{\mathrm{E}} = \frac{\hbar c}{\imath}\frac{\partial}{\partial ct}$, $c\hat{\mathbf{p}} = -\frac{\hbar c}{\imath}\boldsymbol{\nabla}$
Minimal substitution $(E, c\mathbf{p}) - q(V, -\mathbf{A})$	$(E, c\mathbf{p}) - q(V, -\mathbf{A})$	$(\frac{\hbar c}{\imath}\frac{\partial}{\partial ct}, -\frac{\hbar c}{\imath}\boldsymbol{\nabla}) - q(V, \mathbf{A})$
$\Downarrow$		$\Downarrow$
Operator definitions $(\hat{\mathrm{E}} - q\hat{V}, c\hat{\mathbf{p}} - \hat{\mathbf{A}})$		$\hat{V} = V$, $\hat{\mathbf{A}} = \mathbf{A}$

formalism also automatically highlights the invariant nature of $Et - \mathbf{p} \cdot \mathbf{r}$. Hence, $(Et - \mathbf{p} \cdot \mathbf{r})\,\mathbb{1} - (E\mathbf{r} - c^2\mathbf{p}t){\cdot}\boldsymbol{\sigma}/c$ can be rewritten as $\gamma m_0 c^2\,[\,(t - \mathbf{v} \cdot \mathbf{r}/c^2)\mathbb{1} - (\mathbf{r} - \mathbf{v}t){\cdot}\boldsymbol{\sigma}/c\,]$. The Hermitian quantity $\mathbf{L}_{ct,\mathbf{r}} = (\,E\mathbf{r} - c\mathbf{p}\,ct\,){\cdot}\boldsymbol{\sigma}/c$ contains the (j,k)-components, for which $j = 0 \vee k = 0$, and describes the motion of the centre of mass of the energy, while the anti-Hermitian quantity $\imath\mathbf{L}_{\mathbf{r},\mathbf{r}} = \imath(\mathbf{r} \wedge \mathbf{p}){\cdot}\boldsymbol{\sigma}$ contains the (j,k)-components, for which $j \neq 0\,\&\,k \neq 0$. Just as with the electromagnetic field we thus have a symmetric and an anti-symmetric part in this tensor. A Lorentz transformation leaves $\det(ct\mathbb{1} - \mathbf{r}{\cdot}\boldsymbol{\sigma})$ invariant. It also leaves $\det(E\mathbb{1} + c\mathbf{p}{\cdot}\boldsymbol{\sigma})$ invariant. It must thus also leave $\det[\,(ct\mathbb{1} - \mathbf{r}{\cdot}\boldsymbol{\sigma})\,(E\mathbb{1} + c\mathbf{p}{\cdot}\boldsymbol{\sigma})/c\,]$ invariant. Hence, it can be seen that the electromagnetic field, angular momentum, and the Lorentz transformations all have a six-component structure, and that they all transform with Lorentz symmetry. However, the six components of a Lorentz transformation are independent. They are obtained from eight real parameters subject to the constraint that the determinant of the complex matrix they build (and which represents the Lorentz transformation they correspond to) must be 1. The matrix **RP** built with the angular momentum $\mathbf{r} \wedge \mathbf{p}$ has only five independent parameters. In fact, the four-vector $(0, \mathbf{r})$ that can be built from $\mathbf{r}$ contains only three independent components, and this is also true for the four-vector $(E, c\mathbf{p})$, as its length has the fixed value m_0c^2. But the same angular momentum can be obtained from other couples $(\mathbf{r}, \mathbf{p})$. One can argue that the choice of a specific couple corresponds to a phase factor. It can be seen then that the information content of a Lorentz transformation is equivalent to an angular momentum and an additional phase factor. In fact, the angular momentum can be characterized by $(\mathbf{r}, \mathbf{p})$ (i.e. six parameters) or by its length $rp \sin\alpha$ and a pair of orthogonal unit vectors (i.e. five parameters) both normal to the plane defined by $(\mathbf{r}, \mathbf{p})$. It is then the choice of this specific pair of orthogonal unit vectors that contains the phase vector, as the pair can be rotated, leaving the plane invariant, or alternatively, one can make a rotation in the plane without changing the two orthogonal unit vectors. Another way of presenting this could be to define the angular momentum by a unit vector $\mathbf{e}_1$ for $\mathbf{r}$ (three parameters), and a vector $rp \sin\alpha\,\mathbf{e}_2$ orthogonal to it (three more parameters). The choice of the orientation of $\mathbf{e}_1$ in the plane defined by the couple then embodies the phase factor. This phase factor $e^{\imath\varphi}$ then presents itself as related to the rotation angle φ when the isotropic vector $\mathbf{e}_1 + \imath\mathbf{e}_2$ is turned to $e^{\imath\varphi}\,(\mathbf{e}_1 + \imath\mathbf{e}_2)$. To really bring out the full tensor symmetry of the angular momentum, its phase factor can thus be included as a sixth component within the formalism.

It must be stressed that the notation $(\mathbf{E} + \imath\mathbf{B})\cdot\boldsymbol{\sigma}$ is misleading in that it could create the impression that the electromagnetic field behaves as a complex vector. The electromagnetic field is a tensor, not a complex vector. It has been demonstrated that in SL(2,$\mathbb{C}$), a four-vector with representation matrix $\mathbf{V}_1$ transforms according to $\mathbf{V}_1 \to \mathbf{L}\mathbf{V}_1\mathbf{L}^\dagger$. A bi-vector will be of the form $\mathbf{V}_2^\star\mathbf{V}_1$, where $\mathbf{V}_2^\star$ transforms according to $\mathbf{V}_2^\star \to \mathbf{L}^{\dagger-1}\mathbf{V}_2^\star\mathbf{L}^{-1}$, because $\mathbf{V}_2^\star$ behaves like $\mathbf{V}_2^{-1}$. The quantity $\mathbf{V}_2^\star\mathbf{V}_1$ transforms then according to $\mathbf{V}_2^\star\mathbf{V}_1 \to \mathbf{L}^{\dagger-1}\mathbf{V}_2^\star\mathbf{V}_1\mathbf{L}^\dagger$, where $\mathbf{L}^{\dagger-1} \neq \mathbf{L}$. In general, in the Lorentz group $\mathbf{L}^\dagger \neq \mathbf{L}^{-1}$. This shows that the transformation properties for a bi-vector are different from those for a complex vector, e.g. $\mathbf{V}_1 + \imath\mathbf{V}_2$, for which the transformation is still of the form: $\mathbf{V}_1 + \imath\mathbf{V}_2 \to \mathbf{L}(\mathbf{V}_1 + \imath\mathbf{V}_2)\mathbf{L}^\dagger$. This leads to the idea of introducing notations v^μ for the components of the four-vector coded by $\mathbf{V}$ and v_μ for the components of the four-vector coded by $\mathbf{V}^\star$. The components are really linked by the metric tensor: $v_\mu = \sum_\nu g_{\mu\nu}v^\nu$. The components v_μ are called covariant, while the components v^μ are called contravariant. The use of notations F_μ^ν, $F_{\mu\nu}, \ldots$ for the components of the electromagnetic field tensor are well known. The fact that these components have two indices illustrates that the electromagnetic field tensor is not a vector of the type $A_\mu + \imath B_\mu$.

Appendix D

The Analogy Between Electromagnetism and Gravitation

(The contents of this appendix are speculative.)[1]

D.1 Influence of the potential on the rest mass

Here, the analogy between an electron within the attractive electric potential of the nucleus and a planet within the attractive gravitational potential of its star will be used. Consider a planet of rest mass m_0 in the gravitational field of a star at rest with rest mass M_0. Due to the gravitational field of the star, the rest mass of the planet will change. In fact, the contribution of the potential energy to the rest mass must be taken into account. The closer the planet is to the star, the less its potential energy will be. At a distance $r < \infty$ we will have thus: $m_0^* = m_0 + E_{pot}(r)/c^2$, where E_{pot} is the potential energy. The rest-mass of the star will also change due to the interaction: $M_0^* = M_0 + E_{pot}(r)/c^2$. In a first approach, one may think that for the potential energy $E_{pot}(r)$ one should just take $E_{pot} = -GM_0 m_0/r$, but this not self-consistent. The new masses lead to new, modified potentials. Within the modified potentials, there could again be an effect of the potentials on the rest masses. The procedure could be iterated until it converges to a self-consistent solution that could be called the effective potential. The self-consistent approach consists in immediately putting the converged solutions: $m_0^* = m_0 - GM_0^* m_0^*/r$ and $M_0^* = M_0 - GM_0^* m_0^*/r$.

Here, the same difficulty arises as when trying to derive Dirac's substitution rule. It is not possible to go from a situation at rest in the limit

[1]The aim is to make a link between rest mass as containing rotational energy, and potential energy as some deformation energy. The calculations are not rigorous.

$r \to \infty$ to a situation at rest at r in a purely kinematic way, as one cannot satisfy simultaneously the conservation of momentum and the conservation of energy in such a scenario. It can only be achieved by allowing for a simultaneous emission of radiation (which for electromagnetic radiation would require several photons). For a planet, which has no net charge, the radiation would have to be gravitational (but there could also be dissipation of heat as with Io). This is why being at rest within a potential corresponds to an acceleration. The acceleration that brought a particle to rest in a gravitational potential must have been obtained by emitting radiation, for example.

Let us now see how the effects of changes in mass can be taken into account. The aim is to find a new potential $V'(M_0, r)$ where the old rest mass for the star can be entered, and the potential energy calculated as $m_0 V'(M_0, r)$, using the old mass of the planet. In other words, the intention is to make the calculations with m_0 as though the rest mass of the planet would not change due to the potential. It could be claimed that a varying rest mass $M_0^*(r)$ that is expressed in terms of M_0 and r is used by using the effective potential $V'(M_0, r)$. This use of the quantities M_0 and m_0 to construct an effective potential is for practical purposes: they are the only such quantities known. With such an effective potential V', it is not necessary to claim that we have a varying mass M_0; it can equally be claimed that we have an effective potential in terms of the unchanged masses M_0 and m_0, that is different from V.

The total mass of the planet within the effective potential will now be calculated. In terms of the effective potential, it will be $mc^2 + mV'(M_0, r)$ taking into account that the particle is moving by replacing m_0 by $m = \gamma m_0$. This is the way to proceed in special relativity, for example, if one makes calculations of an orbit using the relativistic generalization of Newton's law $\mathbf{F} = d\mathbf{p}/dt$. Here, m_0 is simply replaced by m within $\mathbf{p}$. That the total mass is $mc^2 + mV'(M_0, r)$ within the effective potential is based on the definition of the effective potential: the intention is to perform calculations with unaltered rest masses m_0 and M_0. It will be necessary to create a new effective potential V' such that the calculations yield the correct result when m_0 and M_0 are used. The total energy is thus $E = \gamma m_0 c^2 + \gamma m_0 V'(M_0, r)$. The total energy at infinite distance would be $m_0 c^2$. Hence, the difference in energy between the situation at distance r and the situation at infinite distance is:

$$\Delta E = \gamma m_0 c^2 + \gamma m_0 V'(M_0, r) - m_0 c^2. \tag{D.1}$$

Here $\Delta E = \gamma m_0 c^2 + \gamma m_0 V'(M_0, r) - m_0 c^2 - m_0 V'(M_0, \infty)$, could also be written as $V'(M_0, \infty) = 0$. For a particle that has not radiated in travelling from $r \to \infty$ to r, the energy difference $\Delta E = E(r) - E(\infty)$ would just be zero.

Let us now also calculate the same difference in total energy in terms of the conventional $1/r$ potential $V(M_0, r)$. Here it is necessary to adopt the viewpoint that the potential is just a change of the rest masses. If the rest masses within the total energy are adjusted, according to the rule $m_0^* = m_0 + E_{pot}(r)/c^2$, $M_0^* = M_0 + E_{pot}(r)/c^2$ all the effects of the potential on the mass will have been taken into account. Once this modified "effective" rest mass m_0^* has been introduced, the potential energy can be forgotten. There is no potential energy, there are only changes in rest mass. Within this viewpoint, $\Delta E = (\gamma - 1)m_0^* c^2 = (\gamma - 1)(m_0 + \Delta m)c^2$, where $m_0^* = m_0 + \Delta m$. Of course, the idea is that the changes in rest mass are due to the classical $1/r$-potential $V(r)$. In a first approach the changes in mass could thus be written as: $\Delta m = m_0 V/c^2$. This is not entirely correct, as Δm is due to a modified potential where $V(M_0, r)$ must be replaced by $V(M_0^*, r)$. But the relative difference between M_0^* and M_0 is very small. In fact, $M_0^*(1 + m_0 G/(rc^2)) = M_0$. For Mercury, $m_0 G/(rc^2) = 4 \times 10^{-11}$. Furthermore, we should actually have $\Delta m = m_0^* - m_0 = m_0^* V/c^2 = (m_0 + \Delta m)V/c^2$, in order to have self-consistency. But here again Δm can be ignored with respect to m_0. We obtain then:

$$\Delta E' = (\gamma - 1)(m_0 c^2 + m_0 V(M_0, r)). \tag{D.2}$$

There is a crucial point here: the idea is that the planet has radiated, such that the particle can now be at rest at some distance $r_{max} < \infty$. The quantity $(m_0 c^2 + m_0 V(M_0, r)) = m_0^* c^2$ is then actually the rest mass in r. The quantity $\gamma(m_0 c^2 + m_0 V(M_0, r))$ is the mass at a different point at distance $r < r_{max}$ where the velocity v corresponds to γ. At r_{max}, the velocity is 0, such that $\gamma = 1$.

Once again, the quantity $\Delta E'$ is then zero. However, it must be clear that $\Delta E'$ does no longer correspond to $E(r) - E(\infty)$, but to $E(r) - E(r_{max})$. However, if one does not take into account that the rest mass has changed due to radiation, such that $m_0^* c^2 \neq m_0 c^2$, then one would obtain:

$$\Delta E' = (\gamma - 1)(m_0 c^2) + \gamma m_0 V(M_0, r)). \tag{D.3}$$

This shows that the term $-m_0 V(M_0, r)$ is lacking. The "rest mass" has been subtracted from the mass in motion according to

$\gamma(m_0c^2) + m_0V(M_0, r)) - m_0c^2$, but it has not been realized in the process that the rest mass contains an additional term $-m_0V(M_0, r)$. The correct expression is (D.2), the incorrect one is (D.3), which has the same form as D.1. As the latter form is being used, it is necessary to replace the correct $1/r$-potential V with an effective potential V' that makes up for the error. Hence, $(\gamma - 1)(m_0c^2 + m_0V(M_0, r))$ must be made equal to $(\gamma - 1)(m_0c^2) + \gamma m_0V'(M_0, r))$. By carrying out this procedure, we obtain:

$$V'(M_0, r) = \frac{\gamma - 1}{\gamma} V(M_0, r). \tag{D.4}$$

This looks odd, because it implies that $V'(r) = 0$ when $v = 0$. But it should not be forgetten that $v = 0$ implies that $r = r_{max}$ (and that potentials are defined up to a constant). If it is desired that $v = 0$ at $r < r_{max}$, there must again have been radiation. The point is that for a planet orbiting around the sun, the velocity is not the escape velocity. In other words, the rest mass of the planet will not be given by the condition that it must be zero at infinity. If the mass of the planet initially came from infinity, it must have radiated. The velocity will become zero at a much smaller distance r_{max} from the sun. The approach has taken a singular twist. It started from the idea of calculating a correct effective potential to take into account that the potential changes the rest mass. But in the end, a correct way to take into account this effect was found, and a logically wrong effective potential V' introduced to make up for the error in the rest mass.

It is easy to obtain a good estimation of v/c for Mercury. The term v/c in the Newtonian potential can be calculated non-relativistically, by using the Hamiltonian formalism where the total energy is expressed as $E = T + V$. Note that $T = p^2/(2m)$ follows from expanding $\gamma m_0c^2 \approx m_0c^2 + p^2/(2m)$, and dropping the term m_0c^2. The term of higher order contains v^4/c^4, which is of the order of 10^{-16} for Mercury. In the non-relativistic limit, the first-order approximation $T = p^2/2m$ can thus be taken. Applying this both at infinite distance and at distance r we obtain in this first order approximation: $p^2/(2m) = mv^2/2 = GMm/r$, from which it is easy to see that $v^2/c^2 = 2GM/(rc^2)$. Substituting this value within $V'(r) = (\gamma - 1)V(r)/\gamma$, we obtain:

$$V'(r) = V(r) \left[1 - \sqrt{1 - \frac{2GM}{rc^2}} \right], \tag{D.5}$$

which is just the expression for the effective potential as derived from general relativity and leads to a correct value for the calculation of the perihelion shift of Mercury. Note that it has been assumed here that there is no equivalent of magnetic potential in the frame "at rest", otherwise the equation would contain $(\mathbf{p}-e\mathbf{A})^2/(2m)$ rather than $p^2/2m$. It must also be stressed once more that the first order approximation $T = p^2/2m$ is only valid for small velocities.

There is a big difference between the gravitational case and the case of the electron in the hydrogen atom. For the electron it is necessary to calculate $E(r) - E(\infty)$ in order to be able to calculate the radiation that takes place in the transitions. For Mercury, it is requisite to calculate $E(r) - E(r_{max})$, and we do not observe any radiation of an energy corresponding to $E(r_{max}) - E(\infty)$, even if a lot of energy might have been radiated away in the past.

D.2 Polarization effects

In a sense, this shows that space-time is curved because the rest mass of the planet is changing. The presentation in Section D.1 is phenomenological. It does not explain why the rest mass of the planet changes. It must be that there is work done on the planet by the potential. It could be that the potential changes the distribution of mass or energy within the planet, deforming it. This connects to Einstein's approach in terms of the stress-energy tensor, which describes how a spherical swarm of particles is deformed when it is in free fall within a potential. Note that the changes in the electromagnetic potential corresponding to these effects are different. For gravitation the mass occurs both in the energy $E = mc^2$ and in the interaction. There is only one parameter, due to the equivalence principle. For the electromagnetic case, there are two parameters e and m. The parameter m intervenes within the total energy $E = mc^2$, the parameter e intervenes only within the contributions due to the interaction $V(r) = e^2/(4\pi\epsilon_0 r)$. In the gravitational case, the effect of the change in potential energy on the interaction is direct and scalar. In the electromagnetic case, it is not direct. From the equation $V(r) = e^2/(4\pi\epsilon_0 r)$, one might think that a change of mass does not have any effect on the potential. Nevertheless, the change of mass is a change in electromagnetic energy. The change of mass corresponds to a change in rotational energy, which itself corresponds to a change of the magnetic potential of the electron. This must correspond to redistribution of the currents. When the electron moves, this leads also to a change of the electric potential and a redistribution of the charges.

For the electromagnetic case, it could be considered that the whole works like a polarization process. If a neutral atom is placed within an electric field, the centre of mass of the negative charges will shift with respect to the centre of mass of the positive charges. This will create an electric dipole. The field of this electric dipole will reduce the original field. It could be possible to reason the same way for a planet within the gravitational field of a star as for an electron within the electric field of a nucleus; it would be some kind of electric-dipole effect. Of course, there may be no positive charge within the electron. But there might be more negative charge inwards than outwards, which could decrease the energy. It would not be an absolute dipole, but a relative dipole. Reasoning self-consistently, this must lead to a multi-pole expansion of the charge-current distributions. This represents some analogy with quantum electrodynamics, where the vacuum is polarized.

There is radiation loss within the process of coming in from infinity to the distance r. Ultimately, the whole rest mass could be lost if it were possible to come close enough to the star. By equating the rest mass with the energy loss $2GM/rc^2$, in the simple non-relativistic case we obtain the Schwarzschild radius. By analogy this would lead to an electromagnetic annihilation radius. At this radius $v = c$.

Equating $m_0c^2 = e^2/r$ in the electromagnetic case, could define a radius of closest possible approach. Below this radius, the electron could no longer exist at rest, like a photon which also has no rest mass. This minimum radius conjures up the idea of an analogy with the Schwarzschild radius in general relativity. In the gravitational case, the same reasoning would lead to $m_0c^2 = GMm_0/r$, which with respect to the correct result is off by a factor of 2.

Bibliography

Aspect, A., Dalibard, J., and Roger, G. (1982). *Phys. Rev. Lett.* **49**, 1804–1807.

Bée, M. (1988). *Quasielastic Neutron Scattering* (Adam Hilger, Bristol).

Bell, J.S. (1965). *Physics* **1**, 195.

Biedenharn L.C. and Louck J.D. (1985). *Angular Momentum in Quantum Mechanics, Theory and Application*, Encyclopedia of Mathematics and its Applications, Vol. 8 (Addison-Wesley, Reading, MA, 1981) and (Cambridge University Press).

Blumenthal, O. (1935). "Lebensgeschichte", in David Hibert (ed.), *Gesammelte Abhandlungen* (Julius Springer, Berlin), pp. 388–429.

Bohm, D. (1951). *Quantum Theory* (Prentice Hall, Upper Saddle River, NJ).

Cartan, E. (1981). *The Theory of Spinors* (Dover, New York).

Chaichian, M. and Hagedorn, R. (1998). *Symmetries in Quantum Mechanics, From Angular Momentum to Supersymmetry* (IOP, Bristol).

Clauser, J.F. and Shimony, A. (1978). *Rep. Progr. Phys.* **41**, 1881; Freeman, S.J., and Clauser, J.F. (1972). *Phys. Rev. Lett.* **28**, 938; Clauser, J.F., Horne, M.A., Shimony, A., and Holt, R.A. (1969). *Phys. Rev. Lett.* **23**, 880; Clauser, J.F. and Horne, M.A. (1974). *Phys. Rev. D.* **10**, 526.

Coddens, G. (2002). *Eur. J. Phys.* **23**, 549.

Coddens, G. (2008). *Eur. Phys. J. C* **55**, 145–157; Erratum in (2008). *Eur. Phys. J. C* **56**, 163. This paper must be considered as superseded by the present monograph, which contains corrections for many details and misunderstandings. For example, large portions of Section 4 of this paper are wrong, and this applies also to the Appendices A and C. Also Section 5 needed heavy reworking.

Cornwell, J.F. (1984). *Group Theory in Physics* (Academic Press, London).

Coxeter, H.S.M. (1963). *Regular Polytopes* (Dover, New York).

Dieudonné, J. (1987). *Pour l'honneur de l'esprit humain — les mathématiques aujourd'hui* (Hachette, Paris).

Duneau, M. (1994). *Lectures on Quasicrystals*, Hippert, F. and Gratias, D. (eds.) (Les Editions de Physique, Les Ulis), pp. 153–186.

Einstein, A., Podolsky, B., and Rosen, N. (1935). *Phys. Rev.* **47**, 777.

Farmelo, G. (2009). *The Strangest Man. The Hidden Life of Paul Dirac: Quantum Genius* (Faber & Faber), p. 430.

Feyerabend, P. (1975). *Against method: Outline of an anarchistic theory of knowledge* (Humanities Press, Atlantic Highlights, New Jersey).

Feynman, R.P. Leighton, R.B., and Sands, M. (1964). *The Feynman Lectures on Physics, Vol. 3, Quantum Mechanics* (Addison-Wesley, Reading, MA).

Feynman, R.P. and Weinberg, S. (1987). *Elementary Particles and the Laws of Physics* (Cambridge University Press, New York). Note that Feynman discusses the problem in terms of exchange of particles rather than in terms of exchange of spins.

Frankel, T. (1997). *The Geometry of Physics* (Cambridge University Press, New York).

Galilei, G. (1623). *Il Saggiatore* (Rome).

Gel'fand, I.M., Kapranov, M.M., and Zelevinsky, A.V. (1994). *Discriminants, Resultants and Multidimensional Determinants*, Mathematics: Theory & Applications, R.V. Kadison and I.M. Singer (eds.) (Birkhäuser, Boston).

Ghirardi, G.C., Rimini A., and Weber, T. (1980). *Lett. Nuovo Cim.* **27**, 293.

Greiner, W. (1990). *Relativistic Quantum Mechanics* (Springer, Berlin).

Gross, E.K.U., Runge, E., and Heinonen, O. (1991). *Many-particle Theory* (Adam Hilger, Bristol).

Halberstadt, E. (1980). *Cube hongrois et Théorie des Groupes, Pour la Science* **34**(8), 23–31. The number of different combinations of the Rubik's cube is: $8! \times 3^7 \times 12! \times 2^{10}$, which is 42 252 003 274 489 856 000. The minimal number of moves to solve the group has been proved to be 20.

Harter, W.G. and Weeks, D.E. (1989). *J. Chem. Phys.* **90**, 4744–4771.

Hladik, J. (1996). *Les Spineurs en Physique* (Masson, Paris).

Inui, T., Tanabe, Y., and Onodera, Y. (1990). *Group Theory and its Applications in Physics* (Springer, Heidelberg).

Jackson, J.D. (1963). *Classical Electrodynamics* (Wiley, New York).

Jones, H.F. (1990). *Groups, Representations and Physics* (Adam Hilger, Bristol).

Kuhn, T.S. (1996). *The Structure of Scientific Revolutions* (University of Chicago Press, Chicago).

Klein, F. (1884). *Vorlesungen über das Ikosaheder und die Aufösung der Gleichungen vom fünften Grade* (Teubner, Leipzig). English translation (1956) *Lectures on the icosahedron and the solution of equations of the fifth degree* (Dover, New York).

Lyonnard, S. (1997). Ph.D. thesis (Université d'Orsay). The heuristic idea that delivers some intuition about what is going on behind the scenes of the abstract general formalism of projection operators is described in this Ph.D. thesis. From this basic idea one can develop the whole formalism as it is presented abstractly in advanced textbooks, but it is quite elaborate to render it completely general. Without the heuristics, the abstract theory may look rather impenetrable.

Messiah, A. (1965). *Quantum Mechanics, Vol. I* (North Holland, Amsterdam).

Misner, C.W., Thorne, K.S., and Wheeler, J.A. (1970). *Gravitation* (Freeman, San Francisco).

Naimark, M.A. (1964). *Linear Representations of the Lorentz group* (Pergamon Press, Oxford).

Newman, E. and Penrose, R. (1962). *J. Math. Phys.* **3**, 566.

Penrose, R. and Rindler, W. (1984). *Spinors and Space-Time, Vol. I, Two-spinor Calculus and Relativistic Fields* (Cambridge University Press, Cambridge).

Rhodes, J.A. and Semon, M.D. (2004). *Am. J. Phys.* **72**, 943.

Sagan, B.E. (2001). *The Symmetric Group, Representations, Combinatorial Algorithms, and Symmetric Functions*, 2nd edition (Springer, New York).

Salam, A. (1963). *Theoretical Physics* (International Atomic Energy Agency, Vienna), pp. 173–196.

Schwartz, L. (1973). *Théorie des Distrubitions* (Hermann, Paris).

Shimony, A. (1983). *The New Physics*, Davies, P. (ed.) (Cambridge University Press, Cambridge), pp. 373–395; see also Wheeler, J.A. and Zurek, W.H. (1983). *Quantum Theory and Measurement* (Princeton University Press, New Jersey).

Smirnov, V. (1972). *Cours de Mathémathiques Supérieures*, Vols. 2 and 3 (Mir, Moscow).

Sternberg, S. (1994). *Group Theory in Physics* (Cambridge University Press, Cambridge).

Ziman, J.M. (1972). *Principles of the Theory of Solids* (Cambridge University Press, Cambridge).

Index

www.ingramcontent.com/pod-product-compliance
Ingram Content Group UK Ltd.
Pitfield, Milton Keynes, MK11 3LW, UK
UKHW021820130526
5754IPUK00004B/30

9 781783 266364